Stream Ciphers and Number Theory

Revised Edition

North-Holland Mathematical Library

VOLUME 66

ELSEVIER

Amsterdam – Boston – Heidelberg - London - New York - Oxford – Paris – San Diego –San Francisco – Singapore – Sydney - Tokyo

Stream Ciphers and Number Theory

Revised Edition

Thomas W. Cusick
The State University of New York
at Buffalo, NY, U.S.A.

Cunsheng Ding
The Hong Kong University of
Science and Technology
China

Ari Renvall
The University of Turku
Finland

ELSEVIER

2004

Amsterdam – Boston – Heidelberg - London - New York - Oxford –
Paris – San Diego –San Francisco – Singapore – Sydney - Tokyo

ELSEVIER B.V.
Sara Burgerhartstraat 25
P.O. Box 211, 1000 AE
Amsterdam, The Netherlands

ELSEVIER Inc.
525 B Street
Suite 1900, San Diego
CA 92101-4495, USA

ELSEVIER Ltd
The Boulevard
Langford Lane, Kidlington,
Oxford OX5 1GB, UK

ELSEVIER Ltd
84 Theobalds Road
London WC1X 8RR
UK

First edition 2004

Library of Congress Cataloging in Publication Data
A catalog record is available from the Library of Congress.

British Library Cataloguing in Publication Data
A catalogue record is available from the British Library.

ISBN: 0-444-51631-X
ISSN: 0924-6509 (Series)

∞ The paper used in this publication meets the requirements of ANSI/NISO Z39.48-1992 (Permanence of Paper).
Printed in The Netherlands.

To Kathy, Alan and Laura

—T. W. C.

To Weijuan, Xiang and Lili

—C. D.

To Pirjo and Marko

—A. R.

Preface to the Revised Edition

Since the publication of this monograph in 1998, a considerable amount of advances on interactions between stream ciphers and number theory has been made. The objective of this revised edition is to report the recent advances and correct typos and errors in the original version. Most chapters are revised. In particular, Chapter 6 is completely rewritten.

We thank Y.-H. Park, D. Hong and E. Chun for pointing out an error in computing the linear complexity of the prime-square generator in Chapter 8 of the original edition. We are grateful to S.S. Bedi and R. Pillai for pointing an error in Section 14.6, and for providing us with the source code of their C-implementation of the 2-RA algorithm.

August 2003 Thomas W. Cusick Cunsheng Ding Ari Renvall

Preface to the First Edition

The goal of cryptography is the concealment of messages in such a way that only authorized people can read them. A *cipher* or *cryptosystem* is an algorithm for carrying out this concealment. If a message M is represented as a string of characters $m_1, m_2, \ldots$ from some fixed character set or *alphabet*, then a cipher consists of two processes: *encryption*, a method for converting the message or plaintext into a ciphertext meant to be unreadable by unauthorized people; and *decryption*, a method for recovering the message from the ciphertext.

Broadly speaking, cryptosystems can be classified as either *block ciphers* or *stream ciphers*. A block cipher breaks up a message M into successive blocks $M_1, M_2, \ldots$ of elements from the alphabet. There is a *key set* K such that each key k in the set corresponds to an encryption algorithm E_k which acts on the blocks of plaintext. Thus a plaintext $M_1, M_2, \ldots$ is encrypted as $E_k(M_1), E_k(M_2), \ldots$. There is a decryption algorithm D_k for each key k such that $D_k(E_k(M_i)) = M_i$; thus ciphertext can be converted back into plaintext if the key k and D_k are known. A stream cipher breaks up a message M into its component characters $m_1, m_2, \ldots$. Each character m_i is enciphered with the element k_i of a *keystream* $K = k_1, k_2, \ldots$. If we let $E_{k_i}(m_i)$ denote the encipherment of message character m_i by keystream character k_i (in many cases this encipherment will simply be the sum of k_i and m_i in some suitable Abelian group), then the ciphertext stream is $E_{k_1}(m_1), E_{k_2}(m_2), \ldots$. There is a decryption procedure D_{k_i} for each keystream character such that $D_{k_i}(E_{k_i}(m_i)) = m_i$; thus ciphertext can be converted back into plaintext if the needed characters k_i of the keystream and the corresponding D_{k_i} are known.

Both block ciphers and stream ciphers are in common use today. Stream ciphers are especially prevalent in business, military and diplomatic settings. One advantage that stream ciphers have is that typically they can be implemented very efficiently in computing hardware. Since the security of a stream cipher depends on the randomness properties of the keystream, it is often easier to carry out a mathematical analysis of a stream cipher instead of a block cipher.

This book is almost entirely concerned with stream ciphers. We concentrate on a particular mathematical model for such ciphers which we call *additive natural stream ciphers*. These ciphers use a *natural sequence generator* to produce a periodic keystream. Full definitions of these concepts are given in Chapter 2.

In this book we focus on keystream sequences which we can analyze using number theory. It turns out that we can deduce a great deal of information about the cryptographic properties of many classes of sequences by applying the terminology and theorems of number theory. We make these connections explicit by describing three kinds of *bridges* between stream ciphering problems and number theory problems. A detailed summary of these ideas is given in the introductory Chapter 1.

This is the first book devoted to the study of the extensive crossfertilization between stream ciphers and number theory. Many results in the book are new, and over seventy percent of the results described in this book are based on our recent research results. On the one hand, there are numerous instances where results from number theory are used to answer questions from cryptography. On the other hand, there are many cryptographic problems which suggest new avenues of research in number theory. A few dozen questions of this type, with greatly varying levels of difficulty, are scattered through the book and labelled as Research Problems. For the convenience of the reader, a list of brief summaries of these Research Problems is given in Appendix D.

Launched in 1992, this project has taken us several years to complete. During the whole process, we have benefited from discussions with and comments from several colleagues. We thank Mark Goresky, Tor Helleseth, Andrew Klapper, and Arto Salomaa for reading some parts of this book manuscript and providing us with valuable comments and suggestions. We are grateful to Harald Niederreiter for helpful suggestions and comments. We appreciate the excellent working conditions provided by the State University of New York at Buffalo, University of Turku, Turku Centre for Computer Science, and the National University of Singapore. We acknowledge good cooperation with Elsevier, especially with Drs. Arjen Sevenster, Ms. Claudette van Daalen, and Ms. Titia Kraaij. Finally, we thank all members of our families for their support.

December 1997 Thomas W. Cusick Cunsheng Ding Ari Renvall

Contents

Chapter 1

Introduction

Number theory, which Gauss called the queen of mathematics, has fascinated the human mind since the beginning of recorded history. Arithmetic was already quite sophisticated in Mesopotamia at the end of the third millennium B.C. [178]. Number theory has a great influence not only on other branches of mathematics, but also on many other sciences, such as physics, biology, digital signal processing, coding, computing, and public-key cryptology as well as stream ciphers.

Ciphers, which are usually divided into block and stream ciphers, have been used for millennia to safeguard military and diplomatic communications. Today stream ciphers are still the most used ones in practice for these purposes. The main reason may be that the theory of stream ciphers is much more analytical, while only relatively few aspects of typical block ciphers can be mathematically measured and analyzed. Another important reason may be that well-designed stream ciphers can destroy statistical properties of the plaintext, while block ciphers may not.

It is not strange that number theory and stream ciphers are closely related, since many ciphers actually manipulate numbers. One aim of this book is to set up bridges between number theory and stream ciphers, and to stimulate the interaction between the two fields. Another one is to design some promising keystream generators based on number theory.

This introductory chapter is organized as follows: Section 1.1 presents a number of other sciences upon which number theory has an important impact. Section 1.2 gives a brief introduction to the book.

1.1 Applications of Number Theory

Number theory is an ancient field of study and its content is vast. It has several branches, such as elementary number theory, algebraic number theory, analytic number theory, the geometry of numbers, etc. Sometimes it may be difficult to say whether some mathematical topics should belong to number theory or not. For example, it is difficult to draw a strict line between number theory and algebra. One might regard class field theory as a part of "abstract algebra" rather than "number theory".

Many number-theoretic problems had a great impact upon the development of entire branches of mathematics. For example, the study of the distribution of primes sparked the development of the theory of functions of a complex variable and, in particular, that of the theory of entire functions. Fermat's Last Theorem led to the creation of the theory of algebraic numbers, one of the most important and flourishing branches of modern number theory. It also happens that some of the most fundamental concepts of modern algebra (actually, of all modern mathematics) such as groups, rings, fields, modules, to name only a few, are obtained by the processes of abstraction and generalization from situations we meet in elementary number theory [178].

Number theory has wide applications in digital signal processing, where fast algorithms for digital data processing are of great significance. The Galois fields in which algorithms for the computation of a Fourier transform are simplest are those of the form $GF(2^m+1)$, which is a field whenever $p = 2^m + 1$ is a Fermat prime. In a Galois field $GF(q)$ when q is a Fermat prime, any factor of $q-1$ is a power of 2. The Fourier transform

$$V_k = \sum_{i=0}^{n-1} \omega^{ik} v_i, \quad k = 0, ..., n-1$$

exists whenever n is a divisor of 2^m and ω is an element of order n. By using the Cooley-Tukey algorithm (see [105] for a discussion), a Fourier transform over $GF(2^m+1)$ can be broken down into a sequence of radix-two transforms, which can be implemented rather neatly using only $(n/2)\log_2 n$ multiplications and $(n/2)\log_2 n$ additions.

The Galois fields in which the operation of multiplication is most straightforward are those of the form $GF(2^m-1)$. Arithmetic in the field $GF(2^m-1)$ is quite convenient if the integers are represented as m-bit binary numbers. In the prime field $GF(2^m-1)$ a Fourier transform of block length n exists for every n dividing 2^m-2. These are sometimes called *Mersenne number transforms*. There are also other number transforms [24, 298, 387].

Coding is an important field of communications theory and engineering.

There are quite a number of number-theoretic concepts which are related to codes. Among the most notable ones are group characters and character sums, and Diophantine equations. Character sums are powerful tools for the control of the correlation property of codes for code-division multiple-access systems [200, 207], and for the computation of the weight distribution of some linear codes [198]. Cyclotomy also has applications in coding theory [301].

The Chinese Remainder Theorem developed in the first century B.C. has wide applications in algorithms, modular computation, computer systems, coding theory, digital signal processing, and cryptology [134]. The discrete Fourier transform has an important impact on many fields, but it is only a special case of the Chinese remainder transform which includes quite a number of popular transform techniques [134].

The applications of number theory in public-key cryptography and authentication are well-known. The RSA system, based on number theory, remains one of the most promising public-key cryptosystems [111, 113, 318, 337, 312]. On the other hand, RSA is one of the main motivations for the investigation of factorization, as the security of RSA depends on the assumption that factoring large integers is difficult. Modular hashing is also an important topic of cryptography [357].

Number theory also has applications in physics and biology as well as in other sciences. It is even possible for the theory of quadratic partitions to have applications in chemistry [387]

The application of number theory to stream ciphers looks quite natural, because many ciphers manipulate numbers. The design of pseudorandom number generators is one of the main issues in stream ciphering, where pseudorandom number sequences are often employed as keystreams.

The study of linear recurring sequences has a very long history which was traced from the year 1202 to 1918 by Dickson [110]. One of the most famous linear recurring sequences is the Fibonacci sequence defined by $F_{n+2} = F_{n+1} + F_n$ for $n = 0, 1, \ldots$, with $F_0 = 0, F_1 = 1$. These *Fibonacci numbers* occur in counting the number of leaves, petals and seed grains of many plants [387]. During the 19th century, recurring series of rational integers were frequently studied. Lucas sequences, which were used for primality testing by Lucas, are one such example.

The first systematic investigation of a linear recurrence modulo m was done in 1920 by Carmichael. Nine years later Carmichael gave an account of earlier results, and showed that the recurrence problem is intimately related with many other parts of elementary number theory. Important developments in the study of linear recurrence sequences can be found in Engstrom [147], Ward [434]–[444], Hall [185]–[188], Zieler [472], Selmer [390], and Golomb [169]. A lot of references about linear recurrence sequences can be found in [276, 390].

Though the investigation of sequences dates back at least to 1202, those early investigations were not done from the cryptographic point of view. In spite of the recent intensive investigations of cryptographic sequences, the cryptographic theory of sequences is far from mature. Some sequences were known centuries ago, but their cryptographic properties still remain unknown. One important problem for modern cryptologists is to analyze the cryptographic properties of those sequences. It is interesting to note there are indeed some cryptographically attractive sequences developed centuries ago. Among them are some Legendre sequences which may have more ideal cryptographic attributes than many keystream sequences designed today [123]. It seems that recently much more attention has been given to sequences based on the theory of finite fields and algebra than to sequences based on number theory.

The distribution problem of quadratic and other power residues and nonresidues has been attacked by many mathematicians for centuries with only limited success. The "irregular" distributions of power residues, primes and primitive roots are not only useful in constructing some cryptographic building blocks, but also in other fields. For example, quadratic residues are used in a proposal for designing a concert hall ceiling in [387].

Primes can also be defined in fields other than the rational numbers, for instance, in some algebraic number fields. In the complex number field we have the Gaussian primes defined in the Gaussian ring $Z[i]$. Those primes form an interesting pattern [387], which has been used in weaving tablecloths and tiling floors. We also have the Eisenstein primes defined in the ring $Z[\omega]$, where ω is the cube root $\omega = (1 - \sqrt{-3})/2$. Some primes defined in the integer ring Z remain prime in these two rings, others not. The Eisenstein primes form also an interesting pattern with hexagonal symmetry [387]. These primes can also be used to construct keystream generators.

Some purely number-theoretic problems could also have a deep cryptographic significance. Our first example is the distribution of primitive roots. Concerning this problem Artin conjectured that every integer a, not equal to -1 or to a square, is a primitive root of infinitely many primes. Recently an important progress about Artin's conjecture has been made: Let p_1, p_2 and p_3 be three distinct primes, then at least one of them is a primitive root for infinitely many primes [195]. If Artin's conjecture is proven to be true, this means that building "good" cryptographic sequences over any field is possible. Our second example is Fermat's Last Theorem, which led to algebraic number theory. A proof of Fermat's Last Theorem means that building good cryptographic sequences with periods equal to powers of primes over many fields is theoretically possible. Our third example is the theory of class fields. Some results from class field theory can be used to prove some cryptographically meaningful results. In this book we will need the quadratic partition of primes. Thus, we need to know whether the set

$$B(n) = \{x^2 + ny^2 : (x, y) \in Z \times Z\}$$

contains infinitely many primes. If it does, what is their density? These questions are cryptographically important for some applications. Class field theory can give us some answers.

The word cyclotomy means "circle-division." This problem and cyclotomic polynomials together with cyclotomic numbers were investigated by many mathematicians for a long time. Now these problems have been found to be very useful in designing keystream sequences. Cyclotomies and generalized cyclotomies constitute in fact one of the theoretical bases for many of the generators presented in this book. The stability of (generalized) cyclotomic numbers leads to cryptographically useful properties for many cyclotomic generators. It is useful in cyclotomy to have the Riemann Hypothesis for curves over finite fields, which was proved by Weil in 1948 [449]. This implies the cryptographic significance of the Riemann Hypothesis. Thus, the cryptographic importance of the genus of algebraic curves and of algebraic function fields follows.

In summary, a considerable number of number-theoretic problems are related to the design and analysis of stream ciphers. It seems that no other field has as many applications to stream ciphers as does number theory. This may explain why the secret agencies of some countries require a prereview of papers of certain types about number theory. Indeed, some cryptographic problems can be settled only if progress in some number-theoretic problems can be made.

We do not claim to cover all of the cryptographic applications of number theory. In this book we intend only to give examples to illustrate that number theory is not only a pure mathematical science, but also a very applicable science.

1.2 An Outline of this Book

In this book some bridges between the design of stream ciphers and some number-theoretic problems are built for the first time. With those bridges the applications of number theory to the design and analysis of stream ciphers are then investigated. In our approach, old number-theoretic problems are stressed and new ones are proposed from the cryptographic point of view. Those cryptography-related number-theoretic problems call for further developments in number theory. The main purpose in writing this book is to invoke an interaction between number theory and stream ciphers. This book is organized as follows.

Chapter 2 is an introduction to synchronous and self-synchronous stream ciphers, as well as some keystream generators and some cryptographic factors of sequences. It is intended to introduce only some basic notions. A number of keystream generators are briefly described.

Chapter 3 is about primes and primitive roots as well as sequences, and aims at finding pairs of primes and primitive roots for the purpose of designing keystream sequences. The main cryptographic idea of this chapter is the search for a good partnership, as defined in the chapter, in order to get sequences with both large linear complexity and good linear complexity stability. A main bridge between number theory and stream ciphers is also set up.

Chapter 4 is devoted to cyclotomy and its applications to the design of cryptographic functions. The cryptographic significance of cyclotomic numbers is illustrated. A number of cryptographic functions whose nonlinearity depends on (generalized) cyclotomic numbers are introduced.

Based on the results of Chapter 4, Chapter 5 is devoted to the search for special primes for stream ciphering purposes. The cryptographic values of various kinds of special primes with respect to specific finite fields are analyzed. The sexes of twin primes are introduced and their distributions are investigated. A comparison between primes for stream ciphering and those for RSA is made. It is shown that the distribution problem for special primes is of cryptographic importance.

Chapter 6 is devoted to functions with optimum nonlinearity, and gives a well-rounded treatment of non-Boolean functions with perfect and almost perfect nonlinearity.

Chapter 7 is about the differential analysis of sequences. A natural sequence generator realization of sequences is given and the differential analysis of sequences is introduced. The linear complexity of difference-set characterized and almost difference-set characterized sequences is calculated. The cryptographic importance of Barker sequences is shown.

Based on the results of foregoing chapters, Chapter 8 builds some binary cyclotomic generators. Many security aspects of the generators are analyzed. Chapter 9 gives a detailed analysis of a particular type of cyclotomic generator.

In Chapter 10 a nonbinary generator based on cyclotomy is designed and its security problems are discussed. Some cryptographic ideas behind the cyclotomic generators are also analyzed.

Chapter 11 is dedicated to generators based on permutations. One important cryptographic technique employed in this chapter is that for ensuring "good + bad = good." Cryptographic permutations and some design problems of cyclic-key generators are discussed. Two generators based on permutations are given.

Chapter 12 investigates the application of the theory of quadratic partitions to stream ciphers. There are two motivations for our interest in quadratic partitions: One is the design and analysis of some cryptographic functions, and the other is that we need them to give us some primes in integer domains other than Z for the purpose of designing key stream generators based on the arithmetic of the integer domains. Some cryptographic quadratic forms are discussed, and some elementary results about quadratic forms are introduced.

Chapter 13 introduces the theory of group characters and shows the importance of group characters in designing keystream generators. One of the most important cryptographic ideas in this chapter is that sometimes linear functions are cryptographically more attractive than nonlinear functions. This shows that "linear" mappings with respect to some operations could give the "best nonlinear" cryptographic mappings in another context.

Chapter 14 is mainly about the 2-adic approach to the design and analysis of binary sequences. The relation between the Blum-Blum-Shub sequences [26] and the class numbers of imaginary quadratic fields is another main topic of this chapter.

Chapter 15 describes some fast stream ciphering algorithms based on primes and gives some theoretical results obtained using material from some previous chapters. Compared with other software-oriented algorithms, the ciphering algorithms presented in this chapter are more amenable to analysis.

Chapter 16 discusses some cryptographic problems and philosophies. Among the topics are nonlinearity and linearity of cryptographic mappings, stability and instability, stability and diffusion, the stability of local nonlinearities and differences, correlation stability, pattern stability and mutual information stability, localness and globalness, and goodness versus badness.

The design and analysis of the natural sequence generators, which have been systematically investigated only recently, are among the main topics of this book. One aim of this book is to bridge stream ciphers and number theory. Another aim is to design and analyze a number of cryptographic generators by making use of the bridges. The third aim is to stress old number-theoretic problems and to propose new ones from cryptographic viewpoints. In this book, over thirty cryptography-related research problems are presented. A list of them is given in Appendix D.

There are gaps between some number-theoretic problems and stream-cipher problems because bridges between them have not been set up yet. In this book we try to bridge some of these gaps. Naturally, there are more gaps which remain to be bridged. Thus, the cryptographic meaning and importance of some number-theoretic problems mentioned in this book may be only partially elucidated.

The first set of bridges between some stream ciphering problems and number-theoretic problems is depicted in Figure 1.1.a, where the main cryptographic problems are the linear and sphere complexity of sequences, and the number-theoretic problems include cyclotomic polynomials, special primes, primitive roots and their distribution, and some topics related to these topics.

The second set of bridges considered in this book is depicted in Figure 1.1.b, where the bridge between p-ary sequences and p-adic numbers has been known for hundreds of years. The 2-adic span of binary sequences is closely related to the factorization of $2^N - 1$, and the balance of some sequences is closely related to the class numbers of some quadratic fields. The theory of permutations is the main issue for permutation generators. The third set of bridges is depicted in Figure 1.2, where quite a number of cryptographic and number-theoretic problems are involved.

This book by no means covers all number-theoretic generators. There are also other very interesting ones. Among them are, for example, the quadratic residue generator and the index generator which have been proven to possess some interesting cryptographic properties. For details about these generators we refer to [26, 25, 249], where a nice treatment has been given. However, we will come to some properties of the Blum-Blum-Shub generator in Section 14.8.

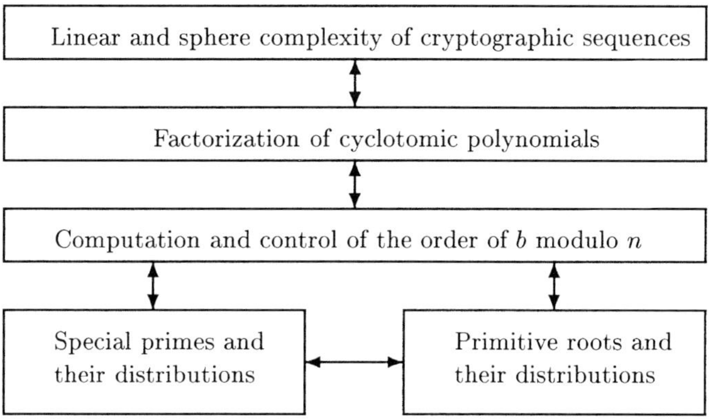

(a) The first set of bridges

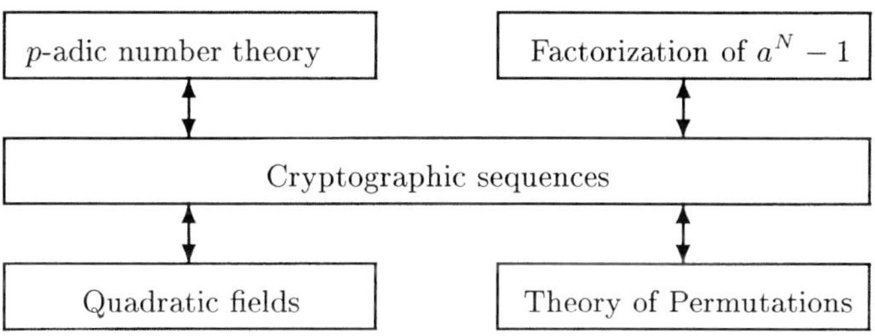

(b) The second set of bridges

Figure 1.1: The first two sets of bridges.

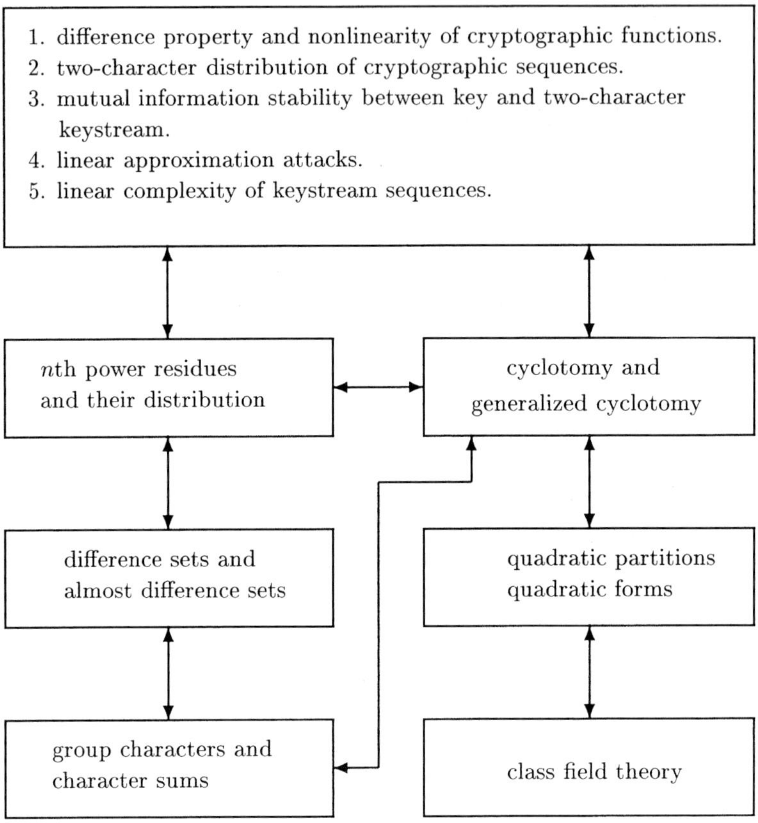

Figure 1.2: The third set of bridges.

Chapter 2

Stream Ciphers

This chapter introduces some basic notions about stream ciphers and describes some keystream generators. Section 2.1 is devoted to the description of synchronous and self-synchronous stream ciphers, and some approaches to the construction of stream ciphers based on block ciphers. Section 2.2 introduces some keystream generators. Section 2.3 considers some cryptographic aspects of sequences, such as linear complexity, weight complexity, sphere complexity, autocorrelation and crosscorrelation functions, pattern distribution, quadratic span and maximum order complexity. Section 2.4 shows the consistency and harmony of the binary natural sequence generator which is the main topic of this book. Section 2.5 considers the security of and attacks on stream ciphers generally.

2.1 Stream Cipher Systems

Ciphering systems are generally classified into block and stream ciphers, in analogy to error-correcting codes which are subdivided into block and convolutional codes. The essential distinction between block and stream ciphers is the memory, as is shown in Figures 2.1(a) and 2.1(b).

A block cipher breaks each plaintext message into successive blocks and enciphers each block M under the control of a key k into a ciphertext block $C = (c_1, \cdots, c_n)$, where the plaintext and ciphertext alphabet are usually identical. Each block is typically several characters long. Simple substitution and homophonic substitution ciphers [103] are examples of block ciphers, even though the unit of encipherment is a single character. This is because the same key is used for each character. A stream cipher specifies a device with internal memory that enciphers the jth digit m_j of the message stream into the jth digit c_j of the ciphertext stream by means of

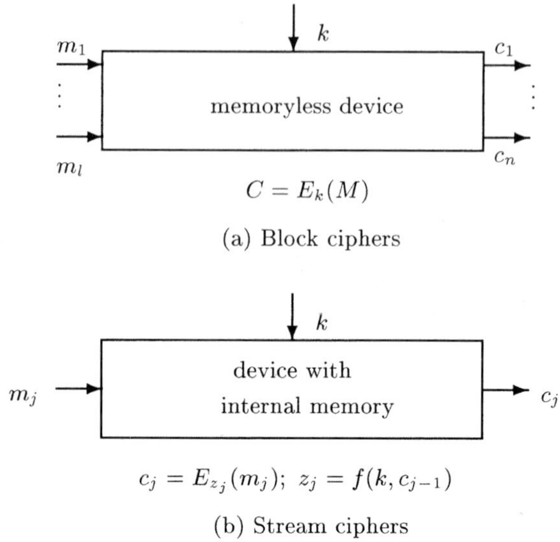

Figure 2.1: The difference between block and stream ciphers.

a function which depends on both the secret key k and the internal state of the stream cipher at time j. The sequence $z^\infty = z_0z_1\cdots$ which controls the enciphering is called the *key stream* or *running key*. The deterministic automaton which produces the key stream from the actual key k and the internal state is called the *running-key generator* or *keystream generator*.

A stream cipher is *periodic* if the keystream repeats after d characters for some fixed d; otherwise it is nonperiodic. Ciphers generated by Rotor and Hagelin machines [103] are periodic stream ciphers. The Vernam cipher is an example of a nonperiodic stream cipher.

There are two different approaches to stream encryption: synchronous methods and self-synchronous methods. In a *synchronous stream cipher*, the next state depends only on the previous state and not on the input so that the succession of states is independent of the message stream. The key stream is therefore generated independently of the message stream. Consequently, the enciphering transformation is memoryless, but time-varying. Thus, if a ciphertext character is lost during transmission, the sender and receiver must resynchronize their generators before they proceed further. Furthermore, this must be done in a way which ensures that no part of the key stream is repeated (thus the keystream generator should not be reset to an earlier state). It is therefore natural, in a synchronous stream cipher,

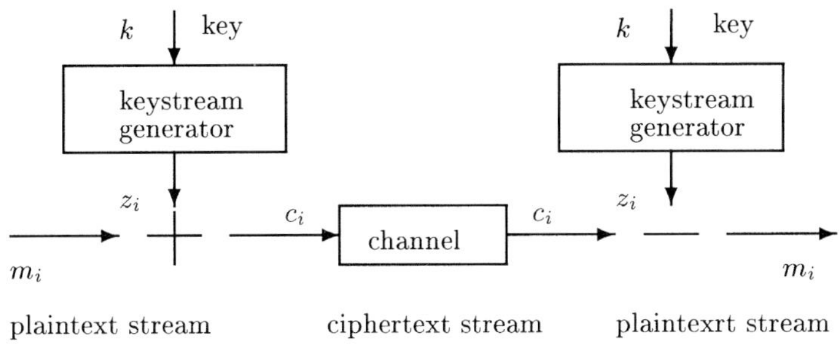

Figure 2.2: Additive synchronous stream ciphers.

to separate the enciphering transformation from the generation process of time-varying parameters which control the deciphering transformation.

In a *self-synchronous stream cipher*, each keystream character is derived from a fixed number n of preceding cipher characters. Thus, if a ciphertext character is lost or altered during transmission, the error propagates forward for n characters, but the cipher resynchronizes itself after n correct ciphertext characters have been received. Self-synchronous stream ciphers are nonperiodic because each key character is functionally dependent on the entire preceding message stream.

Figures 2.1.a and 2.1.b depict block and stream ciphers respectively. In Figure 2.1.a M and C stand for plaintext and ciphertext block respectively, and E_k is the encryption transformation specified by a key k. In Figure 2.1.b m_j and c_j are the plaintext and ciphertext character respectively, z_j the keystream character at time j, f a function for producing the keystream, and E_{z_j} a function for combining the keystream character z_j and the plaintext character m_j. A practical difference between a block and a stream cipher is that the redundancy of a natural language may remain in the ciphertext under a block cipher, while it has been usually made very small with a well-designed stream cipher. This may explain why stream ciphers are still popular in practice.

2.1.1 Additive Synchronous Stream Ciphers

As mentioned above, in a synchronous stream cipher, the key stream, $z^\infty = z_0 z_1 \cdots$, is independent of the message stream. The algorithm that generates the stream must be deterministic so that the stream can be reproduced for decipherment. One important kind of synchronous stream ciphers is the *additive synchronous stream ciphers* depicted by Figure 2.2, where the

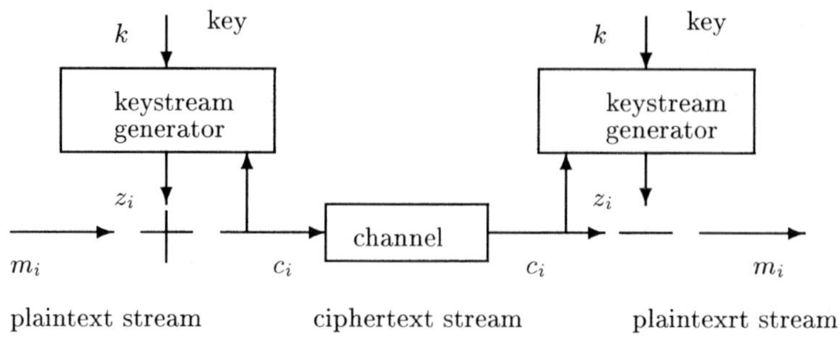

Figure 2.3: Additive self-synchronous stream ciphers.

characters of the key stream are from an Abelian group $(G, +)$ and the ciphertext character c_i is the addition of the key stream character z_i and message stream character m_i, and "$-$" denotes the inverse operation of "$+$".

The main design problem for this kind of stream cipher is the design of the keystream generator. Because the way to combine plaintext and ciphertext characters is very simple, keystream generators for additive synchronous stream ciphering should be strong enough.

2.1.2 Additive Self-Synchronous Stream Ciphers

In a self-synchronous stream cipher each keystream character is derived from a fixed number n of preceding ciphertext characters. The idea of this kind of cipher traces back to the time of Vigenère in the 16th Century. Autokey ciphers and cipher feedback systems are examples of additive self-synchronous stream ciphers [103].

An *autokey cipher* is one in which the key is derived from the message it enciphers. Another important class of self-synchronous stream ciphers consists of those where the cipher is fed back to the keystream generator as depicted in Figure 2.3. The main problems concerning this kind of stream ciphers are the design of the keystream generator and the way in which the feedback ciphertext character is used in the keystream generator. This kind of stream cipher is rather difficult to design and analyze because of the feedback approach.

2.1.3 Nonadditive Synchronous Stream Ciphers

There are advantages and disadvantages in both block ciphering and additive stream ciphering. Additive synchronous stream ciphers have the disadvantage that a ciphertext-plaintext character pair immediately reveals the corresponding keystream character under which the plaintext character is encrypted. This makes possible various kinds of key-recovering attacks such as correlation attacks and collision attacks, equivalent-machine attacks such as the one based on the Berlekamp-Massey algorithm, approximate-machine attacks such as those based on linear approximations. One of their advantages is that the keystream is time-varying, which ensures that the same plaintext character usually corresponds to different ciphertext characters at different times. This usually conceals some statistical properties of the plaintext. Block ciphers have the disadvantage that their keys cannot be changed very frequently due to the problem of key management. In addition, the same block of message corresponds always to the same ciphertext block if one key is selected and fixed. This may make many attacks such as differential attacks on some block ciphers applicable. One of their advantages is that the detection of the modification of messages may be possible owing to the fact that messages are encrypted block by block.

To keep the merits of both additive stream ciphering and block ciphering, but to get rid of the demerits of both approaches, a dynamic block ciphering approach is described as follows. With this approach a keystream generator and a conventional (one-key) block cipher are combined in such a way that some output characters of the keystream generator are employed to serve as the dynamic key of the block cipher for each message block.

For a block cipher of plaintext block length n, let $E_k(.)$ and $D_k(.)$ denote respectively the encryption and decryption transformation specified by a key k. To use the block cipher to encipher and decipher dynamically, a dynamic key k_i for the block cipher is produced by a sequence generator SG as $(z_{ti}, z_{ti+1}, \cdots, z_{ti+t-1})$, where t is a positive integer, and z^∞ denotes the sequence produced by the SG. The parameter t could be 1 or another assigned constant. Thus, the encryption and decryption are done respectively by

$$\begin{aligned} c_i &= E_{k_i}(m_i), \\ m_i &= D_{k_i}(c_i), \end{aligned}$$

where m_i is the plaintext block, c_i the ciphertext block at time i. Since the key k_i is time-varying, this is a dynamic block ciphering, and therefore a nonadditive synchronous stream ciphering approach. The key of the system consists of that of the keystream generator SG.

In this ciphering system it is not necessary to require large linear complexity for the output sequence of the SG if the underlying block cipher is properly designed. One cryptographic idea behind the design is cooperation. The SG and the block cipher should be designed so that they can protect each other. This kind of ciphering approach is intended to thwart as many attacks on block and/or additive synchronous stream ciphers as possible. Indeed, if the system is well designed, it seems that known attack approaches to additive stream ciphers and block ciphers do not apply to the system. To attack the system, one needs to develop new approaches.

Another aim of this system is to get fast ciphering algorithms. It is possible to use fast sequence generators and fast block ciphers in this system to get fast and secure ciphering algorithms.

Additive synchronous stream ciphers and all block ciphers can be regarded as special cases of the system. If the underlying block cipher of the above dynamic ciphering system is chosen as the term-wise addition of the key and the plaintext block, then the system is the usual additive synchronous stream cipher. In this case, we use one of the worst block ciphers in the system. If the SG is chosen such that the keystream is a constant sequence, then it is the usual block ciphering approach. In this case, we employ the worst keystream generator. Thus, the usual block ciphers and additive stream ciphers are special cases of the above approach, and in fact two extreme cases of the system.

2.1.4 Stream Ciphering with Block Ciphers

There are several kinds of modes of using block ciphers. The most studied four are the Electronic Codebook (ECB) mode, the Cipher Block Chaining (CBC) mode, the Cipher Feedback Chaining (CFB) mode, and the Output Feedback Chaining (OFB) mode [216].

In the ECB mode, a block cipher is applied block by block independently. Let $M = M_1 M_2 \cdots M_t$ be the plaintext, then the encryption is carried out as

$$C_i = E_k(M_i) \quad \text{for } i = 1, 2, \cdots, t.$$

Thus, the corresponding ciphertext is $C = C_1 C_2 \cdots C_t$. The decryption is then described by

$$M_i = D_k(C_i) \quad \text{for } i = 1, 2, \cdots, t,$$

where $D_k(x)$ is the inverse transformation of $E_k(x)$. This is a rather straightforward way to use block ciphers.

In the CBC mode the blocks are chained together with an initial value IV. In this mode we assume that the plaintext and ciphertext block space are identical, and that this block space is an Abelian group with an operation $+$. The first ciphertext block is defined as

$$C_1 = E_k(M_1 + IV),$$

where IV is an initial value from the block space. The other ciphertext blocks are then computed as follows:

$$C_i = E_k(M_i + C_{i-1}) \text{ for } i = 2, 3, \cdots, t.$$

To decrypt, the first plaintext block is obtained as

$$M_1 = D_k(C_1) - IV,$$

where "−" is the inverse operation of "+". The other plaintext blocks are then calculated as

$$M_i = D_k(C_i) - C_{i-1} \text{ for } i = 2, 3, \cdots, t.$$

Clearly, the CBC mode makes a block cipher into a stream cipher which has internal memory.

The CFB mode also uses a block cipher for stream ciphering. Assume that we have a block cipher with both plaintext and ciphertext block space A^n, where the alphabet $(A, +)$ is an Abelian group. Let $E_k(x)$ be the encryption transformation, rchop_u denote the function that drops the u rightmost characters of its argument, and lchop_u denote the function that drops the u leftmost characters of its argument. A simple variant of the CFB mode is described as follows. Choose m to be any integer between 1 and n. The stream cipher based on the block cipher has then the alphabet $(A^m, +)$, where the operation "+" of A^m is a natural extension of the operation of A, i.e.,

$$(x_1, \cdots, x_m) + (y_1, \cdots, y_m) = (x_1 + y_1, \cdots, x_m + y_m),$$

where $(x_1, \cdots, x_m) \in A^m$ and $(y_1, \cdots, y_m) \in A^m$. Under the choice of an initial value X_1, the encryption of the ith plaintext character $M_i \in A^m$ is carried out as

$$C_i = M_i + \text{rchop}_{n-m}(E_k(X_i)), \quad X_{i+1} = \text{lchop}_m(X_i) \,\|\, C_i,$$

where $\|$ denotes the concatenation. The decryption is as follows:

$$M_i = C_i - \text{rchop}_{n-m}(E_k(X_i)), \quad X_{i+1} = \text{lchop}_m(X_i) \,\|\, C_i.$$

An internal register is needed to update X_i.

The OFB mode uses also a block cipher for stream ciphering. As in the CFB mode, we have first a block cipher with both plaintext and ciphertext block space A^n, where the alphabet $(A, +)$ is an Abelian group. The stream cipher based on the block cipher is described as follows. The plaintext and ciphertext alphabet of the stream cipher are A^m, where m can be arbitrarily chosen between 1 and n. The stream cipher has an internal register for updating the values $X_i \in A^n$. Let X_1 be the initial value of the register. The encryption of the ith plaintext character $M_i \in A^m$ is carried out as

$$C_i = M_i + \text{rchop}_{n-m}(E_k(X_i)), \quad X_{i+1} = E_k(X_i).$$

The decryption is defined by

$$M_i = C_i - \text{rchop}_{n-m}(E_k(X_i)), \quad X_{i+1} = E_k(X_i).$$

Note that the only difference between the CFB and OFB is the updating of the internal register.

Among the above four modes of operations for block ciphers three of them result in stream ciphers. Naturally, there are many other ways to use block ciphers for stream ciphering. A nonadditive synchronous stream cipher based on block ciphers was described in the previous section. Another approach to the construction of stream ciphers based on block ciphers will be described in the following section.

2.1.5 Cooperatively Distributed Ciphering

There are advantages and disadvantages in both block and additive stream ciphering, as made clear in Section 2.1.3. To keep the advantages of both block and additive stream ciphering and to get rid of their disadvantages, a cooperatively distributed (briefly CD) ciphering system was described by Ding and Salomaa in [135].

The cooperatively distributed ciphering system consists of s components: s conventional block ciphers of the same block length, and a control device which is a sequence generator with internal memory, SG for short, which produces sequences over the alphabet $Z_s = \{0, 1, \cdots, s-1\}$.

Let $k_0, \cdots, k_{s-1}$ be the keys respectively; $E_0(k_0, \cdot), \cdots, E_{s-1}(k_{s-1}, \cdot)$ the encryption transformations specified by the keys; $D_0(k_0, \cdot), \cdots, D_{s-1}(k_{s-1}, \cdot)$ the decryption transformations specified by the keys respectively. Let k_{sg} be the key of the sequence generator, z_i be the output character of the SG at time i. The key of the CD cipher system is $k = (k_{sg}, k_0, \cdots, k_{s-1})$. At each time unit only one of the block ciphers is active, i.e., doing the encryption (respectively decryption). So we have

$$c_i = E_{z_i}(k_{z_i}, m_i),$$

where m_i and c_i are the ith plaintext block and ciphertext block. Similarly, the decryption is defined by

$$m_i = D_{z_i}(k_{z_i}, c_i).$$

In this CD cipher system the SG determines the action of each component block cipher, and it is possible for the encryption algorithms $E_0, \cdots, E_{s-1}$ to be the same, but in this case the keys $k_0, k_1, \cdots, k_{s-1}$ should be pairwise different.

The security of the system can be analyzed as follows. First we consider attacks on block ciphers. All the attacks on block ciphers are done under the assumption that the key is fixed and there is only one encryption (respectively decryption) algorithm. Among such attacks are differential attacks and linear attacks. All of those attacks could not apply in a simple way to this CD cipher system, since we have at least two different encryption (resp. decryption) algorithms or at least two different keys for the underlying block ciphers. Second, though there are a number of attacks on stream ciphers, most of them apply only to additive ones, and consequently to those keystream generators for additive stream ciphers. If the CD cipher system is designed properly, those attacks should not apply.

The CD cipher system is a stream ciphering one, though it is a combination of block and stream ciphers, since a message usually corresponds to different ciphertexts at different times. The purpose of cooperation and distribution is to make infeasible as many known attacks on both block and additive stream ciphers as possible. Given a piece of ciphertext, it is usually difficult for the enemy to know how many times a component block cipher has contributed and where it has distributed.

If the system is designed properly, it is possible to get a very strong cipher by choosing some very weak block ciphers and a weak sequence generator. This shows again the power of cooperation and distribution.

The components and the control device in the CD system should be chosen carefully. In what follows we consider the system consisting of two component block ciphers.

Let K_0 and K_1 be the key spaces of the two block ciphers respectively. Assume that each key of K_0 (resp. K_1) is equally likely. Let $p_0 = \Pr(z = 0)$, $p_1 = \Pr(z = 1)$ and

$$n_i(m, c) = |\{k_i \in K_i | E_i(k_i, m) = c\}|, \; i = 0, 1.$$

Also let $\Pr(m, c)$ denote the probability that c is a corresponding ciphertext block of the plaintext block m. Then it is not difficult to see that

$$\Pr(m, c) = p_0 \frac{n_0(m, c)}{|K_0|} + p_1 \frac{n_1(m, c)}{|K_1|},$$

$$\Pr(z = i; (m, c)) = p_i \frac{n_i(m, c)}{|K_0|}, \; i = 0, 1.$$

It follows that we have the conditional probabilities

$$\Pr(z=0|(m,c)) = \frac{|K_1|p_0n_0(m,c)}{|K_1|p_0n_0(m,c)+|K_0|p_1n_1(m,c)}$$
$$\Pr(z=1|(m,c)) = \frac{|K_0|p_1n_1(m,c)}{|K_1|p_0n_0(m,c)+|K_0|p_1n_1(m,c)}$$

Hence, we have the following expression for the average mutual information

$$\begin{aligned} I(z;(m,c)) &= -\frac{|K_1|p_0n_0(m,c)}{|K_1|p_0n_0(m,c)+|K_0|p_1n_1(m,c)} \\ &\quad \times \log \frac{|K_1|p_0n_0(m,c)}{|K_1|p_0n_0(m,c)+|K_0|p_1n_1(m,c)} \\ &\quad -\frac{|K_0|p_1n_1(m,c)}{|K_1|p_0n_0(m,c)+|K_0|p_1n_1(m,c)} \\ &\quad \times \log \frac{|K_0|p_1n_1(m,c)}{|K_1|p_0n_0(m,c)+|K_0|p_1n_1(m,c)}. \end{aligned}$$

To minimize the above average mutual information, we have to ensure that

$$p_0 \frac{n_0(m,c)}{|K_0|} = p_1 \frac{n_1(m,c)}{|K_1|}. \tag{2.1}$$

Note that

$$\sum_{c\in C} \frac{n_0(m,c)}{|K_0|} = \sum_{c\in C} \frac{n_1(m,c)}{|K_1|} = 1.$$

It follows that

$$p_0 = \sum_{c\in C} p_0 \frac{n_0(m,c)}{|K_0|} = \sum_{c\in C} p_1 \frac{n_1(m,c)}{|K_1|} = p_1.$$

Hence, $p_0 = p_1 = 1/2$, and furthermore

$$\frac{n_0(m,c)}{|K_0|} = \frac{n_1(m,c)}{|K_1|}. \tag{2.2}$$

With the above analysis, we have obtained the following design principle. For the CD cipher system with two component block ciphers, the parameters should be chosen such that

1. $p_0 \approx \frac{1}{2}$;

2. $\frac{n_0(m,c)}{|K_0|} \approx \frac{n_1(m,c)}{|K_1|}$, and if one of $n_0(m, c)$ and $n_1(m, c)$ is zero, so must be the other.

Clearly, a cipher is secure against ciphertext-only attacks if it is secure against known plaintext attacks. Given some plaintext-ciphertext block pairs, a cryptanalyst may first try to get a piece of keystream and then try to recover the key of the SG or to construct a generator which produces the same control sequence, by analyzing the parameters $n_0(m, c)$ and $n_1(m, c)$ of the two block ciphers for the given plaintext-ciphertext pairs. If the two block ciphers are not well designed, and the cryptanalyst gets to know $n_0(m, c) = 0$, then he/she knows immediately that the control digit under which a block cipher is selected is 1. If an attack on the SG is successful, then it remains only to attack the two block ciphers in the usual sense. At this stage the meaning of cooperation is lost. The above design principle is intended to make infeasible this kind of divide-and-conquer attack.

On the other hand, the SG should be designed so that its output sequences have good pattern distributions. If the control sequence is $111 \cdots 1000 \cdots 0$, then the cooperation is obviously very bad.

A CD system can be much more secure than the underlying block ciphers. If the SG is well designed, some weak block ciphers can be employed. It is important that the two block ciphers should have many similarities, just like "twins". This indicates that using only a two-key cooperation within one well-designed algorithm seems better, but in this approach one has to guarantee that the two keys do not specify the same encryption transformation; otherwise there is no cooperation within the system.

2.2 Some Keystream Generators

Finite state machines are important mathematical objects for modeling electronic hardware. Furthermore, due to their recursiveness finite state machines are convenient means for realizing infinite wordfunctions built over finite alphabets. Many keystream generators can be modeled by finite state machines [112, 343]. In a synchronous stream cipher, the running-key generator may generally be viewed as an autonomous finite state machine as depicted in Figure 2.4.

The keystream generator as a finite state machine consists of an output alphabet and a state set, together with two functions and an initial state. The *next state function* f_s maps the current state S_j into a new state S_{j+1} from the state set, and the output function f_0 maps the current state S_j into an output symbol z_j from the output alphabet. The key may determine the next state function and the output function as well as the initial state.

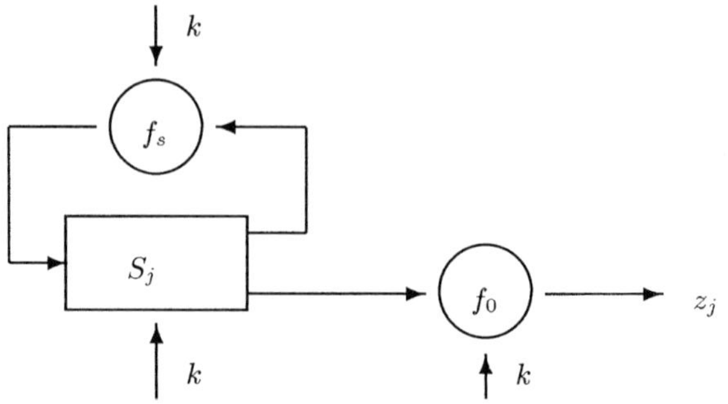

$S_{j+1} = f_s(S_j)$ $z_j = f_0(S_j)$

Figure 2.4: Keystream generators as autonomous finite state machines.

The fundamental problem of designing a keystream generator is to find a next state function f_s and an output function f_0 which are guaranteed to produce a running key z^∞ that satisfies certain cryptographic requirements such as large linear complexity and good linear complexity stability, good autocorrelation, uniform pattern distribution, etc. In some cases, the output function f_0 should possess the good difference property with respect to some binary operation of the state vector space and good nonlinearity with respect to the binary operation of the state vector space and that of the output alphabet space. These binary operations depend on the realization of the next state function f_s. The actual specific requirements for the next state and output function depend on the system in which the generator is used.

In order to meet certain requirements, special classes of finite state machines have been employed as running-key generators. Unfortunately, the theory of autonomous automata whose change of state function is nonlinear has not been well developed. There are many kinds of proposed keystream generators. Some are easy to implement, but their security may be difficult to control; some are secure against certain kinds of attacks, but may have a relatively slow implementation. In what follows we shall give a brief description of some number-theoretic generators and counter generators.

2.2.1 Generators Based on Counters

A counter, one of the simplest automata, has a period, which is often taken to be q^n, where q is a positive integer. A counter of period N counts the numbers $0, 1, ..., N-1$ cyclically. Diffie and Hellman suggested applying

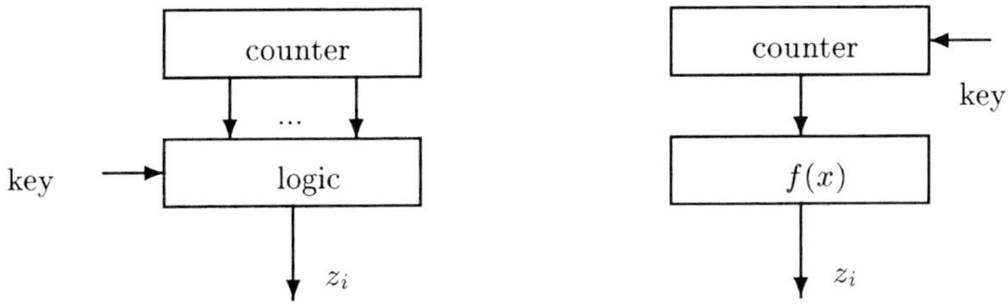

(a) Diffie-Hellman (b) Natural sequence generator

Figure 2.5: Some counter generators.

a nonlinear function to a counter to construct keystream generators [112, p.416], as depicted in Figure 2.5(a). In this kind of generator, the key is used to control the function. The initial values of the counter may be taken as part of the key or as a random value sent as an indicator. A specific proposal given by Diffie and Hellman is to use a fixed component of a block cipher algorithm as the function for the generator of Figure 2.5(a) [112].

If we consider counters of arbitrary period N and use a fixed function $f(x)$ from Z_N to an Abelian group G, we have the generator of Figure 2.5(b). In this generator, the key k is one of the integers $0, 1, ..., N-1$, and the counter begins its cyclical counting at the key value. The arguments x of $f(x)$ are the successive integer values provided by the counter. Thus the output sequence or keystream in G is given by

$$z_i = f((i+k) \bmod N),$$

where the residue modulo N is taken to be an integer between 0 and $N-1$. There are slight differences between the two generators. In the generator of Figure 2.5(a), the key or part of the key is used to control the function, while in the generator of Figure 2.5(b) the function $f(x)$ is fixed and the key is simply the initial value of the register. The generator of Figure 2.5(b) is called the *natural sequence generator* (briefly, NSG) because every periodic sequence can be realized by this generator in a natural way and many security aspects of the generator can be analyzed and controlled. Synchronous additive stream ciphers based on this kind of generator are called *additive natural stream ciphers* [122].

In [122] the differential cryptanalysis and design of the additive natural stream ciphers were studied. It was shown that an improperly designed natural keystream generator could be broken by a differential attack. Other possible attacks, such as key determining attacks based on decision trees, partial-key attacks, linear approximation attacks with respect to the additions of Z_N and the Abelian group over which the key stream is built and key (key stream) correlation attacks, were also possible for this generator if the design parameters are not chosen properly [122]. If the generator is properly designed, the NSG may resist all possible attacks mentioned above. This book is mainly concerned with the design and analysis of this generator.

2.2.2 Some Number-Theoretic Generators

"Pseudorandom" numbers are needed not only in cryptography, but also in numerical simulations for Monte Carlo methods, sampling, numerical analysis, testing computer chips for defects, decision making, and programming slot machines [256, 245]. However, different applications require different random properties of the numbers. For instance, pseudorandom numbers for simulations are different from those for cryptographic purposes in a number of senses. There are several proposed number-theoretic generators. One of these is the *multiplicative generator*, which is described by

$$x_{n+1} = ax_n + b \bmod M,$$

where $0 \leq x_n \leq M - 1$. Here (a, b, M) are the parameters describing the generator and x_0 is the seed. More generally, one can consider polynomial recurrences (mod$N$), or vector-valued polynomial recurrences as done by Lagarias and Reeds [255]. These linear congruential generators are widely used in practice in Monte Carlo methods [245, 375, 288] but they are cryptographically weak. It has been pointed out by Lagarias [256] that for the above linear congruential generator, the parameters a and b can be recovered from three consecutive iterates x_1, x_2, x_3 if M is known. When a, b and M are not known, there is a polynomial algorithm which, given the output $(x_1, \cdots, x_n)$ of a linear congruential generator (mod$M$), will generate a prediction x'_{n+1}. This algorithm has the property that if this is done for $n = 1, 2, 3, \cdots$, it will make at most $3 + \log M$ mistakes. For details about the algorithm we refer to Lagarias [256] and Boyar [30].

If we expand the rational

$$\frac{1}{p} = .d_0 d_1 d_2 \cdots d_j d_{j+1} \cdots$$

in base d, we have the $1/p$ *generator*. Here (p, d) are parameters describing the generator, and the seed is a specified position j of the initial digit; i.e., set $x_n = d_{j+n}$. Details about the generator can be found in [26, 256], however we shall consider this generator in Chapter 14.

There is also the so-called *power generator* described by

$$x_{n+1} = x_n^d \bmod N,$$

where (d, N) are parameters describing the generator and x_0 is the seed. There are two special cases of the power generator, both occurring when $N = p_1 p_2$ is a product of two distinct odd primes. If d is chosen such that $\gcd(d, \phi(N)) = 1$, then the map $x \to x^d$ is a permutation of Z_N^*, and the generator is called the *RSA generator* by Lagarias [256]. If we choose $d = 2$ and $N = p_1 p_2$ with $p_1 = p_2 = 3 \bmod 4$, this is the *square generator.* Properties of these generators can be found in Lagarias [256] and Blum, Blum, and Shub [26]. We shall consider the Blum-Blum-Shub generator in Section 14.8.

A number-theoretic generator based on the exponential operation is the following one described by

$$x_{n+1} = g^{x_n} \bmod N,$$

where (g, N) are parameters describing the generator and x_0 is the seed. For more about this generator, one should consult [256].

The generators of this section could be quite slow, when the modulus is large. By modifying the above generators, one may obtain some number-theoretic bit generators. Among them are the RSA bit generator [3, 256], the modified Rabin bit generator [3, 359, 256], the discrete exponential generator [25, 278, 256].

2.3 Cryptographic Aspects of Sequences

Sequences for stream ciphering purposes are very different from those for other purposes. It is often the case that sequences used in one stream cipher are required to have some properties which are different from those required for some other sequences employed in another stream cipher. Thus, cryptographic sequences may be different in some aspects. However, for keystream sequences for additive synchronous stream ciphers there are some common cryptographic measures of their strength such as the linear complexity (linear span or linear equivalence), sphere complexity, pattern distribution, and autocorrelation property. This section introduces some of these measures and illustrates their cryptographic importance. Here only sequences over finite fields are discussed.

2.3.1 Minimal Polynomial and Linear Complexity

To introduce linear complexity, we need the shift operator on sequences. A left shift operator E is defined by $Es_i = s_{i-1}$ for all possible i. In this way we can define recursively the operators E^l for $l > 1$. Thus, for a polynomial $f(x)$ of $GF(q)[x]$ the polynomial operator $f(E)$ is well defined, if we write $E^0 = 1$, the identity operator. If a sequence is over a finite field $GF(q)$ and $f(x)$ is a polynomial with coefficients in $GF(q)$ given by

$$f(x) = c_0 + c_1 x + \cdots + c_{L-1} x^{L-1},$$

then we define

$$f(E)s_j = c_0 s_j + c_1 s_{j-1} + \cdots + c_{L-1} s_{j-L+1}.$$

Let s^n denote a sequence $s_0 s_1 \cdots s_{n-1}$ of length n over a finite field $GF(q)$. For a finite sequence, the n is finite; for a semi-infinite sequence the n is ∞. A polynomial $f(x) \in GF(q)[x]$ of degree $\leq l$ with $c_0 \neq 0$ is called a *zero polynomial* or *characteristic polynomial* of the sequence s^n if

$$f(E)s_j = 0, \text{ for all } j \text{ with } j \geq l. \tag{2.3}$$

If the above equations hold for l, then they hold also for $l+1$. Thus, for every zero polynomial there is a least $l \geq \deg(f)$ such that the above equations hold. We call the smallest l the *associated recurrence length* of $f(x)$ with respect to the sequence. It is easy to see that there are zero polynomials of a sequence such that their associated recurrence length is minimal. Such a zero polynomial is called a *minimal polynomial* of the sequence, and the associated recurrence length is called the *linear span* or *linear complexity* of the sequence, which is denoted by $\mathrm{L}(s^n)$ hereafter. It follows immediately from this definition that $\mathrm{L}(s^n) = 0$ iff $s^n = 0^n$, where 0^n denotes the all-zero sequence of length n. Another immediate consequence is that, for $0 < n < \infty$, $\mathrm{L}(s^n) = n$ iff $s^{n-1} = 0^{n-1}$ and $s_{n-1} \neq 0$. If a semi-infinite s^∞ is periodic, then its minimal polynomial is unique if we require that $c_0 = 1$. The linear complexity of a periodic sequence is equal to the degree of its minimal polynomial.

The engineering interpretation of linear complexity is as the length of the shortest linear feedback shift register (LFSR) that generates the sequence, see Figure 2.6. Such an LFSR is said to be non-singular when $c_1 \neq 0$, i.e., when it corresponds to a linear recursion of order $\mathrm{L}(s^n)$. The minimal polynomial is actually a feedback polynomial of the LFSR. In the LFSR of length L in Figure 2.6 the boxes containing $s_{j-1}, s_{j-2}, \cdots, s_{j-L+1}, s_{j-L}$ are memory units, and with each clock tick the content of the right-most memory unit is output, while the contents of other memory units are shifted to their right-hand neighboring memory units respectively, the left-most memory unit is then occupied by the new element

$$s_j = -c_1 s_{j-1} - \cdots - c_L s_{j-L}.$$

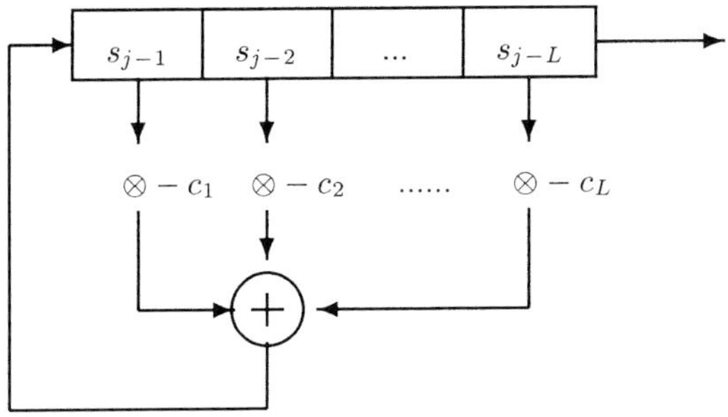

Figure 2.6: The linear feedback shift register interpretation of recursion (2.3).

The polynomial $c(x) = c_0 + c_1 x + \cdots + c_L x^L$ is called the *connection polynomial* or *feedback polynomial* of the LFSR, and the sequence s^n is said to be produced by the LFSR.

The cryptographic significance of the linear complexity of keystream sequences is well known due to the Berlekamp-Massey algorithm. If the linear complexity of a key stream is L, then $2L$ consecutive characters of the sequence could be used to construct the whole key stream with the Berlekamp-Massey algorithm. Thus, it is cryptographically necessary but not sufficient to require keystream sequences for additive stream ciphers to have large linear complexity. This will be illustrated when we introduce the sphere complexity. It should be mentioned, however, that for some nonadditive stream ciphers large linear complexity of the keystream is not a necessary cryptographic requirement (see Section 2.1.3).

Recall that in the definition of linear complexity above, the coefficients of the feedback polynomial and the entries of the sequence are required to be in the same field. Let $s^n = s_0 s_1 \cdots s_{n-1}$ be a sequence over the field $GF(q^m)$. The linear complexity of s^n with respect to the subfield $GF(q)$, here and hereafter denoted as $\mathrm{L}_{GF(q)}(s^n)$, is defined as the smallest nonnegative integer L such that there exist $c_1, c_2, \cdots, c_L \in GF(q)$ for which

$$s_j + c_1 s_{j-1} + \cdots + c_L s_{j-L} = 0, \text{ for all } L \leq j < n. \tag{2.4}$$

In this definition [138], the coefficients $c_1, c_2, \cdots, c_L$ are required to be in $GF(q)$. If $m = 1$, then the two complexities are identical. Thus, the linear complexity $\mathrm{L}_{GF(q)}$ is a generalization of the usual linear complexity. The following inequality clearly holds:

$$\mathrm{L}(s^n) \leq \mathrm{L}_{GF(q)}(s^n). \tag{2.5}$$

We now turn to the cryptographic importance of this generalized linear complexity. It is well known that $GF(q^m)$ can be regarded as a linear space of dimension m over $GF(q)$. Let $u_1, u_2, \cdots, u_m$ be a basis of $GF(q^m)$ over $GF(q)$, then each u of $GF(q^m)$ can be expressed as

$$u = \sum_{i=1}^{m} a_i u_i, \ a_1, \cdots, a_m \in GF(q).$$

Assume that for every j we have

$$s_j = \sum_{i=1}^{m} s_{i,j} u_i, \quad s_{i,j} \in GF(q),$$

then recursion (2.4) is equivalent to

$$\begin{array}{l} s_{i,j} + c_1 s_{i,j-1} + \cdots + c_L s_{i,j-L} = 0, \\ \text{for all } L \leq j \leq n, \text{ and for all } i = 1, \cdots, m. \end{array} \tag{2.6}$$

This means that the linear complexity $\mathrm{L}_{GF(q)}(s^n)$ is the shortest length of the LFSRs which can generate the lower field sequences $s_{i,0} s_{i,1} \cdots s_{i,n-1}, i = 1, 2, \cdots, m$, at the same time only with different initial states. The determination of the shortest LFSR which generates the m sequences is called the LFSR synthesis of multisequences, which is useful in decoding cyclic codes. Several algorithms for the LFSR synthesis of multisequences have been developed [72, 154, 155, 156, 118, 138]. If we use multisequences to encipher a message stream in parallel or to use a matrix sequence to encipher, then this kind of generalized linear complexity is cryptographically important.

Now we turn to the usual linear complexity of periodic sequences. As already mentioned, the linear complexity of periodic sequences over finite fields is precisely the degree of their minimal polynomials. Cryptographically, we need to know not only the linear complexity of a sequence, but the minimal polynomial also. To introduce some results about the minimal polynomials of periodic sequences, we need the concept of generating functions.

The *formal power series* or *generating function* of a semi-infinite sequence s^∞ over $GF(q)$ is defined by

$$s(x) = \sum_{i=0}^{\infty} s_i x^i.$$

If s^∞ is periodic with period N, then we have

$$(1 - x^N) s(x) = s^N(x) = \sum_{i=0}^{N-1} s_i x^i.$$

It follows that the following proposition holds.

Proposition 2.3.1 *The generating function of each periodic sequence* s^∞ *can be expressed as*

$$s(x) = g(x)/f(x) \tag{2.7}$$

with $f(0) \neq 0$ *and* $\deg(g) < \deg(f)$.

The expression in (2.7) is called a *rational form* of the generating function $s(x)$ and of the sequence s^∞. If $\gcd(g(x), f(x)) = 1$, then it is called a *reduced rational form.* Let us stipulate that f_s will denote the minimal polynomial of a sequence s^∞. The following two classic propositions are very useful; their proofs can be found, for example, in [276, 390, 138].

Proposition 2.3.2 *Let* s^∞ *be a periodic sequence over* $GF(q)$ *and*

$$s(x) = r(x)/f(x), \ f(0) = 1$$

a rational form of the generating function of s^∞. *Then* $f(x)$ *is the minimal polynomial of the sequence iff* $\gcd(r(x), f(x)) = 1$.

Concerning the minimal polynomial of the sum of two periodic sequences we have the following conclusion, which follows easily from Proposition 2.3.2.

Proposition 2.3.3 *Let the reduced rational forms of two periodic sequences* s^∞ *and* t^∞ *be respectively*

$$s(x) = r_s(x)/f_s(x), \ t(x) = r_t(x)/f_t(x).$$

Then the minimal polynomial of the sum sequence of the two sequences is given by

$$f_{s+t} = f_s f_t / \gcd(f_s f_t, r_s f_t + r_t f_s).$$

2.3.2 Pattern Distribution of Key Streams

Let s^∞ be a sequence of period N over $GF(q)$, where N is not necessarily the least period. The vector $(s_t, s_{t+\tau_1}, \cdots, s_{t+\tau_{k-1}})$ is called a *pattern of length* k with distances $(\tau_1, \tau_2 - \tau_1, \cdots, \tau_{k-1} - \tau_{k-2})$. A pattern $(s_t, s_{t+\tau})$ was also called a τ*-bigram* by Selmer [390]. The τ-bigrams were first introduced by Zierler under another name for the purpose of studying the autocorrelation of maximum period length sequences [472]. We adopt the usual terminology and call these maximum-length sequences or m-sequences.

Consider the following sequence s^∞ of period 7:

$$s^\infty = \underbrace{0111001}\underbrace{0111001}\ldots$$

and the pattern $0*1**0$, where $*$ indicates an arbitrary element. It is easily seen that this pattern appears only once in a period of the sequence.

The notion of a multiplier was first introduced by Carmichael [66]. If s^∞ is a sequence of least period N, an element $M \neq 1$ in $GF(q), q > 2$, such that

$$M \cdot s^\infty = s_\tau^\infty, \ 0 < \tau < N, \tag{2.8}$$

is called a *multiplier* of the sequence, where s_τ^∞ is the τ-shift version of s^∞. This multiplier is related to the multiplier of residue difference sets [15]. The τ here was called the *span* of M by Ward [434]. For maximum-length sequences over $GF(q)$, Zierler showed that the τ-bigrams are evenly distributed if τ is not the span of some multiplier $M \neq 1$. It is important to observe that no such multiplier exists for binary sequences.

Bigram, trigram and τ-gram are terms from linguistic studies. We will use the term pattern instead. The distribution of some special patterns was investigated by Golomb [167, 169], i.e., the runs of 0's and 1's, which were called *gap* and *block* respectively. He proved that in a binary maximum-length sequence of period $2^n - 1$, there are 2^n runs. Half the runs have length 1, one fourth have length 2, one-eighth have length 3, etc., until two runs of length $n-2$; for each of these lengths, there are equally many gaps and blocks. Finally, there is one gap of length $n-1$ and one block of length n. These are Golomb's three randomness postulates.

Why is the pattern distribution property of a keystream cryptographically important? Some intuitive facts can only give us some superficial reasons. For example, the pattern distribution of length 1 is in fact the distribution of the elements of the field in the sequence. Thus, if there are many more 1's than 0's in a binary sequence, then the sequence is not cryptographically good. To see the cryptographic importance of a roughly equally likely distribution for certain patterns, we first prove a conservation law of patterns.

Consider now the patterns of length k with distances $(\tau_1, \tau_2 - \tau_1, \cdots, \tau_{k-1} - \tau_{k-2})$. For a sequence with least period N over $GF(q)$, the vector variable $(s_t, s_{t+\tau_1}, \cdots, s_{t+\tau_{k-1}})$ takes on vectors of $GF(q)^k$ when t ranges from 0 to $N-1$. Let $n((s_t, s_{t+\tau_1}, \cdots, s_{t+\tau_{k-1}}) = a)$ denote the number of times with which the vector variable $(s_t, s_{t+\tau_1}, \cdots, s_{t+\tau_{k-1}})$ takes on $a \in GF(q)^k$ when t ranges from 0 to $N-1$. It is straightforward to see that the following theorem holds.

Theorem 2.3.4 (The conservation law of patterns) *Let the symbols be the same as before, then*

$$\sum_{a \in GF(q)^k} n((s_t, s_{t+\tau_1}, \cdots, s_{t+\tau_{k-1}}) = a) = N.$$

Clearly, this theorem means that these $n((s_t, s_{t+\tau_1}, \cdots, s_{t+\tau_{k-1}}) = a)$ are conservative. The constant $n((s_t, s_{t+\tau_1}, \cdots, s_{t+\tau_{k-1}}) = a)/N$, denoted as $\Pr(a)$, is the probability that $(s_t, s_{t+\tau_1}, \cdots, s_{t+\tau_{k-1}})$ takes on a. It follows that

$$\sum_{a \in GF(q)^k} \Pr(a) = 1. \tag{2.9}$$

In general, *bad patterns* refer to those which appear with small probability in the key streams. If in a sequence of period N over $GF(q)$ the almost equally likely distribution of a pattern of length k with distances $(\tau_1, \tau_2 - \tau_1, \cdots, \tau_{k-1} - \tau_{k-2})$ is required, this means that $n((s_t, s_{t+\tau_1}, \cdots, s_{t+\tau_{k-1}}) = a)$ is approximately a constant, namely N/q^k. Now the question is why such a uniform pattern distribution should be cryptographically required. This can be shown by the differential attack on the natural sequence generator of Figure 2.5(b) [122]. The idea behind the attack is that bad patterns give much more information about the key than other patterns. That differential cryptanalysis implies the following general randomness requirement for keystream sequences of the NSG:

Randomness requirements: In a sequence of least period N over $GF(q)$ for each k with $1 \leq k \leq \lfloor \log_q N \rfloor$, the pattern $((s_t, s_{t+\tau_1}, \cdots, s_{t+\tau_{k-1}}) = a)$ of length k appears approximately $\lfloor N/q^k \rfloor$ times when t ranges from 0 to $N-1$.

This requirement may be reasonable only for some applications. It is known that uniform pattern distributions in a sequence result in good autocorrelation properties of the sequence, which will be seen in the next subsection.

2.3.3 Correlation Functions

Many problems in ranging systems, radar systems, spread-spectrum communication systems, multiple-terminal system identification, and code-division multiple-access communications systems require sets of signals which have one or both of the following properties:

- Each signal in the set is easy to distinguish from a time-shifted version of itself;

- each signal in the set is easy to distinguish from (a possibly time-shifted version of) every other signal in the set.

This leads to an intensive study of the periodic autocorrelation and crosscorrelation functions of sequences (see Sarwate [381], Helleseth and Kumar [200], also [417, 419, 420]).

Let $GF(q)$ be a finite field. We need the idea of an additive character of $GF(q)$ (see [276], p. 190). Let χ be an additive character of $GF(q)$, and s^∞ and t^∞ be two sequences of period respectively N and M and $P = \text{lcm}\{M, N\}$. Then the *periodic crosscorrelation function* of the two sequences is defined by

$$\text{CC}_{s,t}(l) = \sum_{i=0}^{P-1} \chi(s_i - t_{i+l}) = \sum_{i=0}^{P-1} \chi(s_i)\overline{\chi(t_{i+l})}. \tag{2.10}$$

If the two sequences are identical, then $P = M = N$ and the crosscorrelation function is the so-called *periodic autocorrelation function* of s^∞ described by

$$\text{AC}_s(l) = \sum_{i=0}^{N-1} \chi(s_i - s_{i+l}) = \sum_{i=0}^{N-1} \chi(s_i)\overline{\chi(s_{i+l})}. \tag{2.11}$$

If $q = 2$, then $\chi(a) = (-1)^a$ is an additive character of $GF(2)$, here we identify $GF(2)$ with Z_2. Then (2.10) and (2.11) are the usual crosscorrelation and autocorrelation functions of binary sequences.

For two sequences s^∞ and t^∞ of period respectively N and M and $P = \text{lcm}\{M, N\}$, the *aperiodic crosscorrelation function* of the two sequences is defined by

$$\text{ACC}_{s,t}(l, u, v) = \sum_{i=u}^{v} \chi(s_i - t_{i+l}) = \sum_{i=u}^{v} \chi(s_i)\overline{\chi(t_{i+l})}, \tag{2.12}$$

where $\overline{x}$ denotes the complex conjugate of x. If the two sequences are identical, then $P = M = N$ and the crosscorrelation function is the so-called *aperiodic autocorrelation function* of s^∞ described by

$$\text{AAC}_s(l, u, v) = \sum_{i=u}^{v} \chi(s_i - s_{i+l}) = \sum_{i=u}^{v} \chi(s_i)\overline{\chi(s_{i+l})}. \tag{2.13}$$

Here our definitions of aperiodic functions are slightly different from those used for communication purposes [381], which are defined to be

$\mathrm{AAC}_{s,t}(l,0,v)$ and $\mathrm{AAC}_s(l,0,v)$ respectively. Aperiodic autocorrelation results may be much more important cryptographically than periodic autocorrelation results, since the former reflect local randomness and the latter reflect global randomness: what we actually use for stream ciphering is only a small piece of periodic sequences. Generally speaking, the periodic autocorrelation is relatively much easier to control than the aperiodic autocorrelation. The connections between the autocorrelation function and other cryptographic notions will be described in detail for the binary case in Section 2.4

2.3.4 Sphere Complexity and Linear Cryptanalysis

Let x be a finite sequence of length n over $GF(q)$. The *weight complexity* [137, 138] of the finite sequence is defined by

$$\mathrm{WC}_u(x) = \min_{\mathrm{WH}(y)=u} \mathrm{L}(x+y), \tag{2.14}$$

where $\mathrm{WH}(y)$ denotes the Hamming weight of y, i.e., the number of components of y that are different from zero.

Consider now the space $GF(q)^n$ with Hamming distance d_H. Denoting $S(x,u) = \{y : \ d_H(x,y) = u\}$, by definition we have

$$\mathrm{WC}_u(x) = \min_{y \in S(x,u)} \mathrm{L}(y).$$

This means that the weight complexity is the maximum lower bound of linear complexities of all the sequences of length n on the sphere surface $S(x,u)$. The name of this kind of complexity comes from this geometrical meaning.

Let $O(x,u) = \{y : \ 0 < d_H(x,y) \leq u\}$ be the sphere with center x. The *sphere complexity* [137, 138] is defined by

$$\mathrm{SC}_u(x) = \min_{y \in O(x,u)} \mathrm{L}(y) = \min_{0<v\leq u} \mathrm{WC}_v(x). \tag{2.15}$$

Similarly, let s^∞ be a sequence of period N (not necessarily least period) over $GF(q)$. The weight and sphere complexity of periodic sequences with respect to N are defined respectively by

$$\mathrm{WC}_u(s^\infty) = \min_{\mathrm{WH}(t^N)=u, \mathrm{per}(t^\infty)=N} \mathrm{L}(s^\infty + t^\infty) \tag{2.16}$$

$$\mathrm{SC}_u(s^\infty) = \min_{0<v\leq u} \mathrm{WC}_v(s^\infty), \tag{2.17}$$

where $\mathrm{per}(t^\infty) = N$ denotes that t^∞ has period N.

These two complexities were introduced to measure the stability of the linear complexity function, in analogy to the derivative of functions in Euclidean spaces. The cryptographic background of these complexities is that some key streams with large linear complexity can be approximated by some sequences with much lower linear complexity [137, 138]. The sphere and weight complexity are based on the LFSR approximation model. In contrast to the linear complexity which is based on the shortest LFSR that produces a sequence, the sphere complexity $\mathrm{SC}_k(s^\infty)$ is based on the shortest LFSR that produces another sequence with a probability of agreement no less than $(1 - k/N)$, where N is a period of the sequence s^∞ with which the sphere complexity is concerned. The weight complexity $\mathrm{WC}_k(s^\infty)$ is based on the shortest LFSR that produces another sequence with a probability of agreement equal to $(1 - k/N)$.

To illustrate the difference between the linear complexity and sphere complexity, we consider the binary sequence of period N:

$$s^\infty = \underbrace{0....01}_{N}\underbrace{0....01}_{N}...$$

The linear complexity of the sequence is N by definition since no LFSR of length less than N can produce it. However its sphere complexity $\mathrm{SC}_1(s^\infty) = 1$ by definition since there is an LFSR of length one that produces the all-zero sequence having the probability $(1 - 1/N)$ of agreement with the sequence s^∞.

These LFSR approximation model complexities are cryptographically important only if there is an efficient algorithm to find the LFSR for approximating the original generator. To see the cryptographic importance of these complexities, we describe an attack on all synchronous additive stream ciphers [123].

Suppose that a cryptanalyst has a number of consecutive ciphertext-plaintext pairs of a synchronous additive stream cipher which enable him to derive a piece of key stream, say $z_0 z_1 \cdots z_{n-1}$. Suppose also that he knows nothing else but the plaintext source code of the enemy's messages. What can he do under these assumptions in order to decipher the enemy's ciphertext? The best the cryptanalyst can do may be the construction of a new generator which produces a sequence with a large probability of agreement with the original key stream.

To make things simple, we assume that the key stream is binary. Under the assumption that the linear complexity of the enemy's key stream is very unstable, the cryptanalyst can try to construct an LFSR to approximate the original keystream generator according to the following procedure:

Step 1 Use the Berlekamp-Massey algorithm to construct an LFSR which produces the sequence $z^n = z_0 z_1 \cdots z_{n-1}$. Then use the constructed LFSR to decipher a large piece of ciphertext. If only ϵ percent (this constant can be flexible, say less than 15) of the deciphered ciphertext makes no sense, then accept the LFSR and stop; otherwise, go to Step 2.

Step 2 For $i = 0$ to $n-1$, do the following: Change z_i into $z_i \oplus 1$. Apply the Berlekamp-Massey algorithm to the new sequence to construct an LFSR which produces the new sequence. Then use the constructed LFSR to decipher a large piece of ciphertext. If only ϵ percent (this constant can be flexible, say less than 15) of the deciphered ciphertext makes no sense, then accept the LFSR and stop; otherwise, repeat this step for $i+1$ if $i < n-1$, and go to Step 3 if $i = n-1$.

Step 3 For a possible pair (i, j) with $i < j$ and $i, j \in \{0, 1, ..., n-1\}$, change z_i into $z_i \oplus 1$ and z_j into $z_j \oplus 1$. Then apply the Berlekamp-Massey algorithm to the new sequence to construct an LFSR which produces the new sequence. Then use the constructed LFSR to decipher a large piece of ciphertext. If only ϵ percent (this constant can be flexible, say less than 15) of the deciphered ciphertext makes no sense, then accept the LFSR and stop; otherwise, repeat this step for the next pair (i, j) with $i < j$ if there is a remaining pair, and print "fail" and stop if there is no pair remaining.

Since the complexity of Berlekamp-Massey algorithm for sequences of length n is $O(n^2)$, the complexity of this attack is $O(n^4)$. Thus, if s^∞ is a key stream such that its linear complexity is very large (say, for example, 2^{40}) and $\mathrm{SC}_k(s^\infty)$ is small enough (say less than 1000 for example) for some very small k, then this attack must succeed. The basic idea of this attack is that, we expect that the keystream sequence can be expressed as

$$z^\infty = u^\infty + v^\infty$$

such that u^∞ and u^∞ are of period N, the $\mathrm{WH}(v^N)/N$ is very small and the sequence u^∞ has small linear complexity. This can be done when the linear complexity of the key stream is very unstable. In this case, we expect that the known key stream z^n can be expressed as

$$z^n = u^n + v^n$$

with $\mathrm{WH}(v^n) \leq 2$ if $n < 2N/k$. Furthermore, we may also use the regular decimation sequences of z^n to replace z^n, then derive the minimal polynomial of u^n from the decimated sequences. Thus, it follows that the designer of a synchronous additive stream cipher must ensure that for very small

k's, the sphere complexity $\mathrm{SC}_k(s^\infty)$ is large enough. In other words, the designer of an additive synchronous stream cipher should make sure that his key stream cannot be well approximated by a sequence with small linear complexity, since the above polynomial-time algorithm can be used to find an LFSR to approximate the original keystream sequence if its linear complexity is very unstable. This shows why sphere complexity is cryptographically important. For the purpose of measuring the linear complexity stability of sequences, fixed-complexity distance and variable-complexity distance were introduced. The connection between these measures and some lower bounds on these measures for some sequences can be found in [138].

The linear complexity stability problem for sequences was also considered by Stamp and Martin [412] under the name of k-error linear complexity which is defined to be $\min\{\mathrm{SC}_k(s), \mathrm{L}(s)\}$ and is essentially the same as sphere complexity.

Note that the motivation behind the sphere complexity is linear cryptanalysis on two kinds of stream ciphers introduced by Ding, Xiao and Shan in 1988 [137], see also [119, 138]. The basic idea of linear cryptanalysis for stream ciphers is to use a related linear system to approximate the original highly nonlinear system, or in other words to use linear circuits to approximate nonlinear circuits. For details about the linear cryptanalysis of two kinds of stream ciphers and specific examples we refer to [138]. We note that the linear cryptanalysis for stream ciphers was done earlier than that for block ciphers.

2.3.5 Higher Order Complexities

Linear complexity (also called *linear span*) of a sequence is defined to be the length of the shortest LFSR that generates the sequence. A sequence with very large linear span may be generated by a much shorter FSR (feedback shift register) if nonlinear terms are allowed in the feedback function. If only quadratic terms and linear terms are allowed, then we have the *quadratic span* [157, 69]. If general terms are allowed, then we have the *maximum order complexity* [219] or generally *span* [69].

The quadratic span and other nonlinear spans are cryptographically important only when there are efficient algorithms for finding the shortest nonlinear FSRs or an FSR that is short enough. Suppose there is an efficient algorithm for finding the shortest quadratic FSR that generates any given sequence; then we need to investigate further the relations between the linear span and quadratic span. It is obvious that the linear span of any sequence is greater than or equal to the quadratic span of the sequence. If the relation

$$\mathrm{L}(s^{\infty}) \geq \mathrm{Q}(s^{\infty}) \geq \sqrt{\mathrm{L}(s^{\infty}) + a},$$

holds for every periodic binary sequence, where $\mathrm{Q}(s^{\infty})$ denotes the quadratic span of the sequence s and a is a constant, then control of the linear span results also in control of the quadratic span. Thus investigation of the following research problem is interesting.

Research Problem 2.3.5 *Investigate whether there are constants a and b such that*

$$\mathrm{Q}(s^{\infty}) \geq \sqrt{\mathrm{L}(s^{\infty}) \times b + a}$$

for each sequence of period N over $GF(q)$, where the constants depend only on the period and q.

2.4 Harmony of Binary NSGs

The NSG of Figure 2.5(b) is cryptographically attractive because not only can every periodic sequence be produced with a proper choice of the parameters of the generator, but also many of its security aspects are consistent. For the binary natural sequence generator the following cryptographic analyses are equivalent:

1. differential analysis of the cryptographic function $f(x)$;
2. nonlinearity analysis of the cryptographic function $f(x)$;
3. autocorrelation analysis of the cryptographic function $f(x)$;
4. autocorrelation analysis of the output sequence;
5. two-bit pattern distribution analysis of the output sequence;
6. stability analysis of the mutual information $I(i;\ z_i z_{i+t-1})$ (here and hereafter z^{∞} denotes the output sequence of the NSG);
7. transdensity analysis of the additive stream cipher system with this NSG as the keystream generator (by which we mean the analysis of the probability of agreement between two encryption resp. decryption transformations specified by two encryption resp. decryption keys [122]).

Equivalence is understood in the sense that one analysis gives another analysis, and conversely.

We now prove the equivalence between the above seven analyses and show that the "ideal difference property" of the cryptographic function $f(x)$ ensures automatically

- ideal nonlinearity of the cryptographic function $f(x)$,
- ideal autocorrelation property of $f(x)$,
- ideal autocorrelation property of the output sequence z^∞,
- ideal two-bit pattern distribution property of the output sequence z^∞, and
- ideal balance between the mutual information $I(i;\ z_i z_{i+t-1})$ for all possible pairs $(z_i, z_{i+t-1}) \in Z_2 \times Z_2$, where t is arbitrary.

In what follows Z_N denotes the residue class ring modulo an integer N. Our notation for the autocorrelation function in this section is different from the one in the last section for the sake of simplicity.

Consider now the NSG of Figure 2.5(b). Assume that $(G, +)$ is the Abelian group over which the keystream sequence is constructed, and $|G| = n$. For each $g_i \in G$ let

$$C_i = \{x \in Z_N : f(x) = g_i\}.$$

The ordered set $\{C_0, C_1, \cdots, C_{n-1}\}$ is called the *characteristic class.* For any ordered partition $\{C_0, C_1, \cdots, C_{n-1}\}$ of Z_N, there exists a function $f(x)$ with this partition as its characteristic class. The differential analysis of the system is the analysis of the following *difference parameters*:

$$d_f(g_i, g_j; w) = |C_i \cap (C_j - w)|, \quad (g_i, g_j) \in G \times G, \ w \in Z_N.$$

We say that f has the *ideal difference property* if the values $d_f(g_i, g_j; w)$ are approximately the same for all possible $(g_i, g_j; w)$.

To see why the analysis of the difference parameters can be regarded as a kind of differential analysis, we take $(G, +) = (Z_2, +)$. Consider the input pairs (x, y) such that $x - y = a$, and consider the difference of the corresponding output pairs. Then we have the following expressions

$$\frac{|\{(x,y): f(x) - f(y) = 1,\ x - y = a\}|}{|\{(x,y): x - y = a\}|} = \frac{d_f(0,1;a)}{N} + \frac{d_f(1,0;a)}{N}$$
$$\frac{|\{(x,y): f(x) - f(y) = 0,\ x - y = a\}|}{|\{(x,y): x - y = a\}|} = \frac{d_f(0,0;a)}{N} + \frac{d_f(1,1;a)}{N},$$

These two expressions show that the difference parameters can be regarded as partial differentials or directional differentials of the function $f(x)$.

In what follows we prove the equivalence between the above seven analyses for the binary NSG (natural sequence generator).

Between differential and nonlinearity analysis

Let $g(x)$ be a mapping from an Abelian group $(G,+)$ to another one $(H,+)$. The nonlinearity of g is measured by

$$P_g = \max_{0\neq a\in G} \max_{b\in H} \Pr(f(x+a)-f(x)=b),$$

where $\Pr(A)$ denotes the probability of the occurrence of event A. Here $P_g(a)$ could be called the differential of $g(x)$ at a. However, elementary calculus shows that differentials are ideal measures for nonlinearities. We shall deal with highly nonlinear functions in details in Chapter 6.

The nonlinearity analysis of the cryptographic function $f(x)$ refers to the analysis of the probability $\Pr(f(x+a)-f(x)=b)$. It can be easily seen that

$$\begin{aligned} N\Pr(f(x)-f(x-a)=1) &= d_f(0,1;a)+d_f(1,0;a),\\ N\Pr(f(x)-f(x-a)=0) &= d_f(0,0;a)+d_f(1,1;a) \end{aligned} \tag{2.18}$$

and

$$\begin{aligned} 2d_f(0,0;-a) &= |C_0|-|C_1|+N\Pr(f(x+a)-f(x)=0),\\ 2d_f(1,1;-a) &= |C_1|-|C_0|+N\Pr(f(x+a)-f(x)=0),\\ 2d_f(1,0;-a) &= 2d_f(0,1;-a) = N-N\Pr(f(x+a)-f(x)=0). \end{aligned} \tag{2.19}$$

Then formulae (2.18) and (2.19) show the equivalence.

Between differential and autocorrelation analysis

The autocorrelation analysis of $f(x)$ refers to the analysis of the (normalized) autocorrelation function

$$\mathrm{AC}_f(a) = \frac{1}{N}\sum_{x\in Z_N}(-1)^{f(x+a)-f(x)}.$$

It is easily verified that

$$N\mathrm{AC}_f(a) = N-4d_f(1,0;a) \tag{2.20}$$

and

$$\begin{aligned} 4d_f(0,0;a) &= 4|C_0|-N+N\mathrm{AC}_f(a),\\ 4d_f(1,1;a) &= 4|C_1|-N+N\mathrm{AC}_f(a),\\ 4d_f(1,0;a) &= 4d_f(0,1;a) = N-N\mathrm{AC}_f(a). \end{aligned} \tag{2.21}$$

Combining formulae (2.20) and (2.21) proves the equivalence between the differential and autocorrelation analysis of $f(x)$.

The autocorrelation analysis of the output binary sequence z^∞ refers to the analysis of the autocorrelation function

$$\mathrm{AC}_z(a) = \frac{1}{N} \sum_{i \in Z_N} (-1)^{z_{i+a} - z_i}.$$

Clearly by the definition of the NSG we have

$$\mathrm{AC}_z(a) = \mathrm{AC}_f(a), \quad \text{for each } a.$$

Thus, the above formulae (2.20) and (2.21) are also true if we replace $C_f(a)$ with $C_z(a)$. This fact shows the equivalence between the differential analysis and the autocorrelation analysis of the output sequence z^∞.

Between differential and two-bit pattern distribution analysis

The two-bit pattern distribution analysis of z^∞ is concerned with how the two-bit patterns are distributed (see Section 2.3.2). For each fixed t with $0 < t \leq N - 1$ the vector (z_i, z_{i+t}) takes on elements of $Z_2 \times Z_2$ when i ranges from 0 to $N-1$. Let $n[(z_i, z_{i+t}) = (a, b)]$ denote the number of times which the vector (z_i, z_{i+t}) takes on $(a, b) \in Z_2 \times Z_2$ when i ranges from 0 to $N - 1$. Then we have obviously

$$n[(z_i, z_{i+t}) = (a, b)] = d_f(a, b; -t). \tag{2.22}$$

Thus, for the binary NSG each difference parameter represents in fact the number of times with which a two-bit pattern appears in a segment of length N of the binary output sequence z^∞.

Between differential and mutual information analysis

We are given two bits z_i and z_{i+t} of the output sequence of the binary NSG. It is cryptographically interesting to know how much information these two bits give to the content of the register of the counter in the binary NSG at the time the output bit z_i was produced. It is easy to verify

$$I(i; z_i z_{i+t}) = \log_2 N - \log_2 d_f(z_i, z_{i+t}; -t) \quad \text{bits} \tag{2.23}$$

and

$$d_f(z_i, z_{i+t}; -t) = N 2^{-I(i; z_i z_{i+t})}, \tag{2.24}$$

where the mutual information $I(i; z_i z_{i+t})$ is measured in bits. Formulae (2.23) and (2.24) clearly show the equivalence. In addition they show that the difference parameters are in fact a measure of uncertainty.

Between differential and transdensity analysis

In a cipher system it is possible for two keys to determine the same encryption (resp. decryption) transformation. Even if the two transformations are distinct, it is cryptographically interesting to know the probability of agreement between the ciphertexts given by the two transformations. The control of this probability of agreement may protect a cipher from a key approximation attack, that is, the use of one key to decrypt a message encrypted by another key. Let E_k (resp. D_k) denote the encryption (resp. decryption) transformation specified by the key k. The analysis of the density (briefly, transdensity analysis) of a cipher system refers to the analysis of the probability of agreement $\Pr(E_k(m) = E_{k'}(m))$, where m can be restricted to plaintext blocks or without restriction [122].

For the additive binary stream cipher with the binary NSG as its keystream generator this probability can be expressed easily as

$$\Pr(E_k = E_{k'}) = \mathrm{AC}_z(k - k' \bmod N) = \mathrm{AC}_f(k - k' \bmod N), \qquad (2.25)$$

because of the additive structure of the additive stream cipher and the fact that the keystream sequences specified by all keys are shift versions of each other. Thus, the equivalence follows easily from formula (2.25).

So far we have proved the equivalence between differential analysis and the other six analyses. Thus, equivalence among the seven analyses follows. In addition, there is no trade-off between all the above seven properties and the linear complexity and its stability for this generator (we will see this fact in later chapters). This means that it is possible to design the NSG so that it is not only ideal with respect to all seven properties, but also has large linear complexity and ideal linear complexity stability for the output sequence. It is because of these facts and because every periodic sequence can be produced by the natural sequence generator that the generator is called a natural one [122].

Formulae 2.18–2.25 clearly show that to ensure ideal behavior with respect to all seven properties, it suffices to control the difference property of the cryptographic function $f(x)$. Thus, in later chapters we will concentrate on the control of the difference property of $f(x)$, of the linear complexity, and of the sphere complexity of the output sequence for each specific NSG.

2.5 Security and Attacks

Without attacks on cipher systems there would be no problem of security. Security is associated with attacks and is usually relative to attacks. Attacks are also relative to the assumptions about a cryptanalyst's knowledge of a cipher system.

Attacks can be classified according to the assumed available information about a cipher that the cryptanalyst has. This kind of classification results in three types of attacks: (A) *ciphertext-only attacks* under the assumption that only pieces of ciphertext are known to a cryptanalyst; (B) *known-plaintext attacks* under the assumption that a piece of ciphertext with corresponding plaintext is known; (C) *chosen-plaintext attacks* under the assumption that a cryptanalyst has a chosen piece of plaintext with corresponding ciphertext.

Suppose that a cryptanalyst has got a piece of keystream sequence, then there are two further assumptions concerning attacks on stream ciphers:

B1: It is assumed that the cryptanalyst knows only a piece of keystream. In this case there are the following possible attacks: (1) *equivalent-machine attacks*, which make use of the piece of key stream to construct a new generator which produces the same key stream. For example, if the linear complexity of the key stream is not very large, the Berlekamp-Massey algorithm can be used to construct an LFSR which produces the same key stream. (2) *Approximate-machine attacks*, which make use of the known piece of key stream to construct another generator to approximate the original generator. One example of these attacks is the attack based on the linear complexity stability of the key stream described in Section 2.3.4.

B2: Apart from a piece of keystream it is assumed that the cryptanalyst also knows the type of generator and the cryptographic algorithm. Under these assumptions attacks are flexible in forms and in techniques, for example, those in [4, 5, 138, 122, 471]. Attacks under these assumptions depend on specific systems and on the technique a cryptanalyst uses. In this case, there is one more type of attack than in the case (B1), i.e., *key recovering attacks*, which aim to recover the original key or equivalent keys.

For ciphertext-only attacks, some other information about a cipher is usually assumed to be known to a cryptanalyst. For example, under the assumption that the structure of the keystream generator is known there are several kinds of correlation attacks [401, 306] and key-recovering attacks.

The general idea of the key-recovering attacks is to make use of known data about a cipher to get information about the key. The techniques of attack vary with the structure of the ciphers and with the known data. The following "mother problem" may illustrate the flexibility of such attacks.

Suppose that there are ten children in such an order $z_1 z_2 \cdots z_{10}$ that z_i is older than or has the same birthday as z_{i+1}. We are told that these children have the same father F which is known and also the same mother which could be any one of the 100 mothers $\{M_1, \cdots, M_{100}\}$. If we have further information about the pairs (F, M_i), then we can make use of it to make the set containing the mother smaller or to determine the mother. For instance, by studying the known information about all (F, M_i), suppose we know that twins are only possible for (F, M_1). Then if twins are found in the 10 children, the mother must be M_1. Of course, it may be technically difficult to find this character of (F, M_1). In fact we can identify the keys with mothers, the algorithm with the father, and the ciphertext with the children.

To determine the mother, one person may try to study the distribution of sexes among the children because she has some technique to study the sex of possible children of each (F, M_i). Another person may try to study the color of the hair of the mothers, father and children. Others may study the blood types of father, mothers and children and the age distribution of the mothers and children. Every method must use the idea of "consistency" or "correlation" either in an obvious way or in a hidden way. Perhaps everyone will have her own technique for getting information about the mother.

Keystream generators are flexible and diverse. Each generator may have its own special security aspects, although some common requirements exist (e.g., linear and sphere complexity). Some cipher systems are easy to implement, but may have tradeoffs between known security parameters; some are relatively difficult to implement, but their security may be easy to control; others may have both an easy implementation and ideal security, but be slow. Of course, fewer tradeoffs make the design easier. In designing secure cipher systems the most important problems are:

1. How can we build systems which have as few security tradeoffs as possible?
2. What are the tradeoffs or conflicts in a given system?
3. How do we manage tradeoffs and conflicts?
4. How do we coordinate security and performance?

We consider such questions in later chapters of this book.

Chapter 3

Primes, Primitive Roots and Sequences

In this chapter we search for those pairs (N, $GF(q)$) such that every sequence of period N over $GF(q)$ has both large linear and sphere complexity when the Hamming weight of one period of the sequence is neither too large nor too small. Such pairs (N, $GF(q)$) are called *good partner pairs* since they work in harmony. This is why we write the title of this chapter as "primes, primitive roots and sequences". In this chapter we consider only the linear and sphere complexity aspect of sequences. Other aspects will be studied in later chapters. This chapter is mainly based on Ding [123] and Ding [125].

The cryptography-related topics of number theory discussed in this chapter are: cyclotomic polynomials, Euler's function, Carmichael function, primitive roots, least primitive roots, common primitive roots, Artin's conjectures, Fermat's Last Theorem, order, Wieferich and non-Wieferich primes, Stern primes, Sophie Germain primes, o-primes, e-primes, Tchebychef primes and primes of other forms as well as the Chinese Remainder Theorem.

3.1 Cyclotomic Polynomials

For every integer $n \geq 1$, Euler's function $\phi(n)$ is defined to be the number of integers a such that $\gcd(a, n) = 1$, where $1 \leq a < n$. This function has the following properties:

1. If p is a prime, then $\phi(p) = p - 1$.

2. For any prime p, $\phi(p^k) = p^{k-1}(p-1)$.

3. If $m, n \geq 1$ and $\gcd(m,n) = 1$, then $\phi(mn) = \phi(m)\phi(n)$, that is, ϕ is a *multiplicative function*.

4. For any integer $n = \prod_p p^k$, $\phi(n) = \prod_p p^{k-1}(p-1)$.

Proofs of these properties are easy and can be found in most books about number theory.

Cyclotomic polynomials have close relations with coding theory [279]. It will be seen in the following sections that the linear complexity and period of sequences as well as their stability are also closely related to cyclotomic polynomials. So we summarize now some known results which are needed in later sections.

Let K be a field of characteristic p, n a positive integer not divisible by p, and ξ an nth primitive root of unity over K. Then the polynomial

$$Q_n(x) = \prod_{s=1,\gcd(s,n)=1}^{n} (x - \xi^s)$$

is called the nth *cyclotomic polynomial* over K. References about cyclotomic polynomials can be found, for example, in [276, p.64].

Proposition 3.1.1 Basic Facts [276]:

1. *$Q_n(x)$ is independent of the choice of ξ.*

2. *$\deg(Q_n(x)) = \phi(n)$.*

3. *The coefficients of $Q_n(x)$ belong to the prime subfield of K.*

4. *$x^n - 1 = \prod_{d|n} Q_d(x)$.*

5. *If $K = GF(q)$ with $\gcd(q,n) = 1$, then Q_n factors into $\phi(n)/d$ distinct monic irreducible polynomials in $K[x]$ of the same degree d, where d is the least positive integer such that $q^d \equiv 1 \pmod{n}$, i.e., d is the order (or exponent) of q modulo n, denoted as $\mathrm{ord}(q)$ modulo n or $\mathrm{ord}_n(q)$.*

With the help of Propositions 3.2.1 and 3.1.1, it is not difficult to arrive at the following result, which will play an important role in designing some keystream sequences.

Proposition 3.1.2 *Assume that $\gcd(n,q) = 1$. Then Q_n is irreducible over $GF(q)$ if and only if $n = r^k$, $2r^k$ or 4, where r is an odd prime and $k \geq 0$, and q is a primitive root modulo n.*

3.2 Two Basic Problems from Stream Ciphers

For sequences of period N over the field $GF(q)$, their linear and sphere complexity are closely related with the factorization of cyclotomic polynomials $Q_n(x)$ over $GF(q)$ for all factors n of N. Proposition 3.1.1 says that $Q_n(x)$ factors into $\phi(n)/d$ distinct monic irreducible polynomials in $GF(q)$ of the same degree d, where d is the least positive integer such that $q^d \equiv 1 \pmod{n}$. It follows that, to design sequences with both large linear and sphere complexity, we should find pairs (N, q) such that

1. N has as few factors as possible; and
2. for each factor n of N, $d = \text{ord}_n(q)$ should be as large as possible.

This leads to the following two basic problems in designing cryptographic sequences for certain applications.

Basic Problem 1 *Find large positive integers N and small positive integers q which are powers of primes such that*

1. $\gcd(N, q) = 1$;
2. $\text{ord}_n(q) = \phi(n)$ *for any factor* $n \neq 1$ *of* N.

Basic Problem 2 *Find large positive integers N and small positive integers q, q a power of a prime, such that*

1. $\gcd(N, q) = 1$;
2. *N has few factors;*
3. $\text{ord}_n(q)$, *a factor of* $\phi(n)$, *is as large as possible for any factor* $n \neq 1$ *of* N.

An integer q is said to be a *primitive root* of (or modulo) n if $\text{ord}_n(q) = \phi(n)$. If $g \equiv g' \pmod{N}$, then g is a primitive root of N if and only if g' is a primitive root of N. So for our cryptographic purposes, we discuss here and hereafter primitive roots modulo N only in the range between 2 and $N - 1$. To study the two problems further, we need the following important result of Gauss whose proof can be found in most books about number theory.

Proposition 3.2.1 *If p is a prime, then there exist $\phi(p-1)$ primitive roots of p. The only integers having primitive roots are p^e, $2p^e$, 1, 2 and 4, with p being an odd prime.*

This proposition shows that Basic Problem 1 has a solution if and only if $N = r^k$, or $2r^k$, with r being an odd prime. We shall investigate this basic problem in detail in Sections 3.4 and 3.5.

Before dealing with Basic Problem 2, we present some basic results about the order of integers modulo n. If $\gcd(a,n) = 1$, Euler's theorem states that $a^{\phi(n)} \equiv 1 \pmod{n}$. This implies that $\mathrm{ord}_n(a)$ divides $\phi(n)$. The order of a has a close relation to the *Carmichael function* $\lambda(n)$, which is defined by

$$\begin{aligned}
&\lambda(1) = 1,\ \lambda(2) = 1,\ \lambda(4) = 2,\\
&\lambda(2^r) = 2^{r-2} \text{ (for } r \geq 3).\\
&\lambda(p^r) = p^{r-1}(p-1) = \phi(p^r) \text{ for any odd prime } p \text{ and } r \geq 1,\\
&\lambda(2^r p_1^{r_1} p_2^{r_2} \cdots p_s^{r_s}) = \mathrm{lcm}(\lambda(2^r), \lambda(p_1^{r_1}), \cdots, \lambda(p_s^{r_s})),
\end{aligned}$$

where lcm denotes the least common multiple. It is not difficult to see that the order of a modulo n is at most equal to $\lambda(n)$, and that $\lambda(n)$ divides $\phi(n)$.

It seems difficult to solve Basic Problem 2 completely. However, for those N's which are a product of two distinct primes, it is possible to find the associated q's such that (N, q) is a solution of Basic Problem 2. We shall deal with this problem in Section 3.8.

Before ending this section, we make some preparations for the following two sections. Specifically, we introduce now the concept of *negative order* of an integer a modulo an integer N, and discuss the relation of the negative order with the order.

Definition 3.2.2 *Let N and a be positive integers. If there is a positive integer m such that $a^m \equiv -1 \pmod{N}$, then we call the smallest such m the negative order of a modulo N (we coin the word "negord" to denote the negative order), and denote it as* $\mathrm{nord}_N(a)$.

An integer a may have a negord modulo an integer N or not. As an example, we consider $N = 23$. It is easily checked that 1, 2, 4, 8, 16, 9, 18, 13, 36 and 12 have no negord, but 17, 11, 22, 21, 19, 15, 7 and 14 have a negord. It is for the purpose of investigating the order that we introduce the concept of the negord.

The relation of the order and negord is stated in the following theorem.

Theorem 3.2.3 *Let N be a positive integer. If an integer a, where $1 \leq a \leq N-1$ and $gcd(a, N) = 1$, has a negord modulo N, then*

$$\mathrm{ord}_N(a) = 2\mathrm{nord}_N(a).$$

Proof: By definition $a^{\text{nord}_N(a)} \equiv -1 \pmod{N}$. It follows that $a^{2\text{nord}_N(a)} \equiv 1 \pmod{N}$. Hence, $\text{ord}_N(a)$ divides $2\text{nord}_N(a)$. We now prove that $\text{ord}_N(a) \geq 2\text{nord}_N(a)$. If not so, then there are two possibilities: $\text{ord}_N(a) < \text{nord}_N(a)$ and $\text{nord}_N(a) < \text{ord}_N(a) < 2\text{nord}_N(a)$. It is easily verified that in both cases there must exist an integer l, where $1 \leq l < \text{nord}_N(a)$, such that $a^l \equiv -1 \pmod{N}$. This is contrary to the minimality of the negord of a modulo N. Thus, $\text{ord}_N(a)$ must be equal to $2\text{nord}_N(a)$. □

A simple property of negord, which is similar to that of order, is the following conclusion.

Theorem 3.2.4 *If $a^m \equiv -1 \pmod{N}$ for a positive integer m, then $\text{nord}_N(a)|m$ and $m/\text{nord}_N(a)$ is odd.*

Proof: Let $m = \text{nord}_N(a)h+l$, where $0 \leq l < \text{nord}_N(a)$. We first prove that h must be odd. From $a^m \equiv (a^{\text{nord}_N(a)})^h a^l \pmod{N}$ we get $a^l \equiv (-1)^{h+1} \pmod{N}$. By the definition of the negord h is odd.

If $l \neq 0$, then $l \geq 1$. The equation $a^l \equiv 1 \pmod{N}$ gives that $\text{ord}_N(a) < \text{nord}_N(a)$, which is contrary to Theorem 3.2.3. Therefore, $l = 0$. This completes the proof. □

Now we give a characterization of primitive roots in terms of negord. This characterization is useful in searching for primitive roots.

Theorem 3.2.5 *Let N be a positive integer > 4 which has primitive roots. Then a is a primitive root modulo N if and only if $\text{nord}_N(a) = \phi(N)/2$.*

Proof: If a is a primitive root modulo N, by Proposition 3.2.1 N must be of the form p^e or $2p^e$, where p is an odd prime. Thus $\phi(N)$ must be even. Since $a^{\phi(N)} \equiv 1 \pmod{N}$, we get

$$(a^{\phi(N)/2} + 1)(a^{\phi(N)/2} - 1) \equiv 0 \pmod{N}.$$

This gives $a^{\phi(N)/2} \equiv -1 \pmod{N}$. Thus, the negord of a modulo N exists. Now by Theorem 3.2.3 we have $\text{nord}_N(a) = \phi(N)/2$. The remaining part then follows from Theorem 3.2.3. □

This theorem shows that a necessary condition for a to be a primitive root is $a^{\phi(N)/2} \equiv -1 \pmod{N}$. It can be used as a criterion for primitivity. As an example, we take $N = 43$. Then we have $2^{\phi(N)/2} = 2^{(N-1)/2} = 2^{3\times 7} \equiv -1 \pmod{N}$. But 2 is not a primitive root of 43. This is because $\text{nord}_{43}(2) = 7 \neq 21$.

3.3 A Basic Theorem and Main Bridge

As linear and sphere complexity are important security criteria for keystream sequences for additive stream ciphering, the control of these two parameters becomes one of the key issues in designing keystream generators. For this purpose the following Basic Theorem 3.3.1 is useful [123].

Basic Theorem 3.3.1 (Ding [123]) *Suppose* $N = p_1^{e_1} \cdots p_t^{e_t}$, *where* $p_1, \cdots, p_t$ *are* t *pairwise distinct primes, and* q *is a positive integer such that* $\gcd(q, N) = 1$. *Then for each nonconstant sequence* s^∞ *of period* N *over* $GF(q)$,

$$\mathrm{L}(s^\infty) \geq \min\{\mathrm{ord}_{p_1}(q), \cdots, \mathrm{ord}_{p_t}(q)\}$$

and

$$\begin{aligned} \mathrm{SC}_k(s^\infty) &\geq \min\{\mathrm{ord}_{p_1}(q), \cdots, \mathrm{ord}_{p_t}(q)\}, \\ \textit{if } k &< \min\{\mathrm{WH}(s^N), N - \mathrm{WH}(s^N)\}. \end{aligned}$$

To prove this theorem, we need the following two propositions.

Proposition 3.3.2 *Let* $n_1, n_2, ..., n_t$ *be pairwise relatively prime positive integers, and* g *an integer with* $\gcd(g, n_i) = 1$ *for each* $1 \leq i \leq t$. *Then*

$$\mathrm{ord}_{n_1 n_2 \cdots n_t}(g) = \mathrm{lcm}\{\mathrm{ord}_{n_1}(g), \mathrm{ord}_{n_2}(g), ..., \mathrm{ord}_{n_t}(g)\}.$$

Proof: By the Chinese Remainder Theorem

$$Z_n \cong Z_{n_1} \times \cdots \times Z_{n_t},$$

where $n = n_1 \cdots n_t$, and the isomorphism is given by

$$f(x \bmod n) = (x \bmod n_1, ..., x \bmod n_t).$$

If $g^d \equiv 1 \pmod n$, then $f(g^d \bmod n) = f(1)$. It follows that

$$(g^d \bmod n_1, ..., g^d \bmod n_t) = (1, ..., 1).$$

Thus, each $\mathrm{ord}_{n_i}(g)$ divides d, and therefore

$$\mathrm{lcm}\{\mathrm{ord}_{n_1}(g), ..., \mathrm{ord}_{n_t}(g)\} | \mathrm{ord}_{n_1 \cdots n_t}(g).$$

On the other hand, let $d = \mathrm{lcm}\{\mathrm{ord}_{n_1}(g), ..., \mathrm{ord}_{n_t}(g)\}$. Then $f(g^d \bmod n) = f(1) = (1, ..., 1)$. Thus, $g^d \bmod n = 1$. It follows that $\mathrm{ord}_{n_1 n_2 \cdots n_t}(g)$ must be d. □

Proposition 3.3.3 *Let p be a prime, $k \geq 1$ and $a \geq 1$ be integers. Then* $\text{ord}_{p^k}(a) \geq \text{ord}_p(a)$.

Proof: Assume $\text{ord}_{p^k}(a) = m$. Then $a^m - 1 \equiv 0 \pmod{p^k}$. Hence $a^m - 1 \equiv 0 \pmod{p}$. It follows that $\text{ord}_p(a)$ divides $\text{ord}_{p^k}(a)$ and the conclusion follows. □

Finding conditions for the equality $\text{ord}_{p^k}(a) = \text{ord}_p(a)$ seems to be a complicated, but cryptographically useful problem. For the special case $\text{ord}_p(a) = p - 1$, Proposition 3.4.1 below gives useful information.

Research Problem 3.3.4 *Find conditions which ensure the equality*

$$\text{ord}_{p^k}(a) = \text{ord}_p(a),$$

where p is a prime, k and a are integers no less than 2.

Proof of Basic Theorem 3.3.1: By assumptions and part (5) of Proposition 3.1.1

$$x^N - 1 = \prod_{n|N} Q_n(x)$$

and the polynomial $Q_n(x)$ is equal to the product of $\phi(n)/d$ distinct monic irreducible polynomials over $GF(q)[x]$ of the same degree d, where $d = \text{ord}_n(q)$.

If n divides N, there are integers $h_{i_1}, ..., h_{i_s}$ such that $n = p_{i_1}^{h_{i_1}} \cdots p_{i_s}^{h_{i_s}}$, where $1 \leq h_{i_j} \leq e_{i_j}$ for $j = 1, 2, ..., s$, and $1 \leq s \leq t$. By Propositions 3.3.2 and 3.3.3

$$\begin{aligned} \text{ord}_n(q) &= \text{lcm}\{\text{ord}_{p_{i_1}^{h_{i_1}}}(q), \cdots, \text{ord}_{p_{i_t}^{h_{i_t}}}(q)\} \\ &\geq \max\{\text{ord}_{p_{i_1}^{h_{i_1}}}(q), \cdots, \text{ord}_{p_{i_t}^{h_{i_t}}}(q)\} \\ &\geq \min\{\text{ord}_{p_1}(q), \cdots, \text{ord}_{p_t}(q)\}. \end{aligned}$$

Since the minimum polynomial of each sequence of period N over $GF(q)$ divides $x^N - 1$, and s^∞ is a nonconstant sequence, the conclusion of this theorem follows. □

If $t = 1$ and $e_1 = 1$, Basic Theorem 3.3.1 gives a general lower bound for the linear and sphere complexity of sequences with a prime period. Theorem 3.3.1 is called basic because it gives most of the theorems of Chapters 3 and 4 as special cases.

We say that Theorem 3.3.1 is a bridge between number theory and stream ciphers because it makes a clear connection between the linear and sphere complexity of sequences and many number-theoretic problems such as primes of special forms (e.g., twin primes and Sophie German primes, i.e., primes p with $2p+1$ being also prime) and their distributions, primality testing, primitive roots and their distributions, and primitivity testing. Some of these connections will be made clear in this chapter and Chapter 4.

This basic theorem shows that it is usually quite easy to control the global linear and sphere complexity. However, it seems fairly difficult to control the local linear and sphere complexity.

The condition $\gcd(N, q) = 1$ in Basic Theorem 3.3.1 makes this theorem not applicable for many sequences. The following theorem is a generalization of the basic theorem.

Theorem 3.3.5 *Let $N = p^k p_1^{e_1} \cdots p_t^{e_t}$, where $p, p_1, \cdots, p_t$ are $t+1$ pairwise distinct primes and $k \geq 0, e_1 \geq 1, ..., e_t \geq 1$ are integers. Let $q = p^m$, where $m \geq 1$ is an integer. Then for each sequence s^∞ of period N over $GF(q)$*

1. *either $0 \leq \mathrm{L}(s^\infty) \leq p^k$ in which case the minimal polynomial of s^∞ is $(x-1)^{\mathrm{L}(s^\infty)}$ or*

$$\mathrm{L}(s^\infty) \geq \min\{\mathrm{ord}_{p_1}(q), \cdots, \mathrm{ord}_{p_t}(q)\};$$

2. *either $0 \leq \mathrm{SC}_k(s^\infty) \leq p^k$ or*

$$\mathrm{SC}_k(s^\infty) \geq \min\{\mathrm{ord}_{p_1}(q), \cdots, \mathrm{ord}_{p_t}(q)\},$$
$$\textit{if } k < \min\{\mathrm{WH}(s^N), N - \mathrm{WH}(s^N)\}.$$

Proof: Since the minimal polynomial of s^∞ divides $x^N - 1$, we consider the factorization of $x^N - 1$ over $GF(q)$. Let $N' = p_1^{e_1} \cdots p_t^{e_t}$. Since $GF(q)$ has characteristic p and $\gcd(N', q) = 1$, By Proposition 3.1.1 we have

$$x^N - 1 = (x^{N'} - 1)^{p^k} = (x-1)^{p^k} \prod_{n|N',\ n \neq 1} Q_n(x)^{p^k},$$

where $Q_n(x)$ is the product of $\phi(n)/d$ monic irreducible polynomials over $GF(q)$ with degree $d = \mathrm{ord}_n(q)$. By the proof of Basic Theorem 3.3.1 the degree of the minimal polynomial of s^∞ is no less than

$$\min\{\mathrm{ord}_{p_1}(q), \cdots, \mathrm{ord}_{p_t}(q)\}$$

if the minimal polynomial has a factor of a form other than $(x-1)^h$, where h is an integer. Thus, the conclusions of this theorem follow. □

Theorem 3.3.5 is practical when k is small. The following related result is also useful.

Theorem 3.3.6 *Let $N = p^k p_1^{e_1} \cdots p_t^{e_t}$, where $p, p_1, \cdots, p_t$ are t pairwise distinct primes and $k \geq 0, e_1 \geq 1, ..., e_t \geq 1$ are integers. Let $q = p^m$, where $m \geq 1$ is an integer. Then for each sequence s^∞ of period N over $GF(q)$*

$$\mathrm{L}(s^\infty) \geq \min\{\mathrm{ord}_{p_1}(q), \cdots, \mathrm{ord}_{p_t}(q)\}$$

if $x^{p^k} - 1$ divides $s^N(x) = s_0 + s_1 x + \cdots + s_{N-1}x^{N-1}$, where $s^N = s_0 s_1 \cdots s_{N-1}$ is the first periodic segment of s^∞.

Proof: Note that the minimal polynomial of s^∞ is

$$\frac{x^N - 1}{\gcd(x^N - 1, s^N(x))} = \frac{(x^{N'} - 1)^{p^k}}{\gcd((x^{N'} - 1)^{p^k}, s^N(x))},$$

where $N' = p_1^{e_1} \cdots p_t^{e_t}$. By assumption the minimal polynomial of s^∞ has no factor of the form $(x-1)^h$. The conclusion of this theorem follows from Theorem 3.3.5. □

3.4 Primes, Primitive Roots and Binary Sequences

We begin this section with two definitions. It is obvious that every odd prime p must be in one of the forms $4t \pm 1$. For simplicity we call primes of the forms $4t \pm 1$ with t odd *o-primes*, those with t even *e-primes*. We shall see later that it is necessary to distinguish between these two kinds of primes in designing binary keystream sequences.

A major aim of this section is to search for prime periods p such that all nonconstant sequences of period p over $GF(2)$ have both large linear and sphere complexity. This leads us to the case $q = 2$ of Basic Problem 1 in Section 3.2.

Before going further we have to return to the topic of primitive roots of integers modulo n. For our cryptographic purposes, as will be seen later, we need to know whether g is a primitive root of a prime power p^e for $e \geq 2$ if it is a primitive root of p. For example, 10 is a primitive root of 487, but not a primitive root of 487^2. The following old result ([361], for a proof see [6] for example) clarifies the situation.

Proposition 3.4.1 *Let p be a prime, then the following three assertions are equivalent:*

1. *g is a primitive root of p and $g^{p-1} \not\equiv 1 \pmod{p^2}$;*

2. *g is a primitive root of p^2;*

3. *for every $e \geq 2$, g is a primitive root of p^e.*

Now we investigate the linear and sphere complexity of sequences with period equal to a prime power.

Theorem 3.4.2 *Let r be an odd prime, $N = r^k$ with $k \geq 1$, and let q be a primitive root modulo N. Then for any nonconstant sequence s^∞ of period N over GF(q),*

1. *there exist coefficients $c_i = 0$ or 1 for each i with $0 \leq i \leq k$, such that*

$$\mathrm{L}(s^\infty) = \sum_{i=1}^{k} c_i(r^i - r^{i-1}) + c_0 \geq r - 1;$$

2. *if $k < \min\{\mathrm{WH}(s^N), N - \mathrm{WH}(s^N)\}$, then $\mathrm{SC}_k(s^\infty) \geq r - 1$.*

Proof: From Proposition 3.1.1 it follows that over GF(q)

$$x^N - 1 = (x-1)\prod_{i=1}^{k} Q_{r^i}(x).$$

Then by Proposition 3.1.1, part 5, Proposition 3.4.1 and the assumptions of the theorem, $Q_{r^i}(x)$ is irreducible over GF(q) for each i with $1 \leq i \leq k$. On the other hand, we have $\phi(r^i) = r^i - r^{i-1}$ and the minimal polynomial of each sequence of period N divides $x^N - 1$. It follows from the above facts and the definitions of linear and sphere complexity that the conclusions of the theorem are true. □

This theorem shows that, if k is small and r is large enough, the linear and sphere complexity of any sequence of period r^k without a bad distribution of the elements of $GF(q)$ in the sequence are both good. From Proposition 3.4.1 we get another version of Theorem 3.4.2 as follows:

Theorem 3.4.3 *Let r be an odd prime, $N = r^k$ with $k \geq 2$, and let q be a primitive root modulo r with $\gcd(r, q) = 1$ and $q^{r-1} \not\equiv 1 \pmod{r^2}$. Then for any nonconstant sequence s^∞ of period N over GF(q),*

1. *there exist coefficients $c_i = 0$ or 1 for each i with $0 \leq i \leq k$, such that*

$$\mathrm{L}(s^\infty) = \sum_{i=1}^{k} c_i(r^i - r^{i-1}) + c_0 \geq r - 1;$$

2. *if* $k < \min\{\mathrm{WH}(s^N), N - \mathrm{WH}(s^N)\}$, *then* $\mathrm{SC}_k(s^\infty) \geq r - 1$.

If we take $k = 1$, i.e., sequences with prime periods, in the proof of Theorem 3.4.2, then the proof gives the following:

Theorem 3.4.4 *Let N be an odd prime with* $\gcd(N, q) = 1$, *and let q be a primitive root modulo N. Then for any nonconstant sequence s^∞ of period N over $GF(q)$,*

1. $\mathrm{L}(s^\infty) = N$ *or* $N - 1$;

2. $$\mathrm{SC}_k(s^\infty) = \begin{cases} N \text{ or } N-1, & \text{if } k < \min\{\mathrm{WH}(s^N), N - \mathrm{WH}(s^N)\}; \\ 0, & \text{otherwise.} \end{cases}$$

Theorem 3.4.4 is also a special case of the following theorem.

Theorem 3.4.5 *If N is prime, then for any nonconstant sequence s^∞ of period N over $GF(q)$ and over $GF(q^s \bmod N)$ with* $\gcd(s, N-1) = 1$ *and with $q^s \bmod N$ being a power of a prime,*

1. $\mathrm{L}(s^\infty) \geq \mathrm{ord}_N(q)$;

2. $$\mathrm{SC}_k(s^\infty) = \begin{cases} \geq \mathrm{ord}_N(q), & \text{if } k < \min\{\mathrm{WH}(s^N), N - \mathrm{WH}(s^N)\}; \\ 0, & \text{otherwise.} \end{cases}$$

Proof: Setting $t = 1$ and $e_1 = 1$ in Basic Theorem 3.3.1 proves the theorem. □

We now turn to applications of o-primes in the design of cryptographic binary sequences. Before doing so, we show the importance of the classification of primes into o-primes and e-primes. Let N be a prime > 4. Theorem 3.2.5 shows that a necessary condition for a to be a primitive root modulo N is that a is a quadratic nonresidue modulo N. On the other hand, we have the Legendre symbol formula

$$\left(\frac{2}{N}\right) = (-1)^{(N^2-1)/8} = \begin{cases} +1, & \text{if } N = 8k \pm 1 \\ -1, & \text{if } N = 8k \pm 3. \end{cases} \tag{3.1}$$

Thus, it is possible for 2 to be a primitive root of an o-prime, but not of an e-prime. This is the significance of the classification.

Now we are ready to search for those o-primes which have primitive root 2. The following two propositions give some special primes having primitive root 2.

Proposition 3.4.6 *If* $N = 4t + 1$ *is prime and t is prime, then 2 is a primitive root modulo N.*

Proof: Since t is odd, N is of the form $8k-3$. By (3.1) and Theorem 3.2.4 the negord of 2 modulo N exists and divides $(N-1)/2 = 2t$. Since t is an odd prime, $N > 5$ and therefore $\text{nord}_N(2) \neq 2$. Then by Theorem 3.2.4 $\text{nord}_N(2) = 2t = (N-1)/2$. It follows from Theorem 3.2.5 that 2 is a primitive root of N. □

Proposition 3.4.7 *Let $N = 4t-1$ be a prime with t odd. If $2t-1$ is prime (i.e., $(N-1)/2$ is a Sophie Germain prime), then 2 is a primitive root of N.*

Proof: It is straightforward to give an argument similar to the proof of Proposition 3.4.6. □

For many o-primes of the two forms $4t \pm 1$ with t odd, neither t nor $2t-1$ is prime. So it is necessary to investigate other conditions which can ensure the primitivity of 2 modulo an o-prime.

Proposition 3.4.8 *Let $N = 4t+1$ be an o-prime and $t = t_1t_2$, where t_1 and t_2 are primes. Then 2 is a primitive root modulo N if and only if*

$$2^{2t_1t_2} \equiv -1 \pmod{N}, \; 2^{2t_1} \not\equiv -1 \pmod{N}, \; 2^{2t_2} \not\equiv -1 \pmod{N},$$

Proof: Note that $\text{ord}_N(2)$ divides $\phi(N) = N-1 = 4t_1t_2$ and that t_1 and t_2 are primes. It then follows from the hypotheses that the order of 2 modulo N must be equal to $N-1$. Now Theorem 3.2.5 completes the proof. □

A similar argument proves the following result.

Proposition 3.4.9 *Let $N = 4t-1$ be an o-prime and $2t-1 = t_1t_2$, where t_1 and t_2 are primes. Then 2 is a primitive root modulo N if and only if*

$$2^{t_1} \not\equiv -1 \pmod{N}, \quad 2^{t_2} \not\equiv -1 \pmod{N}, \quad 2^{t_1t_2} \not\equiv 1 \pmod{N}.$$

We note that the above two propositions can be further generalized to the cases in which t and $2t-1$ have square factors. From the above discussions, we get immediately the following corollaries.

Corollary 3.4.10 *If $N = 4t+1$ is prime and t is an odd prime, then for any nonconstant sequence s^∞ of period N over $GF(2)$ and over $GF(2^s \bmod N)$ with $\gcd(s, N-1) = 1$ and with $2^s \bmod N$ being a power of a prime,*

1. $\text{L}(s^\infty) = N$ *or* $N-1$;

2. $\mathrm{SC}_k(s^\infty) = \begin{cases} N \text{ or } N-1, & \text{if } k < \min\{\mathrm{WH}(s^N), N - \mathrm{WH}(s^N)\}; \\ 0, & \text{otherwise.} \end{cases}$

Proof: Combining Theorem 3.4.4 and Proposition 3.4.6 gives this corollary. □

Corollary 3.4.11 *Let* $N = 4t - 1$ *be a prime with* t *odd. If* $(N-1)/2$ *is prime (i.e., it is a Sophie Germain prime), then for any nonconstant sequence* s^∞ *of period* N *over* $GF(2)$ *and over* $GF(2^s \bmod N)$ *with* $\gcd(s, N-1) = 1$ *and with* $2^s \bmod N$ *being a power of a prime,*

1. $\mathrm{L}(s^\infty) = N$ *or* $N - 1$*;*

2. $\mathrm{SC}_k(s^\infty) = \begin{cases} N \text{ or } N-1, & \text{if } k < \min\{\mathrm{WH}(s^N), N - \mathrm{WH}(s^N)\}; \\ 0, & \text{otherwise.} \end{cases}$

Proof: Combining Theorem 3.4.4 and Proposition 3.4.7 gives this corollary. □

Corollary 3.4.12 *Let* $N = 4t+1$ *be an o-prime and* $t = t_1 t_2$*, where* t_1 *and* t_2 *are primes. If*

$$2^{2t_1t_2} \equiv 1 \pmod N, \; 2^{2t_1} \not\equiv -1 \pmod N, \; 2^{2t_2} \not\equiv -1 \pmod N,$$

then for any nonconstant sequence s^∞ *of period* N *over* $GF(2)$ *and over* $GF(2^s \bmod N)$ *with* $\gcd(s, N-1) = 1$ *and with* $2^s \bmod N$ *being a power of a prime,*

1. $\mathrm{L}(s^\infty) = N$ *or* $N - 1$*;*

2. $\mathrm{SC}_k(s^\infty) = \begin{cases} N \text{ or } N-1, & \text{if } k < \min\{\mathrm{WH}(s^N), N - \mathrm{WH}(s^N)\}; \\ 0, & \text{otherwise.} \end{cases}$

Proof: Combining Theorem 3.4.4 and Proposition 3.4.8 gives this corollary. □

Corollary 3.4.13 *Let* $N = 4t - 1$ *be an o-prime and* $2t - 1 = t_1 t_2$*, where* t_1 *and* t_2 *are primes. If*

$$2^{t_1} \not\equiv -1 \pmod N, \; 2^{t_2} \not\equiv -1 \pmod N, \; 2^{t_1t_2} \not\equiv 1 \pmod N,$$

then for any nonconstant sequence s^∞ *of period* N *over* $GF(2)$ *and over* $GF(2^s \bmod N)$ *with* $\gcd(s, N-1) = 1$ *and with* $2^s \bmod N$ *being a power of a prime,*

1. $\mathrm{L}(s^{\infty}) = N$ *or* $N-1$;

2. $\mathrm{SC}_k(s^{\infty}) = \begin{cases} N \text{ or } N-1, & \text{if } k < \min\{\mathrm{WH}(s^N), N - \mathrm{WH}(s^N)\}; \\ 0, & \text{otherwise.} \end{cases}$

Proof: Combining Theorem 3.4.4 and Propositon 3.4.9 gives this corollary. □

The foregoing corollaries are cryptographically useful in designing binary and nonbinary keystream sequences. They tell us that it is easy to control the linear and sphere complexity of sequences of period N over fields GF(2) and GF($2^s \bmod N$) with $\gcd(s, N-1) = 1$ and $2^s \bmod N$ being a power of a prime, when the o-prime is properly chosen. Some binary keystream sequences based on the results of Corollaries 3.4.10 and 3.4.11 will be discussed in later chapters.

3.5 Primes, Primitive Roots and Ternary Sequences

To design cryptographic ternary sequences, we are interested in knowing which primes have primitive root 3. First, we look at some necessary conditions for 3 to be a primitive root.

Theorem 3.5.1 *Let $N = 4t+1$ be a prime. If 3 is a primitive root modulo N, then $t = 3k+1$ for some positive integer k.*

Proof: Clearly $t \not\equiv 2 \pmod 3$, for then 3 divides N. If $t = 3k$ for some k, then by quadratic reciprocity

$$\left(\frac{3}{N}\right) = (-1)^{[(3-1)/2]\times[(N-1)/2]}\left(\frac{N}{3}\right) = \left(\frac{N}{3}\right) = 1.$$

Thus 3 cannot be a primitive root of N if $t = 3k$. Only the case $t = 3k+1$ remains. □

Similarly, we can prove the following theorem:

Theorem 3.5.2 *Let $N = 4t-1$ be a prime. If 3 is a primitive root modulo N, then $t = 3k+2$ for some positive integer k.*

Summarizing the above results, we see that it is only possible for the primes of forms $12k+7$ and $12k+5$ to possess primitive root 3. Now we investigate which primes in the two classes have primitive root 3. First we have an analog (proved by Stern in 1830 [110]) of Proposition 3.4.6.

Proposition 3.5.3 *If $N = 4t + 1$ is prime and $t = 3k + 1$ is prime, then 3 is a primitive root modulo N.*

Proof: The proof of Proposition 3.4.6 applies to this proposition. □

The following result is quite useful in finding primes having primitive root 3.

Theorem 3.5.4 *(The basic theorem about the order of 3 modulo $N = 4t + 1$)*
Let $N = 4t + 1$ be a prime with $t = 3k + 1 = 2^m t'$, where t' is odd. Then $\mathrm{ord}_N(3) = 2^{m+2}t_1$, where t_1 is a factor of t'.

Proof: By quadratic reciprocity, 3 is a quadratic nonresidue mod N, so $3^{(N-1)/2} \equiv -1 \pmod{N}$ and, since $(N-1)/2 = 2^{m+1}t'$, this gives $3^{t'} \not\equiv 1 \pmod{N}$. By Theorem 3.2.4 we obtain $\mathrm{nord}_N(3) = 2^{m+1}t_1$, where t_1 is a factor of t'. Now Theorem 3.2.3 gives $\mathrm{ord}_N(3) = 2\mathrm{nord}_N(3) = 2^{m+2}t_1$. This proves the theorem. □

An important case of the basic theorem above is when $t' = 1$. In this case, the same proof gives the following conclusion obtained by Richelot in 1832 [110]:

Proposition 3.5.5 *If $p = 2^m + 1$ is a prime, every quadratic nonresidue (in particular, 3) is a primitive root of p.*

Combining this proposition and Theorem 3.4.4 yields the following conclusion.

Theorem 3.5.6 *If $N = 2^m + 1$ is a prime, then for any nonconstant sequence s^∞ of period N over $GF(3)$,*

1. $\mathrm{L}(s^\infty) = N$ *or* $N - 1$;

2. $$\mathrm{SC}_k(s^\infty) = \begin{cases} N \text{ or } N-1, & \text{if } k < \min\{\mathrm{WH}(s^N), N - \mathrm{WH}(s^N)\}; \\ 0, & \text{otherwise.} \end{cases}$$

If $p = 2^m + 1$ is a prime, it is an easy exercise to prove that m must be of the form 2^k. Such primes are called *Fermat primes*, and the numbers $F_n = 2^{2^n} + 1$ are called *Fermat numbers*. We shall discuss these numbers later.

Another important special case of Theorem 3.5.4 occurs when t' is a prime. In this case we have the following corollary, which can be easily derived from Basic Theorem 3.5.4.

Corollary 3.5.7 *Let $N = 4t + 1$ be a prime with $t = 3k + 1 = 2^m t'$ even, where t' is an odd prime. If $3^{2^{m+1}} \not\equiv -1 \pmod{N}$, then 3 is a primitive root modulo N.*

Furthermore, if t' is an odd prime $> 3^{2^{m+1}}/2^{m+2}$, it is obvious that $N > 3^{2^{m+1}} + 1$. Thus, it follows that 3 is a primitive root of N. This proves the following result obtained by Tchebychef in 1849 [110].

Proposition 3.5.8 *If $m > 0$ and n is an odd prime $> 9^{2^m}/2^{m+2}$, then 3 is a primitive root of $4n2^m + 1$.*

Primes of the form $k2^m + 1$ are cryptographically attractive. Such primes are related to Fermat numbers, because Euler showed that every factor of F_n (with $n \geq 2$) must be of the form $k \times 2^{n+2} + 1$. Owing to the cryptographic importance of such primes $4n2^m + 1$ with $m > 0$ and n being an odd prime $> 9^{2^m}/2^{m+2}$, we call them *Tchebychef primes*. With Proposition 3.5.8 and Theorem 3.4.4 we arrive at the following conclusion.

Theorem 3.5.9 *If N is a Tchebychef prime. Then for any nonconstant sequence s^∞ of period N over $GF(3)$,*

1. $\mathrm{L}(s^\infty) = N$ *or* $N - 1$;

2. $$\mathrm{SC}_k(s^\infty) = \begin{cases} N \text{ or } N - 1, & \text{if } k < \min\{\mathrm{WH}(s^N), N - \mathrm{WH}(s^N)\}; \\ 0, & \text{otherwise.} \end{cases}$$

A special kind of Tchebychef primes is those of the form $q = 8p + 1$ with $p > 5$ prime. The integer 3 is, of course, a primitive root of such primes q by Proposition 3.5.8.

Applying Theorem 3.2.5 yields the following corollary:

Corollary 3.5.10 *Let $N = 4t - 1$ be a prime with $t = 3k + 2$. If $2k + 1$ is prime, then 3 is a primitive root modulo N if and only if*

$$3^{2k+1} \not\equiv -1 \pmod{N}, \quad 3^{3(2k+1)} \not\equiv 1 \pmod{N}.$$

It is clear that the above results can be further generalized to the cases in which t and $2t - 1$ have square factors. It is easily seen from the above discussions that the following corollaries, which are similar to Corollaries 3.4.11-3.4.13 respectively, are true.

Corollary 3.5.11 *If $N = 4t + 1$ and $t = 3k + 1$ are odd primes, then for any nonconstant sequence s^∞ of period N over $GF(3)$ and over $GF(3^s \bmod N)$ with $\gcd(s, N - 1) = 1$ and with $3^s \bmod N$ being a power of a prime,*

1. $L(s^\infty) = N$ *or* $N-1$;

2. $$SC_k(s^\infty) = \begin{cases} N \text{ or } N-1, & \text{if } k < \min\{WH(s^N), N - WH(s^N)\}; \\ 0, & \text{otherwise.} \end{cases}$$

Corollary 3.5.12 *Let $N = 4t - 1$ be a prime with $t = 3k + 2$ (t odd or even). If $(N-1)/6$ is prime, $3^{2k+1} \not\equiv -1 \pmod N$ and $3^{3(2k+1)} \not\equiv 1 \pmod N$, then for any nonconstant sequence s^∞ of period N over $GF(3)$ and over $GF(3^s \bmod N)$ with $\gcd(s, N-1) = 1$ and with $3^s \bmod N$ being a power of a prime,*

1. $L(s^\infty) = N$ *or* $N-1$;

2. $$SC_k(s^\infty) = \begin{cases} N \text{ or } N-1, & \text{if } k < \min\{WH(s^N), N - WH(s^N)\}; \\ 0, & \text{otherwise.} \end{cases}$$

Corollary 3.5.13 *Let $N = 4t + 1$ be a prime with $t = 3k + 1 = 2^m t_1 t_2$, where t_1 and t_2 are odd primes. If*

$$\begin{aligned} 3^{2^{m+1}t_1} &\not\equiv -1 \pmod N, \\ 3^{2^{m+1}t_2} &\not\equiv -1 \pmod N, \\ 3^{2^{m+1}t_1 t_2} &\not\equiv 1 \pmod N, \end{aligned}$$

then for any nonconstant sequence s^∞ of period N over $GF(3)$ and over $GF(3^s \bmod N)$ with $\gcd(s, N-1) = 1$ and with $3^s \bmod N$ being a power of a prime,

1. $L(s^\infty) = N$ *or* $N-1$;

2. $$SC_k(s^\infty) = \begin{cases} N \text{ or } N-1, & \text{if } k < \min\{WH(s^N), N - WH(s^N)\}; \\ 0, & \text{otherwise.} \end{cases}$$

The foregoing corollaries, which show how to control the linear and sphere complexity of sequences of period N over fields GF(3) and GF($3^s \bmod N$), are cryptographically quite useful in designing ternary keystream sequences. Some ternary keystream generators based on these results will be constructed in later chapters.

3.6 Primes, Negord and Sequences

As shown in the foregoing sections, large primes having certain small primitive roots are useful in constructing cryptographic sequences. However, such primes may not be easy to find. We now show that some primes can also be used to construct cryptographic ternary sequences, even if they do not have primitive root 3.

In Section 3.5 we have seen that primes of the form $k2^m+1$ are cryptographically valuable, when k is a large prime and m is absolutely small, i.e., the Tchebychef primes, which have primitive root 3. But two things should be made clear. First, primes of the forms $4p+1$, $8p+1$ and $16p+1$ seem hard to find, where p is also prime. Second, most of the known large primes of the form $k2^m+1$ have a very small k which is not a prime. It seems difficult to say whether such large primes have a small primitive root other than 2. However, we will prove that some of them are cryptographically valuable, even though they may have no small prime primitive root.

Theorem 3.6.1 *Let $N=4t+1$ be a prime with $t=3k+1=2^m t'$, where t' is odd. Then for any nonconstant sequence s^∞ of period N over $GF(3)$,*

1. $\mathrm{L}(s^\infty) \geq 2^{m+2}$;
2. $\mathrm{SC}_k(s^\infty) = \begin{cases} \geq 2^{m+2}, & \text{if } k < \min\{\mathrm{WH}(s^N), N-\mathrm{WH}(s^N)\}; \\ 0, & \text{otherwise.} \end{cases}$

Proof: By Theorem 3.5.4 $\mathrm{ord}_N(3) \geq 2^{m+2}$. Then the conclusion follows from Theorem 3.4.5. □

This theorem demonstrates that every sequence of such a period N over $GF(3)$ without bad balance has both large linear and sphere complexity, if t' is very small. Similarly, we can prove the following results for sequences over $GF(5)$, $GF(7)$, $GF(11)$, $GF(13)$ and $GF(17)$.

Theorem 3.6.2 *Let $N=4t+1$ be a prime with t being one of the forms $5k+3$ and $5k+4$ and $t=2^m t'$, where t' is odd. Then for any nonconstant sequence s^∞ of period N over $GF(5)$,*

1. $\mathrm{L}(s^\infty) \geq 2^{m+2}$;
2. $\mathrm{SC}_k(s^\infty) = \begin{cases} \geq 2^{m+2}, & \text{if } k < \min\{\mathrm{WH}(s^N), N-\mathrm{WH}(s^N)\}; \\ 0, & \text{otherwise.} \end{cases}$

Theorem 3.6.3 *Let $N=4t+1$ be a prime with t being of one of the forms $7k+1$, $7k+3$ and $7k+4$ and $t=2^m t'$, where t' is odd. Then for any nonconstant sequence s^∞ of period N over $GF(7)$,*

1. $\mathrm{L}(s^\infty) \geq 2^{m+2}$;
2. $\mathrm{SC}_k(s^\infty) = \begin{cases} \geq 2^{m+2}, & \text{if } k < \min\{\mathrm{WH}(s^N), N-\mathrm{WH}(s^N)\}; \\ 0, & \text{otherwise.} \end{cases}$

Theorem 3.6.4 *Let $N = 4t+1$ be a prime with t being of one of the forms $11k+3$, $11k+4$, $11k+5$, $11k+7$ and $11k+10$, and with $t = 2^m t'$, where t' is odd. Then for any nonconstant sequence s^∞ of period N over $GF(11)$,*

1. $\mathrm{L}(s^\infty) \geq 2^{m+2}$;
2. $\mathrm{SC}_k(s^\infty) = \begin{cases} \geq 2^{m+2}, & \text{if } k < \min\{\mathrm{WH}(s^N), N - \mathrm{WH}(s^N)\}; \\ 0, & \text{otherwise.} \end{cases}$

Theorem 3.6.5 *Let $N = 4t+1$ be a prime with t being one of the forms $13k+1$, $13k+5$, $13k+8$,$13k+10$, $13k+11$ and $13k+12$, and with $t = 3k+1 = 2^m t'$, where t' is odd. Then for any nonconstant sequence s^∞ of period N over $GF(13)$,*

1. $\mathrm{L}(s^\infty) \geq 2^{m+2}$;
2. $\mathrm{SC}_k(s^\infty) = \begin{cases} \geq 2^{m+2}, & \text{if } k < \min\{\mathrm{WH}(s^N), N - \mathrm{WH}(s^N)\}; \\ 0, & \text{otherwise.} \end{cases}$

Theorem 3.6.6 *Let $N = 4t+1$ be a prime with t being one of the forms $17k+1$, $17k+7$, $17k+9$,$17k+10$,$17k+11$, $17k+14$, $17k+15$ and $17k+16$, and with $t = 2^m t'$, where t' is odd. Then for any nonconstant sequence s^∞ of period N over $GF(17)$,*

1. $\mathrm{L}(s^\infty) \geq 2^{m+2}$;
2. $\mathrm{SC}_k(s^\infty) = \begin{cases} \geq 2^{m+2}, & \text{if } k < \min\{\mathrm{WH}(s^N), N - \mathrm{WH}(s^N)\}; \\ 0, & \text{otherwise.} \end{cases}$

3.7 Prime Powers, Primitive Roots and Sequences

Cryptographically, we are interested in sequences with period equal to a square of a prime because their linear and sphere complexity are easy to control. We investigate now sequences of period $N = r^2$, with r an odd prime, over some fields. As a corollary of Theorems 3.4.2 or 3.4.3 we have the following results:

Corollary 3.7.1 *Let r be an odd prime, $N = r^2$ and q a primitive root modulo r. Assume that r^2 does not divide $q^{r-1}-1$, then for any nonconstant sequence of period N over $GF(q)$,*

1. $\mathrm{L}(s^\infty)$ *must be equal to one of* $\{\sqrt{N}, \sqrt{N}-1, N-\sqrt{N}, N-\sqrt{N}+1, N-1, N\}$;
2. $\mathrm{SC}_k(s^\infty) \geq \sqrt{N}-1$, *if* $k < \min\{\mathrm{WH}(s^N), N - \mathrm{WH}(s^N)\}$.

Proof: Since q is a primitive root of r and r^2 does not divide $q^{r-1}-1$ by assumptions, by Proposition 3.4.1 q must be a primitive root of r^2. Thus, by Proposition 3.1.1 the cyclotomic polynomials $Q_r(x)$ and $Q_{r^2}(x)$ are irreducible over $GF(q)$. Again from the properties of cyclotomic polynomials it follows that

$$x^N - 1 = (x-1)Q_r(x)Q_{r^2}(x).$$

Note that $\deg(Q_r(x)) = r-1$ and $\deg(Q_{r^2}(x)) = r(r-1)$ since q is a common primitive root of r and r^2. Combining these facts and the fact that the minimum polynomial of each sequence of period N over $GF(q)$ divides x^N-1 proves this theorem. □

Corollary 3.7.1 can also be proved with the following Proposition 3.7.2 and the facts that

$$x^N - 1 = (x-1)Q_r(x)Q_{r^2}(x)$$

and

$$Q_{r^2}(x) = Q_r(x^r).$$

The assumption that, r^2 does not divide $q^{r-1}-1$, ensures that $Q_r(x^r)$ is irreducible over GF(q).

Proposition 3.7.2 *Let $f_1(x)$, $f_2(x)$, $\cdots$, $f_N(x)$ be all the distinct monic irreducible polynomials in GF(q)[x] of degree m and order e, and let $t \geq 2$ be an integer whose factors divide e but not $(q^m-1)/e$. Assume also that $q^m \equiv 1 \pmod 4$ if $t \equiv 0 \pmod 4$. Then $f_1(x^t)$, $f_2(x^t)$, $\cdots$, $f_N(x^t)$ are all the distinct monic irreducible polynomials in GF(q)[x] of degree mt and order et.*

For proof of this proposition, we refer to [276, pp. 97-98].

To apply Corollary 3.7.1 to the design of keystream sequences over $GF(q)$, we should find large primes r such that r^2 does not divide $q^{r-1}-1$. A prime p satisfying the congruence

$$a^{p-1} \equiv 1 \pmod{p^2}$$

is called a *Wieferich prime* with base a. Other primes are called *non-Wieferich primes* with base a. Concerning the Wieferich primes, the following two problems are open [361]:

1. Given base $a \geq 2$, do there exist infinitely many Wieferich primes?

2. Given base $a \geq 2$, do there exist infinitely many non-Wieferich primes?

For our applications, we are mostly interested in finding some large non-Wieferich primes with small bases a equal to a prime or a prime power, especially $a = 2, 3, 5, 7, 11$ and some small powers of these primes. Lehmer showed in 1981 that, with the exceptions of 1093 and 3511, there are no other Wieferich primes $p < 6 \times 10^9$ with base 2 [266]. With base 3, it has been proven that, there are only two Wieferich primes 11 and 10006003, for $p < 2^{30}$ [396, 361, 36]. A table of the Wieferich primes with bases up to 99 and $p < 2^{32}$ has been given in [313].

The quotient

$$q_p(a) = \frac{a^{p-1} - 1}{p}$$

is called the *Fermat quotient* of p with base a. It is interesting to see that the residue modulo p of the Fermat quotient behaves like a logarithm: If p does not divide ab, then

$$q_p(ab) \equiv q_p(a) + q_p(b) \pmod{p}.$$

Also

$$q_p(p-1) \equiv 1 \pmod{p}, \qquad q_p(p+1) \equiv -1 \pmod{p}.$$

This logarithm property may be useful in designing cryptosystems.

It is also interesting that Wieferich primes and Fermat quotients have connections with the first case of Fermat's last theorem [361]. It seems easy to find non-Wieferich primes $N = 4t \pm 1$ with bases 2 and 3 and with t being odd [361]. Indeed, Wieferich primes are almost certainly rare. Thus, to construct sequences with period equal to a prime square, we can find a primitive root q of some prime r and test whether r^2 divides $q^{r-1} - 1$. Of course, theoretical results can avoid such a test. Specific sequence generators of this kind will be discussed in later chapters.

3.8 Prime Products and Sequences

In this section we examine cryptographic sequences with period equal to the product of two distinct primes over some fields. We show that there are many cryptographically good sequences of this kind.

Let $N = rs$ be the product of two distinct odd primes, so there is no primitive root modulo N. However, we have

$$x^N - 1 = \prod_{d|N} Q_d(x) = (x-1)Q_r(x)Q_s(x)Q_{rs}(x),$$

and we show that there exist integers N such that many sequences of period N over some fields have both large linear and sphere complexity. First, we have the following theorem.

Theorem 3.8.1 *Let $N = rs$ be a product of two distinct primes, q a common primitive root of both r and s. Then for every nonconstant sequence s^∞ over $GF(q)$,*

1. $\mathrm{L}(s^\infty) \geq \min\{r-1, s-1\}$;
2. $\mathrm{SC}_k(s^\infty) \geq \min\{r-1, s-1\}$; *if* $k < \min\{\mathrm{WH}(s^N), N - \mathrm{WH}(s^N)\}$.

Proof: This is a special case of Basic Theorem 3.3.1. □

More generally, we have the following theorem:

Theorem 3.8.2 *Let $r_1, \cdots, r_t$ be t pairwise distinct primes, $N = r_1 \cdots r_t$, q a positive integer such that* $\gcd(q, N) = 1$. *Then for each nonconstant sequence s^∞ of period N over $GF(q)$,*

1. $\mathrm{L}(s^\infty) \geq \min\{\mathrm{ord}_{r_1}(q), \cdots, \mathrm{ord}_{r_t}(q)\}$;
2. $\mathrm{SC}_k(s^\infty) \geq \min\{\mathrm{ord}_{r_1}(q), \cdots, \mathrm{ord}_{r_t}(q)\}$, *if* $k < \min\{\mathrm{WH}(s^N), N - \mathrm{WH}(s^N)\}$.

Proof: This is a special case of Basic Theorem 3.3.1. □

Theorems 3.8.1 and 3.8.2 clearly show that to design sequences with large linear and sphere complexity, it suffices to find primes r and s such that $\min\{\mathrm{ord}_r(q), \mathrm{ord}_s(q)\}$ is large enough.

3.8.1 Binary Sequences and Primes

Basing on Theorem 3.8.1 or Theorem 3.8.2, we can easily prove the following corollaries:

Corollary 3.8.3 *Let $r = 4t_1 + 1$, $s = 4t_2 + 1$, $r \neq s$. If r, s, t_1 and t_2 are odd primes, then for any nonconstant binary sequence s^∞ of period $N = rs$,*

1. $\mathrm{L}(s^\infty) \geq \min\{r-1, s-1\}$;
2. $\mathrm{SC}_k(s^\infty) \geq \min\{r-1, s-1\}$, *if* $k < \min\{\mathrm{WH}(s^N), N - \mathrm{WH}(s^N)\}$.

Proof: By Proposition 3.4.6, 2 is a common primitive root of r and s. Then the conclusion of this corollary follows from Theorem 3.8.1 or 3.8.2. □

Corollary 3.8.4 *Let $r = 4r_1 - 1$, $s = 4s_1 - 1$, and let $(r-1)/2$ and $(s-1)/2$ be odd primes. Then for each binary nonconstant sequence s^∞ of period $N = rs$,*

1. $\mathrm{L}(s^\infty) \geq \min\{r-1, s-1\}$;
2. $\mathrm{SC}_k(s^\infty) \geq \min\{r-1, s-1\}$; *if* $k < \min\{\mathrm{WH}(s^N), N - \mathrm{WH}(s^N)\}$.

Proof: By Proposition 3.4.7, 2 is a common primitive root of r and s. Then the conclusion of this corollary follows from Theorem 3.8.1 or 3.8.2. □

Corollary 3.8.5 *Let $r = 4r_1 + 1$, $s = 4s_1 - 1$. If $r, r_1, s, (s-1)/2$ are odd primes, then for each binary nonconstant sequence s^∞ of period $N = rs$,*

1. $\mathrm{L}(s^\infty) \geq \min\{r-1, s-1\}$;
2. $\mathrm{SC}_k(s^\infty) \geq \min\{r-1, s-1\}$; *if* $k < \min\{\mathrm{WH}(s^N), N - \mathrm{WH}(s^N)\}$.

Proof: By Propositions 3.4.6 and 3.4.7, 2 is a common primitive root of r and s. Then the conclusion of this corollary follows from Theorem 3.8.1 or 3.8.2. □

We can also use Propositions 3.4.8 and 3.4.9 to get four kinds of binary sequences with period equal to a product of two primes, which have large linear and sphere complexity if they do not have bad balance between the number of 1's and 0's in one periodic segment.

3.8.2 Ternary Sequences and Primes

To design ternary sequences, we need prime pairs (r, s) which have the common primitive root 3 or prime pairs such that the orders of 3 modulo r and s are large enough. Propositions 3.5.3-3.5.8 enable us to to find such prime pairs having common primitive root 3.

For example, Proposition 3.5.3 and Theorem 3.8.1 give the following corollary.

Corollary 3.8.6 *Let $r = 4r_1 + 1$, $s = 4s_1 + 1$, where r, r_1, s, s_1 all are primes, and $r_1 \equiv s_1 \equiv 1 \pmod 3$. Then for each nonconstant ternary sequence s^∞ of period $N = rs$, we have*

1. $\mathrm{L}(s^\infty) \geq \min\{r-1, s-1\}$;
2. $\mathrm{SC}_k(s^\infty) \geq \min\{r-1, s-1\}$; *if* $k < \min\{\mathrm{WH}(s^N), N - \mathrm{WH}(s^N)\}$.

Since the primes of form $q = 8p + 1$ with $p > 5$ prime are Tchebychef primes, we obtain the following two corollaries from Theorem 3.8.1 plus Propositions 3.5.3 and 3.5.8.

Corollary 3.8.7 *Let* $r = 4r_1 + 1$, $s = 8s_1 + 1$, *where* r, r_1, s, s_1 *all are primes, and* $r_1 \equiv 1 \pmod 3$. *If* $s > 41$, *then for each nonconstant ternary sequence* s^∞ *of period* $N = rs$,

1. $\mathrm{L}(s^\infty) \geq \min\{r-1, s-1\}$;
2. $\mathrm{SC}_k(s^\infty) \geq \min\{r-1, s-1\}$; *if* $k < \min\{\mathrm{WH}(s^N), N - \mathrm{WH}(s^N)\}$.

Corollary 3.8.8 *Let* $r = 8r_1 + 1$, $s = 8s_1 + 1$, *where* r, r_1, s, s_1 *all are primes. If* $r > 41$ *and* $s > 41$, *then for each nonconstant ternary sequence* s^∞ *of period* $N = rs$,

1. $\mathrm{L}(s^\infty) \geq \min\{r-1, s-1\}$;
2. $\mathrm{SC}_k(s^\infty) \geq \min\{r-1, s-1\}$; *if* $k < \min\{\mathrm{WH}(s^N), N - \mathrm{WH}(s^N)\}$.

In some later chapters we will construct generators which can realize the above binary and ternary sequences. Theorems about sequences with period equal to a product of two distinct primes over $GF(q)$ can also be similarly established.

3.9 On Cryptographic Primitive Roots

One role of primitive roots in stream ciphers has already been made clear in Sections 3.4 to 3.8. Primes serve as periods or as factors of periods for keystream sequences, while primitive roots determine the base fields over which the sequences are constructed.

We call primitive roots which are small powers of small primes *cryptographic primitive roots.* Without small primitive roots which are a prime power, a prime may have little cryptographic value for stream ciphers. Thus the distribution of primitive roots has cryptographic importance. This distribution has been investigated by many scholars, to mention a few, Carlitz [65], Vegh [428, 429, 430], Szalay [416] and Shoup [399].

What we need for stream ciphers is small primitive roots which are primes or powers of primes. Investigations of the least primitive root have been done by Bach [9], Burgess and Elliott [48], Elliott [146], Wang [432], Heath-Brown [195] and Murata [315]. To discuss some cryptographically interesting results in this field, we introduce now two notations following Murata. If p is an odd prime number, let $g(p)$ denote the least positive integer which is a primitive root of p, and let $G(p)$ denote the least prime which is a primitive root of p. We use the notation $a(x) \ll b(x)$ to mean that $a(x) \leq cb(x)$ for some constant c. It has been conjectured that $g(p) \ll p^{\epsilon}$ for any $\epsilon > 0$, but only much weaker or conditional results have been obtained so far: Burgess proved that $g(p) \ll p^{(1/4)+\epsilon}$ for any $\epsilon > 0$ [47]. Under the assumption of the Generalized Riemann Hypothesis Wang proved that $g(p) \ll \omega(p-1)^6(\log p)^2$, where $\omega(m)$ denotes the number of distinct prime factors of the integer m [432]. A similar result for the average value of $g(p)$ was established without any unproven hypothesis by Burgess and Elliott [48].

As for the magnitude of $G(p)$, Linnik showed that $G(p) \ll p^A$ for some positive constant A [315]; Elliott obtained that $G(p) \leq 475(\log p)^{8/5}$ holds for infinitely many primes [146]; Heath-Brown proved that $G(p) \leq 5$ for infinitely many primes [195]. The result of Heath-Brown implies that there are infinitely many primes which are cryptographically valuable for stream ciphers. According to Murata [315], "numerical examples show that, in most cases, $g(p)$ is very small. Among the first 19,862 odd primes up to 223,051, $g(p) = 2$ happens for 7429 primes (37.4%), $g(p) = 3$ happens for 4518 primes (22.8%), and $g(p) \leq 6$ holds for about 80% of these primes."

Murata surmised that,

> for almost all primes p, $g(p)$ is not very far from $(p-1)/\phi(p-1)$.

The function $(p-1)/\phi(p-1)$ fluctuates irregularly, but Murata gave the following asymptotic formula:

$$\pi(x)^{-1} \sum_{p \leq x;\ p\ prime} \frac{p-1}{\phi(p-1)} = C + O\left(\frac{\log\log x}{\log x}\right),$$

$$C = \prod_{p\ prime} \left(1 + \frac{1}{(p-1)^2}\right) \doteq 2.827,$$

where $\pi(x)$ denotes the number of primes no larger than x. So he conjectured also that

> for almost all primes p, $(p-1)/\phi(p-1)$ is not very far from the constant C.

Summarizing Murata's above argument, we may expect that, for almost all p, $g(p)$ is not far from the constant C.

If a similar argument applies for $G(p)$ as above, then we could draw the following cryptographic conclusion:

> for most of the large primes p we can expect that we can generate sequences of period p over a field $GF(q)$, where q is a small prime and is a primitive root mod p.

We mention two versions of a famous 1927 conjecture of E. Artin [396]:

Artin's Conjecture 3.9.1 *Every integer a, not equal to -1 or to a square, is a primitive root for infinitely many primes.*

Artin's Conjecture 3.9.2 *If $a \neq -1$ and $a \neq b^n$ with $n > 1$, and if $v_a(N)$ is the number of primes $\leq N$ for which a is a primitive root, then*

$$v_a(N) \sim 0.3739558\pi(N).$$

The conjectures are still unproved, but if certain generalized Riemann hypotheses are assumed, then a modified version of the second one was proved by Hooley [209].

The investigation of primitive roots predates the work of Gauss [159], for example, the eighteenth century work of Lambert and Euler. Since 1800 Gauss, and many others, have devised efficient techniques for finding primitive roots, but no general, explicit, deterministic method has been devised. This remains an important open problem.

In Section 3.8 we noted that it is cryptographically useful to find primes which have common primitive roots. This is also useful in designing twin-prime difference sets and generators based on those difference sets (see Chapters 7 and 8). The Chinese Remainder Theorem can be used to compute the common primitive roots [361]. Assume that $n = p_1p_2 \cdots p_t$ is a product of distinct primes, and g_i is a primitive root of p_i. If g is such that $1 \leq g \leq n-1$ and $g \equiv g_i \pmod{p_i}$ for every $i = 1, ..., t$, then g is a common primitive root of every p_i.

Sometimes we may be interested in sequences with period N which is not of the form p^e or $2p^e$, where p is an odd prime. Such an N has no primitive root. However, it would also be cryptographically attractive if there is a small prime or small prime power g such that the order of g modulo every factor n of N, $\text{ord}_n(g)$, is about as large as $\phi(n)$. The linear and sphere complexity of sequences with such a period over $GF(q)$ are also relatively easy to control.

3.10 Linear Complexity of Sequences over Z_m

The definition of the linear complexity of sequences over fields was given before. The linear complexity for sequences over commutative rings can be similarly defined as follows. Let R be a commutative ring with multiplicative identity 1, and $s^N = s_0s_1 \cdots s_{N-1}$ be a sequence of length N over R, where $s_i \in R$. If s^N satisfies a linear recurrence relation

$$s_i = a_1s_{i-1} + a_2s_{i-2} + \cdots + a_ls_{i-l}, \quad i \geq l, \ a_i \in R,$$

then there exists a shortest such linear recurrence relation. The least l is called the *linear complexity* or *linear span* of the sequence and is denoted by $\mathrm{L}(s^N)$. The linear complexity of a finite sequence s^N is defined to be N if s^N does not satisfy such a linear recurrence relation. For semi-infinite sequences the linear complexity is defined to be $+\infty$ if they satisfy no finite linear recurrence relation. For ultimately periodic sequences the linear complexity is finite. If the linear complexity of a sequence over a field is l, then $2l$ successive terms of the sequence can be used to determine a linear recurrence relation of length l satisfied by the sequence by applying the Berlekamp-Massey algorithm [291], which has complexity $O(l^2)$. Consequently, $2l$ successive characters of the sequence are sufficient to determine the whole sequence. Thus, sequences over fields for additive stream ciphering and for some code-division multiple-access systems should have large linear complexity.

Various results about the period of linear recurrence sequences over Z_m are known, but few results about the linear complexity of sequences over Z_m have been proved. Sequences over fields are easy to construct, and their properties are easy to control. But it seems harder to do this for sequences over residue class rings. In this section we show how to construct sequences over Z_m from those over finite fields Z_p, where p's are primes, and how to control their linear complexity with the help of the Chinese Remainder Theorem.

An important result we need is the following theorem whose second part was implied in the work of Reeds and Sloane [360] without giving a proof.

Lemma 3.10.1 *Let s^∞ be a sequence over Z_m, where $m = m_1 m_2 \cdots m_t$ and m_i are pairwise relatively prime, and let*

$$s(i)^\infty = s^\infty \bmod m_i, \quad i = 1, 2, ..., t,$$

i.e., $s(i)_j = s_j \bmod m_i$ for all possible j.

1. *If s^∞ is (ultimately) periodic, then each sequence $s(i)^\infty$ must be (ultimately) periodic, and $\mathrm{per}(s^\infty) = \mathrm{lcm}\{\mathrm{per}(s(1)^\infty), \cdots, \mathrm{per}(s(t)^\infty)\}$, where $\mathrm{per}(s^\infty)$ denotes the least period.*

2. $\mathrm{L}(s^\infty) = \max\{\mathrm{L}(s(1)^\infty), \cdots, \mathrm{L}(s(t)^\infty)\}$.

Proof: Let φ be the mapping from Z_m to $Z_{m_1} \times \cdots \times Z_{m_t}$ given by

$$\varphi : x \bmod m \mapsto (x \bmod m_1, ..., x \bmod m_t).$$

By the Chinese remainder theorem φ is an isomorphism.

Assume that s^∞ is periodic and $N = \mathrm{per}(s^\infty)$. Then $\varphi(s_{N+j}) = \varphi(s_j)$ for all $j \geq 0$. It follows that $s(i)_{N+j} = s(i)_j$ for all $j \geq 0$ and all $i = 1, ..., t$. Thus, each $s(i)^\infty$ is periodic and $N' := \mathrm{lcm}\{\mathrm{per}(s(1)^\infty), \cdots, \mathrm{per}(s(t)^\infty)\}$ divides N. On the other hand, since $s(i)_{N'+j} = s(i)_j$ for all j and i, we have $\varphi(s_{j+N'}) = \varphi(s_j)$ for all j. Since φ is one-to-one, $s_{j+N'} = s_j$ for all j. It follows that N divides N'. Combining the above results gives $N = N'$.

Now we prove part two. Let $l = \mathrm{L}(s^\infty)$ and let

$$s_j = a_1 s_{j-1} + a_2 s_{j-2} + \cdots + a_l s_{j-l}, \quad j \geq l \tag{3.2}$$

be a shortest linear recurrence relation for s^∞, where $a_i \in Z_m$. Let $\varphi(a_i) = (a(1)_i, \cdots, a(t)_i)$ for $i = 1, ..., t$. Applying the isomorphism φ to (3.2) gives

$$\varphi(s_j) = \varphi(a_1)\varphi(s_{j-1}) + \cdots + \varphi(a_l)\varphi(s_{j-l}), \quad j \geq l$$

from which it follows that for each i with $1 \leq i \leq t$

$$s(i)_j = a(i)_1 s(i)_{j-1} + \cdots + a(i)_l s(i)_{j-l}, \quad j \geq l,$$

where $a(i)_j \in Z_{m_i}$. Thus, $l_i := \mathrm{L}(s(i)^\infty) \leq l$.

On the other hand, let $l' = \max\{l_1, ..., l_t\}$. Assume that

$$s(i)_j = a(i)_1 s(i)_{j-1} + \cdots + a(i)_{l_i} s(i)_{j-l_i}, \quad j \geq l_i$$

is a shortest linear recurrence relation the sequence $s(i)^\infty$ satisfies. Define $a(i)_j = 0$ for all j with $l_{i+1} \leq j \leq l'$, where $1 \leq i \leq t$. Then

$$\begin{aligned}(s(1)_j, \cdots s(t)_j) &= (a(1)_1, \cdots a(t)_1)(s(1)_{j-1}, \cdots s(t)_{j-1}) + \cdots + \\ &+ (a(1)_{l'}, \cdots a(t)_{l'})(s(1)_{j-l'}, \cdots s(t)_{j-l'})\end{aligned} \tag{3.3}$$

holds for each $j \geq l'$. Let $a_j = \varphi^{-1}(a(1)_j, \cdots, a(t)_j) \in Z_m$. Applying the inverse isomorphism φ^{-1} to (3.3) yields

$$s_j = a_1 s_{j-1} + a_2 s_{j-2} + \cdots + a_{l'} s_{j-l'}, \quad j \geq l'.$$

Thus, $l \leq l'$. Hence $l = l'$. This proves part two. □

From now on in this section we stipulate that the period mentioned does not necessarily refer to the least one, i.e., the period is a multiple of the least one. To set up lower bounds on the linear complexity of sequences over Z_m, we need Basic Theorem 3.3.1.

A general lower bound for sequences over Z_m is the following.

Theorem 3.10.2 (Ding [123]) *Let $N = N_1 \cdots N_r$, $m = p_1 \cdots p_t$, where N_1, ..., N_r, p_1, ..., p_t are pairwise distinct primes. If s^∞ is a nonconstant sequence of period N over Z_m, then*

$$\mathrm{L}(s^\infty) \geq \min\{\mathrm{ord}_{N_1}(p_1), ..., \mathrm{ord}_{N_r}(p_1), ..., \mathrm{ord}_{N_1}(p_t), ..., \mathrm{ord}_{N_r}(p_t)\}.$$

Proof: Let $s(i)^\infty = s^\infty \bmod p_i$ for $i = 1, 2, ..., t$. Then $s(i)^\infty$ is a sequence of period N over Z_{p_i}. Since s^∞ is nonconstant, there must exist an integer j such that $s(j)^\infty$ is nonconstant by the Chinese Remainder Theorem. Since $\gcd(p_j, N) = 1$, by Theorem 3.3.1 and Lemma 3.10.1

$$\begin{aligned} \mathrm{L}(s^\infty) &\geq \min\{\mathrm{ord}_{N_1}(p_j), ..., \mathrm{ord}_{N_r}(p_j)\} \\ &\geq \min\{\mathrm{ord}_{N_1}(p_1), ..., \mathrm{ord}_{N_r}(p_1), ..., \mathrm{ord}_{N_1}(p_t), ..., \mathrm{ord}_{N_r}(p_t)\}. \end{aligned}$$

□

A better bound is given by the following theorem.

Theorem 3.10.3 (Ding [123]) *Let $N = N_1 N_2 \cdots N_r$, $m = p_1 \cdots p_t$, where N_1, ..., N_r, p_1, ..., p_t are pairwise distinct primes. Assume that s^∞ is a nonconstant sequence of period N over Z_m. Let*

$$s(i)^\infty = s^\infty \bmod p_i$$

for $i = 1, ..., t$, and let $i_1, ..., i_u$ be integers such that $s(i_1)^\infty, ..., s(i_u)^\infty$ are nonconstant, where $1 \leq i_1 < i_2 < \cdots < i_u \leq t$. Then

$$\begin{aligned} \mathrm{L}(s^\infty) \geq\ & \max\{\min\{\mathrm{ord}_{N_1}(p_{i_1}), ..., \mathrm{ord}_{N_r}(p_{i_1})\}, ..., \\ & \min\{\mathrm{ord}_{N_1}(p_{i_u}), ..., \mathrm{ord}_{N_r}(p_{i_u})\}\}, \end{aligned}$$

Proof: Since s^∞ is nonconstant, one of $s(i)^\infty$ must be nonconstant. By our assumptions and Theorem 3.3.1

$$\mathrm{L}(s(i_k)^\infty) \geq \min\{\mathrm{ord}_{N_1}(p_{i_k}), ..., \mathrm{ord}_{N_r}(p_{i_k})\}$$

for $k = 1, 2, ..., u$. The conclusion then follows from Lemma 3.10.1. □

By Theorems 3.10.2 and 3.10.3, to control the linear complexity of sequences s^∞ over Z_m, we need only to ensure that

$$\min\{\mathrm{ord}_{N_1}(p_i), ..., \mathrm{ord}_{N_r}(p_i)\}$$

is large enough for only one nonconstant sequence $s(i)^\infty$.

A number of practical tight bounds are described by the following five theorems [123].

Theorem 3.10.4 *Let $N, p_1, ..., p_t$ be pairwise distinct primes, and $m = p_1 \cdots p_t$. Assume $p_i^2 + 1 < N$ for $i = 1, 2, ..., t$, and $(N-1)/4$ is an odd prime. For any nonconstant sequence of period N over Z_m,*

1. $\mathrm{L}(s^\infty) \geq (N-1)/4$;

2. *if* $p_1, ..., p_t$ *are quadratic nonresidues modulo* N*, then* $\mathrm{L}(s^\infty) = N-1$ *or* N*.*

Proof: By Theorem 3.10.2

$$\mathrm{L}(s^\infty) \geq \min\{\mathrm{ord}_N(p_1), ..., \mathrm{ord}_N(p_t)\}.$$

Since $\mathrm{ord}_N(p_i)$ divides $N-1$, $\mathrm{ord}_N(p_i)$ must be one of 4, k, $2k$, and $4k$, where $k = (N-1)/4$. Since $N > p_i^2+1 > p_i^2-1$, we have $p_i^4-1 = (p_i^2+1)(p_i^2-1) \not\equiv 0 \pmod{N}$. Thus, $\mathrm{ord}_N(p_i) \geq k = (N-1)/4$. This proves part one.

If $p_1, ..., p_t$ are quadratic nonresidues modulo N, then $p_i^{(N-1)/2} \equiv -1 \pmod{N}$. It then follows that $p_i^{2k} \equiv -1 \pmod{N}$, and that $p_i^k \not\equiv 1 \pmod{N}$. Thus, $\mathrm{ord}_N(p_i) = 4k = N-1$, and $\mathrm{L}(s^\infty) \geq N-1$. This proves part two. □

Theorem 3.10.5 *Let* $N, p_1, ..., p_t$ *be pairwise distinct primes, and* $m = p_1 \cdots p_t$*. Assume* $p_i + 1 < N$ *for* $i = 1, 2, ..., t$*, and* $(N-1)/2$ *is an odd prime. For any nonconstant sequence of period* N *over* Z_m*,*

1. $\mathrm{L}(s^\infty) \geq (N-1)/2$*;*

2. *if* $p_1, ..., p_t$ *are quadratic nonresidues modulo* N*, then* $\mathrm{L}(s^\infty) = N-1$ *or* N*.*

Proof: By Theorem 3.10.2

$$\mathrm{L}(s^\infty) \geq \min\{\mathrm{ord}_N(p_1), ..., \mathrm{ord}_N(p_t)\}.$$

Let $k = (N-1)/2$. Since k is prime and $\mathrm{ord}_N(p_i)$ divides $N-1 = 2k$, $\mathrm{ord}_N(p_i)$ must be one of 2, k and 2k. We first see that $\mathrm{ord}_N(p_i)$ does not equal 2, since $p_i^2 - 1 = (p_i+1)(p_i-1) \not\equiv 0 \pmod{N}$ by the assumption that $p_i + 1 < N$. Thus, $\mathrm{ord}_N(p_i) \geq k = (N-1)/2$. This proves part one. If $p_1, ..., p_t$ are quadratic nonresidues, then $p_i^{(N-1)/2} \equiv -1 \pmod{N}$, so $\mathrm{ord}_N(p_i)$ does not equal k. It follows that $\mathrm{ord}_N(p_i) = N-1$. Thus, $\mathrm{L}(s^\infty) = N-1$ or N. □

In practice the case $N = N_1N_2$ is especially interesting. Sequences of period N_1N_2 over Z_m could also have large linear complexity if the primes $N_1, N_2, p_1, ..., p_t$ are properly designed, as shown by the following three theorems.

Theorem 3.10.6 *Let* $N_1, N_2, p_1, ..., p_t$ *be pairwise distinct primes,* $m = p_1 \cdots p_t$*, and* $N = N_1N_2$*. Assume that* $\max\{p_i^2+1 : i = 1, 2, ..., t\} < \min\{N_1, N_2\}$*, and* $(N_1-1)/4$ *and* $(N_2-1)/4$ *are odd primes. For any nonconstant sequence* s^∞ *of period* N *over* Z_m*,*

1. $L(s^\infty) \geq [\min\{N_1, N_2\} - 1]/4$;
2. *if* $p_1, ..., p_t$ *are quadratic nonresidues modulo both* N_1 *and* N_2, *then*
$$L(s^\infty) \geq \min\{N_1, N_2\} - 1.$$

Proof: By Theorem 3.10.2

$$L(s^\infty) \geq \min\{\text{ord}_{N_1}(p_1), \text{ord}_{N_2}(p_1), ..., \text{ord}_{N_1}(p_t), \text{ord}_{N_2}(p_t)\}.$$

By the proof of Theorem 3.10.4, $\text{ord}_{N_i}(p_j) \geq [\min\{N_1, N_2\} - 1]/4$. The conclusion of part one then follows.

If $p_1, ..., p_t$ are quadratic nonresidues modulo both N_1 and N_2, by the proof of Theorem 3.10.4 we obtain $\text{ord}_{N_i}(p_j) = N_i - 1$. Thus, the conclusion of part two follows. □

Theorem 3.10.7 *Let* $N_1, N_2, p_1, ..., p_t$ *be pairwise distinct primes,* $m = p_1 \cdots p_t$ *and* $N = N_1 N_2$. *Assume that* $\max\{p_i + 1 : i = 1, 2, ..., t\} < \min\{N_1, N_2\}$, *and* $(N_1 - 1)/2$ *and* $(N_2 - 1)/2$ *are odd primes. For any nonconstant sequence* s^∞ *of period* N *over* Z_m,

1. $L(s^\infty) \geq [\min\{N_1, N_2\} - 1]/2$;
2. *if* $p_1, ..., p_t$ *are quadratic nonresidues modulo both* N_1 *and* N_2, *then*
$$L(s^\infty) \geq \min\{N_1, N_2\} - 1.$$

Proof: By Theorem 3.10.2

$$L(s^\infty) \geq \min\{\text{ord}_{N_1}(p_1), \text{ord}_{N_2}(p_1), ..., \text{ord}_{N_1}(p_t), \text{ord}_{N_2}(p_t)\}.$$

By the proof of Theorem 3.10.5, $\text{ord}_{N_i}(p_j) \geq [\min\{N_1, N_2\} - 1]/2$. The conclusion of part one then follows.

If $p_1, ..., p_t$ are quadratic nonresidues modulo both N_1 and N_2, by the proof of Theorem 3.10.5 we obtain $\text{ord}_{N_i}(p_j) = N_i - 1$. Thus, the conclusion of part two follows. □

Theorem 3.10.8 *Let* $N_1, N_2, p_1, ..., p_t$ *be pairwise distinct primes,* $m = p_1 \cdots p_t$, *and* $N = N_1 N_2$. *Assume that* $\max\{p_i + 1 : i = 1, 2, ..., t\} < N_1$, $\max\{p_i^2 + 1 : i = 1, 2, ..., t\} < N_2$ *and* $(N_1 - 1)/2$ *and* $(N_2 - 1)/4$ *are odd primes. For any nonconstant sequence* s^∞ *of period* N *over* Z_m,

1. $L(s^\infty) \geq \min\{(N_1 - 1)/2, (N_2 - 1)/4\}$;

2. if $p_1, ..., p_t$ are quadratic nonresidues modulo both N_1 and N_2, then

$$\mathrm{L}(s^\infty) \geq \min\{N_1, N_2\} - 1.$$

Proof: By Theorem 3.10.2

$$\mathrm{L}(s^\infty) \geq \min\{\mathrm{ord}_{N_1}(p_1), \mathrm{ord}_{N_2}(p_1), ..., \mathrm{ord}_{N_1}(p_t), \mathrm{ord}_{N_2}(p_t)\}.$$

By the proof of Theorem 3.10.4, $\mathrm{ord}_{N_2}(p_j) \geq [N_2 - 1]/4$. By the proof of Theorem 3.10.5, $\mathrm{ord}_{N_1}(p_j) \geq [N_1 - 1]/2$. The conclusion of part one then follows.

If $p_1, ..., p_t$ are quadratic nonresidues modulo both N_1 and N_2, by the proofs of Theorems 3.10.4 and 3.10.5 we obtain $\mathrm{ord}_{N_i}(p_j) = N_i - 1$. Thus, the conclusion of part two follows. □

Lower bounds on the linear complexity of sequences over Z_{p^k} can be developed as follows. Let s^∞ be a sequence of period N over Z_{p^k}, and $s(p)^\infty = s^\infty \bmod p$. Assume that $\mathrm{L}(s^\infty) = l$ and

$$s_i = a_1 s_{i-1} + a_2 s_{i-2} + \cdots + a_l s_{i-l}, \quad i \geq l$$

is a shortest linear recurrence relation for s^∞, then

$$s(p)_i = a(p)_1 s(p)_{i-1} + a(p)_2 s(p)_{i-2} + \cdots + a(p)_l s(p)_{i-l}, \quad i \geq l,$$

where $a(p)_i = a_i \bmod p$, and $s(p)_i = s_i \bmod p$. It follows that

$$\mathrm{L}(s^\infty) \geq \mathrm{L}(s(p)^\infty). \tag{3.4}$$

This inequality will provide a bridge for transferring bounds on the linear complexity of sequences over Z_p to those of sequences over Z_{p^k}.

Theorem 3.10.9 *Let $N = N_1^{n_1} N_2^{n_2} \cdots N_r^{n_r}$, and $m = p^k$, where $N_1, ..., N_r, p$ are pairwise distinct primes. For any sequence s^∞ of period N over Z_m, if $s(p)^\infty$ is not a constant sequence, then*

$$\mathrm{L}(s^\infty) \geq \min\{\mathrm{ord}_{N_1}(p), ..., \mathrm{ord}_{N_r}(p)\}.$$

Proof: By (3.4) $\mathrm{L}(s^\infty) \geq \mathrm{L}(s(p)^\infty)$. The conclusion then follows from Theorem 3.3.1 and the assumption that $s(p)^\infty$ is nonconstant. □

Since we have many ways to control $\mathrm{ord}_{N_i}(p)$, in many cases the linear complexity of sequences over Z_{p^k} is easy to control. As mentioned before, it is necessary to control the linear complexity of sequences over Z_{p^k} due to the Reeds-Sloane algorithm [360].

Since the arithmetic of Z_{2^e} can be efficiently implemented on standard processors, we describe some tight bounds on the linear complexity of sequences over Z_{2^e}. These bounds have already been set up for sequences over fields [123].

Theorem 3.10.10 *Let $N = 8k+3$ and $(N-1)/2$ both be odd primes, and let e be a positive integer. For any sequence s^∞ of period N over Z_{2^e}, if the binary sequence $s(2)^\infty$ is a nonconstant sequence, then*

$$\mathrm{L}(s^\infty) = N-1 \text{ or } N.$$

Proof: Let $N = 2t+1$. By assumption t is prime, so the order of 2 modulo N must be one of 2, t and $2t$. Since $N = 8k+3$, $2^{(N-1)/2} \equiv -1 \pmod{N}$. Because $t \geq 3$, so $N \geq 7$. Thus, the order of 2 modulo N must be $2t = N-1$. The conclusion then follows from Theorem 3.10.9. □

Theorem 3.10.11 *Let $N = 8k-3$ and $(N-1)/4$ both be primes, and let e be a positive integer. For any sequence s^∞ of period N over Z_{2^e}, if the binary sequence $s(2)^\infty$ is a nonconstant sequence, then*

$$\mathrm{L}(s^\infty) = N-1 \text{ or } N.$$

Proof: Let $N = 4t+1$. By assumption t is prime, so the order of 2 modulo N must be one of 4, t, $2t$, and $4t$. Since $N = 8k-3$, we have $2^{(N-1)/2} \equiv -1 \pmod{N}$. Since both t and N are prime, $N \geq 13$. Thus, $\mathrm{ord}_N(2) \neq 4$, since $2^4 - 1 = 15 \not\equiv 0 \pmod{N}$. Since $2^{(N-1)/2} = 2^{2t} \equiv -1 \pmod{N}$, the order of 2 modulo N is not equal to t or $2t$. Hence, the order of 2 modulo N must be $4t = N-1$. The conclusion then follows from Theorem 3.10.9. □

Theorem 3.10.12 *Let $N_1 = 8k_1 + 3$ and $N_2 = 8k_2 + 3$ be primes, where $4k_1 + 1$ and $4k_2 + 1$ are also primes. For any sequence s^∞ of period N_1N_2 over Z_{2^e}, if the binary sequence $s(2)^\infty$ is a nonconstant sequence, then*

$$\mathrm{L}(s^\infty) \geq \min\{N_1, N_2\} - 1.$$

Proof: By the proof of Theorem 3.10.10

$$\mathrm{ord}_{N_1}(2) = N_1 - 1, \quad \mathrm{ord}_{N_2}(2) = N_2 - 1.$$

Then the conclusion follows from Theorem 3.10.9. □

Theorem 3.10.13 *Let* $N_1 = 8k_1 - 3$ *and* $N_2 = 8k_2 - 3$ *be primes, where* $2k_1 - 1$ *and* $2k_2 - 1$ *are also primes. For any sequence* s^∞ *of period* N_1N_2 *over* Z_{2^e}*, if the binary sequence* $s(2)^\infty$ *is a nonconstant sequence, then*

$$\mathrm{L}(s^\infty) \geq \min\{N_1, N_2\} - 1.$$

Proof: By the proof of Theorem 3.10.11

$$\mathrm{ord}_{N_1}(2) = N_1 - 1, \quad \mathrm{ord}_{N_2}(2) = N_2 - 1.$$

Then the conclusion follows from Theorem 3.10.9. □

Theorem 3.10.14 *Let* $N_1 = 8k_1 + 3$ *and* $N_2 = 8k_2 - 3$ *be primes, where* $4k_1 + 1$ *and* $2k_2 - 1$ *are also primes. For any sequence* s^∞ *of period* N_1N_2 *over* Z_{2^e}*, if the binary sequence* $s(2)^\infty$ *is a nonconstant sequence, then*

$$\mathrm{L}(s^\infty) \geq \min\{N_1, N_2\} - 1.$$

Proof: By the proof of Theorems 3.10.10 and 3.10.11

$$\mathrm{ord}_{N_1}(2) = N_1 - 1, \quad \mathrm{ord}_{N_2}(2) = N_2 - 1.$$

Then the conclusion follows from Theorem 3.10.9. □

The bounds of Theorems 3.10.9, 3.10.10, 3.10.11, 3.10.12, 3.10.13, and 3.10.14 show how to control the linear complexity of sequences over Z_{2^e}. For sequences over Z_{p^k}, one can develop similar bounds.

A more general bound on the linear complexity of sequences over Z_m is described by the following theorem [123].

Theorem 3.10.15 *Let* $N = N_1^{n_1}N_2^{n_2}\cdots N_r^{n_r}$*, where* N_i *are distinct primes, and let* $m = p_1^{e_1}p_2^{e_2}\cdots p_t^{e_t}$*, where* p_i *are pairwise distinct primes such that* $\gcd(N, m) = 1$*. For any sequence* s^∞ *of period* N *over* Z_m*, if one of the sequences* $s(p_j)^\infty$ *is nonconstant, then*

$$\mathrm{L}(s^\infty) \geq \min\{\mathrm{ord}_{N_1}(p_1), ..., \mathrm{ord}_{N_r}(p_1), ..., \mathrm{ord}_{N_1}(p_t), ..., \mathrm{ord}_{N_r}(p_t)\}.$$

Proof: Combining Lemma 3.10.1 and Theorem 3.10.9 can prove this theorem. □

All of the bounds presented before are special cases of this more general bound. Whether this bound is tight depends on the parameters N_i, n_i, p_i, e_i. By choosing proper values for these parameters one can easily control the linear complexity of sequences over Z_m. Here we use parameters of special forms to control the linear complexity, instead of using some cryptographic functions to do so.

3.11 Period and its Cryptographic Importance

Let us stipulate that the periods mentioned in this section are least periods. Practical keystream sequences are usually periodic, or at least ultimately periodic, because the proposed sequence generators are usually finite state machines. From the public literature about stream ciphers we could see that little attention has been paid to the particular nature of periods, at most to the size of them. To control the size of the period for sequences, it suffices to control the linear complexity of sequences, since linear complexity is less than or equal to the period.

To design keystream generators, we usually need to consider some of the following problems from both security and implementation viewpoints:

1. the computational complexity of the sequence producing algorithm;
2. the control of the linear complexity of the keystream sequences and of the size of the period;
3. the control of the sphere complexity of the keystream sequences;
4. the control of the frequency distribution of the elements from the ring or field over which the sequence is constructed;
5. the control of the pattern distributions of the output sequences;
6. the control of the difference property of some sequence-producing functions in the generator;
7. the control of the nonlinearity of some sequence-producing functions with respect to some operations in the generator;
8. the control of the correlation property between some output sequences of different stages in the generator.

Traditionally the procedure for designing keystream generators is: first to have an idea about the structure of a generator which is based on some technically simple devices from the viewpoint of fast implementation, then to control the cryptographic properties of the keystream sequences from the security viewpoints by choosing proper parameters for the generator, for example, the choosing of some cryptographic functions.

In fact it is usually difficult to get theoretical results about some items above for many proposed keystream generators. Those commonly known for some generators are the linear and sphere complexity of the output sequences, the frequency distribution of elements of the field over which the sequence is generated, and the nonlinear order of its sequence-producing functions.

Now the question is which of the above problems should be considered first. The order of considering the above problems is cryptographically significant. For instance, many sequence generators cannot generate sequences of arbitrary period. But there do exist some which can produce every periodic sequence by selecting some of the parameters. Thus, if we first consider the performance problem, then we may have a stream cipher system whose security problems are difficult to coordinate. The structure of a generator determines whether there are trade-offs between some cryptographic requirements and the number of trade-offs if there are any.

In this book we consider the design of keystream generators in the following order:

1. choose first cryptographically good periods;

2. design generators which can produce sequences of these periods;

3. control some cryptographic properties above of the generator and at the same time consider the performance of the generator.

The importance of the period for sequences is easily seen from the results in this chapter. Speaking specifically, cryptographically good periods ensure automatically large linear and sphere complexity, provided only that the sequence does not have bad balance of the elements of the field over which the sequence is generated. This approach has the advantage of making the system have as few trade-offs as possible. The importance of periods will be further discussed in some of the following chapters.

As an example, we consider some special periods. Since the order of 2 modulo $2^m - 1$ is m, which is very small, compared with the period $2^m - 1$, the linear and sphere complexity of binary sequences of period 2^m are hard to control. Similarly, since the order of 2 modulo $2^m + 1$ is $2m$, the linear and sphere complexity of binary sequences of period $2^m + 1$ are also hard to control.

Finally, for some generators such as the NSGs the control of the period is easy, while for others it is quite hard. Thus, the structure of a generator determines whether cryptographic aspects of the generator are easy to control or not.

3.12 Recent Advances on the Sphere Complexity

As made clear in Section 2.3.4, the k-error linear complexity introduced by Stamp and Martin [412] in 1993 is defined to be $\min\{\mathrm{SC}_k(s), \mathrm{L}(s)\}$ and is essentially the same as sphere complexity. We mention that the k-error linear complexity was introduced a few years later than the sphere complexity. In this section we provide information on recent results on the sphere complexity. Note that some of these results were described under the name of the k-error linear complexity.

Meidl and Niederreiter considered the expected value of the sphere complexity of periodic sequences [303, 304, 305], and established a lower bound on the expected sphere complexity of periodic sequences. Niederreiter studied periodic sequences with large sphere complexity [323]. Kurosawa, Sato, Sakata and Kishimoto [253] described a relation between the linear complexity and sphere complexity.

Kaida, Uehara and Imamura [227] developed an algorithm for computing the sphere complexity for sequences over $GF(p^m)$ with period p^n. Their algorithm is based on a fast algorithm for computing the linear complexity of sequences over $GF(p^m)$ with period p^n by Ding, Xiao and Shan [138]. Xiao and Wei [467] also developed a fast algorithm for computing the sphere complexity of sequences with period p^n. Lauder and Paterson [261] derived an algorithm for computing the error linear complexity profile of binary sequences of period 2^n.

Chapter 4

Cyclotomy and Cryptographic Functions

The word *cyclotomy* means "circle-division" and refers to the problem of dividing the circumference of the unit circle into a given number, n, of arcs of equal lengths. The ruler-and-compass treatment of this problem was discussed in Euclid's time. Gauss' remarkable result is that, if n is a Fermat prime, then the regular polygons of $2^s n$ sides are constructed with ruler and compass [414].

Our interest in the theory of cyclotomy has stemmed from the rather remarkable fact that the cyclotomic numbers actually represent the difference property and the nonlinearity of some cryptographic functions from Z_p's to some Abelian groups [122] as well as the two-character distributions and autocorrelation property of some cyclotomic sequences. In this chapter we shall construct cryptographic functions based on cyclotomic numbers.

We now fix for this and later chapters the notation $(x \bmod q) \bmod k$, by which we mean that first the number x should be reduced modulo q to give a number between 0 and $q-1$, and then that number should be reduced modulo k to give an integer between 0 and $k-1$.

We make some references to difference sets and almost difference sets in this chapter. The reader not already familiar with these notions should refer to Sections 6.1 and 6.6.

4.1 Cyclotomic Numbers

Let $N = df + 1$ be an odd prime and let θ be a fixed primitive element of Z_N. Denote the multiplicative subgroup (θ^d) as D_0, then the coset decomposition of Z_N^* with respect to the subgroup D_0 is then

$$Z_N^* = \cup_{i=0}^{d-1} D_i,$$

where $D_i = \theta^i D_0$ for $i \geq 0$. The coset D_l is called the *index class* l [15] or *cyclotomic class* l [414]. Let $(l,m)_d$ denote the number of solutions (x,y) of the equation

$$1 = y - x, \quad (x,y) \in D_l \times D_m,$$

or equivalently,

$$(l,m)_d = |(D_l + 1) \cap D_m|.$$

These constants $(l,m)_d$ are called *cyclotomic numbers* of order d [106, 264, 14, 15, 316]. Clearly, there are at most d^2 distinct cyclotomic numbers of order d and these numbers depend not only on N, d, l, m, but also on which of the $\phi(N-1)$ primitive elements of Z_N is chosen.

The following elementary facts about cyclotomic numbers are not hard to prove [106, 15]:

(A) $(l,m)_d = (l',m')_d$ when $l \equiv l' \pmod{d}$ and $m \equiv m' \pmod{d}$;

(B) $(l,m)_d = (d-l, m-l)_d = \begin{cases} (m,l)_d, & f \text{ even} \\ (m+d/2, l+d/2)_d, & f \text{ odd} \end{cases}$

(C) $\sum_{m=0}^{d-1} (l,m)_d = f - n_l$, where

$$n_l = \begin{cases} 1, & l \equiv 0 \pmod{d},\ f \text{ even} \\ 1, & l \equiv d/2 \pmod{d},\ f \text{ odd} \\ 0, & \text{otherwise} \end{cases}$$

(D) $\sum_{l=0}^{d-1} (l,m)_d = f - k_m$, where

$$k_m = \begin{cases} 1, & \text{if } m \equiv 0 \pmod{d}; \\ 0, & \text{otherwise} \end{cases}$$

(E) Diagonal sums (Tze, Chanson, Ding, Helleseth and Park [425])

$$\sum_{l=0}^{d-1} (l, l+m) = \begin{cases} f-1 & \text{if } m = 0, \\ f & \text{if } m \neq 0. \end{cases}$$

(F) $\sum_{l=0}^{d-1} \sum_{m=0}^{d-1} (l,m)_d = df - 1 = N - 2.$

(G) $(l,m)_{d'} = (sl, sm)_d$, where $(l,m)_{d'}$ is based on the primitive root $\theta' \equiv \theta^s \pmod{N}$; necessarily then s is prime to $N-1$.

These elementary facts are very important to our applications, as Properties (C-F) indicate several kinds of conservations between the cyclotomic numbers. They are the theoretical basis for the necessity of keeping the stability of local nonlinearities of some cryptographic functions.

The meaning of the cyclotomic numbers can be seen from another viewpoint. By definition the set $\{(l,m)_d : m = 0, 1, \cdots, d-1\}$ represents how the set $D_l + 1$ is distributed among the cyclotomic classes. Note that

$$|(D_l + \theta^k) \cap D_m| = |(D_{(l+N-1-k) \bmod d} + 1) \cap D_{(m+N-1-k) \bmod d}|$$

for each k, the d sets of numbers $\{(l,m)_d : m = 0, 1, \cdots, d-1\}$ for $l = 0, 1, \cdots, d-1$, represent also the distribution of the elements of any set $D_l + w$ over the d cyclotomic classes, where $w \neq 0$.

As observed above, cyclotomic numbers represent in fact the difference property of the partition $\{D_0, D_1, \cdots, D_{d-1}\}$ of Z_N^*. So they should have connections with difference sets. Necessary and sufficient conditions, that the dth power residues of a prime $N = df + 1$ form a difference set, are that d is even, f is odd and that

$$(l, 0)_d = (f-1)/d \quad \text{for } l = 0, 1, \cdots, d/2 - 1.$$

The existence problem of such difference sets has been solved for $d = 2$ ([338], i.e., the quadratic residues of primes $N = 4t - 1$), $d = 4$ ([76], the biquadratic residue difference set for $N = 2t^2 + 1$, t odd), $d = 8$ ([263], the octic residue difference set for $N = 8a^2 + 1 = 64b^2 + 9$, $k = a^2$, $\lambda = b^2$ with a, b odd).

The best known dth power residue difference sets are the quadratic residue sets of Paley [338], and the biquadratic residue difference set of Chowla [76]. Their applications will be investigated in later chapters. A general theory of the dth power residue difference sets has been developed by Lehmer [263]. Detailed discussions can also be found in Storer [414] and Baumert [15].

4.2 Cyclotomy and Cryptography

Cyclotomic numbers are quite useful in designing cryptographic functions for some stream ciphers. This section will make clear the importance of cyclotomy in the design and analysis of some stream ciphers. This will be done from several points of view. We begin with the additively natural stream ciphers.

4.2.1 Cyclotomy and Difference Parameters

The differential cryptanalysis of the additive natural stream ciphers was studied in [122]. We now give a brief description of the analysis.

Assume that $(G, +)$ is the Abelian group over which the keystream sequence is constructed, and $|G| = n$. For each $g_i \in G$ let $C_i = \{x \in Z_N : f(x) = g_i\}$, where $f(x)$ is the cryptographic function of the NSG in Figure 2.5.b. The ordered set $\{C_0, C_1, \cdots, C_{n-1}\}$ is called the *characteristic class.* For any ordered partition $\{C_0, C_1, \cdots, C_{n-1}\}$ of Z_N, there exists a function $f(x)$ with this partition as its characteristic class. The differential analysis of the system of Figure 2.5.b is the analysis of the following *difference parameters*:

$$d_f(i, j; w) = |C_i \cap (C_j - w)|, \quad (g_i, g_j) \in G \times G, \ w \in Z_N.$$

Thus $d_f(i, j; w)$ is the number of solutions of the equation $w = x_j - x_i$ for $x_j \in C_j, x_i \in C_i$.

The following simple facts are cryptographically important, as they represent some conservation rules between the difference parameters.

$$\sum_j d_f(i, j; w) = |C_i|, \ g_i \in G, \ w \in Z_N;$$

$$\sum_i d_f(i, j; w) = |C_j|; \ g_j \in G, \ w \in Z_N;$$

$$\sum_{i,j} d_f(i, j; w) = N, \ w \in Z_N.$$

When $n = 2$ (so $G = Z_2$) the differential analysis for the additive natural stream ciphers is important because it is equivalent to the following analyses: the nonlinearity analysis of the cryptographic function $f(x)$; the autocorrelation analysis of the keystream sequences; the stability analysis of the mutual information between the key and the two-bit keystream; and the transdensity analysis of the system, by which we mean that the analysis of the probability of agreement between two encryption or decryption transformations specified by two keys [122, 123]. These equivalences have already been proved in Section 2.4.

One cryptographically important aspect of cyclotomic numbers can be shown as follows. Let the notation be the same as in the previous section, so in particular N is an odd prime and $N = df + 1$. What we want to do now is to construct cryptographic functions from Z_N to an Abelian group $(G, +)$ of d elements, where $G = \{g_0, g_1, \cdots, g_{d-1}\}$. Let D_i be the cyclotomic classes of order d defined in the previous section and

$$C_0 = D_0 \cup \{0\}, \quad C_i = D_i, \ i = 1, \cdots, d - 1.$$

Without considering the implementation problem, we define a function from Z_N to $(G, +)$ as: $f(x) = g_i$ iff $x \in C_i$.

If $i \cdot j \neq 0$, then we have

$$d_f(i, j; \theta^k) = (i + N - 1 - k, j + N - 1 - k)_d.$$

On the other hand, we have

$$d_f(0, 0; \theta^k) = |(D_{N-1-k} \cup \{0\}) \cap (D_{N-1-k} \cup \{0\} - 1)|.$$

It follows that

$$0 \leq d_f(0, 0; \theta^k) - (N - 1 - k, N - 1 - k)_d \leq 2.$$

Similarly, we have

$$0 \leq d_f(0, j; \theta^k) - (N - 1 - k, N - k)_d \leq 1.$$

and

$$0 \leq d_f(j, 0; \theta^k) - (N - k, N - 1 - k)_d \leq 1.$$

Thus, we arrive at the conclusion that the difference parameters are almost the same as the cyclotomic numbers. Actually, the nonlinearity of $f(x)$ with respect to additions of Z_N and G is also determined by the cyclotomic numbers. This is clearly shown by the formulae in Section 2.4.

4.2.2 Cyclotomy and the Differential Cryptanalysis

One cryptographic importance of the cyclotomic numbers may be shown by the differential cryptanalysis for the additive natural stream ciphers [122], which can be outlined as follows. Recall that the additive natural stream cipher is an additive one with the NSG of Figure 2.5.b as its keystream generator. Because of the additive structure, knowing a number of plaintext-ciphertext pairs means knowing the same number of keystream characters.

For this cryptanalysis, it is assumed that $f(x)$, N and a piece of keystream sequence $z_0 z_1 \cdots z_{t-1}$ are known to a cryptanalyst, where N is the period of the counter and $f(x)$ is the cryptographic function applied to the counter of the NSG of Figure 2.5.b. The aim of this cryptanalysis is to recover the key of the NSG at the time z_0 was produced. Let $C_i = \{x \in Z_N : f(x) = i\}$ for $i = 0, 1$.

The differential cryptanalysis can be summarized with the following steps [122]:

(a) Find parameters $(i, j; w)$ with $(g_i, g_j) \in G \times G$, $w \in Z_N$ such that

$$d_f(i, j; w) = |C_i \cap (C_j - w)|$$

is as small as possible, and find the corresponding sets

$$D_f(i, j; w) = C_i \cap (C_j - w).$$

(b) Choose one $(i, j; w)$ so that there is a k' such that $(z_{k'}, z_{k'+w}) = (i, j)$. Then

$$k \in D_f(i, j; w) - k'.$$

If $d_f(i, j; w)$ is small, then search the set for k; otherwise, choose another $(i', j'; w')$ and find the corresponding $D_f(i', j'; w') - k''$. Then

$$k \in (D_f(i, j; w) - k') \cap (D_f(i', j'; w') - k'').$$

Continue in this way until the number of the elements of the set which contains k is small enough.

Whether this attack is feasible depends on the difference property of the $f(x)$, the computational complexity of finding some parameters $(i, j; w)$ with both small $d_f(i, j; w)$ and small w, and the determination of the corresponding sets $D_f(i, j; w)$ as well as the known keystream sequence and its length.

Since the difference property of the cryptographic function constructed in the last section is determined by cyclotomic numbers, the importance of cyclotomic numbers to the design of some natural sequence generators which can resist differential cryptanalysis, is clear.

4.2.3 Cryptographic Cyclotomic Numbers

The discussions in the foregoing two subsections clearly show that, to ensure the ideal difference property and nonlinearity of the cryptographic function constructed in Section 4.2.1 with respect to the additions of Z_N and $(G, +)$, the corresponding cyclotomic numbers should be as equal as possible. This condition makes differential cryptanalysis for the additive natural stream ciphers infeasible when the period N of the counter of the NSG is large enough.

By the conservation properties (C-E) in Section 4.1, we need cyclotomic numbers of order d such that all of them are approximately f/d. In other words, we need cyclotomic numbers of order d with good stability. Such cyclotomic numbers are said to be cryptographic.

4.3 Cryptographic Functions from Z_p to Z_d

In Section 4.2 we have made clear the cryptographic significance of cyclotomic numbers and what we need for the construction of cryptographic functions based on cyclotomy. Now we construct some cryptographic functions which we will need in later chapters, i.e., some cryptographic functions from Z_p to an Abelian group $(G, +)$ with d elements.

Before treating some special cases, we first define some cryptographic functions from Z_p to Z_d. Let a prime $p = df + 1$, and let θ be a primitive root modulo p. Setting $u = \theta^f$, we see that u is a primitive dth root in Z_p. Let $U = \{1, u, \cdots, u^{d-1}\}$, then U is a multiplicative subgroup of Z_p^*. First, we define a function from Z_p^* to U by

$$F_1(x) = x^f \bmod p, \quad x \in Z_p^*.$$

It is easy to see that $F_1(x)$ is a surjection and takes on each element of U exactly f times. Then we define a function from U to Z_d by

$$F_2(u^i) = i, \quad 0 \le i \le d-1.$$

Obviously, $F_2(x)$ is well-defined and one-to-one. Thus, the function

$$F^*(x) = F_2(F_1(x)), \quad x \in Z_p^*$$

is a mapping from Z_p^* to Z_d and takes on each element of Z_d exactly f times.

To apply the above functions to stream ciphers, we have to find algorithms to produce the function F_2. First we present a construction of $F_2(x)$ using integer multiplication and integer addition modulo p, i.e., a construction over Z_p.

Our idea is to use polynomial interpolation in the field Z_p. Let $\alpha_0, .., \alpha_{n-1}$ be n points in a field F, and let $V(\alpha_0, ..., \alpha_{n-1})$ denote the $n \times n$ Vandermonde matrix

$$V(\alpha_0, ..., \alpha_{n-1}) = \begin{bmatrix} 1 & \alpha_0 & \alpha_0^2 & \cdots & \alpha_0^{n-1} \\ 1 & \alpha_1 & \alpha_1^2 & \cdots & \alpha_1^{n-1} \\ \vdots & \vdots & \vdots & \vdots & \vdots \\ 1 & \alpha_{n-1} & \alpha_{n-1}^2 & \cdots & \alpha_{n-1}^{n-1} \end{bmatrix}$$

If ω is a primitive nth root of unity, then we write $V([\omega])$ for $V(1, \omega, ..., \omega^{n-1})$.

What we are looking for is one polynomial of degree $d-1$

$$a(x) = a_0 + a_1 x + \cdots + a_{d-1} x^{d-1}$$

in $Z_p[x]$ such that $a(u^i) = i$ for $i = 0, 1, \cdots, d-1$. If the inverse of the $d \times d$ Vandermonde matrix $V([u])$ can be found easily, we can immediately get the polynomial $a(x)$. Thanks to the following Proposition 4.3.1, we obtain the polynomial

$$a(x) = ((d-1)f + 1)(0, 1, \cdots, d-1)V([u^{d-1}])^T(1, x, \cdots, x^{d-1})^T.$$

For small d a polynomial of degree d can be implemented efficiently in hardware or software. For d being a power of 2, an FFT (fast Fourier transform) implementation is also possible.

Proposition 4.3.1 *Let ω be a primitive nth root of unity in a field F in which $n^{-1} = (n \times 1)^{-1}$ exists. Then*

$$V([\omega])^{-1} = n^{-1}V([\omega^{-1}]).$$

Proof: It is straightforward to check that

$$n^{-1}V([\omega^{-1}])V([\omega]) = I_n,$$

where I_n is the n by n identity matrix. □

To design cryptographic functions from Z_p to Z_d, we slightly extend both the above $F_1(x)$ and $a(x)$. Clearly, $F_1 = x^f$ is a well-defined function from Z_p to $U \cup \{0\}$. Thus we define a cryptographic function from Z_p to Z_d by

$$F(x) = a(x^f \bmod p) \bmod d.$$

This function has the following properties:

1. it is balanced to the best possible extent, i.e., it takes on one element of Z_d exactly $(f+1)$ times, and each of the others exactly f times;

2. the nonlinearity and difference property of the function are determined by the cyclotomic numbers of order d; and

3. the linear approximation of the function with respect to the additions of Z_p and Z_d makes no sense, because there are only trivial affine functions from $(Z_p, +)$ to $(Z_d, +)$.

For an effective implementation, only functions derived from small d's are interesting. We see below that the construction of the function for $d = 2$ is especially easy.

4.3.1 The Case $d = 2$

Let $p = 2f + 1$ be a prime. Then it is easy to see that the 4 cyclotomic numbers have the following relations when f is odd:

$$(0,0) = (1,0) = (1,1) = A, \quad (0,1) = B.$$

Employing the elementary facts about cyclotomic numbers described in Section 4.1, we get the following two equations:

$$2A = f - 1, \quad A + B = f.$$

Solving these equations gives the proof of part two in the following proposition. Part one can be similarly proved.

Proposition 4.3.2 *The cyclotomic numbers of order 2 are given by*

1. $(0,0) = (f-2)/2$; $(0,1) = (1,0) = (1,1) = f/2$ *if* f *is even; and*
2. $(0,0) = (1,0) = (1,1) = (f-1)/2$; $(0,1) = (f+1)/2$ *otherwise.*

It follows immediately from this proposition and the definition of difference and almost difference sets that the following proposition is true.

Proposition 4.3.3 *Quadratic residue difference sets and quadratic a.d. sets:*

(i) (Paley [338]) *If* $p = 2f+1 \equiv 3 \pmod 4$, *then the set* D_0 *of quadratic residues modulo p forms a difference set.*

(ii) *If* $p = 2f + 1 \equiv 1 \pmod 4$, *then the set* D_0 *of quadratic residues modulo p forms an a.d. set.*

Using the above facts, we can now give two cryptographic functions from Z_p to Z_2 with optimal difference property and nonlinearity with respect to the additions of the two rings, i.e., the characteristic functions of quadratic residue difference sets and of quadratic residue a.d. sets. The function $f(x)$ is defined by

$$f(x) = (x^f \bmod p) \bmod 2.$$

Since $x^f \bmod p$ takes on only two possible values 1 and $p - 1$, the above function is well defined. This function is needed in designing the DSC and ADSC generators in Chapter 8.

4.3.2 The Case $d = 3$

Cyclotomic numbers of order 3 were calculated by Dickson in 1935 [106]. Given a prime $p = 3t + 1$, t even, by the theory of binary quadratic forms, there are integers L^2 and M^2 which are uniquely determined by

$$4p = L^2 + 27M^2, \quad L \equiv 1 \pmod 3.$$

The sign of L has been chosen so that the congruence $L \equiv 1 \pmod 3$ holds. But the sign of M depends on the primitive root θ employed. In this case the nine cyclotomic numbers (i, j) reduce to $(0, 0)$, $(0, 1)$, $(0, 2)$, $(1, 2)$ and

$$\begin{array}{lll} (1,0) = (0,1), & (1,1) = (0,2), & (2,0) = (0,2), \\ (2,1) = (1,2), & (2,2) = (0,1). & \end{array}$$

The four different cyclotomic numbers are

$$\begin{aligned} (1,2) &= \frac{p+1+L}{9}; \\ (0,0) &= \frac{p-8+L}{9}; \\ (0,1) &= \frac{2p-4-L+9M}{18}; \\ (0,2) &= \frac{2p-4-L-9M}{18}. \end{aligned}$$

The stability of these cyclotomic numbers depends on the actual values of L and M. However, note that

$$1 \le |L| \le \sqrt{4p-27}, \quad 1 \le |M| \le \sqrt{\frac{4p-1}{27}},$$

the stability of the cyclotomic numbers of order 3 is cryptographically ideal for large primes.

A simple formula for the polynomial $a(x)$ can be derived in this case due to the fact $u^2 + u + 1 = 0$. It follows from the discussion at the beginning of Section 4.3 that

$$a(x) = (2f + 1)[3 + (u - 1)x - (u + 2)x^2].$$

Thus, the cryptographic function $F(x)$ defined at the beginning of Section 4.3 can be realized efficiently for the case $d = 3$.

4.3.3 The Case $d = 4$

For cyclotomic numbers of order 4, two sets of formulas were given for the cases of f even and odd respectively by Dickson [106]. We first consider the case f even.

Let $p = x^2 + 4y^2, \ x \equiv 1 \pmod 4$. Here y is two valued, depending on the choice of the primitive root [106]. There are five possible different cyclotomic numbers in this subcase; in fact, we have $(k,h) = (h,k)$, $(0,0)$, $(1,3) = (2,3) = (1,2)$, $(1,1) = (0,3)$, $(2,2) = (0,2)$, $(3,3) = (0,1)$ and

$$
\begin{aligned}
(0,0) &= (p - 11 - 6x)/16 = \frac{p}{4^2} - \frac{6x + 11}{16},\\
(0,1) &= (p - 3 + 2x + 8y)/16 = \frac{p}{4^2} + \frac{2x + 8y - 3}{16},\\
(0,2) &= (p - 3 + 2x)/16 = \frac{p}{4^2} + \frac{2x - 3}{16},\\
(0,3) &= (p - 3 + 2x - 8y)/16 = \frac{p}{4^2} + \frac{2x - 8y - 3}{16},\\
(1,2) &= (p + 1 - 2x)/16 = \frac{p}{4^2} - \frac{2x - 1}{16}.
\end{aligned}
$$

For the case f odd, there are at most five distinct cyclotomic numbers, which are

$$
\begin{aligned}
(0,0) = (2,2) = (2,0) &= (p - 7 + 2x)/16 = \frac{p}{4^2} + \frac{2x - 7}{16},\\
(0,1) = (1,3) = (3,2) &= (p + 1 + 2x - 8y)/16 = \frac{p}{4^2} + \frac{1 + 2x - 8y}{16},\\
(1,2) = (0,3) = (3,1) &= (p + 1 + 2x + 8y)/16 = \frac{p}{4^2} + \frac{1 + 2x + 8y}{16},\\
(0,2) &= (p + 1 - 6x)/16 = \frac{p}{4^2} + \frac{1 - 6x}{16},\\
\text{the rest} &= (p - 3 - 2x)/16 = \frac{p}{4^2} - \frac{3 + 2x}{16},
\end{aligned}
$$

where $p = x^2 + 4y^2$ and $x \equiv 1 \pmod 4$.

The stability of the cyclotomic numbers depends on the actual values of x and y. Nevertheless, since

$$
1 \le |x| \le \sqrt{p - 4}, \quad 1 \le |y| \le \sqrt{\frac{p - 1}{4}},
$$

the stability of the cyclotomic numbers of order 4 is cryptographically ideal for large primes.

Now we investigate the function $a(x)$ for this case. Note that

$$\begin{aligned} a(x) &= (3f+1)(0,1,2,3)V([u^3])^T(1,x,x^2,x^3)^T \\ &= (3f+1)(0,1,2,3)\begin{bmatrix} 1 & 1 & 1 & 1 \\ 1 & u^3 & u^2 & u \\ 1 & u^2 & 1 & u^2 \\ 1 & u & u^2 & u^3 \end{bmatrix}(1,x,x^2,x^3)^T. \end{aligned}$$

Since $u^4 = 1$ we get $u^2 = -1$. This gives

$$a(x) = 2(3f+1)[3 + (u-1)x - x^2 - (u+1)x^3].$$

With those cyclotomic numbers of order 4, we can easily prove the following theorem of Chowla [75], which is useful in designing some keystream sequences.

Proposition 4.3.4 *The biquadratic residues (D_0) of primes $p = 4x^2+1$, x odd, form a difference set with parameters* (N,k,λ)=$(4x^2+1, x^2, (x^2-1)/4)$.

Concerning the difference property of the set $D_0 \cup \{0\}$ for $d = 4$, it has also been proven that the following proposition holds [414, p. 50].

Proposition 4.3.5 *When $d = 4$, $D_0 \cup \{0\}$ forms a difference set of Z_p if and only if $p = 9 + 4x^2$ with x odd.*

Proof: It follows easily from the above cyclotomic constants of order 4 and the definition of difference sets. □

4.3.4 The Case $d = 5$

This case was also treated by Dickson in 1935. In the subcase f even the twenty-five (i,j)'s reduce to (0,0), (0,1), (0,2), (0, 3), (0,4), (1,2), (1,3), and we have also

$$\begin{aligned} &(4,4) = (0,1),\ (3,3) = (0,2),\ (2,2) = (0,3),\ (1,1) = (0,4), \\ &(3,4) = (1,4) = (1,2),\ (2,4) = (2,3) = (1,3). \end{aligned}$$

Furthermore $(k,h) = (h,k)$. Specifically, the cyclotomic numbers depend on the decomposition of the prime p or d^2p into the integral linear combination of integer squares. In this case the prime p can be decomposed as

$$16p = x^2 + 50u^2 + 50v^2 + 125w^2 \tag{4.1}$$

with $x \equiv 1 \pmod 5$ and

$$v^2 - 4uv - u^2 = xw. \tag{4.2}$$

It has been proven that such a decomposition is not unique [106]. Actually, there are exactly eight such decompositions, which satisfy (4.1) and (4.2). If (x, u, v, w) is one, also $(x, -u, -v, w)$ and $(x, \pm v, \mp u, -w)$ are decompositions. The remaining four are derived from these four by changing all signs. Choosing one decomposition (x, u, v, w), Dickson has proved that the seven possible distinct cyclotomic numbers are determined by the following equations

$$\begin{cases} (0,0) + (0,1) + (0,2) + (0,3) + (0,4) = f - 1 \\ (0,1) + (0,4) + 2(1,2) + (1,3) = f \\ (0,2) + (0,3) + (1,2) + 2(1,3) = f \\ 25(1,2) + 25(1,3) - 10t - 4 = x \\ (1,3) - (1,2) = w \\ (0,1) - (0.4) = v \\ (0,2) - (0,3) = u \end{cases}$$

By solving the linear equations, we get

$$\begin{aligned} (0,0) &= (p + 3x - 26)/25, \\ (0,1) &= (4p - 3x + 25w + 50v - 4)/100, \\ (0,2) &= (4p - 3x - 25w + 50u - 4)/100, \\ (0,3) &= (4p - 3x - 25w - 50u - 4)/100, \\ (0,4) &= (4p - 3x + 25w - 50v - 4)/100, \\ (1,2) &= (2p + x - 2 - 25w)/50, \\ (1,3) &= (2p + x - 2 + 25w)/50 \end{aligned}$$

Because of (4.1) and (4.2) as well as the form of these cyclotomic numbers, the cyclotomic numbers of order 5 for the case f even have ideal stability, though it depends on the quadratic decomposition of $16p$.

Similar to the above cases, we can easily get the $a(x)$ for this case:

$$\begin{aligned} a(x) &= (4f - 1)[10 + (u^3 + 2u^2 + 3u - 1)x + (-2u^3 + u^2 - u - 3)x^2 \\ &\quad +(2u^3 - u^2 + u - 2)x^3 - (u^3 + 2u^2 + 3u + 4)x^4]. \end{aligned}$$

Thus, the corresponding function $F(x)$ defined in Section 4.3 has a simple realization.

Table 4.1: The relations of the cyclotomic numbers of order 6.

(h,k)	0	1	2	3	4	5
0	(0,0)	(0,1)	(0,2)	(0,3)	(0,4)	(0,5)
1	(0,1)	(0,5)	(1,2)	(1,3)	(1,4)	(1,2)
2	(0,2)	(1,2)	(0,4)	(1,4)	(2,4)	(1,3)
3	(0,3)	(1,3)	(1,4)	(0,3)	(1,3)	(1,4)
4	(0,4)	(1,4)	(2,4)	(1,3)	(0,2)	(1,2)
5	(0,5)	(1,2)	(1,3)	(1,4)	(1,2)	(0,1)

4.3.5 The Case $d = 6$

The cyclotomic numbers of order 6 were investigated by Dickson [106] and Whiteman [455]. It has been proven that, when $d = 6$, the 36 cyclotomic constants (k, h) depend solely upon the decomposition $A^2 + 3B^2$ of the prime $p = 6f + 1$ [106].

In the subcase f even, there are three sets of cyclotomic numbers, depending on the choice of the primitive element θ of Z_p. Specifically, there are ten possible distinct cyclotomic numbers. The relations of these numbers are given in Table 4.1.

The values of the 10 basic constants are expressible in terms of p, A, B and depend on the cubic character of 2 modulo p. Select the integer m so that $\theta^m \equiv 2 \pmod{p}$, then the three sets of cyclotomic numbers are given in Table B.1.

For the case f odd, the ten basic constants are also expressible in terms of p, A, B and depend also on the cubic character of 2 modulo p (see Table B.2).

From $p = A^2 + 3B^2$ it follows that

$$1 \leq |A| \leq \sqrt{p-3}, \quad 1 \leq |B| \leq \sqrt{(p-1)/3}.$$

These facts together with the two sets of cyclotomic numbers indicate ideal stability of the cyclotomic constants of order 6, though this depends on the actual values of A and B. The derivation of the actual $a(x)$ for this case is also an easy task by hand. Thus, the corresponding function $F(x)$ defined in Section 4.3 can be easily determined.

4.3.6 The Case $d = 8$

The cyclotomic constants of order 8 were given by E. Lehmer [264]. In this case the 64 constants (i, j) have at most 15 different values for a given prime

Table 4.2: The relations of cyclotomic numbers of order 8 in subcase I.

(j,i)	0	1	2	3	4	5	6	7
0	(0,0)	(0,1)	(0,2)	(0,3)	(0,4)	(0,5)	(0,6)	(0,7)
1	(0,1)	(0,7)	(1,2)	(1,3)	(1,4)	(1,5)	(1,6)	(1,2)
2	(0,2)	(1,2)	(0,6)	(1,6)	(2,4)	(2,5)	(2,4)	(1,3)
3	(0,3)	(1,3)	(1,6)	(0,5)	(1,5)	(2,5)	(2,5)	(1,4)
4	(0,4)	(1,4)	(2,4)	(1,5)	(0,4)	(1,4)	(2,4)	(1,5)
5	(0,5)	(1,5)	(2,5)	(2,5)	(1,4)	(0,3)	(1,3)	(1,6)
6	(0,6)	(1,6)	(2,4)	(2,5)	(2,4)	(1,3)	(0,2)	(1,2)
7	(0,7)	(1,2)	(1,3)	(1,4)	(1,5)	(1,6)	(1,2)	(0,1)

$p = 8t + 1$. These values are expressible in terms of p, x, y, a and b in

$$p = x^2 + 4y^2 = a^2 + 2b^2, \ (x \equiv a \equiv 1 \pmod 4).$$

There are two subcases: the cases $p = 16t+1$ and $p = 16t+9$. The relations between the cyclotomic constants in the first subcase are given in Table 4.2 and the 15 fundamental constants are given by Table B.3.

In the second subcase the relations of the cyclotomic numbers are given in Table 4.3, and the fundamental constants are given in Table B.4.

Table 4.3: The relations of cyclotomic numbers of order 8 in subcase II.

(j,i)	0	1	2	3	4	5	6	7
0	(0,0)	(0,1)	(0,2)	(0,3)	(0,4)	(0,5)	(0,6)	(0,7)
1	(1,0)	(1,1)	(1,2)	(1,3)	(0,5)	(1,3)	(0,3)	(1,7)
2	(2,0)	(2,1)	(2,0)	(1,7)	(0,6)	(1,3)	(0,2)	(1,2)
3	(1,1)	(2,1)	(2,1)	(1,0)	(0,7)	(1,7)	(1,2)	(0,1)
4	(0,0)	(1,0)	(2,0)	(1,1)	(0,0)	(1,0)	(2,0)	(1,1)
5	(1,0)	(0,7)	(1,7)	(1,2)	(0,1)	(1,1)	(2,1)	(2,1)
6	(2,0)	(1,7)	(0,6)	(1,3)	(0,2)	(1,2)	(2,0)	(2,1)
7	(1,1)	(1,2)	(1,3)	(0,5)	(0,3)	(1,6)	(1,3)	(1,0)

The cyclotomic numbers of order 8 given in the two tables together with the facts that

$$1 \le |x| \le \sqrt{p-4}, \quad 1 \le |y| \le \sqrt{(p-1)/4},$$
$$1 \le |a| \le \sqrt{p-2}, \quad 1 \le |b| \le \sqrt{(p-1)/2},$$

show that the cyclotomic numbers of order 8 also have ideal stability. This means that the function $F(x)$ defined in Section 4.3 is cryptographically ideal for the natural sequence generator in the case $d = 8$. It is also cryptographically attractive from the viewpoint that the corresponding $a(x)$ is relatively simple because $u^4 = -1$. By calculation we get

$$\begin{aligned} a(x) &= 4(7f+1)[7+(u^3+u^2+u-1)(x-x^7)+(u^2-1)(x^2-x^6) \\ &\quad +(u^3-u^2+u-1)x^3-(u^3-u^2+u+1)x^5-x^4]. \end{aligned}$$

There are not so many dth power difference sets. However, the cyclotomic numbers of order 8 show that it is possible for the octic residues to form a difference set. In fact we have the following two results due to Lehmer [263].

Proposition 4.3.6 *If $p = 8f + 1$, then D_0 forms a difference set of Z_p if and only if p admits the simultaneous representations*

$$p = 9 + 64y^2 = 1 + 8b^2 \quad \textit{where } y \equiv b \equiv 1 \pmod 2.$$

Proof: It is a straightforward application of the cyclotomic constants of order 8 described in Tables B.3 and B.4. □

In this case the octic residue difference set has parameters $k = b^2$, $\lambda = y^2$. One example is the $(N, k, \lambda) = (73, 9, 1)$ difference set

$$\{1, 2, 4, 8, 16, 32, 37, 55, 64\},$$

the next such prime N is 140,411,704,393.

Proposition 4.3.7 *If $p = 8f + 1$, then $D_0 \cup \{0\}$ forms a difference set of Z_p if and only if p admits the simultaneous representation $p = 441 + 64y^2 = 49 + 8b^2$.*

Proof: It is a straightforward application of the cyclotomic constants of order 8 described in Tables B.3 and B.4. □

If the octic residues and zero form a difference set, it has parameter $k = b^2 + 7$, $\lambda = y^2 + 7$ with b odd and y even. It is known that $p = 26041$ admits such a pair of representations, and there is no other $p < 34,352,398,777$ which does.

4.3.7 The Case $d = 10$

The cyclotomic numbers of order ten were attacked by Dickson [106], Bruck [43] and Whiteman [455] with different approaches. The complete tables of the cyclotomic constants of order ten have been given by Whiteman [455].

Dickson showed that if p is a prime of the form $5k + 1$, then there are exactly four integral simultaneous solutions of the pair of diophantine equations

$$16p = x^2 + 50u^2 + 50v^2 + 125w^2, \quad xw = v^2 - 4uv - u^2, \tag{4.3}$$

with x uniquely determined by the condition $x \equiv 1 \pmod 5$. The four solutions are given by (x, u, v, w), $(x, v, -u, -w)$, $(x, -u, -v, w)$, $(x, -v, u, -w)$.

The 100 constants (h, k) have at most 22 different values for a given p, which are expressible in terms of p, x, u, v. Tables 4.4 and 4.5 summarize the relations of the constants in two cases, where (ij) denotes (i, j). There are ten sets of formulas depending on the parity of f and the quintic residue character of 2 modulo p. The 22 essentially different formulas of each set are given in the accompanying four tables, i.e., Tables B.5-B.8.

Table 4.4: The relations of the cyclotomic numbers of order 10 for even f.

	0	1	2	3	4	5	6	7	8	9
0	(00)	(01)	(02)	(03)	(04)	(05)	(06)	(07)	(08)	(09)
1	(01)	(09)	(12)	(13)	(14)	(15)	(16)	(17)	(18)	(12)
2	(02)	(12)	(08)	(18)	(24)	(25)	(26)	(27)	(24)	(13)
3	(03)	(13)	(18)	(07)	(17)	(27)	(36)	(36)	(25)	(14)
4	(04)	(14)	(24)	(17)	(06)	(16)	(26)	(36)	(26)	(15)
5	(05)	(15)	(25)	(27)	(16)	(05)	(15)	(25)	(27)	(16)
6	(06)	(16)	(26)	(36)	(26)	(15)	(04)	(14)	(24)	(17)
7	(07)	(17)	(27)	(36)	(36)	(25)	(14)	(03)	(13)	(18)
8	(08)	(18)	(24)	(25)	(26)	(27)	(24)	(13)	(02)	(12)
9	(09)	(12)	(13)	(14)	(15)	(16)	(17)	(18)	(12)	(01)

It has been proven that the set of tenth power residues or the set of tenth power residues together with zero modulo a prime $p = 10f + 1$ cannot form a difference set [455]. However, (4.3) and the constants in Tables B.5-B.8 show that the cyclotomic numbers of order 10 are roughly flat in each of the cases. This means that the corresponding $F(x)$ defined in Section 4.3 in this case is cryptographically attractive from the difference property and nonlinearity viewpoints. The actual $a(x)$ can be easily calculated by hand.

Table 4.5: The relations of the cyclotomic numbers of order 10 for odd f.

	0	1	2	3	4	5	6	7	8	9
0	(00)	(01)	(02)	(03)	(04)	(05)	(06)	(07)	(08)	(09)
1	(10)	(11)	(12)	(13)	(14)	(06)	(04)	(14)	(18)	(19)
2	(20)	(21)	(22)	(23)	(18)	(07)	(14)	(03)	(13)	(23)
3	(22)	(31)	(31)	(20)	(19)	(08)	(18)	(13)	(02)	(12)
4	(11)	(21)	(31)	(21)	(10)	(09)	(19)	(23)	(12)	(01)
5	(00)	(10)	(20)	(22)	(11)	(00)	(10)	(20)	(22)	(11)
6	(10)	(09)	(19)	(23)	(12)	(01)	(11)	(21)	(31)	(21)
7	(20)	(19)	(08)	(18)	(13)	(02)	(12)	(22)	(31)	(31)
8	(22)	(23)	(18)	(07)	(14)	(03)	(13)	(23)	(20)	(21)
9	(11)	(12)	(13)	(14)	(06)	(04)	(14)	(18)	(19)	(10)

4.3.8 The Case $d = 12$

In this case the 144 cyclotomic constants (i, j) depend solely upon the decompositions $p = x^2 + 4y^2$ and $p = A^2 + 3B^2$ of the prime $p = 12f + 1$, where $x \equiv 1 \pmod 4$ and $A \equiv 1 \pmod 6$. Some partial results in this case were obtained by Dickson [106], but the complete calculation was done by Whiteman [455].

The 144 constants have at most 31 different values for a given p. There are two sets of relations for the cyclotomic constants corresponding to the case f even and odd respectively. The formulas for the cyclotomic numbers are expressible in terms of p, x, y, A, B where the signs of x and A are such that $A \equiv 1 \pmod 4$ and $A \equiv 1 \pmod 6$. It has been proven by Whiteman that there are essentially twelve different sets of formulas depending on the parity of f, the sextic residue character of 2 modulo p, and $\psi(\beta^3, \beta)/\psi(\beta^5, \beta)$, where β is a primitive 12th root of unity and $\psi(\beta^m, \beta^n)$ is the Jacobi sum [455].

For our applications, we are only interested in the values of the cyclotomic constants, in order to estimate their stability. Practically, only cyclotomic numbers of relatively small orders are interesting to us. Other cyclotomic numbers of larger orders could be cryptographically needed, but are too long to present here. We shall describe some of them in Appendix A.

4.4 Cryptographic Functions from Z_{pq} to Z_d

In the foregoing section we constructed some cryptographic functions from Z_p to Z_d by employing the theory of cyclotomy over the prime fields Z_p, where these p are prime. The idea of the approach is first to get a proper partition $\{D_0, \cdots, D_{d-1}\}$ of the multiplicative group Z_p^* such that the partition has good difference property. Then we put the zero of the field Z_p into one class D_i to get a partition of Z_p. The assignment of the zero element only slightly changes the difference property of the original partition of Z_p^*. Finally, the characteristic function of the partition of Z_p is the constructed cryptographic function.

To construct cryptographic functions from Z_{pq} to Z_d, where p and q are distinct primes, we will follow the same idea. However, it should first be noted that Z_{pq} is only a ring, not a field. So we shall first get a partition of the multiplicative group $(Z_{pq}^*, \cdot)$ with good difference property. Then we assign the $p+q-1$ elements of $Z_{pq} \setminus Z_{pq}^*$ to the partition to get a partition of Z_{pq} with ideal difference property. To this end, we need Whiteman's theory of generalized cyclotomy.

4.4.1 Whiteman's Generalized Cyclotomy and Cryptography

We first consider the motivation for the theory of generalized cyclotomy. In 1958 Stanton and Sprott [411] published a generalization of the following result.

Proposition 4.4.1 *Let g be a common primitive root of p and $p+2$, where p and $p+2$ are both prime. Then the numbers*

$$1, g, g^2, \cdots, g^{(p^2-3)/2}; 0, p+2, 2(p+2), \cdots, (p-1)(p+2)$$

form a difference set with parameters $(N, k, \lambda) = (p(p+2), (N-1)/2, (N-3)/4)$, i.e., a Hadamard difference set, where $N = p(p+2)$.

According to [15, p.131] and [457], these so-called twin prime sets were already known, although in slightly different guise. They had been independently discovered by Stanton and Sprott, Kesava Menon [230], Brauer [31], Chowla [76], perhaps first by Gruner [181]. Motivated by the above proposition, Whiteman had generalized the theory of cyclotomy for the purpose of investigating residue difference sets modulo pq [457].

Whiteman's approach is first to get a proper representation of the elements of the multiplicative group Z_{pq}^* as described in the following proposition [457].

Proposition 4.4.2 *Let g be a fixed common primitive root of both primes p and q; let $d = \gcd(p-1, q-1)$ and $de = (p-1)(q-1)$. Then there exists an integer x such that*

$$Z_{pq}^* = \{g^s x^i : s = 0, 1, \cdots, e-1;\ i = 0, 1, \cdots, d-1\}.$$

Proof: Let $N = pq$. By the Chinese Remainder Theorem the common primitive root g exists since p and q are primes. Let x and y be a pair of integers satisfying the simultaneous congruences

$$\begin{array}{ll} x \equiv g \pmod{p}, & y \equiv 1 \pmod{p} \\ x \equiv 1 \pmod{q}, & y \equiv g \pmod{q}. \end{array} \tag{4.4}$$

The existence and uniqueness of such x, y are guaranteed by the Chinese Remainder Theorem. Clearly we have $xy \equiv g \pmod{N}$. Since g is a common primitive root of p and q, by the Chinese Remainder Theorem

$$\begin{aligned} \mathrm{ord}_N(g) &= \mathrm{lcm}\{\mathrm{ord}_p(g), \mathrm{ord}_q(g)\} \\ &= \mathrm{lcm}\{p-1, q-1\} \\ &= (p-1)(q-1)/d = e. \end{aligned}$$

Now we prove that the integer x defined by (4.4) satisfies the assertion of the proposition. To this end, we first show that no power g^s $(s = 0, 1, ..., e-1)$ of g is congruent modulo N to a power x^i $(i = 0, 1, ..., d-1)$ of x except when $s = i = 0$. This is true because the congruence $x^s y^s \equiv x^i \pmod{N}$ together with (4.4) implies that $p-1$ divides $s-i$ and $q-1$ divides s. Consequently d divides i and so $i \geq d$ unless $i = 0$. It follows that the congruence

$$g^s x^i \equiv g^t x^j \pmod{N} \quad (s, t = 0, 1, ..., e-1,\ i, j = 0, 1, ..., d-1)$$

is impossible unless $s = t$ and $i = j$. This completes the proof. □

The set Z_{pq}^* is also called the reduced residue system modulo $N = pq$. In Whiteman's generalized cyclotomy the *index class* or *cyclotomic class* D_i consists of e numbers and is defined by

$$D_i = \{g^s x^i : s = 0, 1, \cdots, e-1\},\ i = 0, 1, \cdots, d-1$$

and the generalized cyclotomic number $(i, j)_d$ is defined by

$$(i, j)_d = |(D_i + 1) \cap D_j|.$$

There are d cyclotomic classes $D_0, \cdots, D_{d-1}$, which form a partition of Z_{pq}^*.

The integer x of Proposition 4.4.2 is not unique, and the integer y defined by (4.4) could serve equally well in the same role. It is an immediate consequence of Proposition 4.4.2 and its proof that

$$x^d \equiv g^u \pmod{N}$$

for some fixed u such that $0 \le u \le e-1$. We note that $u \ne 1$ because the order of x modulo N is not equal to $\phi(N)$.

We now analyze the relation between the difference property of the partition of Z_{pq}^* and the generalized cyclotomic numbers. It is obvious that $x \in Z_{pq}^*$. By the definition of x and the proof of Proposition 4.4.2 the order of x modulo N is $p-1 \ge d$. Let $w \in Z_{pq}^*$. Then there must exist two integers s and t with $0 \le s \le e-1$, $0 \le t \le d-1$ such that $w = g^s x^t$. Because $x^d = g^u$ for some fixed u such that $0 \le u \le e-1$, the difference parameter can be expressed as

$$\begin{aligned} d(i,j;w) &= |(D_i + g^s x^t) \cap D_j| \\ &= |(D_{(p-1-t+i) \ mod \ d} + 1) \cap D_{(p-1-t+j) \ mod \ d}| \\ &= ((p-1-t+i) \ mod \ d, (p-1-t+j) \ mod \ d)_d, \end{aligned}$$

where $0 \le i, \ j \le d-1$; $w \in Z_N^*$. This means that for each $(i,j;w)$ with $0 \le i,j \le d-1$, $w \in Z_{pq}^*$, the difference parameter $d(i,j;w)$ is in fact one cyclotomic number. We will discuss the case for $w \notin Z_{pq}^*$ later.

Similar to the cyclotomy in prime fields Z_p, there are some elementary properties of this generalized cyclotomy. Clearly, d is even. Let $p-1 = df$, $q-1 = df'$, $e = dff'$ for some relatively prime integers f, f' (in particular f, f' are not both even). Then it is easy to prove

$$-1 \equiv \begin{cases} g^{e/2} \pmod{N} & \text{when } ff' \text{ is odd;} \\ g^v x^{d/2} \pmod{N} & \text{when } ff' \text{ is even,} \end{cases}$$

where v is some fixed integer, $0 \le v \le e-1$. The following properties of Whiteman's generalized cyclotomic numbers are fundamental and their proofs are left to the reader:

(A) $(l,m)_d = (l',m')_d$ when $l \equiv l' \pmod{d}$ and $m \equiv m' \pmod{d}$;

(B) $(l,m)_d = (d-l, m-l)_d = \begin{cases} (m,l)_d, & ff' \text{ odd} \\ (m+d/2, l+d/2)_d, & ff' \text{ even} \end{cases}$

(C) $\sum_{m=0}^{d-1}(l,m)_d = \frac{(p-2)(q-2)-1}{d} + n_l$, where

$$n_l = \begin{cases} 1, & l \equiv 0 \pmod{d}, & ff' \text{ odd,} \\ 1, & l \equiv d/2 \pmod{d}, & ff' \text{ even,} \\ 0, & \text{otherwise.} \end{cases}$$

(D) $\sum_{l=0}^{d-1}(l,m)_d = \frac{(p-2)(q-2)-1}{d} + k_m$, where

$$k_m = \begin{cases} 1, & \text{if } m \equiv 0 \pmod{d}; \\ 0, & \text{otherwise.} \end{cases}$$

(E) $\sum_{l=0}^{d-1}\sum_{m=0}^{d-1}(l,m)_d = (p-2)(q-2)$.

The elementary properties above can be easily proved from the definitions [457, 15].

As seen above, the index classes $D_0, \cdots, D_{d-1}$ form a partition of Z_N^*. Define

$$R = \{0\}, \quad P = \{p, 2p, \cdots, (q-1)p\}, \quad Q = \{q, 2q, \cdots, (p-1)q\},$$

so the sets $D_0, \cdots, D_{d-1}$; R; P; Q form a partition of Z_N. To extend the partition of Z_N^* into one of Z_N having ideal difference property, we have to study the difference property among the above sets. The following result [457, p.112] is useful for our application.

Proposition 4.4.3 *For any* $r \in P \cup Q$,

$$d(0,1;r) = |(D_0 + r) \cap D_1| = (p-1)(q-1)/d^2.$$

For our cryptographic purpose, we need to know the value of $d(i,j;r)$. Actually, the following more general proposition is true.

Theorem 4.4.4 *For any* $r \in P \cup Q$ *and any* $1 \le k \le d-1$,

$$d(0,k;r) = |(D_0 + r) \cap D_k| = (p-1)(q-1)/d^2.$$

Proof: Let $r \in P$ and let g, x generate the reduced residue system modulo N as in Proposition 4.4.2. Then $x \not\equiv 1 \pmod{N}$ and there is some fixed integer v with $0 \le v \le p-1-k$ such that

$$g^v x^k \equiv 1 \pmod{p}.$$

By definition the difference parameter $d(0,k;r)$ is the number of solutions of the congruence

$$y - z \equiv r \pmod{N} \tag{4.5}$$

with $y \in D_k$ and $z \in D_0$, which is equivalent to the congruence

$$g^t x^k - g^s \equiv r \pmod{N} \tag{4.6}$$

with $t, s \in \{0, 1, ..., e-1\}$. In order for (4.6) to be solvable, it is thus necessary that $t - s \equiv v \pmod{p-1}$. Note that $1 \le k \le d-1$, and therefore $0 \le v \le p-2$. Thus for each s with $0 \le s \le e-1$ there are precisely $(q-1)/d$ values of t with

$$t = v + s + m(p-1),$$

where $0 \le m < (q-1)/d$, for which the right side of (4.6) is divisible by p. Fix m and consider (4.6) for any $q-1$ consecutive values of s. The $q-1$ differences $g^{m(p-1)+s+v}x^k - g^s$ are $a, ag, ..., ag^{q-2}$, where a is an integer with $a \not\equiv 0 \pmod{q}$. Since g is a primitive root of q, they are congruent modulo N to $p, 2p, ..., (q-1)p$ in some order. Hence, for a fixed m, as s ranges from s_0 to s_0+q-2 the difference in (4.6) represents any fixed r precisely once. For each value of m there are $e/(q-1)$ such ranges of s. Thus a fixed r is represented exactly $(p-1)(q-1)/d^2$ times by (4.5). By symmetry the same result is true when q but not p divides r and therefore the theorem is proved. □

The following proposition [457] is also essential to our application.

Proposition 4.4.5 *For any $r \in P \cup Q$,*

$$\begin{aligned} d(0,0;r) &= |(D_0 + r) \cap D_0| \\ &= \begin{cases} (p-1)(q-1-d)/d^2, & r \in P, r \notin Q; \\ (q-1)(p-1-d)/d^2, & r \in Q, r \notin P. \end{cases} \end{aligned}$$

Proof: By symmetry we need only to prove the first part. The proof for this proposition is only a modification of that of Theorem 4.4.4. Let $r \in P$. A necessary condition for the solvability of $g^t - g^s \equiv r \pmod{N}$, where $s, t \in \{0, 1, ..., e-1\}$, is $t \equiv s \pmod{p-1}$. Recall that the order of g modulo N is e. Therefore, for each integer $m = 1, ..., (q-1-d)/d$ the difference $g^{m(p-1)} - 1$ is divisible by p but not by N. But if $m = 0$, this difference also equals zero and hence is divisible by N. It follows that for a fixed m ($m = 1, ..., (q-1-d)/d$) no two of the $q-1$ differences $g^{m(p-1)+s} - g^s$ with s in

$$\{j(q-1), j(q-1)+1, ..., j(q-1)+q-2\}$$

are congruent modulo N. As a result, these $q-1$ differences are congruent modulo N to the integers $p, 2p, ..., (q-1)p$. Consequently, as m ranges from 1 to $(q-1-d)/d$ and s ranges from 0 to $e-1$, the fixed value of r occurs $(p-1)(q-1-d)/d^2$ times amongst the differences under consideration. This completes the proof. □

Since $x \in Z_N^*$ and $x^d = g^u$ for some u with $0 \le u \le e-1$, for each r we have

$$\begin{aligned} d(i,j;r) &= |(D_i + r) \cap D_j| \\ &= |x^{d-i}(D_i + r) \cap x^{d-i} D_j| \\ &= d(0, (j+d-i) \bmod d; x^{d-i} r \bmod N). \end{aligned}$$

If $r \in P$ (resp. Q), then $x^{d-i}r \in P$ (resp. Q). Combining all the foregoing results in this section, we have completed the proof of the following theorem.

Theorem 4.4.6 *Let the notations be the same as before. Then for the partition $D_0, \cdots, D_{d-1}$ of Z_N^* and $r \neq 0$, we have*

$$d(i,j;r) = \begin{cases} (p-1)(q-1)/d^2, & i \neq j, \ r \in P \cup Q; \\ (p-1)(q-1-d)/d^2, & i = j, \ r \in P, \ r \notin Q; \\ (q-1)(p-1-d)/d^2, & i = j, \ r \in Q, \ r \notin P; \\ (i', j')_d \text{ for some } (i', j'), & \text{otherwise.} \end{cases}$$

In order to put the elements of R, P, Q into some of the D_i's to get a partition of Z_N with good difference property, we need the following result [457].

Proposition 4.4.7 *Let the notations be the same as before. If $r \notin Q \cup R$, then*

$$|D_0 \cap (Q \cup \{0\} + r)| = (p-1)/d.$$

Proof: A proof similar to that of Theorem 4.4.4 and Proposition 4.4.5 is easy to formulate. □

With the help of Propositions 4.4.3, 4.4.5, and 4.4.7, we can easily prove the following result of Whiteman [457].

Proposition 4.4.8 *Let the notations be the same as before. Then the set $D_0 \cup R \cup Q$, which is*

$$\{1, g, g^2, \cdots, g^{e-1}; 0, q, 2q, \cdots, (p-1)q\},$$

forms a difference set with parameters $N = pq$, $k = (N-1)/d$, $\lambda = (N-1-d)/d^2$ if and only if the following conditions are satisfied:

$$\begin{aligned} & q = (d-1)p + 2 \\ & (i, 0)_d = (d-1)[(p-1)/d]^2 \quad (i = 0, 1, \cdots, d-1). \end{aligned}$$

It is important to note that $n = k - \lambda = (p-f)^2$ by the first condition above. It follows that $\gcd(n, N) = 1$. In this case we can prove that the linear complexity of the characteristic sequences of these difference sets is N or $N-1$.

The cryptographically important consequence of the above Whiteman's theorem is that the set $D_0 \cup \{0\} \cup Q$ is a difference set whenever $q = p+2$. An application of the difference sets will be given in the next chapter.

4.4.2 Cryptographic Functions from Z_{pq} to Z_2

To design cryptographic binary sequences, we need functions from Z_{pq} to Z_2 with good nonlinearity with respect to the additions of the two rings. We now consider the characteristic function of the partition $\{R \cup Q \cup D_0, P \cup D_1\} = \{C_0, C_1\}$ of Z_{pq}. In what follows in this subsection we assume that $d = \gcd(p-1, q-1) = 2$. To analyze the function, we need the generalized cyclotomic numbers of order 2 obtained by Whiteman [457]:

Proposition 4.4.9 *Let the notations be the same as before. If ff' is even, we have $(0,0) = (1,0) = (1,1)$ and two different cyclotomic numbers*

$$(0,0) = \frac{(p-2)(q-2)+1}{4}, \quad (0,1) = \frac{(p-2)(q-2)-3}{4}.$$

If ff' is odd, we have $(0,1) = (1,0) = (1,1)$ and

$$(0,0) = \frac{(p-2)(q-2)+3}{4}, \quad (0,1) = \frac{(p-2)(q-2)-1}{4}.$$

With these generalized cyclotomic constants of order 2 we are ready to analyze the difference property of the partition $\{C_0, C_1\}$ of Z_{pq}. Note that

$$d(0,0;r) = |[(R+r) \cup (Q+r) \cup (D_0+r)] \cap [R \cup Q \cup D_0]|.$$

Setting

$$\begin{aligned} a(0,0;r) &= |(Q+r) \cap Q| + |(Q+r) \cap D_0| \\ &+ |(D_0+r) \cap Q| + |(D_0+r) \cap D_0|, \end{aligned}$$

we can prove

$$0 \le d(0,0;r) - a(0,0,r) \le 2.$$

So our task now is to estimate the $a(0,0;r)$ with $r \ne 0$. One simple fact is

$$|(Q+r) \cap Q| = \begin{cases} p-2, & r \in Q; \\ 0, & r \in P \cup Z_{pq}^*; \end{cases}$$

Note that if $r \in P$, then it is possible to have $Q + r \subset D_0$. Thus, for each r we have the following two obvious facts:

$$0 \le |(Q + r) \cap D_0| \le p - 1;$$
$$0 \le |Q \cap (D_0 + r)| \le p - 1.$$

It follows that

$$\begin{aligned} |(D_0 + r) \cap D_0| &\le a(0, 0; r) \\ &\le 3p - 4 + |(D_0 + r) \cap D_0|. \end{aligned}$$

Setting

$$B = \max \left\{ \frac{(p-2)(q-2)+3}{4}, \frac{(p-1)(q-3)}{4}, \frac{(p-3)(q-1)}{4} \right\}$$

and

$$C = \min \left\{ \frac{(p-2)(q-2)-3}{4}, \frac{(p-1)(q-3)}{4}, \frac{(p-3)(q-1)}{4} \right\},$$

we get

$$C \le a(0, 0; r) \le 3p - 4 + B,$$

and therefore

$$C \le d(0, 0; r) \le 3p - 2 + B.$$

We can similarly prove that for each $r \neq 0$,

$$C \le d(1, 1; r) \le 3q - 4 + B.$$

In what follows we analyze $d(1, 0; r)$ and $d(0, 1; r)$. By definition we have

$$\begin{aligned} d(1, 0; r) &= |(C_1 + r) \cap C_0| = |[(P + r) \cup (D_1 + r)] \cap (R \cup Q \cup D_0)| \\ &= |(P + r) \cap R| + |(P + r) \cap Q| + |(P + r) \cap D_0| \\ &\quad + |(D_1 + r) \cap R| + |(D_1 + r) \cap Q| + |(D_1 + r) \cap D_0|. \end{aligned}$$

If $r \in P$, then by Proposition 4.4.3 we have

$$|(D_1 + r) \cap D_0| = (p - 1)(q - 1)/4.$$

In addition we clearly have

$$|(P + r) \cap Q| = |(P + r) \cap D_0| = |(D_1 + r) \cap R| = 0$$
$$|(P + r) \cap R| = 1, \ 0 \le |(D_1 + r) \cap Q| \le p - 1.$$

Hence, we obtain in the case $r \in P$

$$1 + \frac{(p-1)(q-1)}{4} \leq d(1,0;r) \leq \frac{(p-1)(q-1)}{4} + p.$$

If $r \in Q$, we can similarly prove

$$\frac{(p-1)(q-1)}{4} \leq d(1,0;r) \leq \frac{(p-1)(q-1)}{4} + q - 1.$$

If $r \in Z_{pq}^*$, then by Proposition 4.4.9 we get

$$\frac{(p-2)(q-2)-3}{4} \leq |(D_1 + r) \cap D_0| \leq \frac{(p-2)(q-2)+3}{4}.$$

In addition it is easily seen that

$$\begin{aligned}
&|(P+r) \cap R| = 0, \\
&0 \leq |(P+r) \cap Q| \leq \min\{p-1, q-1\}, \\
&0 \leq |(P+r) \cap D_0| \leq q-1, \\
&0 \leq |(D_1+r) \cap Q| \leq p-1, \\
&0 \leq |(D_1+r) \cap R| \leq 1.
\end{aligned}$$

It follows in this case that

$$\begin{aligned}
\tfrac{(p-2)(q-2)-3}{4} &\leq d(1,0;r) \leq \\
&\leq \tfrac{(p-1)(q-1)}{4} + \min\{p-1, q-1\} + \tfrac{3}{4}(p+q) + \tfrac{1}{2}.
\end{aligned}$$

Combining the results for the three cases, we obtain

$$\frac{(p-2)(q-2)-3}{4} \leq d(1,0;r) \leq \frac{(p-1)(q-1)}{4} + E,$$

where

$$E = \max\left\{p, q-1, \min\{p-1, q-1\} + \frac{3}{4}(p+q) + \frac{1}{2}\right\}.$$

Similarly, one can prove

$$\frac{(p-2)(q-2)-3}{4} \leq d(0,1;r) \leq \frac{(p-1)(q-1)}{4} + E,$$

So far we have completed the analysis of the difference property of the partition $\{C_0, C_1\}$ of Z_{pq}. The above results show that this partition has good difference property if $|p-q|$ is small enough. In this case, the facts that

$|C_0| = (p-1)(q-1)/2+q$ and $|C_1| = (p-1)(q-1)/2+p-1$ show that the function also has good balance. This means that the characteristic function of this partition is cryptographically attractive from a number of viewpoints. The best case is $q = p+2$, i.e., twin primes. However, for our applications, the conditions that $|q-p|$ is small enough and $\gcd(p-1, q-1) = 2$, suffice to guarantee many cryptographic properties of the function.

For our applications, we are much concerned with the implementation of the characteristic function. We now prove that the characteristic function of the partition $\{C_0, C_1\}$ has the following expression:

$$F_C(j) = \begin{cases} 0, & j \in R \cup Q; \\ 1, & j \in P; \\ (1 - (\frac{j}{p})(\frac{j}{q}))/2, & \text{otherwise.} \end{cases}$$

Since g is a common primitive root of both p and q, it is easy to see

$$\left(\frac{g^i}{p}\right)\left(\frac{g^i}{q}\right) = (-1)^i \times (-1)^i = 1$$

for each i with $0 \le i \le e-1$. Thus, for each $j \in D_0$ we have $F_C(j) = 0$. By the construction of x, i.e, $x \equiv g \pmod{p}$ and $x \equiv 1 \pmod{q}$, we have

$$\left(\frac{x}{p}\right)\left(\frac{x}{q}\right) = -1 \times 1 = -1.$$

This means that for each $j \in D_1$, we have $F_C(j) = 1$. The remaining parts can be easily seen. We will discuss the implementation of the characteristic function further in Chapter 8.

4.4.3 Cryptographic Functions from Z_{pq} to Z_4

To design cryptographic functions from Z_{pq} to Z_4, we need the generalized cyclotomic numbers of order 4 obtained by Whiteman [457]. When $d = 4$, both primes p and q in the product $N = pq$ are of the form $4t+1$. By a well-known theorem [275, p. 128] there are exactly two representations of N in the form $N = a^2 + b^2$ with $a \equiv 1 \pmod{4}$ and the sign of b indeterminate. Let

$$N = a^2 + 4b^2, \quad N = a'^2 + 4b'^2 \quad (a \equiv a' \equiv 1 \pmod{4})$$

denote these two representations. Let g be a common primitive root of p and q, and x be selected by the Chinese Remainder Theorem as in the foregoing section. Let (i, j) be the cyclotomic number defined as before. Whiteman proved that the sixteen cyclotomic constants depend solely upon one of the two representations. He also showed, for ff' even,

$$\begin{array}{ll} 8(0,0)_4 = -a + 2M + 3, & 8(0,1)_4 = -a - 4b + 2M - 1, \\ 8(0,2)_4 = 3a + 2M - 1, & 8(0,3)_4 = -a + 4b + 2M - 1, \\ 8(0,0)_4 = -a + 2M + 3, & \end{array}$$

and the remaining (i,j)'s are equal to one of the cyclotomic numbers above. Here $M = [(p-2)(q-2) - 1]/4$. For the case ff' odd, he proved

$$\begin{array}{ll} 8(0,0)_4 = 3a + 2M + 5, & 8(0,1)_4 = -a + 4b + 2M + 1 \\ 8(0,2)_4 = -a + 2M + 1, & 8(0,3)_4 = -a - 4b + 2M + 1 \\ 8(1,2)_4 = a + 2M - 1, & \end{array}$$

and the remaining (i,j)'s are equal to one of the cyclotomic numbers above. Note that

$$1 \le |a| \le \sqrt{pq - 4}, \quad 1 \le |b| \le \sqrt{(pq-1)/4}$$

and that M is relatively much larger than a and b. We conclude that the cyclotomic numbers have ideal stability in both cases.

One cryptographically interesting result derived from the cyclotomic numbers of order 4 is the following proposition of Whiteman [457] about difference sets modulo pq.

Proposition 4.4.10 *Let p and q be distinct primes such that $\gcd(p-1, q-1) = 4$ and let $e = (p-1)(q-1)/4$. Let g, g' be distinct common primitive roots of p, q with $g' \not\equiv g^r \pmod{N}$ for any r. Then one (but not both) of the sets*

$$\{1, g, g^2, \cdots, g^{e-1}; 0, q, 2q, \cdots, (p-1)q\}$$
$$\{1, g', g'^2, \cdots, g'^{e-1}; 0, q, 2q, \cdots, (p-1)q\}$$

is a difference set with parameters $N = pq$, $k = (N-1)/4$, $\lambda = (N-5)/16$ if and only if $q = 3p+2$ and k is an odd square.

To design cryptographic functions from Z_{pq} to Z_4, one can follow the same approach as in the foregoing section. Such a function will be described in Section 8.3.

4.5 Cryptographic Functions from Z_{p^2} to Z_2

In Section 3.7 we have seen that some sequences with period the square of an odd prime are cryptographically attractive. To design generators which can produce such binary sequences, we need functions from Z_{p^2} to Z_2 with good nonlinearity with respect to the additions of the two rings. Motivated by the approach in the foregoing sections, we want to follow suit for this case. That is, we first get a partition of the multiplicative group $Z_{p^2}^*$, which has $\phi(p^2) = p(p-1)$ elements. We then extend the partition to get a partition of Z_{p^2} with good difference property [123].

The generalized cyclotomy of order 2 with respect to p^2 was considered for the prime-square generator in [123]. Ding and Helleseth have extended this kind of generalized cyclotomy of order 2 with respect to p^2 into that with respect to general n [129]. For our purpose we introduce here this generalized cyclotomy of order 2 with respect to only p^m, where m is any positive integer.

Let g be a primitive root of p^m, then g is also a primitive root of p^i for $i = 1, 2, \cdots, m-1$. The generalized cyclotomic classes of order 2 are defined by

$$D_0^{(p^m)} = (g^2), \quad D_1^{(p^m)} = gD_0^{(p^m)},$$

where the arithmetic is that of Z_{p^m}. Define

$$R^{(p^m)} = \{0, p, 2p, \cdots, (p^{m-1} - 1)p\}.$$

Then

$$Z_{p^m} = R^{(p^m)} \cup D_0^{(p^m)} \cup D_1^{(p^m)}.$$

As before, the generalized cyclotomic numbers of order 2 with respect to p^m are defined by

$$(i, j)_{p^m} = |(D_i^{(p^m)} + 1) \cap D_j^{(p^m)}|.$$

Lemma 4.5.1

$$\left|R^{(p^m)} \cap \left(D_1^{(p^m)} + 1\right)\right| = \begin{cases} 0, & p \equiv 1 \pmod 4, \\ p^{m-1}, & p \equiv 3 \pmod 4; \end{cases}$$

$$\left|R^{(p^m)} \cap \left(D_0^{(p^m)} + 1\right)\right| = \begin{cases} p^{m-1}, & p \equiv 1 \pmod 4, \\ 0, & p \equiv 3 \pmod 4; \end{cases}$$

Proof: We have $g^{2s} + 1 \in R^{(p^m)}$ if and only if $g^{2s} \equiv -1 \pmod p$. Since g is a primitive root of p, $g^{2s} \equiv -1 \pmod p$ if and only if $2s \equiv (p-1)/2 \pmod{p-1}$. This is impossible if $p \equiv 3 \pmod 4$. If $p \equiv 1 \pmod 4$, then $(p-1)/2$ is even. So $2s = (p-1)/2 + a(p-1)$ for some a. It follows that

$$0 \le 2s = (p-1)\frac{1+2a}{2} \le p^{m-1}(p-1).$$

Hence, $0 \leq a \leq p^{m-1} - 1$. This proves the second part of the lemma, and the first part then follows easily. □

The relations between the cyclotomic numbers $(i, j)_{p^m}$ are described by the following lemma:

Lemma 4.5.2 *1.* $(0,0)_{p^m} + (1,0)_{p^m} = \frac{p^{m-1}(p-3)}{2}$.

2. $(0,1)_{p^m} + (1,1)_{p^m} = \frac{p^{m-1}(p-1)}{2}$.

3. $$(1,0)_{p^m} + (1,1)_{p^m} = \begin{cases} \frac{p^{m-1}(p-1)}{2}, & p \equiv 1 \pmod 4, \\ \frac{p^{m-1}(p-3)}{2}, & p \equiv 3 \pmod 4. \end{cases}$$

4. $$(0,0)_{p^m} + (0,1)_{p^m} = \begin{cases} \frac{p^{m-1}(p-3)}{2}, & p \equiv 1 \pmod 4, \\ \frac{p^{m-1}(p-1)}{2}, & p \equiv 3 \pmod 4. \end{cases}$$

Proof: We prove only part four, and the rest can be similarly proved. Recall that

$$D_0^{(p^m)} \cup D_1^{(p^m)} \cup R^{(p^m)} = Z_{p^m}.$$

Using this, the definitions and Lemma 4.5.1 give

$$\begin{aligned} (0,0)_{p^m} + (0,1)_{p^m} &= \left|D_0^{(p^m)} \cap \left(D_0^{(p^m)} + 1\right)\right| + \left|D_1^{(p^m)} \cap \left(D_0^{(p^m)} + 1\right)\right| \\ &= p^{m-1}(p-1)/2 - \left|R^{(p^m)} \cap \left(D_0^{(p^m)} + 1\right)\right| \\ &= \begin{cases} \frac{p^{m-1}(p-3)}{2}, & p \equiv 1 \pmod 4, \\ \frac{p^{m-1}(p-1)}{2}, & p \equiv 3 \pmod 4. \end{cases} \end{aligned}$$

□

Theorem 4.5.3 *If* $p \equiv 3 \pmod 4$, *then*

$$(1,0)_{p^m} = (0,0)_{p^m} = (1,1)_{p^m} = \frac{p^{m-1}(p-3)}{4}, \quad (0,1)_{p^m} = \frac{p^{m-1}(p+1)}{4}.$$

If $p \equiv 1 \pmod 4$, *then*

$$(0,1)_{p^m} = (1,0)_{p^m} = (1,1)_{p^m} = \frac{p^{m-1}(p-1)}{4}, \quad (0,0)_{p^m} = \frac{p^{m-1}(p-5)}{4}.$$

Proof: Since g is a primitive root of p^m,

$$D_i^{(p^m)} \bmod p = \{\underbrace{x, \cdots, x}_{p^{m-1}} : x \in D_i^{(p)}\}. \tag{4.7}$$

It follows that

$$\begin{aligned}(j,i)_{p^m} &= \left|D_i^{(p^m)} \cap \left(D_j^{(p^m)} + 1\right)\right| \\ &= p^{m-1}\left|D_i^{(p)} \cap \left(D_j^{(p)} + 1\right)\right| \\ &= p^{m-1}(j,i)_p.\end{aligned}$$

The theorem then follows from Proposition 4.3.2. □

The generalized cyclotomic numbers of order 2 for the special case $m = 2$ were conjectured in [123], and the conjecture was proved by Pei with another method [341]. The above general result is due to Ding and Helleseth [129].

Theorem 4.5.4 *For any* $r \in R^{(p^m)}$,

$$\left|D_i^{(p^m)} \cap \left(D_j^{(p^m)} + r\right)\right| = \begin{cases} p^{m-1}(p-1)/2, & \text{if } i = j; \\ 0, & \text{otherwise.}\end{cases}$$

Proof: Note that $r \equiv 0 \pmod p$. It follows from (4.7) that

$$\left|D_i^{(p^m)} \cap \left(D_j^{(p^m)} + r\right)\right| = p^{m-1}\left|D_i^{(p)} \cap D_j^{(p)}\right|.$$

The conclusion then follows. □

Theorem 4.5.3 shows that the generalized cyclotomic numbers of order 2 with respect to p^2 have ideal stability. Thus, the partitions $\{D_0^{(p^2)} \cup R^{(p^2)}, D_1^{(p^2)}\}$ and $\{D_0^{(p^2)}, D_1^{(p^2)} \cup R^{(p^2)}\}$ both have ideal stability. The corresponding characteristic functions are respectively

$$F_C(x) = \begin{cases} 1, & x \in R^{(p^2)}; \\ (x^{p(p-1)/2} \bmod p^2) \bmod 2, & \text{otherwise}\end{cases}$$

and

$$F_C(x) = \begin{cases} 0, & x \in R^{(p^2)}; \\ (x^{p(p-1)/2} \bmod p^2) \bmod 2, & \text{otherwise}\end{cases}$$

The application of the functions will be further discussed in the next chapter. Of course, we can put the elements of $R^{(p^2)}$ into $D_0^{(p^2)}$ and $D_1^{(p^2)}$ arbitrarily. In doing so, we get binary functions which have about the same property as the above $F_C(x)$'s.

4.6 Cryptographic Functions Defined on $GF(p^m)$

In the foregoing sections we have used cyclotomy to construct some cryptographic functions from Z_p, Z_{pq} and Z_{p^2} to some rings Z_d. Those functions have a numerical realization. For cryptography we may need functions from a finite field $GF(p^m)$ to some Abelian group. To this end, we need the extended theory of cyclotomy in finite fields $GF(p^m)$, for example as given by Storer [414]. Let $q = p^m = df + 1$ and let α be a generating element of $GF(q)$. Define $D_0 = (\alpha^d)$ and $D_i = \alpha^i D_0$ for each i, where $1 \leq i \leq d-1$. Then the cyclotomic numbers of order d with respect to $GF(q)$ are defined by

$$(i, j)_d = |(D_i + 1) \cap D_j|,$$

where $0 \leq i \leq d-1$ and $0 \leq j \leq d-1$. Similar results about the cyclotomic numbers have been obtained for the cases $d = 2, 4, 6, 8$ [414]. The construction of such functions is the same as that for the case of Z_p. For instance, for a finite field $GF(p^m)$ with $N := p^m = df + 1$, a cryptographic function from $GF(p^m)$ to Z_d can be defined by

$$f(x) = \begin{cases} a \bmod d, & x = \xi^a; \\ 0, & x = 0, \end{cases}$$

where ξ is a primitive root of N. The above $f(x)$ takes on the zero of Z_d exactly $f+1$ times, and each of the others f times. This means that it has an ideal balance property. The nonlinearity and difference property of the function depend on the stability of cyclotomic numbers of order d defined on the finite field $GF(p^m)$.

4.7 The Origin of Cyclotomic Numbers

The introduction of cyclotomic numbers by Gauss [159] was related to higher reciprocity, the cyclotomic equation, the constructibility of regular polygons and the quadratic partition of primes of the form $3t + 1$ into $x^2 + 27y^2$ [159, 414, 86].

Let $\theta = e^{2\pi i/p}$ be a primitive pth root of unity, and let g be a primitive root modulo p. Now suppose that $p = ef + 1$, and λ is an integer. Gauss defines the *period*[1] $(f; \lambda)$ to be the sum

$$(f; \lambda) = \sum_{j=0}^{f-1} \theta^{\lambda g^{ej}}.$$

[1] Gauss' original symbol for the period is (f, λ) [159]. We use $(f; \lambda)$ instead for the purpose of avoiding confusion with the symbol for cyclotomic numbers.

Let D_0 be the subgroup (g^e) of Z_p^*, and $D_i = g^i D_0$ for $0 \le i \le e-1$ as before. Then it is clear that

$$(f; g^i) = \sum_{r \in D_i} \theta^r,$$

which means that the period is the sum of a set of roots of the cyclotomic equation $x^p - 1 = 0$. It is therefore easily seen that $(f; \lambda)$ is independent of the choice of the primitive element g. Since D_0, D_1, $\cdots$, D_{e-1} are a partition of Z_p^*, it follows from the above formula that

$$\sum_{i=0}^{e-1} (f; g^i) = -1.$$

These periods are related to cyclotomic numbers, higher reciprocity, cyclotomic equations, and the construction of regular polygons. They are the key to Gauss' study of the cyclotomic field $Q(\theta_p)$ [159, 86]. If $p = 3f+1$, then the three periods are $(f;1)$, $(f;g)$ and $(f;g^2)$. By analyzing the products of the periods, he introduced the cyclotomic numbers and got the following remarkable result as a byproduct [86, pp.86 and 94-95]:

Proposition 4.7.1 *If* $4p = a^2 + 27b^2$ *and* $a \equiv 1 \pmod 3$, *then the number of solutions modulo* p *of* $x^3 - y^3 \equiv 1 \pmod p$ *is* $N = p + a - 2$.

Gauss' work [159] implies the following derivation of the solution of the above proposition. Assume $p = 3f + 1$. Let $(f; \lambda)$ and $(f; \mu)$ be periods, and write $(f; \mu) = \theta^{\mu_1} + \cdots, +\theta^{\mu_f}$. The first thing he did is to prove

$$(f; \lambda)(f; \mu) = \sum_{j=1}^{f} (f; \lambda + \mu_j). \tag{4.8}$$

Then for $i, j \in \{0, 1, 2\}$, he defined the cyclotomic number (i, j) to be the number of pairs (m, n), $0 \le m, n \le f-1$, such that

$$1 + g^{3m+i} \equiv g^{3n+j} \pmod p. \tag{4.9}$$

Here the cyclotomic number (m, n) is the $(n, m)_3$ defined at the beginning of this section. With this definition he proved that the number of solutions modulo p of the equation

$$x^3 - y^3 \equiv 1 \pmod p \tag{4.10}$$

is $N = 9(0,0) + 6$. The relations about the products of periods and cyclotomic numbers in this case obtained by Gauss with the help of (4.8) are

$$(f;1)(f;1) = f + (0,0)(f;1) + (0,1)(f;g) + (0,2)(f;g^2) \tag{4.11}$$

and

$$(f;1)(f;g) = (1,0)(f;1) + (1,1)(f;g) + (1,2)(f;g^2), \tag{4.12}$$

which give the following results:

$$(0,0) + (0,1) + (0,2) = f - 1, \tag{4.13}$$

$$(1,0) + (1,1) + (1,2) = f. \tag{4.14}$$

These two relations are special cases of the general conservation laws in Section 4.1. By expanding $(f;g) \cdot (f;1)$ and comparing it to (4.11) and (4.12), he got that $(1,0) = (2,2)$, $(1,1) = (2,0)$ and $(1,2) = (2,1)$. This reduces the 9 cyclotomic numbers to three:

$$\begin{aligned} \alpha &= (1,2) = (2,1) = (0,0) + 1 \\ \beta &= (0,1) = (1,0) = (2,2) \\ \gamma &= (0,2) = (2,0) = (1,1). \end{aligned}$$

Note that $(f;1)(f;g)(f;g^2)$ is an integer. By expanding this quantity in terms of α, β and γ, he obtained

$$\alpha^2 + \beta^2 + \gamma^2 - \alpha = \alpha\beta + \beta\gamma + \alpha\gamma. \tag{4.15}$$

With this result he showed further that

$$(6\alpha - 3\beta - 3\gamma - 2)^2 + 27(\beta - \gamma)^2 = 12(\alpha + \beta + \gamma) + 4. \tag{4.16}$$

Using (4.15), the above result that $\alpha + \beta + \gamma = f$ and $p = 3f + 1$, he obtained

$$4p = a^2 + 27b^2, \tag{4.17}$$

where $a = 6\alpha - 3\beta - 3\gamma - 2$ and $b = \beta - \gamma$. A little more analysis then gave him the result

$$a = 9\alpha - 3(\alpha + \beta + \gamma) - 2 = 9\alpha - p - 1.$$

Finally, the facts that $\alpha = (0,0) + 1$ and $N = 9(0,0) + 6$ gave Gauss the conclusion that

$$a = N - p + 2.$$

This is what Gauss did about the solution of (4.10) by introducing cyclotomic numbers and periods. It should be pointed out that Gauss did not really state Proposition 4.7.1 explicitly in his *Disquisitiones Arithmeticae*, though he proved the result. Our description follows the refinement by Cox [86]. Gauss' method to calculate the cyclotomic numbers is the foundation of later methods extended by Dickson [106, 107, 108, 109] and many others.

In what follows we give the general relation between the product of periods and cyclotomic numbers. Recall that g is a primitive root modulo p. Then for each $a \in Z_p^*$, there must exist an integer $h_a \in \{0, 1, \cdots, p-2\}$ such that $a^{-1} = g^{h_a}$. In particular, we have

$$-1 = (-1)^{-1} = g^{(p-1)/2}.$$

Each element $a \in D_i$ must be expressed as $a = g^{i+ej}$ for some j, where $0 \le j \le f-1$. It follows that

$$a^{-1} = g^{p-1-i-ej} = g^{e(f-j)-i}.$$

Then by the basic properties of cyclotomic numbers we have

$$(u + h_a, v + h_a + (p-1)/2)_e = (u - i, v - i + (p-1)/2)_e$$

for all $a \in D_i$, where $0 \le i \le e-1$. Therefore we get

$$\begin{aligned} H(u,v,\theta) &= \textstyle\sum_{a=1}^{p-1} (u + h_a, v + h_a + (p-1)/2)_e \theta^a \\ &= \textstyle\sum_{i=0}^{e-1} \sum_{a \in D_i} (u + h_a, v + h_a + (p-1)/2)_e \theta^a \\ &= \textstyle\sum_{i=0}^{e-1} (u + h_a, v + h_a + (p-1)/2)_e (f; g^i). \end{aligned}$$

Furthermore for each pair (u, v) with $0 \le u \le p-2$ and $0 \le v \le p-2$, we have

$$\begin{aligned} &(f; g^u)(f; g^v) \\ = \; &\left(\textstyle\sum_{i \in D_u} \theta^i\right) \left(\textstyle\sum_{j \in -D_v} \theta^{-j}\right) \\ = \; &\left(\textstyle\sum_{i \in D_u} \theta^i\right) \left(\textstyle\sum_{j \in D_{v+(p-1)/2}} \theta^{-j}\right) \\ = \; &\begin{cases} f + H(u,v,\theta), & \text{if } u - v - (p-1)/2 \equiv 0 \pmod{e}; \\ H(u,v,\theta), & \text{otherwise.} \end{cases} \end{aligned} \tag{4.18}$$

Formula (4.18) shows that the product of any two periods can be expressed as a linear combination of the periods $(f; 1), (f; g), \cdots, (f; g^{e-1})$ plus a constant, and that the coefficients are cyclotomic numbers of order e.

Iteratively using formula (4.18) two times, we obtain

$$(f; g^u)(f; g^v)(f; g^w) = c + \sum_{i=0}^{e-1} c_i (f; g^i),$$

where $c, c_0, \cdots, c_{e-1}$ are integer coefficients which are linear combinations of cyclotomic numbers and the products

$$(h_1, h_2)(h_3, h_4).$$

For the product of more than three periods, we have similar expressions, but some of the coefficients have contributions from the product of more than two cyclotomic numbers.

Chapter 5

Special Primes and Sequences

In this chapter we are concerned with two topics: the search for large primes which will be needed later in designing stream ciphers, and the statement of some number-theoretic problems which are related to the design of stream ciphers.

Before studying the cryptographic value of various kinds of primes, we make it clear that primes are evaluated only from the following cryptographic points of view: linear complexity, sphere complexity and period stability when primes are used as periods of sequences or used as some factors of periods. Thus, some primes may not be valuable from these viewpoints, but this does not mean they are cryptographically useless. They may be valuable from other cryptographic viewpoints or in some cryptosystems other than stream ciphers.

The cryptographic importance of the following primes are discussed in this chapter: Sophie Germain primes, Mersenne primes, primes of form $k2^n + 1$, prime repunits, primes of the forms $((4u)^n - 1)/(4u - 1)$, $n! \pm 1$ and $p\# \pm 1$, twin primes. Other problems which are related to this chapter are: conjectures about Sophie Germain primes, twin-prime conjectures, Cullen numbers, perfect numbers, Legendre and Jacobi symbols, quadratic residues and nonresidues, the distribution of primes and twin primes, the sexes of twins and their distribution, RSA primes.

5.1 Sophie Germain Primes and Sequences

An odd prime p is called a *Sophie Germain prime* if $2p + 1$ is also a prime. Sophie Germain primes play an important role in designing certain stream ciphers. Their importance in designing sequences with both a large linear complexity and good linear complexity stability was made clear in the foregoing chapter. They are also important in number theory, because a number of famous number-theoretic problems have connections with them.

5.1.1 Their Importance in Stream Ciphers

The importance of these primes in designing sequences with prime period $2p+1$ was shown in Chapter 3. Sophie Germain primes are an excellent partner period for many finite fields. By definition a Sophie Germain prime p can be written as $p = 2p_1+1$, where p_1 could be even or odd. For example, for the Sophie Germain prime $p = 3$, $p_1 = 1$ is odd; for the Sophie Germain prime $p = 5$, $p_1 = 2$ is even. For simplicity, we always denote $2p+1$ with q for any Sophie Germain prime in this section. For any Sophie Germain prime p, $q = 2p+1$ can be written as

$$q = 2p + 1 = 4(p_1 + 1) - 1.$$

This means that for a Sophie Germain prime, q can be an o-prime or an e-prime (see Section 3.4), depending on the parity of p_1. On the other hand, for a Sophie Germain prime p the corresponding q must be of the form $4t-1$. It is easily seen that the following theorem is true.

Theorem 5.1.1 *For a Sophie Germain prime p, q is an o-prime if and only if $p \equiv 1 \pmod 4$; an e-prime if and only if $p \equiv 3 \pmod 4$.*

If $q = 2p+1$ is an o-prime, it is one of the best primes for designing binary sequences of period q, as shown clearly by Corollary 3.4.11, and to construct binary sequences with a period of a product of two o-primes, as shown by Corollaries 3.8.4 and 3.8.5. By Theorem 3.5.2 the prime $2p+1$ cannot have primitive root 3 if p is a Sophie Germain prime. It can also be proven that for a Sophie Germain prime p, the prime $2p+1$ has primitive root 5 if and only if p is one of the forms $10k+1$ and $10k+3$, and that $2p+1$ has primitive root 7 if and only if p is one of the forms $14k+5$ and $14k+11$. Summarizing the results, let p be a Sophie Germain prime, then

1. $2p+1$ has primitive root 2 if and only if p is of the form $4k+1$;
2. $2p+1$ never has primitive root 3;
3. $2p+1$ has primitive root 5 if and only if p is one of the forms $10k+1$ or $10k+3$;
4. $2p+1$ has primitive root 7 if and only if p is one of the forms $14k+5$ or $14k+11$.

So the importance of these primes in designing sequences over $GF(5)$, $GF(7)$ is evident. The corresponding cryptographic values of Sophie Germain primes with respect to $GF(11)$ and other prime fields can be similarly investigated.

Another cryptographic property of such primes is that $q-1$ has only two factors: 2 and p. Therefore there is no known fast algorithm for solving the discrete logarithm problem in the field Z_q. This could also be cryptographically beneficial for some stream ciphers based on such primes.

The most cryptographically important property of Sophie Germain primes may be the following:

Theorem 5.1.2 *Let p be a Sophie Germain prime and $q = 2p+1$. Then for each positive integer a with $2 \le a \le q-2$,*

$$\text{ord}_q(a) = (q-1)/2 \ \textit{or}\ q-1,$$

and for each nonconstant sequence s^∞ of period q over $GF(a)$

1. *$\text{L}(s^\infty)$ must be equal to one of $(q-1)/2$, $(q+1)/2$, $q-1$, and q;*

2. *$\text{SC}_k(s^\infty) \ge (q-1)/2$, if $k < \min\{\text{WH}(s^N), N - \text{WH}(s^N)\}$.*

Proof: Since $\text{ord}_q(a)$ divides $q-1 = 2p$, $\text{ord}_q(a)$ must be one of $2, p, 2p$. Because q does not divide $(a+1)(a-1)$, we have $a^2 - 1 \not\equiv 0 \pmod q$. It follows that $\text{ord}_q(a) = (q-1)/2$ or $q-1$. The remaining conclusions follow easily from Theorem 3.3.1. □

This theorem means that for a Sophie Germain prime p, the prime $q = 2p+1$ is one of the most ideal periods for sequences over any finite field $GF(t)$ with $\gcd(t,q) = 1$ and $2 \le t \le q-2$. Thus, there do exist primes which are good partners for most finite fields.

5.1.2 Their Relations with Other Number-theoretic Problems

Sophie Germain primes are closely related with *Mersenne numbers*, which are numbers $M_q = 2^q - 1$ with q prime, and with the first case of Fermat's last theorem. The following classical result about the relation between Sophie Germain primes and Mersenne numbers was stated by Euler in 1750 and proved by Lagrange (1775) and by Lucas (1878) independently [361].

Proposition 5.1.3 *If p is a prime $p \equiv 3 \pmod 4$, then $2p+1$ divides M_p if and only if $2p+1$ is a prime; in this case, if $p > 3$, then M_p is composite.*

This relation is useful in the search for Sophie Germain e-primes. Note that M_p has factor 23, 47, 167, 263, 359, 383, 479 and 503 respectively for $p = 11$, 23, 83, 131, 179, 191, 239 and 251, we get eight Sophie Germain e-primes.

The relation of Sophie Germain primes with the first case of Fermat's last theorem has been established by Sophie Germain and can be stated as follows (for a proof see [215, p. 275]). For extensions of Sophie Germain's theorem, see [361] for example.

Proposition 5.1.4 *If p is a Sophie Germain prime, then there are no integers x, y, z, different from 0 and not multiples of p, such that $x^p + y^p = z^p$.*

5.1.3 The Existence Problem

It is still an open problem whether there are infinitely many Sophie Germain primes. However, there are two conjectures about this problem [396], which are the following.

Conjecture 5.1.5 *There are infinitely many p such that $q = 2p+1$ is also prime, that is, there are infinitely many Sophie Germain primes.*

Conjecture 5.1.6 *There are infinitely many $p = 4m + 3$ such that $q = 2p + 1$ is also prime, that is, there are infinitely many Sophie Germain e-primes.*

The first conjecture is similar to the famous twin-prime conjecture. The second conjecture is stronger than the first one. Examples are $p =$ 16035002279, 16045032383, 16048973639, 16052557019, 16086619079, etc. There is strong evidence for the validity of these two conjectures; details about this can be found, for example, in [396, 361].

For cryptographic purposes what we are really interested in is whether there are large Sophie Germain primes, and where they are if there are some. The size of primes we need for the design of some keystream generators depends on the system.

5.1.4 A Search for Cryptographic Sophie Germain Primes

In this subsection we search for specific large Sophie Germain primes by making use of some known primes of the form $h2^n - 1$. Four cryptographic Sophie Germain primes will be found among them, of which two are Sophie Germain o-primes, two are Sophie Germain e-primes.

Let $S = \{h2^n - 1 : h, n \text{ are positive integers}\}$. It is possible to find Sophie Germain primes in this set, since $2S + 1 \subseteq S$.

In 1956 Riesel published a table of all primes of the form $M = (6a+1)2^n - 1$ and $M' = (6a-1)2^n - 1$ for $a \leq 9$ and $1 \leq n \leq 150$ [363]. Later in 1968 Williams and Zarnke extended the table for values $a \leq 25$ and $1 \leq n \leq 1000$ [458]. These numbers were tested for primality by using a theorem due to Lehmer [262]. Riesel developed a Lucasian criterion of primality for primes of the form $N = 3A2^n - 1$, and used it to have given a table of all primes for odd $A \leq 35$ and all $n \leq 1000$ [364].

By comparing the primes in the table presented by Williams and Zarnke we have found about 37 Sophie Germain primes. However, cryptographically interesting large ones are

$$p = 3 \times 9 \times 2^{121} - 1 \text{ and } 3 \times 21 \times 2^{758} - 1.$$

For both of the above Sophie Germain primes, it is easily seen that $p \equiv 3 \pmod 4$. This means that they are Sophie Germain e-primes which have no primitive root 2.

Let $q = 2p + 1 = 4t - 1$, then we get the corresponding two t's for the above two Sophie Germain primes

$$t_1 = 3 \times 9 \times 2^{120}, \quad t_2 = 3 \times 21 \times 2^{757}.$$

It is obvious that $t_1 \equiv 0 \pmod 3$ and $t_2 \equiv 0 \pmod 3$. It follows from Theorem 3.5.2 that 3 is not a primitive root of the above two Sophie Germain primes.

Similarly, one can investigate whether other integers are primitive roots of the above primes. It should be pointed out that every large Sophie Germain prime could be cryptographically valuable due to Theorem 5.1.2.

Williams and Zarnke found in 1972 all primes of the form $2A3^n + 1$ and of the form $2A3^n - 1$ for $1 \leq A \leq 50$ and $1 \leq n \leq 325$ [459]. By comparing the primes in the second table by Williams and Zarnke [459], we have found only two cryptographic Sophie Germain primes. They are

$$p = 10 \times 3^{140} - 1 \text{ and } 26 \times 3^{122} - 1.$$

Let $q = 2p + 1$. Then it is easily verified that for the above two Sophie Germain primes p, the corresponding two q's are o-primes. Let $q = 2p+1 = 4t - 1$, then the corresponding

$$t_1 = 5 \times 3^{140}, \quad t_2 = 13 \times 3^{122}.$$

A simple computation gives $t_1 \equiv 0 \pmod 5$ and $t_2 \equiv 2 \pmod 5$. So the second Sophie Germain prime has two small prime primitive roots 2 and 5.

According to [362], the largest Sophie Germain prime known in 1991 was $39051 \times 2^{6001} - 1$, which was discovered by Keller in 1986. It has no primitive roots 2 and 3, but has primitive root 5. There are also two large ones: $296385 \times 2^{4251} - 1$ and $53375 \times 2^{4204} - 1$, which were discovered by Brown, Noll, Parady, G. Smith, J. Smith and Zarantonello [362]. These two Sophie Germain primes have no primitive roots 2, 3, 5 and 7. Anyway, these are too large for our application for the time being.

5.2 Tchebychef Primes and Sequences

Recall that Tchebychef primes are those of the form $4n2^m + 1$ with $m > 0$ and n is an odd prime $> 9^{2^m}/2^{m+2}$. We call them Tchebychef primes owing to the cryptographically important result of Tchebychef (Proposition 3.5.8).

5.2.1 Their Cryptographic Significance

The cryptographic significance of the Tchebychef primes can be strengthened by the following three results, which can be easily derived from Proposition 3.5.8.

Proposition 5.2.1 *If p and $q = 8p + 1$ are both odd primes with $p > 11$, then 3 is a primitive root of q.*

Proposition 5.2.2 *If p and $q = 16p + 1$ are both odd primes with $p > 411$, then 3 is a primitive root of q.*

Proposition 5.2.3 *If p and $q = 32p + 1$ are both odd primes with $p >$ 1345211, then 3 is a primitive root of q.*

Primes like those above can be used to design ternary sequences with period $8p + 1$, $16p + 1$ or $32p + 1$, and with period equal to the product of two such primes. Sequences over $GF(5)$ and $GF(7)$ based on primes of these forms can also be designed.

5.2.2 Existence and Search Problem

Tchebychef primes are of the form $p2^n + 1$ with p being relatively much larger than n. For our cryptographic purposes we are concerned with whether there are large primes of the form $q = lp + 1$ with $l = 8, 16$ and 32.

Dirichlet's theorem on primes in arithmetic progressions says, given $n \geq 1$, there exist infinitely many integers $k \geq 1$, such that $k \times 2^n + 1$ is a prime. This result shows it is possible that there are large Tchebychef primes. However, it is still an open problem whether such primes exist. Many large primes of the form $k \times 2^n + 1$ with k being small have been found [10], but such primes are not Tchebychef primes. For the purpose of designing cryptographic sequences, the investigation into the following problems is important.

Research Problem 5.2.4 *Find large primes p such that $4p+1$ is also a prime.*

Research Problem 5.2.5 *Find large primes p such that $8p+1$ is also a prime.*

Research Problem 5.2.6 *Find large primes p such that $16p+1$ is also a prime.*

Research Problem 5.2.7 *Find large primes p such that $32p+1$ is also a prime.*

A fact of possible cryptographic interest about primes of the form $8p+1$ is the following. Vaughan proved in 1973 that either there are infinitely many primes p such that $8p+1$ is a prime or the product of two distinct primes, or there are infinitely many primes p such that $8p+1$ is the product of three distinct primes [426].

5.3 Other Primes of Form $k \times 2^n + 1$ and Sequences

Tchebychef primes seem hard to find, but many primes of the form $k \times 2^n + 1$ with small k and large n have been found. Much attention has been paid to numbers of this form, because the factors of Fermat numbers are of such a form. A search for such primes was done by Matthew and Williams [295], Robinson [370], Shippee [398] and Baillie [10]. According to [10], the method used to test $k \times 2^n + 1$ for primality was stated originally by Proth [358], and proven in [369]. The idea of the method is: Given $N = k2^n + 1$ with $k < 2^n$, we look for a number D which makes the Jacobi symbol $(D/N) = -1$. If 3 does not divide k, we may take $D = 3$; if 3 divides k, a (usually short) search is conducted for a suitable D. Then N is prime if and only if $D^{(N-1)/2} \equiv -1 \pmod{N}$.

In [10] all primes of the form $k \times 2^n + 1$ for k odd, $1 \leq k \leq 150$, $1 \leq n \leq 1500$, were given. Many large primes of this form with small k were found. We present here a list of large primes of this form obtained by Robinson [370], Matthew and Williams [295] and Baillie [10].

Let $t = k2^{n-2}$. To test whether it is possible for a prime $k2^n + 1$ to have primitive roots 3, 5, 7 and 11, we just calculate the value $T_m = t \bmod m$ for $m = 3, 5, 7, 11$. If $T_3 = 1$, then it is possible for the prime to have primitive root 3; if $T_5 \in \{3, 4\}$, it is possible to have primitive root 5; if $T_7 \in \{1, 3, 4\}$, then it is possible to have primitive root 7; if $T_{11} \in \{3, 4, 5, 7, 10\}$, it is possible to have primitive root 11. For example, for the prime $3 \times 2^{201} + 1$, we have $T_5 = 4$, $T_7 = 6$ and $T_{11} = 7$. In fact, T_m is easy to calculate. We do not even need a computer for moderate values of k and n. We take $7 \times 2^{830} + 1$ as an example. For this prime, $k = 7 \times 2^{828}$. We calculate now T_3 and T_5. It is easy to see that

$$T_3 = (7 \bmod 3)(4 \bmod 3)^{414} \bmod 3 = 1$$
$$T_5 = (7 \bmod 5)(4 \bmod 5)^{414} \bmod 5 = 2(-1)^{414} \bmod 5 = 2.$$

Thus, we just present here a tables of large primes of this form (n is no less than 100). The case $k = 1$ gives Fermat primes, which will be discussed at the end of this section.

According to [362], the largest two primes of the form $k \times 2^n + 1$ with $n \geq 2$ are $8423 \times 2^{59877} + 1$ and $8423 \times 2^{55157} + 1$, which were discovered by Buell and Young in 1988 and 1987. The largest known prime of the form $k^2 \times 2^n + 1$ was discovered by Keller in 1984: $17^2 \times 2^{18502} + 1 = (17 \times 2^{9251})^2 + 1$. This is also the largest known prime of the form $n^2 + 1$. The largest known prime of the form $k^4 \times 2^n + 1$ is $6954^4 \times 2^{9952} + 1$.

The numbers of the form $Cn = n \times 2^n + 1$ are known as *Cullen numbers.* $C141$ is prime which was given by Robinson in the above table. Keller showed in 1984 that Cn is also prime for n =4713, 5795, 6611, 18497, and for other $n \leq 2000$, Cn is composite. Whether such a special form has cryptographic interest seems to be an open problem. It depends on the finite field $GF(q)$, over which the sequence is constructed, and on $\text{ord}_p(q)$.

For stream cipher purposes, the known primes of this form seem to be large enough. On the other hand, almost all of the large primes in the two tables are valuable in one of the prime fields $GF(3)$, $GF(5)$, $GF(11)$ and $GF(13)$. To further investigate their cryptographic value in sequence designing over a finite field $GF(q)$, we have to know the order of q modulo these primes or develop a tight lower bound for the order.

Research Problem 5.3.1 *For large primes of the form $k \times 2^n + 1$ with $k < 2^n$, find positive integers q such that the order of q modulo the prime is large enough.*

If $k = 1$, then every quadratic nonresidue of a Fermat prime is a primitive root modulo this prime. On the other hand if 2^n is very small and k is a large prime, Tchebychef proved that 3 is a primitive root of this prime. So we might conjecture that there are infinitely many primes $N = k2^n + 1$ such that the integer 3 is a primitive root. This is true for 5, 17, 97 and 113. Further research on the distribution of these primes needs to be done. Proth's theorem is the basis for testing the primality of integers of the form $k \times 2^n + 1$.

Table 5.1: First table of large primes of the form $k \times 2^n + 1$.

k	n
3	189, 201, 209, 276, 353, 408, 438, 534
5	127, 1947
7	120, 174, 180, 190, 290, 320, 390, 432, 616, 830
9	134, 162, 206, 211, 366, 663, 782, 1305, 1411, 1494
11	125, 127, 209, 211
13	188, 308, 316, 1000
15	112, 168, 229, 297, 339, 517, 522, 654, 900
17	147, 243, 267, 347, 471, 747
19	366, 1246
21	124, 128, 129, 187, 209, 276, 313, 397, 899
23	341, 381, 389, 649
25	184, 232, 268, 340, 448, 554, 664, 740, 748, 1280, 1328
27	175, 215, 275, 407, 455, 1076, 1090
29	103, 143, 185, 231, 245, 391, 1053, 1175
31	140, 216, 416
33	118, 289, 412, 453, 525, 726, 828, 1420
35	147, 245, 327, 355, 663, 1423, 1443
37	106, 110, 166, 236, 254, 286, 290, 712, 1240
39	251, 370, 375, 389, 407, 518, 818, 865, 1057
41	215, 289, 379
43	104, 144, 158, 252, 778, 1076
45	189, 200, 333, 372, 443, 464, 801, 1374
47	583, 1483
49	118, 390, 594, 1202
51	119, 175, 187, 257, 263, 267, 321, 333, 695, 825, 1485
53	105, 133, 485, 857
55	220, 244, 262, 286, 344, 356, 392
57	190, 398, 456, 502, 719, 1312, 1399
59	291, 1085
61	168
63	133, 153, 228, 280, 314, 326, 334, 340, 410, 429, 626, 693, 741, 768, 1150, 1290, 1441
65	129, 151, 205, 239, 257, 271, 307, 351, 397, 479, 553, 1317
67	102, 134, 214, 236, 238, 342, 354, 382, 454, 470, 598, 726, 870, 1148, 1366
69	145, 515, 842, 1450
71	119, 299, 417, 705
73	110, 212, 230
75	102, 163, 222, 247, 312, 397, 430, 675, 831, 984, 1018, 1054
77	287, 483, 559, 655, 667
79	206, 538, 970, 1330

Table 5.2: Second table of large primes of the form $k \times 2^n + 1$.

k	n
81	104, 121, 125, 148, 152, 267, 271, 277, 296, 324, 344, 396, 421, 436, 447, 539, 577, 592, 711, 809, 852, 1384
83	157, 181, 233, 373
85	148, 200, 624, 1300
87	104, 134, 207, 518, 602, 1268, 1302
89	589, 711
91	168, 260, 696
93	108, 122, 164, 170, 226, 298, 398, 686, 1020, 1110, 1478
95	111, 167, 175, 237, 533, 621, 661, 753, 993, 1039
97	266, 400, 652, 722
99	126, 143, 162, 170, 186, 189, 206, 211, 270, 319, 369, 410, 433, 631, 894
101	117, 123, 143, 173, 387, 389, 513, 633, 827, 971, 1103
103	138, 250, 616, 622, 736
105	107, 155, 182, 215, 273, 382, 392, 413, 434, 490
107	291, 303, 311, 479, 567
109	318
111	128, 137, 193, 676
113	145, 365, 409, 509, 553, 673, 733, 961, 1045
115	114, 228, 396, 456, 482, 1298
117	156, 382, 454, 643, 867, 1416
119	553, 1115
121	228, 264, 320, 732, 788
123	128, 141, 268, 333, 476, 742, 832, 1173
125	281, 331, 491, 581, 941, 1205, 1279, 1411
127	114, 180, 214, 504, 558, 964, 1098, 1420
129	111, 287, 414, 786, 966, 1071
131	153, 165, 199, 261, 285, 361, 373, 465, 475, 529, 765
133	124, 174, 192, 336, 600, 720, 1092, 1138
135	106, 108, 202, 238, 253, 282, 330, 361, 452, 459, 646, 895, 922, 1201, 1402, 1441, 1462
137	203, 395, 467, 875
139	914
141	103, 117, 133, 137, 141, 160, 291, 303, 343, 488, 535, 555, 556, 640, 756, 897, 917
143	293, 333, 393, 809, 825
145	250, 276, 312, 562, 636, 1366
147	134, 155, 179, 258, 275, 475, 620, 824, 888
149	125, 127, 137, 191, 513, 819, 827, 921, 931, 1047, 1147

Let $N = k \times 2^n + 1$, where $1 \leq k < 2^n$. By Proth's theorem N is prime if $a^{(N-1)/2} \equiv -1 \pmod{N}$ for some a. Since the determination of the primitivity of 3 modulo primes of this form is cryptographically important, we propose now the following problem.

Research Problem 5.3.2 *For primes of the form $N = k \times 2^n + 1$, where $1 \leq k < 2^n$, determine when 3 is a primitive root.*

We mention the following empirical result of Robinson [370], which is cryptographically interesting since only quadratic nonresidues are candidates for primitive roots.

Proposition 5.3.3 *If $N = k \times 2^n + 1$ is prime, where k is odd, $0 < k < 100$, and $0 < n < 512$, then the smallest positive quadratic nonresidue of N does not exceed 23. The smallest nonresidue is 23 in just three cases:*

$$N = 39 \times 2^{13} + 1, \quad 33 \times 2^{28} + 1, \quad 57 \times 2^{90} + 1.$$

Recall that the numbers $F_n = 2^{2^n} + 1$ are called Fermat numbers and primes of such a form are referred to as Fermat primes. So far the primality of the F_n is known for n from 0 up to 22. Among these 23 Fermat numbers only F_0, F_1, F_2, F_3, F_4 are prime, and other F_n's are composite for $5 \leq n \leq 23$ [49]. For those composite F_n some of their prime factors could be cryptographically interesting. It is easily seen that the order of 2 modulo Fermat primes is very small, so they are not suitable for the construction of binary sequences since it is hard to control the linear and sphere complexity of binary sequences with period F_n. However, they might be valuable in the design of non-binary sequences.

5.4 Primes of Form $(a^n - 1)/(a - 1)$ and Sequences

Primes of the form $(a^n - 1)/(a - 1)$ have been investigated for many years. When $a = 2$, numbers $M_p = 2^p - 1$ of this form are called *Mersenne numbers*, and primes of this form are called *Mersenne primes*. Mersenne primes are closely related to perfect numbers. In fact Euler proved that, if q is a prime and $M_q = 2^q - 1$ is a prime, then $n = 2^{q-1}(2^q - 1)$ is a perfect number [184]. What we are concerned with is the cryptographic value of the Mersenne primes. When $a = 10$, numbers of the form are called *repunits*, and primes of this form are called *prime repunits*. Another cryptographically interesting case is $a = 4u$ with u an odd prime.

5.4.1 Mersenne Primes and Sequences

In 1988 Colquitt and Welsh Jr. found the Mersenne prime $2^{110503}-1$, and stated that there are exactly two exponents between 100000 and 139268 [83]. This was the final step in establishing the complete list of the first 31 Mersenne primes. Since then, several more Mersenne primes have been found. However, the task of showing that there are no Mersenne primes between two known ones is computationally intensive and is often left undone for years after the discovery of a new Mersenne prime. For the last two primes in Table 5.3, this task was only completed in 1997. The latest information on Mersenne primes can be obtained from the Web page of Chris Caldwell [50]. For our applications and evaluation purposes, it is convenient to have the specific Mersenne primes of Table 5.3.

Now we turn to the cryptographic value of the Mersenne primes. To this end, we need the *Legendre* and *Jacobi symbols*, whose definition and properties we briefly review here. If $p > 2$ does not divide a and if there exists an integer b such that $a = b^2 \bmod p$, then a is called a *quadratic residue* modulo p; otherwise, it is a *quadratic nonresidue* modulo p.

The Legendre symbol is defined by

$$\left(\frac{a}{p}\right) = (a/p) = \begin{cases} 0, & \text{if } p|a, \\ +1, & \text{if } a \text{ is a quadratic residue modulo } p, \\ -1, & \text{otherwise.} \end{cases}$$

To analyze the order of integers modulo a prime, the following theorem of Euler is sometimes useful, which is

$$\left(\frac{a}{p}\right) \equiv a^{(p-1)/2} \pmod p.$$

Let a be a nonzero integer, and b be an odd integer, such that $\gcd(a, b) = 1$. The Jacobi symbol (a/b) is defined as an extension of Legendre's symbol as follows. Let $|b| = \prod_{p|b} p^{e_p}$ (with $e_p \geq 1$). Then

$$\left(\frac{a}{b}\right) = \left(\frac{a}{-b}\right) = \prod_{p|b} \left(\frac{a}{p}\right)^{e_p}.$$

For a Mersenne prime $p = 2^m - 1$, it is easy to see that $\mathrm{ord}_p(2) = m$. This means that it is difficult to control the linear and sphere complexity for binary sequences with period a Mersenne prime.

It is clear [361] that if n is odd, $n \geq 3$, then $M_n = 2^n - 1 \equiv 7 \pmod{12}$. And if $N \equiv 7 \pmod{12}$, then by quadratic reciprocity the Jacobi symbol

$$\left(\frac{3}{N}\right) = \left(\frac{N}{3}\right)(-1)^{(N-1)/2} = -1.$$

Table 5.3: The first 33 Mersenne primes.

p	*Year*	*Discover*
2	–	–
3	–	–
5	–	–
7	–	–
13	1461	Anonymous
17	1588	P. A. Cataldi
19	1588	P. A. Cataldi
31	1750	L. Euler
61	1883	I. M. Pervushin
89	1911	R. E. Powers
107	1913	E. Fauquembergue
127	1876	E. Lucas
521	1952	R. M. Robinson
607	1952	R. M. Robinson
1279	1952	R. M. Robinson
2203	1952	R. M. Robinson
2281	1952	R. M. Robinson
3217	1957	H. Riesel
4253	1961	A. Hurwitz
4423	1961	A. Hurwitz
9689	1963	D. B. Gillies
9941	1963	D. B. Gillies
11213	1963	D. B. Gillies
19937	1971	B. Tuckerman
21701	1978	C. Noll & L. Nickel
23209	1979	C. Noll
44497	1979	H. Nelson & D. Slowinski
86243	1982	D. Slowinski
110503	1988	W. N. Colquitt & L. Welsh
132049	1983	D. Slowinski
216091	1985	D. Slowinski
756839	1992	D. Slowinski & P. Gage
859433	1994	D. Slowinski & P. Gage

Thus for Mersenne primes M_p, we have $3^{(M_p-1)/2} \equiv -1 \pmod{M_p}$, which corresponds to the congruence $2^{(q-1)/2} \equiv -1 \pmod{q}$ if q is an o-prime. This means that 3 is a candidate to be a primitive root modulo a Mersenne prime. However, this does not ensure the primitivity of 3 modulo a Mersenne prime. For example, 3 is a primitive root of M_3, but not a primitive root of M_5. What we can prove about the order of 3 is

$$\text{ord}(3) = 2u,$$

where u is a factor of $(M_p - 1)/2 = 2^{p-1} - 1$. This is true for every a such that $a^{(M_p-1)/2} \equiv -1 \pmod{M_p}$. To analyze the order of integers modulo a Mersenne prime generally, we have to observe the factors of $2^{p-1} - 1$ for those Mersenne primes M_p. In the book by Brillhart, Lehmer, Selfridge, Tuckerman and Wagstaff, a table of the factorization of $2^n - 1$, $n \leq 310$, was given [39]. Many more factorizations have been done since. According to the tables $2^n - 1$ usually has many small factors. So it seems difficult to design cryptographic sequences with period a Mersenne prime due to the difficulty of controlling the linear and sphere complexity of those sequences. For Mersenne primes M_p for which the factorization of $2^{p-1}-1$ is not known, their cryptographic value is still an open problem.

Research Problem 5.4.1 *Investigate whether Mersenne primes have prime primitive roots or small primitive roots which are a power of a prime.*

Mersenne primes M_p with $2^{p-1} - 1$ having only small factors are *bad cryptographic primes*, since they have no good partner field $GF(q)$ such that the linear and sphere complexity of sequences of period M_p over the field are easy to control. They are quite different from Sophie German primes, which are an excellent partner for many finite fields. However, this evaluation is only based on the ease of controlling the linear and sphere complexity.

5.4.2 Cryptographic Primes of Form $((4u)^n - 1)/(4u - 1)$

Primes of the form $((4u)^n - 1)/(4u - 1)$ with u odd, may be cryptographically useful. We first prove the following result. Let $p = ((4u)^n - 1)/(4u - 1)$ be a prime, then

$$p - 1 = 4u\frac{(4u)^{n-1} - 1}{4u - 1}.$$

Since $4u$ is even, we have the following theorem.

Theorem 5.4.2 *A prime of the form $[(4u)^n - 1]/(4u - 1)$ is an o-prime if and only if u is odd.*

It follows from Section 3.4 that o-primes could be very useful in designing cryptographic binary sequences. For the case $u = 3$, Williams and Seah made a search for all n with $2 \leq n \leq 1000$ [461]. From their table four large primes are found, i.e.,

$$\frac{12^{97}-1}{11}, \quad \frac{12^{109}-1}{11}, \quad \frac{12^{317}-1}{11}, \quad \frac{12^{353}-1}{11}.$$

These primes are of the form $4t + 1$ with t odd. Obviously, t is an odd composite. To see their cryptographic value with respect to $GF(2)$, we need to solve the following problem.

Research Problem 5.4.3 *Study the primitivity of* 2 *and the order of* 2 *modulo the above four primes.*

For the purpose of designing binary keystream sequences, we need large primes of the form $[(4u)^n - 1]/(4u - 1)$ with u odd for which 2 is a primitive root 2 or at least has large order. Thus, we propose the following general problem.

Research Problem 5.4.4 *Find large primes of the form* $[(4u)^n - 1]/(4u - 1)$ *with odd* $u \geq 3$ *for which* 2 *has large order.*

5.4.3 Prime Repunits and their Cryptographic Values

Repunits are the decimal integers 1, 11, 111, 1111, ... Rn is used to denote

$$11...1 = (10^n - 1)/9.$$

The known facts about repunits are

1. if Rn is a prime, then n must be a prime;

2. a repunit ($\neq 1$) cannot be a square;

3. a repunit ($\neq 1$) cannot be a cube.

The known prime repunits are only $R2$, $R19$, $R317$ and $R1031$, of which $R317$ was discovered by Williams [460], $R1031$ by Williams and Dubner [462]. These are the only known prime repunits Rp for $p \leq 10000$. Though it is still an open problem whether there are infinitely many prime repunits, the only cryptographically interesting prime repunit is $R317$ since $R1031$ is too large and the others are too small.

To evaluate the cryptographic value of this prime repunit, we first analyze whether it is an o-prime. Since

$$\frac{Rn+1}{4} = \frac{10^{n-1}+10^{n-2}+\cdots+10^3}{4} + 28,$$

we have $Rn = 4u - 1$ with u even for each $n \geq 3$. This proves the following theorem.

Theorem 5.4.5 *Prime repunits are e-primes, and 2 is therefore never a primitive root of a prime repunit.*

To see the cryptographic value of $R317$ in designing keystream sequences over $GF(a)$, we should solve the following problem.

Research Problem 5.4.6 *For each positive integer a, investigate the order of a modulo $R317$.*

5.5 $n! \pm 1$ and $p\# \pm 1$ Primes and Sequences

Let $p\#$ denote the product of all primes that are no larger than p; for example, $7\# = 2 \times 3 \times 5 \times 7 = 210$. The primality of numbers of the forms $n! + 1$ and $p\# + 1$ was investigated by Borning [29], Templer [418], Buhler, Crandall and Penk [46], and Caldwell [49]. In [46, 49] primes of the forms $n! - 1$ and $p\# - 1$ were also investigated. These investigations have led to the determination of all primes less than 10^{1000} of the forms $n! \pm 1$ and $p\# \pm 1$ [46]. These primes are

- primes $N = n! + 1$ for $n = 1, 2, 3, 11, 27, 37, 41, 73, 77$, 116, 154, 320, 340, 399, 427, 872, 1477;
- primes $N = n! - 1$ for $n = 3, 4, 6, 7, 12, 14, 30, 32, 33, 38, 94, 166, 324$, 379, 469, 546, 974, 1963, 3507, 3610;
- primes $N = p\# + 1$ for $p = 2, 3, 5, 7, 11, 31, 379, 1019, 1021, 2657$, 3229, 4547, 4787, 11549, 13649, 18523, 23801, 24029;
- primes $N = p\# - 1$ for $p = 3, 5, 11, 41, 89, 317, 991, 1873, 2053, 2377$, 4093, 4297, 4583, 6569, 13033, 15877.

Primality was verified by the classic $N^2 - 1$ primality test of [38].

Primes of the forms $n! \pm 1$ are obviously e-primes, so 2 is never a primitive root of these primes. Thus, to design good binary sequences with period of such a prime we have to investigate the orders of 2 modulo these primes.

Comparatively, primes of the form $n!+1$ seem to be worse than those of the form $n!-1$, because $n!$ has many more small factors than $n!-2$. Much more cryptographically interesting is the fact that there may exist large Sophie Germain primes of the form $(n!-2)/2$. Let $N_n = n! - 1$, then

$$\begin{array}{ll} (N_3 - 1)/2 = 2, & (N_4 - 1)/2 = 11, \\ (N_6 - 1)/2 = 359, & (N_7 - 1)/2 = 2519. \end{array}$$

The first three are primes; but $(N_7 - 1)/2$ is not prime. So solving the following problem is cryptographically interesting.

Research Problem 5.5.1 *Analyze whether* $(N_{12} - 1)/2$, $(N_{14} - 1)/2$, $(N_{30} - 1)/2$ *and* $(N_{32} - 1)/2$ *are prime.*

Primes of the forms $p\# \pm 1$ seem also cryptographically interesting. Primes of the form $p\# + 1$ must be of the form $4t - 1$. It is not difficult to get the following results:

$$\begin{array}{rcl cl} 5\# + 1 &=& 31 &=& 4 \times 8 - 1; \\ 7\# + 1 &=& 6211 &=& 4 \times 53 - 1; \\ 11\# + 1 &=& 2311 &=& 4 \times 578 - 1; \\ 31\# + 1 &=& 200560490131 &=& 4 \times 50140122533 - 1. \end{array}$$

These calculations show that primes of the form $p\# + 1$ may be o-primes or e-primes. If some of them are o-primes, we still need to know whether they have primitive root 2 or whether the orders of 2 modulo them are large enough.

Primes of the form $p\# - 1$ must be of the form $4t + 1$. By calculations

$$\begin{array}{l} 5\# - 1 = 29 = 4 \times 7 + 1; \\ 11\# - 1 = 2309 = 4 \times 577 + 1; \\ 13\# - 1 = 30029 = 4 \times 7507 + 1; \\ 41\# - 1 = 4 \times t + 1, \text{ with } t \text{ even.} \end{array}$$

Let $P_p = p\# - 1$. These results show that $(P_p - 1)/4$ is prime for $p = 5, 11, 13$. If P_p and $(P_p - 1)/4$ are both prime, they could be very useful in designing cryptographic sequences over $GF(2)$, $GF(3)$ and $GF(5)$. Thus, the investigation of the following problem is worthwhile.

Research Problem 5.5.2 *Study whether* $(P_p - 1)/4$ *is prime for* $p = 41$, 89, *and* 317.

5.6 Twin Primes and Sequences over $GF(2)$

Twin primes, i.e., pairs of primes of the form p and $p+2$, occur very high up in the integers. Statistical results indicate that the twins tend to thin out compared with the primes. Some theoretical evidence is given by the following theorem of Brun:

$$\begin{aligned} B &= \sum_{(p,p+2)\ twin\ primes} \left(\frac{1}{p} + \frac{1}{p+2}\right) \\ &\approx 1.90216054. \end{aligned}$$

The constant B is now referred to as *Brun's constant*, which was calculated based on intuitive considerations about the distribution of twin primes. For details about the calculations, one may consult Shanks [394] and Brent [33]. Brun's theorem implies that there are not very many twin primes compared with the total number of primes, since $\sum_{p\ prime} 1/p$ diverges. However, for cryptographic purposes what we are concerned with is not whether there are infinitely many twin primes, but whether there are large enough twin primes. Statistical results indicate that there should exist infinitely many twin primes. For example, if we let $\pi_2(x)$ denote the number of primes p such that $p+2$ is also prime and $p+2 < x$, it is known that $\pi_2(10^3) = 35$, $\pi_2(10^4) = 205$, $\pi_2(10^5) = 1224$, $\pi_2(10^6) = 8169$, $\pi_2(10^7) = 58980$, $\pi_2(10^8) = 440312$ and $\pi_2(10^{11}) = 224376048$ [361].

At present the largest known pairs of twin primes are $1706595 \times 2^{11235} \pm 1$ and $571305 \times 2^{7701} \pm 1$, which were found in 1990 by Parady, Smith and Zarantonello [339]. For our cryptographic purposes the pairs of twin primes presented in Table 5.4, which are based on [339, 361], seem too large.

5.6.1 The Significance of Twins and their Sexes

Before evaluating the cryptographic value of twin primes, we prove a cryptographically interesting property of twin primes. To this end, we need some definitions [123].

Definition 5.6.1 *Let $(p, p+2)$ be a pair of twin primes and $p \equiv \Xi(p) \pmod 4$, where $\Xi(p) = \pm 1$. Then we call $\Xi(p)$ the* sex characteristic *of the twins.*

Definition 5.6.2 *If the twins $(p, p+2) = (4t-1, 4t+1)$ for some t, then we say that the twins have the same sex; otherwise, we say that they have different sexes.*

Table 5.4: The known twin primes having more than 1000 digits.

Twin Primes	*Discover*	*Year*
$107570463 \times 10^{2250} \pm 1$	Dubner	1985
$43690485351513 \times 10^{1995} \pm 1$	same	1985
$520995090 \times 10^{6624} \pm 1$	Aktin & Rickert	1984
$519912 \times 10^{1420} \pm 1$	Dubner	1984
$217695 \times 10^{1404} \pm 1$	same	1984
$219649815 \times 10^{4481} \pm 1$	Aktin & Rickert	1983
$1639494 \times (2^{4423} - 1) \pm 1$	Keller	1983
$2445810 \times (2^{4253} - 1) \pm 1$	same	1983
$218313 \times 10^{1068} \pm 1$	Dubner	1985
$499032 \times 10^{1040} \pm 1$	same	1984
$403089 \times 10^{1040} \pm 1$	same	1984
$256200945 \times 2^{3423} \pm 1$	Aktin & Rickert	1980
$663777 \times 2^{7650} \pm 1$	Parady, Smith, Zarantonello	1990
$571305 \times 2^{7701} \pm 1$	Parady, Smith, Zarantonello	1990
$1706595 \times 2^{11235} \pm 1$	Parady, Smith, Zarantonello	1990

In the above definitions, we say that twin primes $(p, p+2)$ have the same sex, because in the expression of the form $4u \pm 1$, the u's for both p and $p+2$ are the same, and have therefore the same parity, if $p = 4t - 1$. If $p = 4t + 1$, then $p + 2 = 4(t + 1) - 1$ and t and $t + 1$ have different parities. That is why we call them twins with different sexes. This discussion has also proved the following two properties of twins [123].

Theorem 5.6.3 *(The Sex Principle of Twins)*
If the smaller of the twins has sex characteristic -1, *then the twins have the same sex; otherwise, they have different sexes.*

Theorem 5.6.4 *If* p *and* $p+2$ *have the same sex, then it is possible for them to have the common primitive root* 2 *(a common best partner); otherwise, they never have.*

We make such a classification for twin primes because of its cryptographic importance. Speaking specifically, twin primes with the same sex can be e-primes or o-primes, and in a pair of twin primes with different sexes there must be one which is an o-prime. The importance of o-primes in binary sequence designing has already been made clear in Chapter 3. In later chapters we will see that twin primes are also of much value in designing good cryptographic functions, which are based on the famous twin-prime difference sets, where a common primitive root is required. Thus, twin primes are cryptographically important from two viewpoints: the control of the linear and sphere complexity of binary sequences; and the designing of good cryptographic functions. If we consider the two aspects together in the design of binary stream cipher systems, we may find that the practically useful twin primes may be those with different sexes, and those $(p, p+2) = (4t-1, 4t+1)$ with same sex and with t odd. This will be shown in later chapters.

What we have mentioned may be only partial cryptographic values of twin primes with respect to the design of binary sequences. To evaluate their values further, we should at least solve part of the following problems:

Research Problem 5.6.5 *Investigate whether there are large twin o-primes which have the common primitive root* 2.

Research Problem 5.6.6 *Investigate for which large twin primes with different sexes there must exist one of the twins which has primitive root* 2.

Research Problem 5.6.7 *Find large twin primes* $(p, p+2)$ *such that* $\text{ord}_p(2)$ *and* $\text{ord}_{p+2}(2)$ *both are large enough.*

5.6.2 Cryptographic Twins and the Sex Distribution

As mentioned above, we are cryptographically interested in twin primes with different sexes and twin o-primes. So it is cryptographically important to know the frequency of occurrence of twin primes with different sexes in the twin-prime series. Let F_t denote the frequency of occurrence of twin primes with different sexes in all the twin primes $(p, p+2)$ such that $p+2 \le t$. By a statistical calculation on all twin primes between 3 and 10094, we get the following results:

$$\begin{aligned} F_{3302} &= 45/87 \approx 0.5172; \\ F_{7952} &= 90/175 \approx 0.5143; \\ F_{10094} &= 107/209 \approx 0.5115. \end{aligned}$$

So from the above statistical results, we can expect that

$$F = f_\infty = 0.5.$$

From the above Brun's theorem, we get an empirical formula that

$$\sum_{(p,p+2)\ with\ different\ sexes} \left(\frac{1}{p} + \frac{1}{p+2}\right) = \frac{1}{2}\sum\left(\frac{1}{p} + \frac{1}{p+2}\right) \approx 0.95108027.$$

<table>
<tr><td rowspan="2">A 1/2</td><td>B 1/4</td></tr>
<tr><td>C 1/4</td></tr>
</table>

A — the set of twin primes with different sexes.
B — the set of twin e-primes with same sex.
C — the set of twin o-primes with same sex.

Figure 5.1: An empirical distribution of the classified twin primes.

Since twin o-primes (in this case they must have the same sex) are more important than twin primes with different sexes, it is also important to know the frequency of occurrence of twin o-primes in twin primes with the same sex. Statistical results also show that this frequency tends to 0.5. Thus, considering the twin primes less than or equal to a given large integer we may conclude

1. about half of the pairs of twin primes has the same sex; the other half has different sexes;

2. approximately 1/4 twin primes are twin o-primes.

Figure 5.1 describes our empirical results about the sex distributions of twin primes.

The above statistical results show that there should exist large twin primes which possess different sexes and large twin o-primes. The only instances in Table 5.4 are the pairs $1639494 \times (2^{4423} - 1) \pm 1$ and $2445810 \times (2^{4253} - 1) \pm 1$ discovered by Keller. For the first twins we have $p = 1639494 \times (2^{4423} - 1) - 1$ and

$$(p - 1)/4 - [1639494 \times 2^{4423} - 1639496]/4 = \text{even} - \text{even} = \text{even}.$$

Thus they are twin e-primes. For the next twins, let $p = 2445810 \times (2^{4253} - 1) - 1$, then

$$(p - 1)/4 = [2445810 \times 2^{4253} - 2445812]/4 = \text{even} - 611453 = \text{odd}.$$

Hence they are twin o-primes. But it seems unknown whether 2 is a primitive root of the twin o-primes.

Research Problem 5.6.8 *Examine whether 2 is a primitive root of the above twin o-primes and investigate the order of 2 modulo the above two pairs of twin primes.*

5.7 Twin Primes and Sequences over $GF(3)$

In the foregoing section the cryptographic usefulness of twin primes with respect to $GF(2)$ was discussed. In this section we are concerned with the cryptographic usefulness of twin primes with respect to $GF(3)$.

We are interested in whether the twin primes have small common primitive roots. Since $t \equiv 1 \pmod 3$ if and only if $t+1 \equiv 2 \pmod 3$, if $(p, p+2) = (4t+1, 4(t+1)-1)$ are twin primes, then it is possible for them to have the common primitive root 3; otherwise it is impossible. Thus we have the following theorem:

Theorem 5.7.1 *If the smaller of a pair of twin primes has the sex characteristic* 1, *then it is possible for them to have the common primitive root* 3*; otherwise, it is impossible.*

The twin primes 29 and 31 have the common primitive root 3. However, 3 is not a common primitive root of 17 and 19. It is therefore cryptographically interesting to solve the following problem.

Research Problem 5.7.2 *Study the following problems:*

1. *What proportion of twin primes has the common primitive root* 3*?*
2. *Find some large twin primes which have the common primitive root* 3.
3. *Find large twin primes such that* $\mathrm{ord}_p(3)$ *and* $\mathrm{ord}_{p+2}(3)$ *both are large enough.*

We note that all the large pairs in Table 5.4 have no primitive root 3, because the smaller all have the sex characteristic -1. Is it really difficult to find large twin primes in which the smaller has sex characteristic 1? This seems to be an open problem.

We have already proved that if the smaller of a pair of twin primes has the sex characteristic -1, it is impossible for it to have primitive root 3. We now prove further that it is also impossible for the other prime to have the primitive root 3. If $(p, p+2) = (4t-1, 4t+1)$ are twin primes, it follows from Theorems 3.5.1 and 3.5.2 that $t \not\equiv 2 \pmod 3$ and $t \not\equiv 1 \pmod 3$, so we must have $t = 3k$ for some k. It follows further from Theorems 3.5.1 and 3.5.2 that it is impossible for p or $p+2$ to have primitive root 3. We therefore have the following result.

Theorem 5.7.3 *If the smaller of a pair of twin primes has the sex characteristic* -1*, then neither prime has primitive root* 3.

It follows from the above theorem that all the twin primes in Table 5.4 cannot have primitive root 3. The cryptographic value of twin primes over $GF(5)$, $GF(7)$ and over other prime fields can be similarly investigated.

5.8 Other Special Primes and Sequences

Two large primes which are related to the Mersenne prime M_{127} are: $114(2^{127}-1)+1$ (41 decimal digits) and $180(2^{127})^2+1$ (79 decimal digits), which were discovered by Miller and Wheeler in 1951 according to Zagier [470]. It seems unknown whether 2 is a primitive root of the two primes. However, it can be seen that if we use the first one as the period of a non-constant binary sequence, then its linear complexity is at least $2^{127}-1$. Further cryptographic properties of the two primes need to be investigated.

5.9 Prime Distributions and their Significance

In Chapter 3, primes were classified into two classes: e-primes and o-primes. This classification is of importance from the viewpoint of constructing binary sequences with both large linear and sphere complexity, since e-primes never have primitive root 2, while it is possible for an o-prime to have primitive root 2. Similarly, for the purpose of designing ternary sequences with both large linear and sphere complexity, primes can be divided into four classes: $\{p \text{ prime} : \ p \equiv a \pmod{12}\}$ for a =1, 3, 5 and 7. Only the two classes corresponding to $a=5$ and $a=7$ may have primitive root 3. For the purpose of designing cryptographic sequences over $GF(5)$, $GF(7)$, etc., similar classifications can also be made.

Cryptographically we need "large" primes which have primitive root a, where a is a small prime or power of a small prime, and those such that the order of a modulo those primes is large enough. However, the meaning of "large" will change over time, and may vary with the development of attack methods and of technology (for example, high-speed special purpose attack machines). Let $\pi(x) = |\{p \text{ prime } | p \leq x\}|$ and $\pi_{d,a}(x) = |\{p \text{ prime } | p \leq x, p \equiv a \pmod{d}\}|$. We see that it is cryptographically interesting to know

1. whether there are infinitely many primes in the arithmetic progression $\{a+kd | k \geq 0\}$, where $\gcd(a,d)=1$; and

2. how the density function $D_{d,a}(x) = \pi_{d,a}(x)/\pi(x)$ behaves.

The first problem was solved by Dirichlet in 1837. Dirichlet's theorem about primes in arithmetic progression states that, if $d \geq 2$, $a \neq 0$ and

$\gcd(a, d) = 1$, then the arithmetic progression $\{a + kd | k = 0, 1, 2, ..., \}$ contains infinitely many primes. The second problem was solved by de la Vallèe Poussin. He proved that

$$\pi_{d,a}(x) \sim \frac{1}{\phi(d)} \cdot \frac{x}{\log x},$$

which is the same, for any a, such that $\gcd(a, d) = 1$. It follows that

$$\lim_{x \to \infty} D_{d,a}(x) = 1/\phi(d).$$

This means that the set of primes in the arithmetic progression $\{a + dk | k \geq 1\}$ has the asymptotic density $1/\phi(d)$ with respect to the set of all primes.

Despite the fact that the asymptotic behavior of $\pi_{d,a}(x)$ is the same, for every a, $1 \leq a < d$, with $\gcd(a, d) = 1$, it is usually different for different a's. It is known that $x = 608,981,813,029$ is the minimum value for which $\pi_{3,1}(x) > \pi_{3,2}(x)$, and that $x = 26861$ is the minimum value for which $\pi_{4,1}(x) > \pi_{4,3}(x)$. Nevertheless, empirical results show that the difference $|\pi_{d,a}(x) - \pi_{d,a'}(x)|$ is usually very small with respect to $\pi(x)$, where $\gcd(a, d) = 1$ and $\gcd(a', d) = 1$.

Although the above two problems are solved, they do not provide us with exactly what we want to know for the design of some stream ciphers. In fact we want to know for cryptographic applications:

1. whether Artin's conjectures in Section 3.9 are true; and
2. if they are true, which primes in the classes $\{8k + 3 | k \geq 1\}$ and $\{8k + 5 | k \geq 1\}$ (resp. $\{12k + 5 | k \geq 1\}$ and $\{12k + 7 | k \geq 1\}$) have primitive root 2 (resp. 3).

These two cryptographically important problems are still open.

5.10 Primes for Stream Ciphers and for RSA

The RSA public-key cryptosystem supports both secrecy and authentication, and hence can provide complete and self-contained support for public-key distribution and signatures. In this system a user chooses primes p and q and computes $n = p \times q$ and $\phi(n) = (p - 1)(q - 1)$. He then chooses e to be an integer in $[1, n - 1]$ such that $\gcd(e, \phi(n)) = 1$. Further, the user finds an integer d such that $e \times d \equiv 1 \pmod{\phi(n)}$. The public parameters are n and e, while d, p, q, and $\phi(n)$ are kept secret.

Based on these parameters the public (encryption) and private (decryption) transformations are respectively defined by

$$E(M) = M^e \bmod n, \qquad D(C) = C^d \bmod n,$$

where $M \in [0, n-1]$ denotes message, and $C \in [0, n-1]$ signed message or enciphered message, and D and E are inverses. Since d is private, so is D; and since n and e are public, so is E. This constitutes a cryptosystem that can be used for both secrecy and authentication. That is, for secrecy, A sends $E_B(M)$ to B as usual; for authentication, A sends $D_A(M)$ as usual. For both secrecy and authentication, suppose first that message digests are not employed. Assuming $n_A \leq n_B$, A computes $C = E_B(D_A(M))$ and sends C to B. Then B recovers M as usual by $M = E_A(D_B(E_B(D_A(M))))$. In the case that $n_A \geq n_B$, A can instead transmit $C' = D_A(E_B(M))$. Then B can recover M as $M = D_B(E_A(D_A(E_B(M))))$.

As usual, the choice of primes p and q is determined by the known attacks on this system. One attack on the RSA system is based on the iteration of the public transformation, another type of attack is based on various methods for the factorization of n. These lead to different restrictions on the choice of p and q.

The iteration attack works like this: If the message M is sent as $E(M) = M^e \bmod n$, let m be the order of e modulo $\phi(n)$. Then applying E successively m times gives $M^{e^m} = M \bmod n$. In this way message is recovered by employing only public information E. For any message M with $\gcd(M, n) = 1$, less than m iterations may be enough. Now the problem is whether such an attack is computationally feasible. This depends on the size of m and the factors of m. To easily protect the system from this attack, Rivest [368] suggested that p and q be chosen as follows: $p = ap' + 1$, $p' = bp'' + 1$, and $q = cq' + 1$, $q' = dq'' + 1$, where p', p'', q' and q'' are distinct primes and a, b, c and d are small integers. Specifically, Riesel suggested that p' and q' be chosen as Sophie Germain primes and thus a and c be chosen as 2 [365].

Another kind of attack is based on the factorization of the modulus n. The quadratic sieve is one of the most efficient general purpose factoring algorithms. It is applicable to composite integers of no special form. References about this topic can be found in [352, 353, 354, 337]. Lenstra's elliptic curve technique [269] and the number field sieve [271] are oriented to integers of special forms. For security against factoring, it is suggested that p and q should have more than 100 decimal digits. There are also some other considerations in choosing the primes, see, for example, the so-called strong primes [170].

It can be seen in the foregoing sections that the choice of primes for the purposes of designing cryptographic sequences differs in the details, because designing sequences over each prime field has its own special features. However, there are some similarities. For example, to design sequences of prime period p over prime field $GF(q)$, it is desirable to require that $\mathrm{ord}_p(q)$ is as large as possible.

The conditions we need to impose on two primes p and q in designing sequences of period $N = pq$ over some prime field are similar to some extent to the conditions needed for primes p and q for RSA, but not exactly the same. For example, in both cases we want $p - 1$ and $q - 1$ to have a very large prime factor. But in the design of sequences, we may require that p and q have a common prime primitive root (for example, 2, 3, 5, or 7). At least for the field $GF(t)$ the orders of t modulo p and q should both be large enough.

Generally, we may say that although there are some similarities between the choices of primes for stream ciphers and for RSA, most of the considerations employed in these two choices are different. It should be mentioned here that the criteria for choosing primes for different kinds of stream ciphers also vary. We have considered the requirements for primes only from the linear and sphere complexity viewpoints. Other requirements will be encountered later. Here we have only made a comparison between primes for RSA and those for stream ciphers. Details about public-key cryptography can be found, for example, in Diffie [113], Koblitz [247] and Salomaa [380].

Chapter 6

Highly Nonlinear Functions

Every cryptographic function for ciphering is directly or indirectly responsible for combining plaintext and keystream characters. It is cryptographic functions that define the encryption and decryption algorithms. So the security of every cryptosystem should depend *mainly* on the design of cryptographic functions.

Cryptographic requirements for those functions vary from system to system. cryptographic goodness and badness of functions depend not only on the structure of the functions, but also on the specific system in which they are used and how they are used. Both linear and highly nonlinear functions are useful in stream ciphers.

Functions with high nonlinearity have important applications in cryptography [19, 58, 294, 308, 332, 333], sequences [349] and coding theory [54, 229, 279, 463]. In cryptography, functions with high nonlinearity are necessary for achieving confusion.

During the last twenty years, there has been a lot of studies of Boolean functions with high nonlinearity. See for example, [53], [56], [57], [58], [59], [62], [63], [64], [139], [140], [141], [142], [333], [373]. Non-Boolean functions have also important applications in cryptography [51, 52, 308], sequences [252, 335] and coding theory [191, 349], but they have been less studied.

This chapter gives a well-rounded treatment of non-Boolean functions with optimum or almost optimum nonlinearity, and is from Carlet and Ding [60]. The reader is referred to Carlet [58] for a survey on Boolean functions with perfect nonlinearity, i.e., bent functions.

6.1 Preliminaries

Let f be a function from an Abelian group $(A,+)$ of order n to another Abelian group $(B,+)$ of order m. f is *linear* if and only if $f(x+y) = f(x)+f(y)$ for all $x, y \in A$. A function g is *affine* if and only if $g = f + b$, where f is linear and b is a constant. Clearly, the zero function is linear. If f is a nonzero linear function from A to B, let $H = \{x \in A | \, f(x) = 0\}$. Then H is a subgroup of A, $f(A)$ is a subgroup of B and, denoting by $|S|$ the size of a set S, $|f(A)| \times |H| = n$. In the case that n is odd and m is a power of 2, the only linear function from A to B is the zero function, since if $f \neq 0$, then $|f(A)|$ is even, a contradiction with the fact that n is odd; thus all affine functions are constant functions.

The (Hamming) distance between two functions f and g from A to B, denoted by $d(f, g)$, is defined to be

$$d(f,g) = |\{x \in A | f(x) - g(x) \neq 0\}|.$$

One way of measuring the nonlinearity of a function f from $(A,+)$ to $(B,+)$ is to use the minimum distance between f and all affine functions from $(A,+)$ to $(B,+)$. With this approach the nonlinearity of f is defined to be

$$N_f = \min_{l \in L} d(f, l), \tag{6.1}$$

where L denotes the set of all affine functions from $(A,+)$ to $(B,+)$. This measure of nonlinearity is related to linear cryptanalysis [294], but it is not useful in some general cases. For example, as pointed out above, in the case $|A|$ is odd and $|B|$ is a power of 2, this measure makes little sense as there are no non-constant affine functions from $(A,+)$ to $(B,+)$.

A robust measure [332] of the nonlinearity of functions is related to differential cryptanalysis [22, 23] and uses the derivatives $D_a f(x) = f(x+a) - f(x)$. It may be defined by

$$P_f = \max_{0 \neq a \in A} \max_{b \in B} \Pr(D_a f(x) = b), \tag{6.2}$$

where $\Pr(E)$ denotes the probability of the occurrence of event E. The smaller the value of P_f, the higher the corresponding nonlinearity of f (if f is linear, then $P_f = 1$). In some cases, it is possible to find the exact relation between the two measures on nonlinearity. We will come back to this later. Note that both nonlinearity measures are relative to the two operations of the two Abelian groups.

6.2 Functions with perfect nonlinearity

Let f be a function from $(A, +)$ to $(B, +)$. For any $b \in B$ define

$$C_b = f^{-1}(b) = \{a \in A | f(a) = b\}. \tag{6.3}$$

We have the following property.

Lemma 6.2.1 *Let f be a function from $(A, +)$ to $(B, +)$. Then, for every $a \in A$ and every $b \in B$*

$$\Pr(D_a f(x) = b) = \frac{\sum_{z \in B} |C_z \cap (C_{z+b} - a)|}{|A|}.$$

Proof: We have

$$\begin{aligned} & |\{x \in A | D_a f(x) = b\}| \\ = & \left| \bigcup_{z \in B} \{x \in A | f(x) = z \text{ and } f(x+a) = z+b\} \right| \\ = & \left| \bigcup_{z \in B} (C_z \cap (C_{z+b} - a)) \right| \\ = & \sum_{z \in B} |C_z \cap (C_{z+b} - a)| . \end{aligned}$$

The conclusion then follows. □

Notice that, for every $a \in A$, the sets $\{x \in A | D_a f(x) = b\}$ constitute a partition of A, and thus we have the following lemma.

Lemma 6.2.2 *For every $a \in A$, we have*

$$|A| = \sum_{b \in B} |\{x \in A | D_a f(x) = b\}| .$$

Note that the maximum of a sequence of numbers is greater than or equal to its mean. It then follows that, for every $a \in A$,

$$\max_{b \in B} [\Pr(D_a f(x) = b)] = \max_{b \in B} \frac{|\{x \in A | D_a f(x) = b\}|}{|A|} \geq \frac{1}{|B|}.$$

Then

$$P_f \geq \frac{1}{|B|}. \tag{6.4}$$

This lower bound can be considered as an upper bound for the nonlinearity of f. For applications in coding theory and cryptography we wish to find functions with the smallest possible P_f.

Definition 6.2.3 *A function $f : A \to B$ has perfect nonlinearity if $P_f = \frac{1}{|B|}$.*

Since the maximum of a sequence of numbers equals its mean if and only if the sequence is constant, inequality (6.4) is an equality if and only if, for every $b \in B$ and every $a \in A^* = A\backslash\{0\}$, the quantity $|\{x \in A | D_a f(x) = b\}|$ has value $\frac{|A|}{|B|}$.

Definition 6.2.4 *A function $g : A \to B$ is balanced if the size of $g^{-1}(b)$ is the same for every $b \in B$ (this size is then $\frac{|A|}{|B|}$).*

Theorem 6.2.5 *A function $f : A \to B$ has perfect nonlinearity if and only if, for every $a \in A^* = A\backslash\{0\}$, the derivative $D_a f$ is balanced (this is possible only if $|B|$ divides $|A|$).*

In the case of Boolean functions (i.e. functions from $GF(2)^n$ to $GF(2)$, where $GF(2)$ is the two-element field), perfect nonlinear functions are also called bent [373]. We recall at Subsection 6.2.6 the definitions and properties of bent functions.

6.2.1 Stability of the set of perfect nonlinear functions under actions of general affine groups

The addition of any perfect nonlinear function from $(A, +)$ to $(B, +)$ and any affine function from $(A, +)$ to $(B, +)$ is clearly a perfect nonlinear function.

Theorem 6.2.6 *Assume that $f(x)$ is a function from $(A, +)$ to $(B, +)$ with perfect nonlinearity and $l(x)$ is a linear or an affine permutation from $(A, +)$ to $(A, +)$, then the composition $f \circ l$ is another function from $(A, +)$ to $(B, +)$ with perfect nonlinearity.*

Proof: If $l(x)$ is a linear permutation, then $f(l(x + a)) - f(l(x))$ is equal to $f(l(x) + l(a)) - f(l(x))$ and is balanced for every $a \neq 0$ since $l(a) \neq 0$ if and only if $a \neq 0$. If $l(x)$ is a translation, say $l(x) = x + u$, then $f(l(x + a)) - f(l(x)) = f(x + u + a) - f(x + u)$ is balanced. The conclusion then follows by composition. □

Theorem 6.2.7 *Let $f : (A, +) \to (B, +)$ have perfect nonlinearity, and let $l : (B, +) \to (C, +)$ be a linear onto function. Then the composition $l \circ f$ is a function from $(A, +)$ to $(C, +)$ with perfect nonlinearity.*

Proof: Since l is linear, we have

$$l(f(x+a)) - l(f(x)) = l(f(x+a) - f(x)).$$

The conclusion then follows from the facts that l is linear and onto and that f has perfect nonlinearity. □

Theorem 6.2.7 leads to a construction of perfect nonlinear functions which is rather useful, as justified by the results of Proposition 6.5.3.

6.2.2 Perfect nonlinear functions and difference partitions

Perfect nonlinear functions are naturally related to a combinatorial notion of difference partition introduced by Carlet and Ding [60]. Let $(A, +)$ and $(B, +)$ be two Abelian groups of orders n and m respectively. Assume that $\{C_b | b \in B\}$ is a partition of A. We call $\{C_b | b \in B\}$ an (n, m, δ) *difference partition* of $(A, +)$ with respect to $(B, +)$ if

$$\sum_{z \in B} |C_z \cap (C_{z+b} - a)| \leq \delta \tag{6.5}$$

for all $b \in B$ and all nonzero elements a of A, and if for at least one pair (a, b) the equality of (6.5) is achieved. Note that for a difference partition $\{C_b | b \in B\}$ some C_b may be empty. The difference partitions defined here are quite different from the difference families that have been studied in combinatorics [20, Chapter VII].

Since $\{C_z \cap (C_{z+b} - a) | z, b \in B\}$ is a partition of A, we have

$$\delta m \geq n. \tag{6.6}$$

The case of equality corresponds to perfect nonlinear functions.

Proposition 6.2.8 *Let $(A, +)$ and $(B, +)$ be Abelian groups of orders n and m respectively. Let $\{C_b | b \in B\}$ be an (n, m, δ) difference partition of $(A, +)$ with respect to $(B, +)$. Let f be the function from A to B defined by $f(x) = b$, for every $x \in C_b$. Then $P_f = \frac{\delta}{n}$. Thus, f has perfect nonlinearity if and only if m divides n and $\{C_b(f) | b \in B\}$ is an $(n, m, n/m)$ difference partition of $(A, +)$ with respect to $(B, +)$.*

Proof: It follows from Lemma 6.2.1. □

If $\{C_b(f) | b \in B\}$ is an $(n, m, n/m)$ difference partition of $(A, +)$ with respect to $(B, +)$, then the equality in (6.5) holds for all $b \in B$ and all nonzero elements a of A.

There are some restrictions on the possible sizes of the sets C_b.

Theorem 6.2.9 (Carlet and Ding [60]) *Let $(A,+)$ and $(B,+)$ be Abelian groups of orders n and m respectively, where m divides n. If an $(n,m,n/m)$ difference partition $\{C_b|b \in B\}$ of A with respect to B exists, then for any nonzero $b \in B$*

$$\left\{ \begin{array}{l} \sum_{z \in B} k_z^2 = \frac{n^2+(m-1)n}{m}, \\ \sum_{z \in B} k_z k_{z+b} = \frac{n(n-1)}{m}, \\ \sum_{z \in B} k_z = n, \end{array} \right. \tag{6.7}$$

where $k_z = |C_z|$ for each $z \in B$.

Proof: If $\{C_b|b \in B\}$ is an $(n,m,n/m)$ difference partition, we have $\sum_{z \in B} k_z = n$ and

$$\sum_{z \in B} |C_z \cap (C_{z+b} - a)| = \frac{n}{m}$$

for all $b \in B$ and all nonzero elements a of A. It then follows that for any nonzero $b \in B$

$$\begin{aligned} \frac{n(n-1)}{m} &= \sum_{a \in A\setminus\{0\}} \sum_{z \in B} |C_z \cap (C_{z+b} - a)| \\ &= \sum_{z \in B} \sum_{a \in A\setminus\{0\}} |C_z \cap (C_{z+b} - a)| \\ &= \sum_{z \in B} |\{x \in A, a \in A^* | f(x) = z \text{ and } f(x+a) = z+b\}| \\ &= \sum_{z \in B} |\{x \in A, a \in A | f(x) = z \text{ and } f(x+a) = z+b\}| \\ &= \sum_{z \in B} k_z k_{b+z}. \end{aligned}$$

Similarly, we obtain

$$\begin{aligned} \frac{n(n-1)}{m} &= \sum_{a \in A\setminus\{0\}} \sum_{z \in B} |C_z \cap (C_z - a)| \\ &= \sum_{z \in B} \sum_{a \in A\setminus\{0\}} |C_z \cap (C_z - a)| \\ &= \sum_{z \in B} |\{x \in A, a \in A^* | f(x) = z \text{ and } f(x+a) = z\}| \\ &= \sum_{z \in B} k_z(k_z - 1) \end{aligned}$$

$$= \sum_{z \in B} k_z^2 - \sum_{z \in B} k_z$$
$$= \sum_{z \in B} k_z^2 - n.$$

This completes the proof. □

Theorem 6.2.10 (Carlet and Ding [60]) *Let $(A, +)$ and $(B, +)$ be Abelian groups of orders n and m respectively, where n is a multiple of m. If f is a function from A to B with perfect nonlinearity $P_f = \frac{1}{m}$, then for any $b \in B$*

$$\frac{n}{m} - \sqrt{\frac{(m-1)n}{m}} \le k_b \le \frac{n}{m} + \sqrt{\frac{(m-1)n}{m}},$$

where $k_z = |\{x \in A | f(x) = z\}|$. Furthermore,

$$\frac{(m-1)n}{m} - \sqrt{\frac{(m-1)n}{m}} \le N_f \le \frac{(m-1)n}{m} + \sqrt{\frac{(m-1)n}{m}}.$$

If B has exponent 2, i.e., $2b = 0$ for any $b \in B$, then for any $b \in B$

$$\frac{n - (m-1)\sqrt{n}}{m} \le k_b \le \frac{n + (m-1)\sqrt{n}}{m},$$

where $k_z = |\{x \in A | f(x) = z\}|$. Furthermore,

$$\frac{(m-1)n - (m-1)\sqrt{n}}{m} \le N_f \le \frac{(m-1)n + (m-1)\sqrt{n}}{m}.$$

Proof: We prove the first conclusion. Set $k_b = n/m + \lambda_b$. It follows from the last equation of (6.7) that $\sum_b \lambda_b = 0$. Combining this equality and the first one of (6.7) yields

$$\sum_b \lambda_b^2 = \frac{(m-1)n}{m}.$$

Hence $|\lambda_b| \le \sqrt{\frac{(m-1)n}{m}}$. This proves the conclusion on k_b. The lower and upper bounds on N_f then follow from the bounds on k_b and the fact that the sum of a function with perfect nonlinearity is again a function with perfect nonlinearity.

We now prove the bounds for the case that B has exponent 2. For any nonzero $b \in B$, by (6.7)

$$\begin{aligned} \textstyle\sum_{z \in B}(k_z - k_{z+b})^2 &= \textstyle\sum_{z \in B} k_z^2 - 2\sum_{z \in B} k_z k_{z+b} + \sum_{z \in B} k_{z+b}^2 \\ &= 2\frac{n^2 + (m-1)n}{m} - 2\frac{n(n-1)}{m} \\ &= 2n. \end{aligned} \tag{6.8}$$

Since B has exponent 2, in the summation

$$\sum_{z \in B} (k_z - k_{z+b})^2$$

both $(k_z - k_{z+b})^2$ and $(k_{z+b} - k_z)^2$ occur as terms. Then by (6.8)

$$2(k_z - k_{z+b})^2 = (k_z - k_{z+b})^2 + (k_{z+b} - k_z)^2 \leq 2n$$

and hence

$$-\sqrt{n} \leq k_z - k_{z+b} \leq \sqrt{n}. \tag{6.9}$$

It follows that

$$-(m-1)\sqrt{n} \leq (m-1)k_z - \sum_{b \neq 0} k_{z+b} \leq (m-1)\sqrt{n}.$$

Note that $\sum_{b \neq 0} k_{z+b} = n - k_z$. We have

$$\frac{n - (m-1)\sqrt{n}}{m} \leq k_z \leq \frac{n + (m-1)\sqrt{n}}{m}.$$

The bounds on N_f follow from those on k_b and the fact that the sum of a function with perfect nonlinearity and any affine function gives also a function with perfect nonlinearity. □

For the existence of functions with perfect nonlinearity, we have the following result.

Theorem 6.2.11 *Assume that there is a function with perfect nonlinearity from an Abelian group of order n to another Abelian group of order m, where m divides n. If m is even, then n is a square. If m is odd, then*

$$z^2 = nx^2 + (-1)^{(m-1)/2} my^2$$

has a nontrivial solution in integers.

Theorem 6.2.11 is a direct consequence of Lemma 6.3.7 below, which was stated in [40, 41] for the existence of generalized Hadamard matrices.

6.2.3 Functions with perfect nonlinearity and difference matrices

It is known that Boolean functions with perfect nonlinearity (i.e. bent functions) are related to Hadamard matrices [373]. More generally, functions with perfect nonlinearity are related to the so-called difference matrices and generalized Hadamard matrices.

Let $(G, +)$ be a group of order m. An $(m, k; \lambda)$ *difference matrix* is a $k \times m\lambda$ matrix $D = (d_{ij})$ with entries from G, so that for each $1 \leq h < j \leq k$, the list

$$\{d_{hl} - d_{jl} | 1 \leq l \leq m\lambda\}$$

contains λ times every element of G. Similarly, difference matrices can be defined over nonAbelian groups [20, 82]. A *generalized Hadamard matrix* GH(m, λ) is a $(m, m\lambda; \lambda)$ difference matrix. Hence Hadamard difference matrices are special difference matrices. In particular, a Hadamard matrix $H(4n)$ is a GH$(2, 2n)$ over the group $(\{1, -1\}, \cdot)$.

Theorem 6.2.12 [60] *Let f be a function from an Abelian group $(A, +)$ of order n to another one $(B, +)$ of order m, where m divides n. Let $A = \{a_0, a_1, \ldots, a_{n-1}\}$, and define an $n \times n$ matrix D as*

$$D = \begin{pmatrix} f(a_0 + a_0) & f(a_0 + a_1) & \cdots & f(a_0 + a_{n-1}) \\ f(a_1 + a_0) & f(a_1 + a_1) & \cdots & f(a_1 + a_{n-1}) \\ \vdots & \vdots & \vdots & \vdots \\ f(a_{n-1} + a_0) & f(a_{n-1} + a_1) & \cdots & f(a_{n-1} + a_{n-1}) \end{pmatrix}.$$

Then f has perfect nonlinearity $P_f = \frac{1}{m}$ if and only if D is a GH$(m, n/m)$, i.e., an $n \times n$ generalized Hadamard matrix.

Proof: By Theorem 6.2.5, f has perfect nonlinearity if and only if $D_a f(x) = f(x + a) - f(x)$ takes on each element of B exactly n/m times for each nonzero element a of A. The conclusion then follows. □

Remarks:

(a) Any k rows of the matrix D of Theorem 6.2.12 gives an $(m, k; n/m)$ difference matrix over B. Theorem 6.2.12 shows that every function with perfect nonlinearity gives generalized Hadamard matrices. But clearly, many generalized Hadamard matrices do not give functions with optimum nonlinearity.

(b) Theorem 6.2.12 is a rather straightforward result, which traces back to at least [101].

Example 6.2.13 Define the function $f(x)$ from $GF(q)^{2t}$ to $GF(q)$ as

$$f(x_1, x_2, \ldots, x_{2t}) = x_1x_2 + x_3x_4 + \ldots + x_{2t-1}x_{2t}.$$

We will show in Theorem 6.5.1 that this function is perfect nonlinear. Then the matrix D of Theorem 6.2.12 is a (q, q^{2t}, q^{2t-1}) difference matrix, i.e., a generalized Hadamard matrix $\mathrm{GH}(q, q^{2t-1})$.

Remark: It is shown by de Launey that for any group G of prime power order q and any integer $t > 0$, there is a $\mathrm{GH}(q, q^{2t-1})$ over G [100]. Here G may not be elementary Abelian. It remains to be checked whether the construction of Corollary 6.2.13 is the same as the one of de Launey [100].

6.2.4 A characterization of perfect nonlinearity by means of Fourier transform

We denote by e the exponent of A; it is the maximum order of elements of A; it is also called the characteristic of A since A is in additive representation. A homomorphism between A and a multiplicative group G is any mapping χ from A to G such that

$$\chi(a + a') = \chi(a)\chi(a') \text{ for all } a, a' \in A.$$

A *character* of A is any homomorphism from A to the multiplicative group of all complex e-th roots of unity. The multiplicative group $\hat{A}$ of characters of A is isomorphic to the group A [205]. We fix some isomorphism from A to $\hat{A}$ and we denote by χ_α the image of $\alpha \in A$ by this isomorphism. χ_0 is the trivial character, i.e. the constant function 1.

For every $a \neq 0$, we have $\sum_{\alpha \in A} \chi_\alpha(a) = 0$; indeed, there exists $\alpha_0 \in A$ such that $\chi_{\alpha_0}(a) \neq 1$; then the equality

$$\sum_{\alpha \in A} \chi_\alpha(a) = \sum_{\alpha \in A} \chi_{\alpha+\alpha_0}(a) = \chi_{\alpha_0}(a) \sum_{\alpha \in A} \chi_\alpha(a)$$

implies $\sum_{\alpha \in A} \chi_\alpha(a) = 0$.

Let E be any subgroup of A. Denote by $E^\perp$ the subgroup of A of elements α such that $\chi_\alpha(a) = 1$ for all $a \in E$. Then

$$\sum_{a \in E} \chi_\alpha(a) = 0; \ \forall \alpha \notin E^\perp \tag{6.10}$$

and

$$\sum_{\alpha \in E^\perp} \chi_\alpha(a) = 0; \ \forall a \notin E. \tag{6.11}$$

The characters satisfy the orthogonality relation

$$\langle \chi_{\alpha_1}, \chi_{\alpha_2} \rangle = \sum_{a \in A} \chi_{\alpha_1}(a)\overline{\chi_{\alpha_2}(a)} = \begin{cases} 0 & \text{if } \alpha_1 \neq \alpha_2 \\ |A| & \text{if } \alpha_1 = \alpha_2 \end{cases},$$

where $\overline{\chi_{\alpha_2}(a)}$ denotes the complex conjugate of $\chi_{\alpha_2}(a)$.

The *Fourier transform* of any complex-valued function φ on A is defined by

$$\widehat{\varphi}(\alpha) = \sum_{a \in A} \varphi(a)\chi_\alpha(a).$$

A direct consequence of property (6.11) is that for every elements α_0 and a_0 in A and for every subgroup E of A, we have

$$\sum_{\alpha \in \alpha_0 + E^\perp} \chi_\alpha(a_0)\widehat{\varphi}(\alpha) = |E^\perp|\, \chi_{\alpha_0}(a_0) \sum_{a \in -a_0 + E} \chi_{\alpha_0}(a)\varphi(a). \tag{6.12}$$

Indeed,

$$\begin{aligned} \sum_{\alpha \in \alpha_0 + E^\perp} \chi_\alpha(a_0)\widehat{\varphi}(\alpha) &= \sum_{\alpha \in E^\perp} \chi_{\alpha_0+\alpha}(a_0)\widehat{\varphi}(\alpha_0 + \alpha) \\ &= \sum_{\alpha \in E^\perp} \sum_{a \in A} \varphi(a)\chi_{\alpha_0+\alpha}(a_0 + a) \\ &= \sum_{a \in A} \varphi(a)\chi_{\alpha_0}(a_0 + a) \left(\sum_{\alpha \in E^\perp} \chi_\alpha(a_0 + a) \right) \\ &= |E^\perp|\, \chi_{\alpha_0}(a_0) \sum_{a \in -a_0 + E} \chi_{\alpha_0}(a)\varphi(a). \end{aligned}$$

The Fourier transform of the product of two functions φ_1 and φ_2 equals the normalized convolution of the Fourier transforms of φ_1 and φ_2:

$$\widehat{\varphi_1\varphi_2}(\alpha) = \frac{1}{|A|}\widehat{\varphi_1} * \widehat{\varphi_2}(\alpha) = \frac{1}{|A|} \sum_{\alpha' \in A} \widehat{\varphi_1}(\alpha')\widehat{\varphi_2}(\alpha - \alpha'). \tag{6.13}$$

Equality (6.13) with $\varphi_2 = \overline{\varphi_1}$ and $\alpha = 0$ gives *Parseval's relation*:

$$\sum_{a \in A} |\varphi(a)|^2 = \frac{1}{|A|} \sum_{\alpha \in A} |\widehat{\varphi}(\alpha)|^2.$$

The inverse Fourier transform is determined by the equality:

$$\varphi(a) = \frac{1}{|A|} \sum_{\alpha \in A} \widehat{\varphi}(\alpha)\overline{\chi_\alpha(a)}.$$

Note that φ satisfies $\varphi(a) = 0$, for every $a \neq 0$, if and only if $\widehat{\varphi}$ is constant and that φ is constant if and only if $\widehat{\varphi}(\alpha) = 0$, for every $\alpha \neq 0$.

Let f be a function from A to a group B. We denote by e' the exponent of B and we fix again an isomorphism between B and $\hat{B}$ (the group of homomorphisms from B to the multiplicative group of all complex e'-th roots of unity); we denote by χ'_β the image of $\beta \in B$ by this isomorphism. For every $\beta \in B$, we denote by f_β the complex-valued function $\chi'_\beta \circ f$ and we have, for every $\alpha \in A$,

$$\widehat{f_\beta}(\alpha) = \sum_{a \in A} \chi'_\beta \circ f(a)\, \chi_\alpha(a).$$

Parseval's relation on f_β gives

$$\sum_{\alpha \in A} |\widehat{f_\beta}(\alpha)|^2 = |A|^2.$$

A characterization of perfect nonlinearity by means of Fourier transform was given by Carlet and Ding [60], and will be presented in Theorem 6.2.16. It generalizes results given in [373] for Boolean functions, in [2] for functions defined over finite fields and in [61] for functions defined over residue class rings. To introduce this characterization, we need first to characterize balanced functions and to recall a classical property of Fourier transform.

Proposition 6.2.14 [60] *Let f be any function from A to B. Then f is balanced if and only if, for every $\beta \in B^*$ we have*

$$\widehat{f_\beta}(0) = 0.$$

Proof: We have

$$\widehat{f_\beta}(0) = \sum_{a \in A} \chi'_\beta \circ f(a) = \sum_{b \in B} |C_b|\, \chi'_\beta(b). \tag{6.14}$$

Thus, if f is balanced and $\beta \neq 0$, then $\widehat{f_\beta}(0) = \frac{|A|}{|B|} \sum_{b \in B} \chi'_\beta(b) = 0$. Conversely, if , for every $\beta \in B^*$ we have $\widehat{f_\beta}(0) = 0$, then, according to relation (6.14), the integer-valued function $b \mapsto |C_b|$ admits as Fourier transform the function $\beta \mapsto \begin{cases} 0 & \text{if } \beta \neq 0 \\ |A| & \text{if } \beta = 0 \end{cases}$, and according to the properties of the Fourier transform recalled above, it is constant. □

Lemma 6.2.15 (Carlet and Ding [60]) *Let $f : A \to B$ and $D_a f(x) = f(x+a) - f(x)$. Let $\mathrm{AC}_{f_\beta}(a)$ be the value at 0 of the Fourier transform of $(D_a f)_\beta$: $\mathrm{AC}_{f_\beta}(a) = \sum_{x \in A} \chi'_\beta(D_a f(x))$. Then, AC_{f_β} has Fourier transform $|\widehat{f_\beta}|^2$.*

Proof:

$$\widehat{\mathrm{AC}_{f_\beta}}(\alpha) = \sum_{a\in A} \widehat{D_a f_\beta}(0)\chi_\alpha(a) = \sum_{a\in A}\sum_{x\in A} \chi'_\beta(f(x+a))\overline{\chi'_\beta(f(x))}\chi_\alpha(a) =$$

$$\sum_{a\in A}\sum_{x\in A} \chi'_\beta(f(x+a))\overline{\chi'_\beta(f(x))}\chi_\alpha(x+a)\overline{\chi_\alpha(x)} = \widehat{f_\beta}(\alpha)\overline{\widehat{f_\beta}(\alpha)}.$$

□

AC_{f_β} is often called the autocorrelation function of f_β. When only one nonzero β exists, i.e. when $B = GF(2)$, it is also called the autocorrelation function of f.

Theorem 6.2.16 (Carlet and Ding [60]) *Let f be any function from an Abelian group A to an Abelian group B. Then f has perfect nonlinearity if and only if, for every $\beta \in B^*$ and every $\alpha \in A$, $\widehat{f_\beta}(\alpha)$ has magnitude $\sqrt{|A|}$.*

Proof: According to Theorem 6.2.5, f has perfect nonlinearity if and only if for every $a \neq 0$ the function $D_a f(x) = f(x+a) - f(x)$ is balanced. Thus, according to Proposition 6.2.14, f has perfect nonlinearity if and only if for every $a \in A^*$ and every $\beta \in B^*$ we have $\mathrm{AC}_{f_\beta}(a) = 0$. Thus, according to the properties of the Fourier transform recalled above, f has perfect nonlinearity if and only if for every $\beta \in B^*$, AC_{f_β} has constant Fourier transform (this constant value must be $|A|$). Lemma 6.2.15 completes the proof. □

Theorem 6.2.16 states that f has perfect nonlinearity if and only if, for every $\beta \in B^*$, f_β is bent in the sense of Logachev, Salnikov and Yashchenko. We recall at Subsection 6.2.6 the original notion of bent functions and its successive generalizations.

6.2.5 Obtaining functions with perfect nonlinearity from known ones

At Subsection 6.2.1, we have seen obvious ways of obtaining perfect nonlinear functions from known ones. Another one is as follows: let A, A' and B be three Abelian groups. Let $f : A \mapsto B$ and $g : A' \mapsto B$ be two perfect nonlinear mappings. Then $f \otimes g : A \times A' \mapsto B$ defined by $(f \otimes g)(x, y) = f(x) + g(y)$ is perfect nonlinear. We give now a non-trivial similar construction. Theorem 6.2.17 and the remark which follows it generalize the most part of the theorem in [56], which was stated for Boolean bent functions.

Theorem 6.2.17 (Carlet and Ding [60]) *Assume that the size of A is a square. Let E be a subgroup of A of size $\sqrt{|A|}$. Assume that $f(x)$ is a function from $(A,+)$ to $(B,+)$ with perfect nonlinearity and that f takes constant value on E. Then every function obtained from f by choosing another constant value for f on E has also perfect nonlinearity.*

Proof: Let b be any element of B. Define $g(x) = f(x)$ if $x \notin E$; $g(x) = f(x) + b$ if $x \in E$. Let β be any nonzero element of B. Denote by ω_β the constant value of f_β on E. Recall that we denote by $E^\perp$ the set of elements α of A such that $\chi_\alpha(a) = 1$ for all $a \in E$.
Let us first prove that $\widehat{f_\beta}(\alpha) = \omega_\beta |E|$ for every $\alpha \in E^\perp$. According to relation (6.12) applied to $\varphi = f_\beta$ and to $a_0 = \alpha_0 = 0$, we have $\sum_{\alpha \in E^\perp} \widehat{f_\beta}(\alpha) = \omega_\beta |E^\perp| |E|$. Since, according to Theorem 6.2.16, $\widehat{f_\beta}(\alpha)$ has magnitude $|E| = \sqrt{|A|}$ for every α, we deduce that $\widehat{f_\beta}(\alpha)$ equals $\omega_\beta \sqrt{|A|}$ for every $\alpha \in E^\perp$.
We have $\widehat{g_\beta}(\alpha) = \widehat{f_\beta}(\alpha) + \omega_\beta(\chi'_\beta(b) - 1) \sum_{a \in E} \chi_\alpha(a)$. Thus $\widehat{g_\beta}(\alpha)$ equals $\widehat{f_\beta}(\alpha)$ for every $\alpha \notin E^\perp$. And for every $\alpha \in E^\perp$ we have $\widehat{g_\beta}(\alpha) = \omega_\beta \sqrt{|A|} + \omega_\beta(\chi'_\beta(b) - 1)\sqrt{|A|} = \omega_\beta \chi'_\beta(b)\sqrt{|A|}$. Thus, $\widehat{g_\beta}(\alpha)$ has magnitude $\sqrt{|A|}$ for every $\alpha \in A$ and every $\beta \in B^*$, and g has therefore perfect nonlinearity. □

Remarks:

(a) The same proof shows that if φ is bent on A in the sense of Logachev, Salnikov and Yashchenko (see Subsection 6.2.6) and if it is constant on E, then $\widehat{\varphi}$ is constant on $E^\perp$ and φ remains bent if we change its constant value on E.

(b) Since $\widehat{f_\beta}$ is constant on $E^\perp$, applying property (6.12) to $\widehat{f_\beta}$ and to $\alpha_0 = 0$ shows that for every $a_0 \notin E$: $\sum_{a \in a_0 + E} f_\beta(a) = 0$. This is equivalent to the fact that f is balanced on every coset of E in A, according to Proposition 6.2.14.

(c) According to property (6.12), we have also $\sum_{\alpha \in \alpha_0 + E^\perp} \widehat{f_\beta}(\alpha) = 0$ for every $\alpha_0 \notin E^\perp$. If there exists a function g from A to B such that $\widehat{f_\beta} = \sqrt{|A|}\, g_\beta$ (using the same terminology as Kumar, Scholtz and Welch in [252], we can say that f is regular-bent), this implies that g is balanced on every coset of $E^\perp$.

(d) Theorem 6.2.17 is still valid if we only assume that the restriction of f to E is affine and if we change the values of f on E by adding a constant (apply Theorem 6.2.17 to $f + l$ where f is affine). It is also valid if E is a coset of a subgroup (change $f(x)$ into $f(x + u)$).

(e) We give after Theorem 6.5.1 an example of application of Theorem 6.2.17. In the case of this example, there exists a function g from A to B such that $\widehat{f_\beta} = \sqrt{|A|}\, g_\beta$.

6.2.6 Bent functions and perfect nonlinearity

Let A be the Abelian group $GF(2)^n$, $B = GF(2)$ and f a function from A to B. Using the notation of Subsection 6.2.4, we have $f_1(a) = (-1)^{f(a)}$ and $\widehat{f_1}(\alpha) = \sum_{a \in GF(2)^n} (-1)^{f(a)+\alpha \cdot a}$ where $\alpha \cdot a = \alpha_1 a_1 + \ldots + \alpha_n a_n$ is the usual inner product in $GF(2)^n$. The Fourier transform of $f_1 = (-1)^f$ is often called the *Walsh transform* of f. The notion of binary *bent function*, introduced by Rothaus in [373], is related to Parseval's relation $\sum_{\alpha \in GF(2)^n} |\widehat{f_1}(\alpha)|^2 = 2^{2n}$: a function $f : GF(2)^n \to GF(2)$ is bent if $\sum_{a \in GF(2)^n} (-1)^{f(a)+\alpha \cdot a}$ has constant magnitude for every $\alpha \in GF(2)^n$, or equivalently if the maximum of $|\widehat{f_1}(\alpha)|^2$ equals its mean 2^n (this is equivalent to say that f lies at maximum Hamming distance from the set of affine functions); this is possible only if n is even. As shown by Rothaus, and also according to Theorem 6.2.16, this notion is equivalent to perfect nonlinearity. More information on binary bent functions can be found in the survey paper [58] and in Canteaut, Carlet, Charpin and Fontaine [53], Carlet [56, 57, 58, 59], Carlet and Guillot [62, 63], Dobbertin [139], Hou and Langevin [212], and Wolfmann [463].

Logachev, Salnikov and Yashchenko have adapted this notion in [277] to the general case of functions φ from any finite Abelian group A to the set of complex numbers of magnitude 1 (see also Hou [211]): φ is bent if $\widehat{\varphi}(\alpha)$ has constant magnitude $\sqrt{|A|}$ for every $\alpha \in A$.

The notion of binary bent function has been generalized to functions from a finite Abelian group A to a finite Abelian group B in two directions:

- Kumar, Scholtz and Welch [252] have generalized it to functions f from Z_q^n to $Z_q = Z/qZ$, where q is any positive number. The function f_1 equals then ${\omega_q}^f$, where $\omega_q = \exp(2i\pi/q)$ (where $i = \sqrt{-1}$) and we have $\widehat{f_1}(\alpha) = \sum_{a \in Z_q^n} {\omega_q}^{f(a)+\alpha \cdot a}$. Kumar, Scholtz and Welch called *generalized bent* any function f from Z_q^n to Z_q such that $\widehat{f_1}$ has constant magnitude $\sqrt{q^n}$, i.e. such that f_1 is bent in the sense of Logachev, Salnikov and Yashchenko. Obviously, a stronger notion could also be considered: for every $\beta \neq 0$, f_β is bent in the sense of Logachev, Salnikov and Yashchenko. But this notion does not deserve a specific denomination since, as shown in [61] and also according to Theorem 6.2.16, it is equivalent to perfect nonlinearity.

- Ambrosimov [2] considers functions f from $GF(q)^n$ to $GF(q)$ where q is a power of a prime p, and $GF(q)$ is the finite field of order q. For every $\beta \in GF(q)$, f_β equals ${\omega_p}^{\mathrm{Tr}(\beta f)}$ where Tr is the trace function from $GF(q)$ to $GF(p)$ and where $\omega_p = \exp(2i\pi/p)$. Then $\widehat{f_\beta}(\alpha)$ equals $\sum_{a\in GF(q)^n} {\omega_p}^{\mathrm{Tr}(\beta f(a)+\alpha\cdot a)}$. The function f is called bent by Ambrosimov if, for every nonzero β, $\widehat{f_\beta}$ has constant magnitude $\sqrt{q^n}$, i.e. if $f_\beta = {\omega_p}^{\mathrm{Tr}(\beta f)}$ is bent in the sense of Logachev, Salnikov and Yashchenko. As shown by Ambrosimov and according to Theorem 6.2.16, this notion is equivalent to perfect nonlinearity.

The notions of bent functions by Kumar, Scholtz and Welch and by Ambrosimov, when they both apply, that is when q is a prime, have different definitions but are in fact equivalent, as shown in [252].

6.3 Binary functions with optimum nonlinearity

In this section, we consider the case $(B, +) = (GF(2), +)$ and functions from A to B. If $(A, +)$ is cyclic, then functions from A to B with optimal nonlinearity are the same as binary sequences with optimal autocorrelation, i.e., *perfect sequences*. The main references for this section are [125, 225].

Let $n = |A|$. For a function f from A to B, the *autocorrelation function* of f is

$$\mathrm{AC}_f(a) = \sum_{x\in A} (-1)^{f(x+a)-f(x)}.$$

The *support* of f is the set

$$S_f = \{x \in A | f(x) = 1\}.$$

The *weight* of f is defined to be $|S_f|$, and denoted by w_f. We also say that f is the *characteristic function* of S_f.

Considering the Fourier transform of $D_a f$ at vector 0, we have, according to Lemma 6.2.15

$$\sum_{a\in A} \mathrm{AC}_f(a) = (n - 2w_f)^2. \tag{6.15}$$

For any subset H of A, we define the *difference function*

$$d_H(a) = |(H+a) \cap H|, \tag{6.16}$$

where $H + a = \{x + a | x \in H\}$.

The following easy result plays an important role in the sequel.

Theorem 6.3.1 *Let f be a function from A to B, and let k be the weight of f. Then for any nonzero $a \in A$,*

$$\Pr(D_a f(x) = b) = \begin{cases} \frac{n-2(k-d_{S_f}(a))}{n}, & b = 0, \\ \frac{2(k-d_{S_f}(a))}{n}, & b = 1. \end{cases}$$

Proof: This is a generalization of Theorem 4.4 in [125]. We have $\Pr(D_a f(x) = 1) = \frac{1}{n} w_{D_a f} = \frac{1}{n}(2\, w_f - 2\, d_{S_f}(a))$ and $\Pr(D_a f(x) = 0) = 1 - \Pr(D_a f(x) = 1)$. □

6.3.1 The case $n \equiv 0 \pmod 4$

Let $(G, +)$ be an Abelian group with v elements, and let D be a k-subset of G. Then D is called a (v, k, λ) *difference set* of G if the equation $x - y = g$ has exactly λ solutions $(x, y) \in D \times D$ for every nonzero element $g \in G$. A trivial necessary condition for the existence of a (v, k, λ) difference set is

$$k(k-1) = (v-1)\lambda. \tag{6.17}$$

Theorem 6.3.2 *Let D be a (v, k, λ) difference set of an Abelian group $(A, +)$ with v elements, and let $f_D(x)$ be the function with support D. Then*

(a) *for any nonzero $a \in A$,*

$$\Pr(f_D(x+a) - f_D(x) = b) = \begin{cases} [v - 2(k-\lambda)]/v, & b = 0, \\ 2(k-\lambda)/v, & b = 1. \end{cases}$$

(b) $P_{f_D} = \max\left\{\frac{v-2(k-\lambda)}{v}, \frac{2(k-\lambda)}{v}\right\}$.

Proof: This is a generalization of Theorem 4.5 in [125]. The conclusion follows from Theorem 6.3.1. □

Theorem 6.3.3 [60] *Let f be a function from A to B. Then the following three conclusions are equivalent:*

(A) $P_f = \frac{1}{2}$;

(B) $\mathrm{AC}_f(a) = 0$ *for every nonzero element a of A;*

(C) *the support* S_f *is a* $(4u^2, 2u^2 \pm u, u(u \pm 1))$ *difference set of A, where* $n = 4u^2$.

Proof: According to Theorem 6.2.5 and Proposition 6.2.14, (A) and (B) are equivalent. By Theorem 6.3.2, (C) implies (A). If (B) is true, then for every nonzero a, the function $f(x)\,f(x+a)$ has constant weight and the support S_f is therefore a difference set. According to Theorem 6.3.2, $v \equiv 0 \pmod 4$. It is well known that a symmetric design with $v = 4u$ can only exist if u is a perfect square and the parameters of S_f have the form $(4u^2, 2u^2 \pm u, u(u \pm 1))$ (see Jungnickel [224, p. 282]). □

It follows from Theorem 6.3.3 that $(4u^2, 2u^2 \pm u, u(u \pm 1))$ difference sets, called *Hadamard difference set*, of an Abelian group A give all binary functions with perfect nonlinearity. Detailed information about Hadamard difference sets can be found in [225]. We just mention the following.

Lemma 6.3.4 [226] *Let G be any group which is a direct product of an Abelian group of order* 2^e *and exponent at most e, where* $e = 2d + 2$ *for some nonnegative integer d, with groups of the type* $Z_{m_i}^2$, *where each* m_i *is a power of* 3, *and groups of the type* $Z_{p_j}^4$, *where the* p_j *are (not necessarily distinct) odd primes. Then G contains a Hadamard difference set.*

Combining Theorem 6.3.3 and Lemma 6.3.4 proves the following.

Theorem 6.3.5 [60] *Let*

$$A = Z_2^{2d+2} \times Z_{m_1}^2 \times \ldots \times Z_{m_t}^2 \times Z_{p_1}^4 \times \ldots \times Z_{p_s}^4, \tag{6.18}$$

where each m_i *is a power of* 3, *the* p_j *are (not necessarily distinct) odd primes,* $s \geq 0$ *and* $t \geq 0$. *Then there are binary functions from A to B with perfect nonlinearity.*

As recalled at Subsection 6.2.6, Boolean functions (i.e. functions from $GF(2)^n$ to $GF(2)$) have perfect nonlinearity if and only if they are bent.

Numerous binary functions with perfect nonlinearity from the set A of (6.18) to $B = GF(2)$ can be constructed as indicated in Theorem 6.3.5 by using the actual constructions of the Hadamard difference sets indicated in Lemma 6.3.4: for details, we refer to Arasu, Davis, Jedwab, Sehgal [8], Chen [73], Kraemer [248], Turyn [424], and Xia [464].

6.3.2 The case $n \equiv 3 \pmod 4$

In this section, let $(A, +)$ be an Abelian group of order $n \equiv 3 \pmod 4$, and $B = GF(2)$. The following theorem is the function version of perfect sequences [225].

Theorem 6.3.6 [60] *Let f be a function from A to B. Then the minimum possible value for P_f is $\frac{1}{2} + \frac{1}{2n}$ and the following two conclusions are equivalent:*

(A) $P_f = \frac{1}{2} + \frac{1}{2n}$;

(B) *the support S_f is an $\left(n, \frac{n-1}{2}, \frac{n-3}{4}\right)$ or $\left(n, \frac{n+1}{2}, \frac{n+1}{4}\right)$ difference set of A.*

Proof: Let k be the weight of f. Note that $[n - 2(k - d_{S_f}(a))] + 2(k - d_{S_f}(a)) = n$. By Theorem 6.3.1, to minimize P_f we need to minimize the maximum magnitude of

$$[n - 2(k - d_{S_f}(a))] - 2(k - d_{S_f}(a)) = n - 4(k - d_{S_f}(a))$$

where a ranges over A^*. Since $n \equiv -1 \pmod 4$, the minimal possible magnitude of $n - 4(k - d_{S_f}(a))$ corresponds to $n - 4(k - d_{S_f}(a)) = -1$. Thus, P_f is minimal if and only if $d_{S_f}(a) = k - \frac{n+1}{4}$ for every nonzero $a \in A$, i.e., if S_f is an $\left(n, k, k - \frac{n+1}{4}\right)$ difference set of A. It then follows from the equation

$$k(k-1) = (n-1)\left(k - \frac{n+1}{4}\right)$$

that $k = \frac{n \pm 1}{2}$, and the minimal value for P_f is $\frac{1}{2} + \frac{1}{2n}$. □

We say that f has *optimum nonlinearity* if P_f achieves the minimum value (here $\frac{1}{2} + \frac{1}{2n}$).

Since the complement of any $\left(n, \frac{n-1}{2}, \frac{n-3}{4}\right)$ difference set is an $\left(n, \frac{n+1}{2}, \frac{n+1}{4}\right)$ difference set and vice versa, we consider only difference sets with parameters $\left(n, \frac{n-1}{2}, \frac{n-3}{4}\right)$. Difference sets of this type are called *Paley-Hadamard difference sets.* Any Paley-Hadamard difference set of A gives a function from A to B with optimum nonlinearity.

Paley-Hadamard difference sets include the following classes:

(1) with parameters $(2^t - 1, 2^{t-1} - 1, 2^{t-2} - 1)$, for description of difference sets with these parameters see Dillon [116], Dillon and Dobbertin [117], Gordon, Mills and Welch [171],Pott [356], Xiang [465];

(2) with parameters $\left(n, \frac{n-1}{2}, \frac{n-3}{4}\right)$, where $n = q(q+2)$ and both q and $q+2$ are prime powers. These are generalizations of the twin-prime difference sets, and may be defined as

$$\{(g,h) \in GF(q) \times GF(q+2) : g,h \neq 0 \text{ and } \chi(g)\chi(h) = 1\}$$
$$\cup\{(g,0) : g \in GF(q)\},$$

where $\chi(x) = +1$ if x is a nonzero square in the corresponding field, and $\chi(x) = -1$ otherwise [226];

(3) with parameters $\left(n, \frac{n-1}{2}, \frac{n-3}{4}\right)$, where $n = q$ is a prime power congruent to 3 (mod 4). They are Paley difference sets and just consist of all the squares in $GF(q)^*$ [226];

(4) with parameters $\left(n, \frac{n-1}{2}, \frac{n-3}{4}\right)$, where $n = q$ is a prime power of the form $q = 4s^2 + 27$. They are cyclotomic difference sets and can be described as [224]

$$D = D_0^{(6,q)} \cup D_1^{(6,q)} \cup D_3^{(6,q)},$$

where $D_0^{(6,q)}$ denotes the multiplicative group generated by α^6, $D_i^{(6,q)} = \alpha^i D_0^{(6,q)}$ denotes the cosets, and α is a primitive element of $GF(q)$.

6.3.3 The case $n \equiv 2 \pmod 4$

As before let $(A, +)$ be an Abelian group of order n. Let C be a k-subset of A. The set C is an (n, k, λ, t) *almost difference set* of A if $d_C(a) = |(C+a) \cap C|$ takes on the value λ altogether t times and the value $\lambda + 1$ altogether $n-1-t$ times when a ranges over all the nonzero elements of A.

Two kinds of almost difference sets were introduced in [99] and [123, 125] (see also [130]). They were generalized and unified in [131].

For (n, k, λ, t) almost difference sets of A we have the following basic relation

$$k(k-1) = t\lambda + (n-1-t)(\lambda+1). \tag{6.19}$$

The following lemma due to Bruck, Chowla and Ryser will be needed later.

Lemma 6.3.7 *Let D be an (n, k, λ) difference set in a group G.*

(i) *If n is even, then $k - \lambda$ is a square.*

(ii) *If n is odd, then the equation*

$$x^2 = (k - \lambda)y^2 + (-1)^{\frac{n-1}{2}} \lambda z^2 \tag{6.20}$$

has a solution in integers x, y, z, not all zero.

We consider now functions f from A to B with optimum nonlinearity. As before, let S_f and k be the support and weight of f respectively. When A is cyclic, the first part of the following theorem is the function version of the corresponding results about perfect sequences [225].

Theorem 6.3.8 [60] *The minimum possible value for P_f is $\frac{1}{2} + \frac{1}{n}$. Furthermore, $P_f = \frac{1}{2} + \frac{1}{n}$ if and only if*

(a) *the support S_f is a difference set with parameters*

$$\left(n, \frac{n \pm \sqrt{3n-2}}{2}, \frac{n+2 \pm 2\sqrt{3n-2}}{4}\right); \tag{6.21}$$

(b) *or the support S_f is an almost difference set with parameters*

$$\left(n, k, k - \frac{n+2}{4}, \frac{4nk - 4k^2 - (n-1)(n-2)}{4}\right). \tag{6.22}$$

Proof: The minimum discrepancy between $n - 2(k - d_{S_f}(\alpha))$ and $2(k - d_{S_f}(\alpha))$ is 2, since $n \equiv 2 \pmod 4$. By Theorem 6.3.1, the nonlinearity measure P_f achieves its minimum value if and only if one of the following three cases happens:

(A) $[n-2(k-d_{S_f}(\alpha))]-2(k-d_{S_f}(\alpha))$ takes on only value 2 when α ranges over all nonzero elements of A;

(B) $[n-2(k-d_{S_f}(\alpha))]-2(k-d_{S_f}(\alpha))$ takes on only value -2 when α ranges over all nonzero elements of A;

(C) $[n-2(k-d_{S_f}(\alpha))]-2(k-d_{S_f}(\alpha))$ takes on both values 2 and -2 when α ranges over all nonzero elements of A.

In all three cases the minimum value for P_f is $\frac{1}{2} + \frac{1}{n}$.

If (A) happens, then S_f is an $\left(n, k, k - \frac{n-2}{4}\right)$ difference set. Hence we obtain

$$k(k-1) = (n-1)\left(k - \frac{n-2}{4}\right).$$

Whence

$$k = \frac{n \pm \sqrt{3n-2}}{2}.$$

Hence S_f is an $\left(n, \frac{n\pm\sqrt{3n-2}}{2}, \frac{n+2\pm2\sqrt{3n-2}}{4}\right)$ difference set.

We now prove that (B) cannot happen. Suppose that (B) happens. Then S_f is an $\left(n, k, k - \frac{n+2}{4}\right)$ difference set. Hence we obtain

$$k(k-1) = (n-1)\left(k - \frac{n+2}{4}\right).$$

Whence

$$\left(k - \frac{n}{2}\right)^2 + \frac{n-2}{4} = 0.$$

This is impossible.

By definition, (C) happens if and only if

$$d_{S_f}(\alpha) = k - \frac{n \pm 2}{4},$$

which is equivalent to S_f being an $\left(n, k, k - \frac{n+2}{4}, t\right)$ almost difference set of A. It then follows from (6.19) that

$$t = \frac{4nk - 4k^2 - (n-1)(n-2)}{4}. \tag{6.23}$$

□

Remarks:

(I) Note that $1 \leq t \leq n-2$. It follows from (6.23) that

$$\frac{n - \sqrt{3(n-2)}}{2} \leq k \leq \frac{n + \sqrt{3(n-2)}}{2} \tag{6.24}$$

if f has optimum nonlinearity. This means that in the case $n \equiv 2 \pmod 4$ the weight k of functions with optimum nonlinearity is more flexible, compared with the two cases $n \equiv 0 \pmod 4$ and $n \equiv 3 \pmod 4$.

(II) The condition of (6.17) and Lemma 6.3.7 cannot be used to rule out the existence of difference sets with parameters of (6.21). For examples, $(66, 40, 24)$ and $(902, 477, 252)$ are such parameters. However, it is known that no difference sets with parameters $(66, 40, 24)$ exist [224]. No difference set with the parameters of (6.21) is known. In the cyclic case, more information on the existence can be found in [225].

Research Problem 6.3.9 *Construct difference sets with the parameters of (6.21) or show that difference sets with such parameters do not exist.*

We describe now the classes of binary functions with optimum nonlinearity which correspond to the known almost difference sets with the parameters of (6.22). To this end, we need to define cyclotomic classes and numbers. Let $GF(q)$ be a finite field, and let d divide $q-1$. For a primitive element α of $GF(q)$, define $D_0^{(d,q)} = (\alpha^d)$, the multiplicative group generated by α^d, and

$$D_h^{(d,q)} = \alpha^h D_0^{(d,q)} \text{ for } h = 1, 2, \ldots, d-1.$$

These $D_h^{(d,q)}$ are called *cyclotomic classes* of order d. The *cyclotomic numbers* of order d with respect to $GF(q)$ are defined as

$$(h, j) = \left| \left(D_h^{(d,q)} + 1 \right) \cap D_j^{(d,q)} \right|.$$

Clearly, there are at most d^2 different cyclotomic numbers of order d.

The cyclotomic classes of order 4 can be used to describe several classes of binary functions with optimum nonlinearity. Consider the finite field $GF(q)$, where $q \equiv 5 \pmod 8$. It is known that q has a quadratic partition $q = s^2 + 4t^2$, with $s \equiv \pm 1 \pmod 4$. Let $D_h^{(4,q)}$ be the cyclotomic classes of order 4.

Theorem 6.3.10 *Let $h, j, l \in \{0, 1, 2, 3\}$ be three pairwise distinct integers, and define*

$$C = \left[\{0\} \times \left(D_h^{(4,q)} \cup D_j^{(4,q)} \right) \right] \cup \left[\{1\} \times \left(D_l^{(4,q)} \cup D_j^{(4,q)} \right) \right].$$

Then C is an $\left(n, \frac{n-2}{2}, \frac{n-6}{4}, \frac{3n-6}{4}\right)$ almost difference set of $A = GF(2) \times GF(q)$ if

(1) $t = 1$ *and* $(h, j, l) \in \{(0, 1, 3), (0, 2, 1)\}$*; or*

(2) $s = 1$ *and* $(h, j, l) \in \{(1, 0, 3), (0, 1, 2)\}$.

Theorem 6.3.10 is a generalization of two results in [131]. The proof given in [131] can be slightly modified to give a proof of Theorem 6.3.10 by using cyclotomic numbers of order 4 for general finite fields [414].

It follows from Theorems 6.3.8 and 6.3.10 that the characteristic functions f_C of the several classes of almost difference sets C described in Theorem 6.3.10 have optimum nonlinearity. Furthermore these functions have weight $\frac{n-2}{2}$, where $n = 2q$. So we say that they are almost balanced.

Theorem 6.3.11 *Let $h, j, l \in \{0, 1, 2, 3\}$ be three pairwise distinct integers, and define*

$$C = \left[\{0\} \times \left(D_h^{(4,q)} \cup D_j^{(4,q)}\right)\right] \cup \left[\{1\} \times \left(D_l^{(4,q)} \cup D_j^{(4,q)}\right)\right] \cup \{0, 0\}.$$

Then C is an $\left(n, \frac{n}{2}, \frac{n-2}{4}, \frac{3n-2}{4}\right)$ almost difference set of $A = GF(2) \times GF(q)$ if

(1) *$t = 1$ and $(h, j, l) \in \{(0, 1, 3), (0, 2, 3), (1, 2, 0), (1, 3, 0)\}$; or*

(2) *$s = 1$ and $(h, j, l) \in \{(0, 1, 2), (0, 3, 2), (1, 0, 3), (1, 2, 3)\}$.*

Theorem 6.3.11 is also a generalization of two results in [131]. The proof given in [131] can also be slightly modified to give a proof of Theorem 6.3.11 by using cyclotomic numbers of order 4 for general finite fields [414].

It follows from Theorems 6.3.8 and 6.3.11 that the characteristic functions f_C of the two classes of almost difference sets C described in Theorem 6.3.11 have optimum nonlinearity. Furthermore these functions have weight $\frac{n}{2}$, where $n = 2q$. Hence they are balanced.

We now describe another class of functions with optimum nonlinearity. Let $q \equiv 3 \pmod 4$. Let $D_h^{(2,q)}$ denote the cyclotomic classes of order 2 with respect to $GF(q)$ and let α be the primitive element employed to define the cyclotomic classes of order 2.

Theorem 6.3.12 *Define a function from $(Z_{q-1}, +)$ to $(GF(2), +)$ as*

$$f(h) = \begin{cases} 1 & \text{if } \alpha^h \in (D_1^{(2,q)} - 1) \\ 0 & \text{otherwise.} \end{cases}$$

Then f has optimum nonlinearity.

Theorem 6.3.12 is the function-oriented version of a result about binary sequences with optimum autocorrelation given in [268]. The support of the function f defined in Theorem 6.3.12 is of course an almost difference set by Theorem 6.3.8.

6.3.4 The case $n \equiv 1 \pmod 4$ and $n > 1$

In this section we assume that $n \equiv 1 \pmod 4$ and consider binary functions f from A to B with optimum nonlinearity. As before, let S_f and k be the support and weight of f respectively.

Theorem 6.3.13 [60] *The possible minimum value for P_f is $\frac{1}{2}+\frac{1}{2n}$. Furthermore, $P_f = \frac{1}{2}+\frac{1}{2n}$ if and only if the support S_f is a difference set with parameters*

$$\left(n, \frac{n \pm \sqrt{2n-1}}{2}, \frac{n+1 \pm 2\sqrt{2n-1}}{4}\right). \tag{6.25}$$

Proof: The proof is similar to that of Theorem 6.3.8 and is omitted. □

Remarks:

(a) For any difference set with parameters of (6.25), the number $\frac{n \pm \sqrt{2n-1}}{2}$ must be a square.

(b) The parameters of (6.25) satisfy the conditions of both (6.17) and Lemma 6.3.7. Note that

$$\left(\sqrt{\frac{n \pm \sqrt{2n-1}}{2}}, 1, 1\right)$$

is a solution to (6.20). Examples of parameters are

$$\begin{array}{lll} (13,9,6), & (25,16,10), & (41,25,15), \\ (61,36,21), & (85,49,28). & \end{array}$$

But it is known that among the parameters above only difference sets with parameters $(13,9,6)$ exist [224]. The set $D = \{2,4,5,6,7,8,10,11,12\}$ is a $(13,9,6)$ difference set in Z_{13}. It is known that no cyclic Abelian difference set of this type exists for $13 < n \le 20201$ [225]

Research Problem 6.3.14 *Construct new difference sets with parameters of (6.25) or show that difference sets with such parameters do not exist for $n > 20201$. (We are interested only in the case $n > 20201$ because of Remark (b) above.)*

Theorem 6.3.15 [60] *$P_f = \frac{1}{2}+\frac{3}{2n}$ if and only if the support S_f is an almost difference set with parameters*

$$\left(n, k, k-\frac{n+3}{4}, \frac{4nk-4k^2-(n-1)^2}{4}\right).$$

Proof: The proof is similar to that of Theorem 6.3.8 and is omitted. □

Similarly, we have the following bounds for the weight of f

$$\frac{n-\sqrt{2n-5}}{2} \leq k \leq \frac{n+\sqrt{2n-5}}{2} \tag{6.26}$$

if f has nonlinearity $P_f = \frac{1}{2} + \frac{3}{2n}$.

Theorem 6.3.16 *Let $q \equiv 1 \pmod 4$ and let $D_h^{(2,q)}$ denote the cyclotomic classes of order 2. Then the function from $(GF(q), +)$ to $(GF(2), +)$ defined by*

$$f(x) = \begin{cases} 1 & \text{if } x \in D_0^{(2,q)} \\ 0 & \text{otherwise} \end{cases}$$

has nonlinearity $P_f = \frac{1}{2} + \frac{3}{2n}$.

Proof: It can be proved with the help of Theorem 6.3.1 and the cyclotomic numbers of order 2 [414]. □

Theorem 6.3.17 *Let $q = 4q' + 1 = x^2 + 4y^2$ be a power of an odd prime with $x \equiv 1 \pmod 4$. Then $D_h^{(4,q)} \cup D_j^{(4,q)}$ is an $\left(q, \frac{q-1}{2}, \frac{q-5}{4}, \frac{q-1}{2}\right)$ almost difference set if and only if q' is odd, $y = \pm 1$, and $(h,j) \in \{(0,1),(1,2),(2,3),(3,0)\}$.*

Theorem 6.3.17 is a slight generalization of a class of almost difference sets in [130]. The proof given in [130] can be slightly modified to give a proof of Theorem 6.3.17 by using cyclotomic numbers of order 4 for general finite fields [414].

It follows from Theorems 6.3.8 and 6.3.17 that the characteristic functions f_C of the class of almost difference sets C described in Theorem 6.3.17 have nonlinearity $P_f = \frac{1}{2} + \frac{3}{2n}$. Furthermore these functions have weight $\frac{q-1}{2}$, and thus are balanced.

6.3.5 Minimum distance from affine functions

In Sections 6.3.1 and 6.3.3, we have described binary functions from A to B with optimum nonlinearity constructed from difference sets in the two cases $n \equiv 0 \pmod 4$ and $n \equiv 2 \pmod 4$, where n is the order of A. In this section we are concerned with the minimum distance of such a function with all affine functions from A to B. We call the two constant functions 0 and 1 *trivial affine functions*.

Theorem 6.3.18 *Suppose D is an (n, k, λ) difference set of A, and $f_D(x)$ is the characteristic function of D. Assume that $l(x)$ is any nontrivial affine function from A to B. Then*

$$\Pr(f_D(x) = l(x)) = \frac{1}{2} \pm \frac{\sqrt{1-c}}{2\sqrt{n}},$$

where $\Pr(f_D(x) = l(x))$ denotes the probability of agreement between $f_D(x)$ and $l(x)$, and $c = \frac{n-4(k-\lambda)}{n}$. Hence the distance between $f_D(x)$ and $l(x)$ is

$$d(f_D(x), l(x)) = \frac{n}{2} \pm \frac{\sqrt{1-c}}{2}\sqrt{n}.$$

Proof: This is a generalization of Theorem 4.8 in [125]. The proof is essentially the same as the one given in [125], and is omitted. □

If D is a Hadamard difference set, then $c = 0$ and

$$d(f_D(x), l(x)) = \frac{n \pm \sqrt{n}}{2}.$$

Hence the minimum distance N_f between $f_D(x)$ and all affine functions is $\frac{n-\sqrt{n}}{2}$ (and is optimal, according to Parseval's relation). This was known for bent functions. It is shown here that this is also true for the characteristic function of any Hadamard difference sets.

6.4 Nonbinary functions with optimum nonlinearity

6.4.1 The case $|B| = 3$

Since the Abelian group of order 3 is unique up to isomorphism, in the case $m = 3$ we assume that $(B, +) = (Z_3, +)$. In this case if $\{C_0, C_1, C_2\}$ is an $(n, 3, n/3)$ difference partition of A with respect to B, then the conditions of (6.7) reduce to

$$\begin{aligned} k_0^2 + k_1^2 + k_2^2 &= \frac{n^2 + 2n}{3}, \\ k_0 + k_1 + k_2 &= n, \end{aligned}$$

since these two equalities imply $k_0k_1 + k_1k_2 + k_2k_0 = \frac{n^2-n}{3}$. For example,

$$(k_0, k_1, k_2) = \left(\frac{n+\sqrt{n}}{3}, \frac{n+\sqrt{n}}{3}, \frac{n-2\sqrt{n}}{3}\right)$$

and

$$(k_0, k_1, k_2) = \left(\frac{n-\sqrt{n}}{3}, \frac{n-\sqrt{n}}{3}, \frac{n+2\sqrt{n}}{3}\right)$$

are solutions to the two equations above. In fact, $(n, 3, n/3)$ difference partitions of some A with respect to B, or equivalently, functions from some A to B with perfect nonlinearity, do exit. When $q = 3$ Theorem 6.5.1 below gives a large class of perfect nonlinear functions with $|B| = 3$.

6.4.2 The case $|B|$=4

When $B = Z_4$, we have the following constraints:

Theorem 6.4.1 *Let $(A, +)$ be an Abelian group of order n and let $(B, +) = (Z_4, +)$, where n is a multiple of 4. If an $(n, 4, n/4)$ difference partition $\{C_b | b \in B\}$ of A with respect to B exists, then*

$$\begin{cases} k_0 + k_2 = \frac{n\pm\sqrt{n}}{2}, \\ k_1 + k_3 = \frac{n\mp\sqrt{n}}{2}, \end{cases} \tag{6.27}$$

where $k_z = |C_z|$ for each $z \in B$.

Proof: If $\{C_b | b \in B\}$ is an $(n, 4, n/4)$ difference partition, then the conditions of (6.7) reduce to

$$\begin{aligned} k_0k_2 + k_1k_3 &= \frac{n(n-1)}{8}, \\ k_0 + k_1 + k_2 + k_3 &= n, \\ k_0^2 + k_1^2 + k_2^2 + k_3^2 &= \frac{n^2+3n}{4}, \end{aligned}$$

since $k_0k_1 + k_1k_2 + k_2k_3 + k_3k_0 = k_0k_3 + k_1k_0 + k_2k_1 + k_3k_2 = (k_0 + k_1 + k_2 + k_3)^2 - (k_0^2 + k_1^2 + k_2^2 + k_3^2) - 2(k_0k_2 + k_1k_3)$. It then follows that

$$\begin{aligned} (k_0 + k_2)^2 + (k_1 + k_3)^2 &= \tfrac{n^2+n}{2}, \\ (k_0 + k_2) + (k_1 + k_3) &= n. \end{aligned} \tag{6.28}$$

Solving the set of equations proves the conclusion. □

We shall see at Subsection 6.5.5 that there exist perfect nonlinear functions from $A = Z_4^n$ to $B = Z_4$, where n is any positive integer greater than 1.

Theorem 6.4.2 [60] *Let $(A,+)$ be an Abelian group of order n and let $(B,+)$ be either $(Z_2 \times Z_2,+)$ or $(GF(2^2),+)$, where n is a multiple of 4. If an $(n,4,n/4)$ difference partition $\{C_b|b \in B\}$ of A with respect to B exists, then the vector $(k_{(0,0)}, k_{(0,1)}, k_{(1,0)}, k_{(1,1)})$ must take on one of the following:*

$$\begin{array}{ll} \left(\frac{n+3\sqrt{n}}{4}, \frac{n-\sqrt{n}}{4}, \frac{n-\sqrt{n}}{4}, \frac{n-\sqrt{n}}{4}\right), & \left(\frac{n-\sqrt{n}}{4}, \frac{n-\sqrt{n}}{4}, \frac{n-\sqrt{n}}{4}, \frac{n+3\sqrt{n}}{4}\right), \\ \left(\frac{n-\sqrt{n}}{4}, \frac{n-\sqrt{n}}{4}, \frac{n+3\sqrt{n}}{4}, \frac{n-\sqrt{n}}{4}\right), & \left(\frac{n-\sqrt{n}}{4}, \frac{n+3\sqrt{n}}{4}, \frac{n-\sqrt{n}}{4}, (\frac{n-\sqrt{n}}{4}\right), \\ \left(\frac{n-3\sqrt{n}}{4}, \frac{n+\sqrt{n}}{4}, \frac{n+\sqrt{n}}{4}, \frac{n+\sqrt{n}}{4}\right), & \left(\frac{n+\sqrt{n}}{4}, \frac{n+\sqrt{n}}{4}, \frac{n+\sqrt{n}}{4}, \frac{n-3\sqrt{n}}{4}\right), \\ \left(\frac{n+\sqrt{n}}{4}, \frac{n+\sqrt{n}}{4}, \frac{n-3\sqrt{n}}{4}, \frac{n+\sqrt{n}}{4}\right), & \left(\frac{n+\sqrt{n}}{4}, \frac{n-3\sqrt{n}}{4}, \frac{n+\sqrt{n}}{4}, (\frac{n+\sqrt{n}}{4}\right), \end{array} \qquad (6.29)$$

where $k_{(i,j)} = |C_{(i,j)}|$ for each $(i,j) \in B$.

Proof: Note that $(GF(2^2),+)$ is isomorphic to $(Z_2 \times Z_2,+)$. We need to consider $B = Z_2 \times Z_2$ only. If $\{C_b|b \in B\}$ is an $(n,4,n/4)$ difference partition of A with respect to B, then the conditions of (6.7) reduce to

$$\left\{\begin{array}{lcl} k_{(0,0)}k_{(0,1)} + k_{(1,0)}k_{(1,1)} & = & \frac{n(n-1)}{8} \\ k_{(0,0)}k_{(1,0)} + k_{(0,1)}k_{(1,1)} & = & \frac{n(n-1)}{8} \\ k_{(0,0)}k_{(1,1)} + k_{(1,0)}k_{(0,1)} & = & \frac{n(n-1)}{8} \\ k^2_{(0,0)} + k^2_{(0,1)} + k^2_{(1,0)} + k^2_{(1,1)} & = & \frac{n^2+3n}{4}. \end{array}\right. \qquad (6.30)$$

Solving the set of equations above gives

$$\left\{\begin{array}{lcl} k_{(0,0)} + k_{(0,1)} & = & \frac{n\pm\sqrt{n}}{2} \\ k_{(1,0)} + k_{(1,1)} & = & \frac{n\mp\sqrt{n}}{2}, \end{array}\right.$$

$$\left\{\begin{array}{lcl} k_{(0,0)} + k_{(1,0)} & = & \frac{n\pm\sqrt{n}}{2} \\ k_{(0,1)} + k_{(1,1)} & = & \frac{n\mp\sqrt{n}}{2}, \end{array}\right.$$

$$\left\{\begin{array}{lcl} k_{(0,0)} + k_{(1,1)} & = & \frac{n\pm\sqrt{n}}{2} \\ k_{(1,0)} + k_{(0,1)} & = & \frac{n\mp\sqrt{n}}{2}. \end{array}\right.$$

So there are eight cases. In each case, we obtain two solutions ($k_{(0,0)}$, $k_{(0,1)}$, $k_{(1,0)}$, $k_{(1,1)}$). Altogether we get the eight solutions of (6.29). It is checked that they are indeed solutions of (6.30). This completes the proof. □

Theorem 6.4.3 [60] *Let $(A,+)$ be an Abelian group of order n and let $(B,+)$ be either $(Z_2 \times Z_2,+)$ or $(GF(2^2),+)$, where n is a multiple of 4. If f is a function from A to B with perfect nonlinearity $P_f = \frac{1}{4}$, then*

$$N_f = \frac{3n - 3\sqrt{n}}{4} \quad or \quad \frac{3n - \sqrt{n}}{4}.$$

Proof: We consider only the case $B = Z_2 \times Z_2$. For any affine function $l(x)$, $g(x) = f(x) - l(x)$ must have perfect nonlinearity $P_g = \frac{1}{4}$ as $f(x)$ has perfect nonlinearity. Let $k_{(i,j)} = |\{x \in A | g(x) = (i,j)\}|$. By Theorem 6.4.2, $(k_{(0,0)}, k_{(0,1)}, k_{(1,0)}, k_{(1,1)})$ must take on one of the eight vectors listed in Theorem 6.4.2. The conclusion of this theorem then follows. □

Remarks:

(1) The nonlinearity N_f measures the minimum distance between f and all affine functions from A to B. Theorem 6.4.2 means that the best affine approximation of any function from A to B with perfect nonlinearity is very poor.

(2) The conditions of (6.28), those of (6.27), and Theorem 6.4.3 may suggest that functions with optimum nonlinearity P_f may not have optimum nonlinearity N_f. In other words the two kinds of measures of nonlinearity are not consistent for nonbinary functions. This is not strange, as sometimes the nonlinearity measure N_f makes little sense.

(3) When $q = 4$, Theorem 6.5.1 below will give a large class of perfect nonlinear functions with $|B| = 4$.

6.5 Constructions of functions with optimum nonlinearity

We give the basic constructions. They can be modified and combined by using the results of Section 6.2.

6.5.1 Functions from $(GF(q)^n, +)$ to $(GF(q), +)$

Let p be a prime and $q = p^l$. We have seen at Subsection 6.2.6 of Section 6.2 that for every $\beta \in GF(q)$, f_β equals ${\omega_p}^{\mathrm{Tr}(\beta f)}$ where Tr is the trace function from $GF(q)$ to $GF(p)$ and where $\omega_p = \exp(2i\pi/p)$. Thus, $\widehat{f_\beta}(\alpha)$ equals $\sum_{a \in GF(q)^n} {\omega_p}^{\mathrm{Tr}(\beta f(a) + \alpha \cdot a)}$.

We extend now the known constructions of perfect nonlinear Boolean functions [115] to this more general framework.

Let $(A, +) = (GF(q)^n, +)$, where n is even. Then the following function f from $(A, +)$ to $(GF(q), +)$

$$f(x_1, x_2, \ldots, x_n) = x_1 x_{n/2+1} + x_2 x_{n/2+2} + \ldots + x_{n/2} x_n$$

has perfect nonlinearity $P_f = \frac{1}{q}$. Hence $\{C_b(f) | b \in GF(q)\}$ is a (q^n, q, q^{n-1}) difference partition, where $C_b(f) = \{x \in A | f(x) = b\}$.

More generally, we have the following result.

Theorem 6.5.1 [60] *Let n be any even positive integer and let π be a bijective mapping from $GF(q)^{n/2}$ to $GF(q)^{n/2}$. We denote its coordinate functions by $\pi_1, \ldots, \pi_{n/2}$. Let g be a function from $GF(q)^{n/2}$ to $GF(q)$. Then*

$$\begin{aligned} f(x_1, x_2, \ldots, x_n) &= x_1\pi_1(x_{n/2+1}, \ldots, x_n) + x_2\pi_2(x_{n/2+1}, \ldots, x_n) + \ldots + \\ &\quad x_{n/2}\pi_{n/2}(x_{n/2+1}, \ldots, x_n) + g(x_{n/2+1}, \ldots, x_n) \end{aligned}$$

has perfect nonlinearity $P_f = \frac{1}{q}$

Proof: Denote $(x_1, x_2, \ldots, x_{n/2})$ by x and $(x_{n/2+1}, x_{n/2+2}, \ldots, x_n)$ by x'. We have $f(x, x') = x \cdot \pi(x') + g(x')$. For every $0 \neq \beta \in GF(q)$ and every $\alpha, \alpha' \in GF(q)^{n/2}$, we have

$$\widehat{f_\beta}(\alpha, \alpha') = \sum_{x, x' \in GF(q)^{n/2}} \omega_p^{\mathrm{Tr}(\beta[x\cdot\pi(x')+g(x')]+\alpha\cdot x+\alpha'\cdot x')},$$

where Tr is the trace function from $GF(q)$ to $GF(p)$.
The partial sum $\sum_{x\in GF(q)^{n/2}} \omega_p^{\mathrm{Tr}(\beta[x\cdot\pi(x')+g(x')]+\alpha\cdot x+\alpha'\cdot x')}$ is null if $\beta\,\pi(x') + \alpha \neq 0$. Thus

$$\widehat{f_\beta}(\alpha, \alpha') = q^{n/2} \sum_{x' \in \pi^{-1}(-\alpha/\beta)} \omega_p^{\mathrm{Tr}(\beta g(x')+\alpha'\cdot x')},$$

and, since $\pi^{-1}(-\alpha/\beta)$ is a singleton, f has perfect nonlinearity according to Theorem 6.2.16. □

This class of functions is often called *Maiorana-McFarland's class*.

The functions f in the class of Maiorana-McFarland functions with constant g can be modified using Theorem 6.2.17: take $E = \{0\} \times GF(q)^{n/2}$ in this theorem; denote by δ_0 the Dirac symbol ($\delta_0(x) = 1$ if $x = 0$, $\delta_0(x) = 0$ otherwise); we have that, for every $\lambda, \mu \in GF(q)$, the function $f(x_1, x_2, \ldots, x_n) = x_1\pi_1(x_{n/2+1}, \ldots, x_n) + x_2\pi_2(x_{n/2+1}, \ldots, x_n) + \ldots + x_{n/2}\pi_{n/2}(x_{n/2+1}, \ldots, x_n) + \lambda\delta_0(x) + \mu$ is perfect nonlinear.

Remark: Let q be an odd prime, then every polynomial function of degree 2 from $GF(q)$ to $GF(q)$ is bent [252] and therefore perfect nonlinear. Let q be a power of 2 and let $b_0, \ldots, b_4$ be elements of $GF(q)$. Then, as shown by Ambrosimov in [2], the function from $GF(q)^2$ to $GF(q)$: $f(x_1, x_2) = b_0 + b_1x_1 + b_2x_2 + b_3x_1^2 + b_4x_2^2 + x_1x_2$ has also perfect nonlinearity.

Another adaptation of a classical construction is the following [60].

Theorem 6.5.2 *Let p be a prime and $q = p^l$. Let $(A,+) = (GF(q)^n,+)$, where n is even. We identify $GF(q)^{n/2}$ with the field $GF(q^{n/2})$. Let g be any balanced function from $GF(q^{n/2})$ to $GF(q)$. Then the following function f from $(A,+)$ to $(GF(q),+)$*

$$f(x,x') = g(x\,x'^{q^{n/2}-2}),\ x,x' \in GF(q^{n/2})$$

has perfect nonlinearity $P_f = \frac{1}{q}$.

Proof: For every $0 \neq \beta \in GF(q)$ and every $\alpha, \alpha' \in GF(q^{n/2})$, we have

$$\widehat{f_\beta}(\alpha,\alpha') = \sum_{x,x' \in GF(q^{n/2})} \omega_p^{\mathrm{Tr}(\beta\, g(x\,x'^{q^{n/2}-2})) + \mathrm{Tr}'(\alpha\, x + \alpha'\, x')},$$

where Tr is the trace function from $GF(q)$ to $GF(p)$ and Tr' is the trace function from $GF(q^{n/2})$ to $GF(p)$. Writing $x = x'\, z$ for every $x' \neq 0$, we have

$$\sum_{x \in GF(q^{n/2}), x' \in GF(q^{n/2*})} \omega_p^{\mathrm{Tr}(\beta\, g(x\,x'^{q^{n/2}-2})) + \mathrm{Tr}'(\alpha\, x + \alpha'\, x')} =$$

$$\sum_{z \in GF(q^{n/2}), x' \in GF(q^{n/2*})} \omega_p^{\mathrm{Tr}(\beta\, g(z)) + \mathrm{Tr}'((\alpha\, z + \alpha')x')} =$$

$$\sum_{z,x' \in GF(q^{n/2})} \omega_p^{\mathrm{Tr}(\beta\, g(z)) + \mathrm{Tr}'((\alpha\, z + \alpha')x')} - \sum_{z \in GF(q^{n/2})} \omega_p^{\mathrm{Tr}(\beta\, g(z))}.$$

Since g is balanced, we have $\sum_{z \in GF(q^{n/2})} \omega_p^{\mathrm{Tr}(\beta\, g(z))} = 0$, according to Proposition 6.2.14. Thus

$$\widehat{f_\beta}(\alpha,\alpha') = \sum_{x \in GF(q^{n/2})} \omega_p^{\mathrm{Tr}(\beta\, g(0)) + \mathrm{Tr}'(\alpha\, x)} + \sum_{z,x' \in GF(q^{n/2})} \omega_p^{\mathrm{Tr}(\beta\, g(z)) + \mathrm{Tr}'((\alpha\, z + \alpha')x')}.$$

The partial sum $\sum_{x' \in GF(q^{n/2})} \omega_p^{\mathrm{Tr}(\beta\, g(z)) + \mathrm{Tr}'((\alpha\, z + \alpha')x')}$ is null if $\alpha\, z + \alpha' \neq 0$. If $\alpha \neq 0$, since the sum $\sum_{x \in GF(q^{n/2})} \omega_p^{\mathrm{Tr}(\beta\, g(0)) + \mathrm{Tr}'(\alpha\, x)}$ is null, we deduce that $\widehat{f_\beta}(\alpha,\alpha')$ has magnitude $q^{n/2}$. And if $\alpha = 0$ and $\alpha' \neq 0$, then $\widehat{f_\beta}(\alpha,\alpha') = q^{n/2}\omega_p^{\mathrm{Tr}(\beta\, g(0))}$ has also magnitude $q^{n/2}$. We deduce that $\widehat{f_\beta}(0,0)$ has magnitude $q^{n/2}$ as well, thanks to Parseval's relation. Thus, f has perfect nonlinearity according to Theorem 6.2.16. □

This class of functions is often called *Dillon's class* or *Partial Spreads class* (when $q = 2$, the support of the function is a partial spread).

6.5.2 Functions from $(GF(q)^n, +)$ to $(GF(q)^n, +)$: perfect and almost perfect nonlinear mappings

We consider now the case of mappings f from $GF(q)^n$ to $GF(q)^n$ where $q = p^l$. Since $GF(q)^n$ can be identified, as a vector space over $GF(p)$ with $GF(q^n) = GF(p^{ln})$, this case reduces to that of mappings f from $GF(p^m)$ to $GF(p^m)$.
If $p = 2$, the minimum possible value of P_f is $\frac{2}{p^m}$, because the characteristic of the field being equal to 2, any solution x of the equation $D_a f(x) = b$ can be paired with the solution $x+a$. If $p > 2$, then the minimum possible value of P_f is $\frac{1}{p^m}$. A function f from $GF(p^m)$ to $GF(p^m)$ is called *almost perfect nonlinear* if $P_f = \frac{2}{p^m}$, and *perfect nonlinear* if $P_f = \frac{1}{p^m}$ [332, 333]. Perfect nonlinear mappings are also called *planar* functions. Perfect and almost perfect nonlinear mappings have important applications in cryptography and coding theory [19, 54, 202, 333]. In this section we summarize known perfect and almost perfect nonlinear functions.

Known almost perfect nonlinear power functions x^s from $GF(2^m)$ to $GF(2^m)$ are the following:

- $s = 2^m - 2$ (Beth and Ding [19], Nyberg [333]).
- $s = 2^h + 1$ with $\gcd(h, m) = 1$, where $1 \le h \le (m-1)/2$ if m is odd and $1 \le h \le (m-2)/2$ if m is even (Nyberg [333], Gold [160]).
- $s = 2^{2h} - 2^h + 1$ with $\gcd(h, m) = 1$, where $1 \le h \le (m-1)/2$ if m is odd and $1 \le h \le (m-2)/2$ if m is even (Kasami [228], Janwa and Wilson [220]).
- $s = 2^{(m-1)/2} + 3$, where m is odd (Dobbertin [141]).
- $s = 2^{(m-1)/2} + 2^{(m-1)/4} - 1$, where $m \equiv 1 \pmod 4$ (Dobbertin [142]).
- $s = 2^{(m-1)/2} + 2^{(3m-1)/4} - 1$, where $m \equiv 3 \pmod 4$ (Dobbertin [142]).

Known perfect nonlinear power functions x^s from $GF(p^m)$ to $GF(p^m)$, where $p > 2$, are the following (Coulter and Matthews [85], see also Helleseth and Sandberg [201]):

- $s = 2$.
- $s = p^k + 1$, where $m/\gcd(m, k)$ is odd.
- $s = (3^k + 1)/2$, where $p = 3$, k is odd, and $\gcd(m, k) = 1$.

The case $s = 2$ was known earlier in [101] under the name of generalized Hadamard matrices.

We deduce that if

- $s = 2$, or
- $s = p^k + 1$, where $m/\gcd(m, k)$ is odd, or
- $s = (3^k + 1)/2$, where $p = 3$, k is odd, and $\gcd(m, k) = 1$,

then the matrix D of Theorem 6.2.12 is a $(q, q, 1)$ difference matrix, i.e., a generalized Hadamard matrix $\mathrm{GH}(q, 1)$.

The following proposition illustrates the idea of constructing new perfect nonlinear functions from known ones.

Proposition 6.5.3 *Define* $f(x) = \mathrm{Tr}_{GF(p^m)/GF(p^h)}(x^s)$, *where* m *and* h *are integers with* $1 \le h|m$, p *is an odd prime, and* $\mathrm{Tr}_{GF(p^m)/GF(p^h)}$ *is the trace function from* $GF(p^m)$ *to* $GF(p^h)$. *If*

- $s = 2$, *or*
- $s = p^k + 1$, *where* $m/\gcd(m, k)$ *is odd, or*
- $s = (3^k + 1)/2$, *where* $p = 3$, k *is odd, and* $\gcd(m, k) = 1$,

then

(a) $f(x)$ *is a function from* $GF(p^m)$ *to* $GF(p^h)$ *with perfect nonlinearity, and*

(b) *the matrix* D *of Theorem 6.2.12 defined by* f *is a generalized Hadamard matrix* $GH(p^h, p^{m-h})$.

Proof: As made clear before, x^s has perfect nonlinearity if s takes on one of the three values above. The conclusion in part (a) then follows from Theorem 6.2.7. The conclusion of part (b) then follows from Theorem 6.2.12. □

Known almost perfect nonlinear power functions x^s from $GF(p^m)$ to $GF(p^m)$, where p is odd, are the following (due to Helleseth and Sandberg [201], and Helleseth, Rong, and Sandberg [202]):

- $s = p^m - 2$, where $p^m \equiv 2 \pmod 3$ [202].
- $s = \frac{p^m - 1}{2} - 1$, where $p \equiv 3, 7 \pmod{20}$, $p^m > 7$, $p^m \neq 27$, and m is odd [201].

- $s = 3$, where $p \neq 3$ [202].
- $s = \frac{p^m+1}{4} + \frac{p^m-1}{2}$, where $p^m \equiv 3 \pmod 8$ [202].
- $s = \frac{p^m+1}{4}$, where $p^m \equiv 7 \pmod 8$ [202].
- $s = p^m - 3$, where $n > 1$ is odd and $p = 3$ [202].
- $s = \frac{2p^m-1}{3}$, where $p^m \equiv 2 \pmod 3$ [202].
- $s = p^{m/2} + 2$, where $p > 3$ is prime and $p^{m/2} \equiv 1 \pmod 3$ [202].
- $s = p^{(m+1)/2} - 1$, where m is odd and $p = 3$ [202].
- $s = \frac{5^k+1}{2}$, where $\gcd(2m, k) = 1$ and $p = 5$ [202].

Functions from $GF(p^m)$ to $GF(p^m)$ with high nonlinearity that are not perfect or almost perfect nonlinear may be found in Beth and Ding [19], Dobbertin [140], Gold [160], Helleseth and Sandberg [201], Helleseth, Rong and Sandberg [202], Kasami [228], and Lachaud and Wolfmann [254].

Note that any power function is a group homomorphism. The perfect and almost perfect nonlinear functions in this section illustrate an idea which will be used again in Subsection 6.5.3.

6.5.3 Functions with optimum nonlinearity from linear functions

One way of getting functions with optimum nonlinearity with respect to a pair of operations is to use linear functions with respect to another pair of operations. The following theorem illustrates this idea [125, p. 125].

Theorem 6.5.4 *Any nonzero linear function f from $(GF(q^m), +)$ to $(GF(q), +)$ is a function from $(GF(q^m)^*, \times)$ to $(GF(q), +)$ with optimum nonlinearity with respect to the two operations $\times$ and $+$ and* $P_f = \frac{1}{q} + \frac{1}{q(q^m-1)}$.

The idea of obtaining highly nonlinear functions from linear functions is by far the most useful tool.

6.5.4 Other functions from $(GF(2^m)^*, \times)$ to $(GF(2), +)$ with optimum nonlinearity

We have obtained at Theorem 6.5.4 functions from $(GF(q^m)^*, \times)$ to $(GF(q), +)$ with optimum nonlinearity. The most interesting practical case is when $q = 2$. Several other examples of functions with optimum nonlinearity are known in this case. Indeed, Boolean functions defined on $GF(2^m)$

and such that, for every $a \neq 1$, the function $f(x) + f(ax)$ is balanced are said to have ideal autocorrelation and present much interest for the construction of good sequences for CDMA communications systems. So much work has been done to obtain such functions. Their restrictions to $GF(2^m)^*$ have optimum nonlinearity $P_f = \frac{2^{m-1}}{2^m-1} = \frac{1}{2} + \frac{1}{2(2^m-1)}$. Thus, as shown at Subsection 6.3.2, their supports are cyclic difference sets with the so-called "Singer parameters" (this strengthens the reasons why these functions have been much studied).

We list now the known constructions. Note that, if $f(x)$ has ideal autocorrelation, $\gcd(2^m - 1, \nu) = 1$ and $a \in GF(2^m)$ is nonzero, then $f(ax^\nu)$ has also ideal autocorrelation.

- Theorem 6.5.4 corresponds to the fact that the Boolean function on $GF(2^m)$ equal to $\mathrm{Tr}(x)$, where Tr denotes the trace function from $GF(2^m)$ to $GF(2)$ has ideal autocorrelation (this can be generalized to any finite field). We have indeed:

$$\sum_{x \in GF(2^m)} (-1)^{\mathrm{Tr}(x)+\mathrm{Tr}(ax)} = \sum_{x \in GF(2^m)} (-1)^{\mathrm{Tr}((1+a)x)} = 0.$$

  The support of this function is called a *Singer cyclic difference set*. This construction is generalized into GMW (Gordon-Mills-Welch) construction:

$$f(x) = \mathrm{Tr}\left[\left(\mathrm{Tr}_{GF(2^m)/GF(2^r)}(x)\right)^t\right]$$

  where r divides m and $\gcd(t, 2^m - 1) = 1$, $\mathrm{Tr}_{GF(2^m)/GF(2^r)}$ is the trace function from $GF(2^m)$ to $GF(2^r)$, and Tr is the trace function from $GF(2^r)$ to $GF(2)$.

- A second way to construct functions with ideal autocorrelation is by using Maschietti's method [116, 290]: find κ such that $\gcd(\kappa, 2^m - 1) = 1$ and such that the map $x \mapsto x + x^\kappa$ is 2 to 1 (i.e. such that for every $y \in GF(2^m)$ there exist either two or no $x \in GF(2^m)$ such that $y = x + x^\kappa$). Then $GF(2^n) \setminus \{x + x^\kappa; x \in GF(2^n)\}$ is the support of a function f with ideal auto-correlation. Singer sets with $\nu = 1$ correspond to $\kappa = 2$. For m odd, $\kappa = 6$ (Segre case) and two other more complex cases also work (see [117]).

- A third way is by using No et al. method [330]: f is then the indicator of the set $\{x^d + (x + 1)^d;\ x \in GF(2^n)\}$ (if the mapping $x \mapsto x^d$ is not a permutation) or of its complement (if it is a permutation), where $\gcd(d, 2^m - 1) = 1$ and where the map $x \mapsto x^d + (x + 1)^d$ is 2 to 1. Take k such that $\gcd(k, m) = 1$ and $d = 2^{2k} - 2^k + 1$ (called Kasami exponent); then as shown by Dillon and Dobbertin in [117] (see also [116]), f has ideal autocorrelation.

- A last way is when $2^m - 1$ is a prime to take for f the indicator of the set of all elements α^t (α a primitive element of $GF(2^n)$) such that t is not a square mod $2^m - 1$.

6.5.5 Functions from Z_q^n to Z_q

If q is not a prime, it has been shown in [61] that only one construction among all known constructions of generalized bent functions can produce perfect nonlinear functions. This construction, due to Hou [210], is a generalization of Dillon's (i.e. Partial Spreads) construction of binary bent functions. It uses the notion of Galois ring and can be specified to produce perfect nonlinear functions from Z_q^n to Z_q where q is a power of a prime and n is even [61].

The question whether functions with perfect nonlinearity exist on Z_q^n for n odd arises. A construction valid for $A = Z_4^n$ where n is any positive integer greater than 1 and $B = Z_4$ has been given in [61]. It uses also Galois rings.

Research Problem 6.5.5 *Construct perfect nonlinear functions from Z_q^n to Z_q for n odd and $q \neq 4$, q being not a prime.*

Theorem 6.5.6 *Define $f : Z_{p^2} \to Z_p$ by $f(h + jp) = hj \bmod p$ for $0 \leq h, j \leq p - 1$. Then f has perfect nonlinearity with respect to $(Z_{p^2}, +)$ and $(Z_p, +)$.*

Theorem 6.5.7 *Let $f : Z_{p^2} \to Z_p$ be a mapping whose restriction to $Z_{p^2}^*$ is a surjective homomorphism with respect to $(Z_{p^2}^*, \cdot)$ and $(Z_p, +)$ and is zero otherwise. Then f has perfect nonlinearity with respect to $(Z_{p^2}, +)$ and $(Z_p, +)$.*

Theorem 6.5.6 and Theorem 6.5.7 are the functional versions of results about generalized Hadamard matrices due to de Launey [102] and Brock [41] respectively. We now give one specific function of the type of Theorem 6.5.7.

Example 6.5.8 Let p be an odd prime, and let α be a primitive root modulo p^2. Define f as

$$f(x) = \begin{cases} h \pmod{p} & \text{if } x = \alpha^h \text{ for some } h \\ 0 & \text{otherwise.} \end{cases}$$

Then f satisfies the conditions of Theorem 6.5.7 and has thus perfect nonlinearity.

Chapter 7

Difference Sets and Sequences

As seen in Chapter 6, the autocorrelation property of a binary periodic sequence is closely related to the difference property of its characteristic set with respect to the addition of Z_N, where N is a period of the sequence. Generally speaking, the better the difference property of its characteristic set, the smaller $\max_{0 \neq w} |\mathrm{AC}_s(w)|$ will be. In particular, for residue difference sets the autocorrelation functions of their characteristic sequences (briefly, DSC sequences) are 2-valued. For almost difference sets of Z_N's the autocorrelation functions of their characteristic sequences (briefly, ADSC sequences) are 3-valued. Furthermore, the characteristic sequences of difference sets and almost difference sets with parameters (N, k, λ) having $k - \lambda \approx N/4$ have good autocorrelation property. The autocorrelation property of sequences is cryptographically important for at least one reason: the control of the transformation density of some stream ciphers [122]. In addition, the autocorrelation property determines the two-digit pattern distributions of binary sequences.

Due to the cryptographic significance of DSC sequences and ADSC sequences this chapter mainly introduces the differential analysis of those sequences and presents some results about their linear complexity. The NSG realization of sequences is also presented to show the significance of the differential analysis of sequences.

7.1 The NSG Realization of Sequences

There are many ways to generate sequences, as shown by the many kinds of proposed generators. In spite of the flexibility of generating binary sequences, every binary sequence generator is equivalent to a natural sequence generator (NSG) described in Chapter 2. We say two generators are *equiv-*

alent if, given any output sequence of one of the generators, the other generator can produce the same output sequence when the parameters of the generator are properly chosen. In this section we search for those NSGs which can produce some given sequences and for the equivalent NSGs of some known generators. To this end, we need the trace representation of sequences.

It is well known that every periodic sequence in $K = GF(q)$ has a trace representation described by the following two propositions [276, pp. 406 and 467].

Proposition 7.1.1 *Let s^∞ be a periodic sequence in $K = GF(q)$ whose characteristic polynomial $f(x)$ of degree k is irreducible over K. Let α be a root of $f(x)$ in the extension field $F = GF(q^k)$. Then there exists a uniquely determined $\theta \in F$ such that*

$$s_n = \mathrm{Tr}_{F/K}(\theta\alpha^n), \ n \geq 0,$$

where $\mathrm{Tr}_{F/K}(x)$ is the trace function.

The characteristic polynomial of a sequence refers to a zero polynomial of the sequence, which is a multiple of the monic minimal polynomial of the sequence. Proposition 7.1.1 gives a trace representation only for periodic sequences whose characteristic polynomials are irreducible over K. Generally we have the following conclusion [276, p. 467].

Proposition 7.1.2 *Let s^∞ be a periodic sequence in $K = GF(q)$ with characteristic polynomial $f(x) = f_1(x) \cdots f_r(x)$, where the $f_i(x)$ are distinct irreducible polynomials over K. For $i = 1, \cdots, r$, let α_i be a root of $f_i(x)$ in its splitting field F_i over K. Then there exist uniquely determined elements $\theta_1 \in F_1, \cdots, \theta_r \in F_r$ such that*

$$s_n = \mathrm{Tr}_{F_1/K}(\theta_1\alpha_1^n) + \cdots + \mathrm{Tr}_{F_r/K}(\theta_r\alpha_r^n), \ n \geq 0.$$

Now we describe an NSG realization of periodic sequences in the finite field $K = GF(q)$. Let s^∞ be the sequence described in Proposition 7.1.1; then one of its NSG realizations is depicted by Figure 7.1. For the sequence s^∞ of Proposition 7.1.2 we have an NSG realization in Figure 7.2. The NSG realization of the maximum-length sequences is easy given the above two propositions.

If one has a characteristic polynomial of a sequence, it is possible to give an NSG realization of the sequence. However the computational complexity could be very large, depending on the sequence. Finding the minimal polynomial of a periodic sequence could be easy as we have the efficient

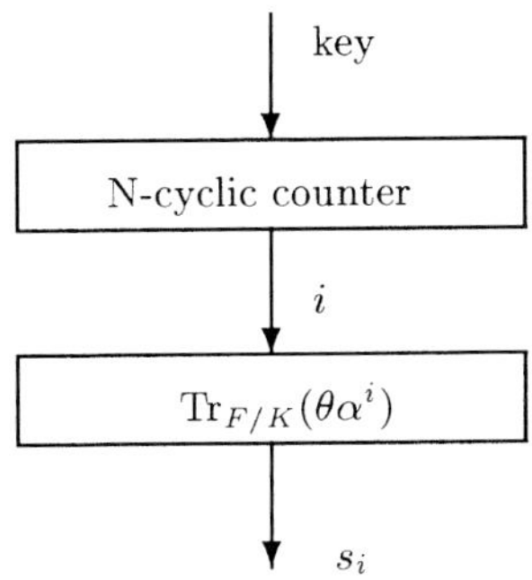

Figure 7.1: The NSG realization of some sequences.

Berlekamp-Massey algorithm. But factoring a polynomial and finding the parameters θ_i and α_i of Proposition 7.1.2 could be hard. We also note that the NSG realization of a sequence is not unique.

7.2 Differential Analysis of Sequences

For any sequence generator (SG), suppose that its output sequence s^∞ over a finite group $(G, +)$ has period N. Let

$$C_s(g) = \{i : \ s_i = g, \ 0 \leq i \leq N-1\}, \ g \in G$$

and f_s be the characteristic function of the partition $\{C_s(g) : g \in G\}$. The analysis of the difference parameters

$$d_s(i, j; w) = |C_s(i) \cap (C_s(j) - w)|, \ (i, j; w) \in G \times G \times Z_N,$$

is called the *differential analysis* of the sequence. The conservation laws between the difference parameters are given in Section 4.2.1. The differential analysis of sequences could be finer than the autocorrelation analysis. However, for binary sequences they are equivalent.

It is clear that the differential analysis is in fact the two-character pattern distribution analysis, since the difference parameters $d_s(i, j; w)$ represent the number of appearances of one two-character pattern in a period of the sequence.

Let ξ be a group character of $(G, +)$. By definition the periodic autocorrelation function of a sequence s^∞ of period N over G is given by

$$\mathrm{AC}_s(l) \quad = \quad \sum_{i=0}^{N-1} \xi(s_i - s_{i+l})$$

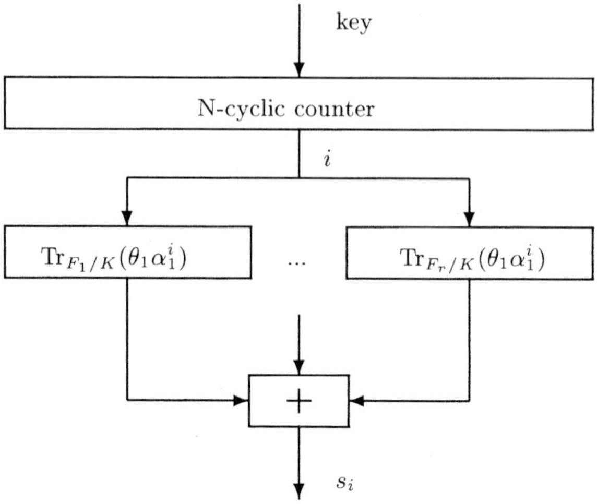

Figure 7.2: The NSG realization of some general sequences.

$$
\begin{aligned}
&= \sum_{v\in G} |\{0 \le i \le N-1 | s_i - s_{i+l} = v\}| \xi(v) \\
&= \sum_{v\in G} \sum_{u\in G} |C_s(u) \cap [C_s(u-v) - l]| \xi(v) \\
&= \sum_{v\in G} \sum_{u\in G} d_s(u, u-v; l) \xi(v).
\end{aligned}
$$

Thus, if the difference parameter $d_s(i, j; w)$ is a constant for all $(i, j) \in G \times G$, the autocorrelation value $\mathrm{AC}_s(l) = 0$ if $l \neq 0$. Generally, the flatter the difference parameters, the smaller the autocorrelation values $|\mathrm{AC}_s(l)|$ for $l \neq 0$. But the converse may not be true when $|G| \geq 3$. In summary, the differential analysis gives the autocorrelation analysis and two-character pattern analysis.

Note that every periodic sequence has an NSG realization and many generators have an equivalent NSG. Thus, if an equivalent NSG of a keystream generator can be constructed, the differential analysis of the NSG is necessary due to the differential attack described in [122]. If we cannot ensure that an equivalent NSG of the keystream generator cannot be constructed, then we should carry out the differential analysis of the keystream. Otherwise, a bad difference property of the keystream sequence could lead to the determination of some parameters of the NSG with which the NSG could produce the same keystream sequence.

7.3 Linear Complexity of DSC (ADSC) Sequences

It is known that for any binary maximum-length sequence s^∞ of period $2^m - 1$, its characteristic set is a $(2^m - 1, 2^{m-1}, 2^{m-2})$ difference set (for example, see [404], p. 314). On the other hand, the m-sequences satisfy also Golomb's three postulates. But these sequences have only linear complexity m, which is very small compared with the period $2^m - 1$. However, there are some DSC sequences with large linear complexity. In fact there do exist DSC sequences having maximum linear complexity, as described by the following proposition [122].

Proposition 7.3.1 *Let D be an (N, k, λ)-difference set of Z_N and s^∞ be its periodic characteristic sequence. Then*

1. *if k is even and λ odd, then* $\mathrm{L}(s^\infty) = N - 1$*;*
2. *if k is odd and λ even, then* $\mathrm{L}(s^\infty) = N$*;*
3. *if k and λ both are even, then*

$$\mathrm{L}(s^\infty) = \deg\left[\frac{\gcd(s^N(x^{-1})x^N,\ x^N-1)}{\gcd(\gcd(s^N(x),\ x^N-1),\ \gcd(s^N(x^{-1})x^N,\ x^N-1))}\right];$$

4. *if k and λ both are odd, then*

$$\mathrm{L}(s^\infty) = \deg\left[\frac{\gcd(s^N(x^{-1})x^N,\ x^N-1)(x+1)}{\gcd(\gcd(s^N(x),\ x^N-1),\ \gcd(s^N(x^{-1})x^N,\ x^N-1))}\right],$$

where $s^N(x) = s_0 + s_1x + \cdots + s_{N-1}x^{N-1}$.

Proof: It is well-known [138], [276, pp. 418-423], that the minimal polynomial of a sequence of period N over $GF(q)$ can be expressed as

$$f_s(x) = \frac{x^N - 1}{\gcd(s^N(x), x^N - 1)}.$$

Since the characteristic sequences are binary, our arithmetic is now on $GF(2)$. Let D be the characteristic set of s^∞. Since D is a difference set

$$\begin{aligned} & s^N(x)s^N(x^{-1})x^N \\ \equiv\ & \sum_{i,j}^{k} x^{d_i - d_j} \pmod{x^N - 1} \\ \equiv\ & (n \bmod 2) + (\lambda \bmod 2)(1 + x + \cdots + x^{N-1}) \pmod{x^N - 1}, \end{aligned}$$

where $n = k - \lambda$.

If k is even and λ is odd, then n is odd, and

$$s^N(x)s^N(x^{-1})x^N \equiv 1 + (1 + x + \cdots + x^{N-1}) \pmod{x^N - 1}.$$

By the difference-set property $k(k-1) = (N-1)\lambda$. Thus N must be odd. It follows further from the assumptions of the proposition that $(x+1)$ but not $(x+1)^2$ divides $s^N(x)$. Hence

$$\gcd(s^N(x), x^N - 1) = x - 1, \quad f_s(x) = (x^N - 1)/(x - 1).$$

Thus the linear complexity of the sequence is $N - 1$. This proves part one.

If k is odd and λ even, then

$$s^N(x)s^N(x^{-1})x^N \equiv 1 \pmod{x^N - 1}.$$

It follows that $\gcd(s^N(x), x^N - 1) = 1$, and $\mathrm{L}(s^\infty) = N$. This proves part two.

If k and λ both are even, then

$$s^N(x)s^N(x^{-1})x^N \equiv 0 \pmod{x^N - 1}$$

and therefore

$$\gcd(s^N(x), x^N - 1)\gcd(s^N(x^{-1})x^N, x^N - 1) \equiv 0 \pmod{x^N - 1}.$$

whence $\gcd(s^N(x), x^N - 1)$ is equal to

$$\frac{(x^N - 1)\gcd(\gcd(s^N, x^N - 1), \gcd(s^N(x^{-1})x^N, x^N - 1))}{\gcd(s^N(x^{-1})x^N, x^N - 1)}.$$

This proves part three. The remaining part four can be proved similarly. □

Set $n = k - \lambda$. The linear complexity of the DSC sequences is optimal for those with parameter n odd. This also shows the cryptographic importance of the parameter n. For those DSC sequences with parameter n even, the linear complexity seems hard to control. As an example, we consider the binary maximum-length sequences. Their characteristic sets form $(2^m - 1, 2^{m-1}, 2^{m-2})$ difference sets. For those difference sets we have $n = k - \lambda = 2^{m-2}$ which is even. When n is even, the formulae for the linear complexity in Proposition 7.3.1 are not practical in general. But in some special cases they might be reduced into practical ones.

Planar difference sets are those with parameters (N, k, λ) having $\lambda = 1$. If we can find planar difference sets with k even, then we get sequences with maximum linear complexity. However, since $k \approx \sqrt{N}$, those sequences are fairly unbalanced. If the prime $p \neq 2$, the periodic characteristic sequences of those $(p^{2j}+p^j+1, p^j+1, 1)$ difference sets have linear complexity $N-1$ and they are also fairly unbalanced. Another family of difference sets is the Singer difference sets with parameters

$$N = \frac{q^{m+1}-1}{q-1}, \quad k = \frac{q^m-1}{q-1}, \quad \lambda = \frac{q^{m-1}-1}{q-1},$$

which exist whenever q is a prime power and $m \geq 2$ [405], [15, pp.99-104], [404, pp.313-314]. Since

$$k - \lambda = q^{m-1}, \quad \lambda = 1 + q + \cdots + q^{m-2},$$

the linear complexity of the periodic characteristic sequences of these difference sets is $N-1$ if q is not a power of 2. However, unfortunatly we have $N/k \approx q$. This kind of unbalance may restrict the cryptographic application of these sequences.

A difference set which is composed of all the mth powers modulo some prime N, or of the mth powers and zero, is called an mth *power residue difference set.* Probably the cryptographically most important periodic characteristic sequences of difference sets are those of the quadratic residue difference sets.

Let D be an (N, k, λ) difference set of Z_N (see Proposition 4.3.3). The polynomial

$$H(x) = x^{d_1} + x^{d_2} + \cdots + x^{d_k}$$

over the ring Z_N is called the *Hall polynomial* of the difference set, the *generating polynomial* of the difference set or the *difference set polynomial* . In terms of this polynomial the difference set property is

$$H(x)H(x^{-1}) = \sum_{i,\ j}^{k} x^{d_i - d_j} \equiv n + \lambda(1 + x + \cdots + x^{N-1}) \pmod{x^N - 1},$$

where $n = k - \lambda$. Let s^∞ be the periodic characteristic sequence of the $(N,\ k,\ \lambda)$ difference set D, then

$$\begin{aligned} s^N(x) &= s_0 + s_1 x + \cdots + s_{N-1} x^{N-1} \\ &= x^{d_1} + x^{d_2} + \cdots + x^{d_k}, \end{aligned}$$

where "+" denotes the modulo 2 addition. Thus, if we consider the Hall-polynomial over GF(2), then we have $s^N(x) = H(x)$. It is by employing the formula

$$\begin{aligned} & s^N(x)s^N(x^{-1}) \\ = & \sum_{i,j}^{k} x^{d_i-d_j} \equiv n+\lambda(1+x+\cdots+x^{N-1}) \pmod{x^N-1} \end{aligned}$$

that the above general conclusions about the linear complexity of DSC sequences have been proved. However, with almost difference sets we do not have such a nice fact to employ. So it seems not easy to control the linear complexity by controlling the parity of n. However, we can control the linear complexity of ADSC sequences by employing the results of Chapter 3. It should be mentioned here that there are ADSC sequences which have optimal linear complexity. Examples are the characteristic sequences of quadratic residues modulo primes of the form $4t+1$ (see Proposition 4.3.3).

Research Problem 7.3.2 *Analyze the linear complexity of the ADSC sequences.*

7.4 Barker Sequences

In some communication systems the value $\max_{1\le j\le N-1}|\mathrm{AAC}_s(j,0,v)|$ should be as small as possible [12], where $\mathrm{AAC}_s(j,0,v)$ denotes the aperiodic autocorrelation function of the sequence. Sequences with $\mathrm{AAC}_s(l,0,v))$ having values from $\{-1,0,1\}$ were called *Barker sequences* [15, p.96], [381, p.611].

According to [15, p.96], only the following Barker sequences are known:

$$\begin{aligned} N &= 2 \quad 00 \\ N &= 3 \quad 001 \\ N &= 4 \quad 0001;\ 0010 \\ N &= 5 \quad 00010 \\ N &= 7 \quad 0001101 \\ N &= 11 \quad 00011101101 \\ N &= 13 \quad 0000011001010 \end{aligned}$$

together with the sequences which may be derived from them by the following transformations:

$$s_i' = (i+s_i) \bmod 2;$$
$$s_i' = (i+1+s_i) \bmod 2;$$
$$s_i' = (1+s_i) \bmod 2.$$

It is known that a binary sequence of period $N > 13$ is a Barker sequence if and only if it is the characteristic sequence of a $(4n^2, 2n^2 - n, n^2 - n)$ difference set of Z_{4n^2} [15, p.97]. Thus, to construct Barker sequences, we have to find difference sets of this type, which are called *Menon difference sets* [7]. It was long known that if any further Barker sequences exist they must have $n \geq 55$, i.e., $N = 4n^2 \geq 12,100$ [15, p.97]. For the next twenty years little was achieved in the search for Menon difference sets of residue rings [7]. Then in 1992 Eliahou and Kervaire [144, p.363] raised the bound on n to $n \geq 689$, so $N \geq 1,898,884$.

Barker sequences are cryptographically interesting from two points of view: On the one hand, a Barker sequence of period $4n^2$ has maximum linear complexity $4n^2$ if n is odd. This can be seen from Proposition 7.3.1 since $k - \lambda = n^2$ is odd. On the other hand, if we use the characteristic function of the corresponding Menon difference set as the cryptographic function for the natural sequence generator and use further this generator as the keystream generator for the binary additive stream cipher, then the stream cipher has optimal local (encryption and decryption) transformation density (see Chapter 16). For our cryptographic applications, we need to consider at least two things: the search for Menon difference sets of Z_{4n^2} with large n's; and the realization of the characteristic functions of them.

Research Problem 7.4.1 *Find Menon difference sets of Z_{4n^2} for large n if there are any.*

The Barker sequences are also closely related to the so-called circulant Hadamard matrices. A matrix is said to be a *circulant* if each successive row is derived from the previous row by shifting it cyclically one position to the right. An example is the following

$$H = \begin{bmatrix} +1 & +1 & +1 & -1 \\ -1 & +1 & +1 & +1 \\ +1 & -1 & +1 & +1 \\ +1 & +1 & -1 & +1 \end{bmatrix}.$$

If a matrix has entries ± 1 and its rows are orthogonal, it is called a *Hadamard matrix*. The above H is a Hadamard matrix and is the only known circulant Hadamard matrix. It is not hard to see that there is a one-to-one correspondence between Barker sequences of even length $N \geq 4$ and circulant Hadamard matrices. Thus, if there exists any further circulant Hadamard matrix its order $N \geq 1,898,884$. Whether there are further circulant Hadamard matrices remains a well-known open problem.

Research Problem 7.4.2 *Investigate whether there are circulant Hadamard matrices of order $N \geq 1,898,884$.*

Chapter 8

Binary Cyclotomic Generators

In Chapter 3 we investigated the relations between primes, primitive roots and sequences, and saw that some sequences with a prime period, prime square period or period the product of two distinct primes over some suitable fields, could have some cryptographically good attributes, i.e., ideal linear and sphere complexity, and ideal period stability. In addition, these cryptographic attributes could be obtained with only a few conditions, i.e., a proper choice of the primes such that the orders of q modulo these primes are large enough, and a choice of the cryptographic function such that the Hamming weight of one period of the sequence is neither too small nor too large compared with the period, where $GF(q)$ is the field over which the sequence is constructed.

In Chapter 5 the cryptographic value of various kinds of primes was analyzed with respect to the construction of sequences with large linear and sphere complexity as well as period stability. In Chapters 4 and 6 many cryptographic functions were constructed and analyzed. Having all this preparation in the foregoing chapters, we shall describe some binary natural sequence generators based on cyclotomy and generalized cyclotomy in this chapter, and analyze some of their properties. These generators were studied in details by Ding [123, 124, 125, 127].

8.1 Cyclotomic Generator of Order 2k

Let $N = 2kf + 1$, and $D_0, D_1, \cdots, D_{2k-1}$ be the cyclotomic classes of order $2k$ defined in Chapter 4. The cyclotomic generator of order $2k$ is described by

$$s_i = [(i_0 + i)^{(N-1)/2k} \bmod N] \bmod 2, \quad i \geq 0,$$

where $0 \le i_0 \le N-1$ is the key of the generator.

If the cyclotomic numbers of order $2k$ are roughly flat and k is very small, we can prove that the cryptographic function

$$F_k(x) = [x^{(N-1)/2k} \bmod N] \bmod 2 \tag{8.1}$$

has good nonlinearity with respect to the additions of Z_N and Z_2. The actual nonlinearity depends on the size of k and the actual quadratic partition of the prime and the cyclotomic numbers. Even if the cyclotomic numbers of order $2k$ are quite flat, a large k may lead to relatively bad nonlinearity of the cryptographic function. From this point of view, only those generators derived from small k are cryptographically attractive.

Let $N = 2kf+1$ be a prime. If $kf \equiv 0 \pmod 4$, then 2 is never a primitive root of N. Thanks to Basic Theorem 3.3.1 the linear and sphere complexity of the output sequence of the cyclotomic generator of order $2k$ can be controlled by choosing a prime $N = 2kf+1$ such that $\text{ord}_N(2)$ is large enough.

Cyclotomic generator of order 4

The cyclotomic generator of order 4 is especially interesting. We now show that the cryptographic function of (8.1) has good nonlinearity when $k=2$. In this case $N = 4f+1$. In Section 4.3.3 we saw that the cyclotomic numbers of order 4 are roughly flat, though the stability depends on the actual decomposition of $N = x^2+4y^2$ with $x \equiv 1 \pmod 4$.

Let D_0, D_1, D_2, D_3 be the cyclotomic classes defined in Section 4.1. Then the set of difference parameters $\{d_D(i,j;r) : i,j = 0,1,2,3, \ r \in Z_N\}$ with respect to the partition of Z_N^* is actually the set of cyclotomic numbers of order 4. Let

$$E_0 = D_0 \cup D_2, \quad E_1 = D_1 \cup D_3.$$

Then $\{E_0, E_1\}$ is a partition of Z_N^*. By simple arguments we have

$$\begin{aligned} d_E(0,1;r) &= |E_0 \cap (E_1 + r)| \\ &= d_D(0,1;r) + d_D(2,1;r) + d_D(0,3;r) + d_D(2,3;r) \end{aligned}$$

and

$$\begin{aligned} d_E(0,0;r) &= d_D(0,0;r) + d_D(0,2;r) + d_D(2,0;r) + d_D(2,2;r), \\ d_E(1,0;r) &= d_D(1,0;r) + d_D(1,2;r) + d_D(3,0;r) + d_D(3,2;r), \\ d_E(1,1;r) &= d_D(1,1;r) + d_D(1,3;r) + d_D(3,1;r) + d_D(3,3;r). \end{aligned}$$

Since the cyclotomic numbers of order 4 are roughly flat, the partition $\{E_0, E_1\}$ has the ideal difference property. Similarly, we have the same conclusion if E_0 is the union of any two cyclotomic classes and E_1 is that of the other two cyclotomic classes. Thus, the function $F_2(x) = (x^{(N-1)/4} \bmod N) \bmod 2$ has good nonlinearity.

Cyclotomic generator of order 2

When $k = 1$ we have the cyclotomic generator of order 2. This is one of the most interesting cyclotomic generators. We now give a brief description of some of the cryptographic properties of this generator. Detailed analysis will be given in Chapter 9.

A Sophie Germain prime p is one such that both p and $2p+1$ are prime. For a Sophie Germain prime p, $N = 2p + 1$ must be of the form $4t - 1$. Let $N = 4t - 1 = 2p + 1$ with t being odd, where p is a Sophie Germain prime. Proposition 3.4.7 shows that 2 is a primitive root of N. Furthermore, by Corollary 3.4.11 for any sequence s^∞ of period N over $GF(2)$, we have

1. $\mathrm{L}(s^\infty) = N$ or $N - 1$;
2. $\mathrm{SC}_k(s^\infty) = \begin{cases} N \text{ or } N-1, & \text{if } k < \min\{\mathrm{WH}(s^N), N - \mathrm{WH}(s^\infty)\}; \\ 0, & \text{otherwise.} \end{cases}$

This ensures large linear and sphere complexity for such sequences.

According to Proposition 4.3.3 the quadratic residues modulo a prime $N = 4t-1$ form an $(N, (N-1)/2, (N-3)/4)$ difference set. When the prime N for this generator is chosen of the form $4t - 1$ the generator is called a DSC (difference-set characterized) generator; otherwise it is referred to as an ADSC (almost difference-set characterized) generator since the set of quadratic residues modulo $N = 4t + 1$ form an almost difference set [122, 123].

If $N = 4t - 1 = 2p + 1$ is chosen such that p is a Sophie Germain prime with t being odd, the DSC generator has the following cryptographic attributes: its output sequences have the maximum linear and sphere complexity, the best autocorrelation property, the best period stability; and the cryptographic function $f(x)$ has the best nonlinearity and the worst linear approximation with respect to the additions of Z_N and Z_2. These properties follow from results of Chapters 3, 4, 5 and 7. It will be proven in Chapter 9 that the generator is computationally secure against some decision-tree based attacks (see Section 9.4).

In 1830 Stern [110] proved that if $N = 4t + 1$ is a prime such that t is also a prime, then 2 is a primitive root modulo N. Due to the cryptographic significance of this result, we call such primes *Stern primes*. If we choose

N to be a Stern prime in the cyclotomic generator of order 2, Corollary 3.4.10 and Proposition 4.3.3 (i.e., the almost-difference-set property) show that the ADSC generator has almost the same cryptographic properties as the DSC generator.

So far only a few classes of almost difference sets have been found, i.e., the quadratic residue almost difference sets for primes of form $4t+1$ and some biquadratic residue almost difference sets as well as some octic residue almost difference sets. For our application, we need other almost difference sets such that their characteristic functions can be realized efficiently.

Research Problem 8.1.1 *Find large Stern primes.*

Research Problem 8.1.2 *Find other almost difference sets with large parameters.*

By Basic Theorem 3.3.1 a prime N such that $\mathrm{ord}_N(2)$ is large enough suffices to control the linear and sphere complexity of the output sequence of the cyclotomic generator of order 2. In Chapter 9 we will analyze the cyclotomic generator of order 2 in detail.

8.2 Two-Prime Generator of Order 2

Suppose that $\gcd(p-1, q-1) = 2$ and recall the definition of Whiteman's generalized cyclotomy of order 2 described in Section 4.4. Let D_0 and D_1 denote the two cyclotomic classes.

Define

$$P = \{p, 2p, \cdots, (q-1)p\}, \quad Q = \{q, 2q, \cdots, (p-1)q\}.$$
$$R = \{0\}, \quad C_0 = R \cup Q \cup D_0, \quad C_1 = P \cup D_1.$$

Then

$$C_0 \cup C_1 = Z_{pq}, \quad C_0 \cap C_1 = \emptyset.$$

The generalized cyclotomic binary sequence s^∞ of order 2 with respect to the primes p and q is defined as

$$s_i = \left\{ \begin{array}{ll} 0, & \text{if } (i \bmod N) \in C_0; \\ 1, & \text{if } (i \bmod N) \in C_1. \end{array} \right\} \text{ for all } i \geq 0.$$

In Section 4.4.2 we have proved that the sequence can be expressed as $s_i = F(i \bmod N)$ with

$$F(i) = \left\{ \begin{array}{ll} 0, & i \in R \cup Q; \\ 1, & i \in P; \\ \left(1 - \left(\frac{i}{p}\right)\left(\frac{i}{q}\right)\right)/2, & \text{otherwise.} \end{array} \right\} \text{ for all } 0 \leq i \leq N-1, \quad (8.2)$$

where $\left(\frac{a}{p}\right)$ denotes the Legendre symbol.

The *two-prime generator* is defined by

$$s(i_0)_i = F((i_0 + i) \bmod N) \text{ for all } i \geq 0,$$

where $0 \leq i_0 \leq N - 1$ is the key of this generator, $N = pq$, and $F(x)$ is defined above. Thus, the output sequences of the two-prime generator are shift-versions of the above generalized cyclotomic binary sequence s^∞. So we need only to compute its linear complexity.

Computation of the Linear Complexity

The following lemma is straightforward by Propositions 2.3.1 and 2.3.2.

Lemma 8.2.1 *Let s^∞ be a sequence of period n over a field F, and*

$$S^n(x) = s_0 + s_1 x + \cdots + s_{n-1} x^{n-1}.$$

Then

1. *the minimal polynomial of s^∞ is given by*

$$(x^n - 1)/\gcd(x^n - 1, S^n(x)); \tag{8.3}$$

2. *the linear complexity of s^∞ is given by*

$$n - \deg(\gcd(x^n - 1, S^n(x))). \tag{8.4}$$

Proof: It is left as an exercise. □

To compute the linear complexity of the sequence s^∞, we need a number of lemmas.

Lemma 8.2.2 *Let g, e, d, D_0 and D_1 be the same as in Section 4.4.1. Let x be defined by the congruences in (4.4). Then*

1. $\mathrm{ord}_N(g) = e$, *where* $\mathrm{ord}_N(g)$ *denotes the order of g modulo N.*

2. *D_0 is a group with respect to the integer multiplication modulo pq.*

3. *If $a \in D_0$ then $aD_1 = D_1$ and $aD_0 = D_0$; if $a \in D_1$ then $aD_1 = D_0$ and $aD_0 = D_1$.*

Proof: Since g is a common primitive root of both p and q, by the Chinese Remainder Theorem

$$\begin{aligned} \mathrm{ord}_N(g) &= \mathrm{lcm}(\mathrm{ord}_p(g), \mathrm{ord}_q(g)) \\ &= \mathrm{lcm}(p-1, q-1) \\ &= (p-1)(q-1)/d = e. \end{aligned}$$

This proves part one.

The second part follows easily from part one and the definition of D_0.

Since $x \in Z_N^*$, there must exist an integer u with $0 \le u \le e-1$ such that $x^2 = g^u$. If $a \in D_1$, there must exist a v such that $a = g^v x$. It follows that

$$\begin{aligned} aD_1 &= \{g^{s+v}x^2 : s = 0, 1, \cdots, e-1\} \\ &= \{g^{s+v+u} : s = 0, 1, \cdots, e-1\} \\ &= D_0. \end{aligned}$$

The remaining parts can be similarly proved. □

Let m be the order of 2 modulo N. Then the field $GF(2^m)$ has an Nth primitive root of unity. Define

$$\begin{aligned} S(x) &= \sum_{i \in C_1} x^i \\ &= \left(\sum_{i \in P} + \sum_{i \in D_1} \right) x^i \in GF(2)[x]. \end{aligned}$$

By (8.3) we now compute $\gcd(x^N - 1, S(x))$. To this end, we need some auxiliary results.

Let θ be a primitive Nth root of unity in $GF(2^m)$. We have then

$$0 = \theta^N - 1 = (\theta^p)^q - 1 = (\theta^p - 1)(1 + \theta^p + \theta^{2p} + \cdots + \theta^{(q-1)p}).$$

It follows that

$$\theta^p + \theta^{2p} + \cdots + \theta^{(q-1)p} = 1. \tag{8.5}$$

By symmetry we get

$$\theta^q + \theta^{2q} + \cdots + \theta^{(p-1)q} = 1. \tag{8.6}$$

Lemma 8.2.3 *Let the symbols be the same as before. Then*

$$\sum_{i\in D_1}\theta^{ai}=\begin{cases}(\frac{p-1}{2} \bmod 2), & \text{if } a\in P;\\ (\frac{q-1}{2} \bmod 2), & \text{if } a\in Q;\end{cases}$$

Proof: Suppose that $a\in Q$. Since g is a common primitive root of both p and q and the order of g modulo N is e, by the definition of x in (4.4) we have

$$\begin{aligned}D_i \bmod p &= \{g^s x^i \bmod p : s=0,1,\cdots,e-1\}\\ &= \{g^{s+i} \bmod p : s=0,1,\cdots,e-1\}\\ &= \{1,2,\cdots,p-1\}.\end{aligned}$$

When s ranges over $\{0,1,\cdots,e-1\}$, $g^s x^i \bmod p$ takes on each element of $\{1,2,\cdots,p-1\}$ $(q-1)/2$ times. It follows from (8.6) that

$$\begin{aligned}\sum_{j\in D_1}\theta^{aj} &= \left(\frac{q-1}{2} \bmod 2\right)\sum_{j\in Q}\theta^j\\ &= \left(\frac{q-1}{2} \bmod 2\right).\end{aligned}$$

The second part follows by symmetry. □

Lemma 8.2.4 *Let the symbols be the same as before. We have*

$$S(\theta^a)=\begin{cases}S(\theta), & a\in D_0;\\ S(\theta)+1, & a\in D_1;\\ 1+(\frac{p-1}{2} \bmod 2), & a\in P;\\ (\frac{q-1}{2} \bmod 2), & a\in Q.\end{cases}$$

Proof: By Lemma 8.2.2, $aD_0=D_0$ if $a\in D_0$. If $a\in D_0$, $aP=P$ since $\gcd(a,q)=1$. Hence

$$\begin{aligned}S(\theta^a) &= \sum_{i\in P}\theta^{ai}+\sum_{i\in D_1}\theta^{ai}\\ &= \sum_{i\in P}\theta^{i}+\sum_{j\in aD_1}\theta^{j}\\ &= \sum_{i\in P}\theta^{i}+\sum_{j\in D_1}\theta^{j}\\ &= S(\theta).\end{aligned}$$

If $a \in D_1$, by Lemma 8.2.2 $aD_1 = D_0$. Note that

$$\left(\sum_{i\in D_0} + \sum_{i\in D_1} + \sum_{i\in P} + \sum_{i\in Q}\right)\theta^i + 1 = \sum_{i=0}^{N-1}\theta^i = 0.$$

From this, (8.5), (8.6) and Lemma 8.2.2 we obtain

$$\begin{aligned} S(\theta^a) &= \sum_{i\in P}\theta^{ai} + \sum_{i\in D_1}\theta^{ai} \\ &= \sum_{i\in P}\theta^i + \sum_{j\in aD_1}\theta^j \\ &= \sum_{i\in P}\theta^i + \sum_{j\in D_0}\theta^j \\ &= S(\theta) + 1 + \sum_{i\in Q}\theta^i + \sum_{i\in P}\theta^i \\ &= S(\theta) + 1. \end{aligned}$$

If $a \in P$, then $aP = P$ since $\gcd(p, q) = 1$. Then by Lemma 8.2.3

$$\begin{aligned} S(\theta^a) &= \sum_{i\in P}\theta^{ai} + \sum_{i\in D_1}\theta^{ai} \\ &= \sum_{i\in P}\theta^i + \sum_{i\in D_1}\theta^{ai} \\ &= 1 + \sum_{i\in D_1}\theta^{ai} \\ &= 1 + \left(\frac{p-1}{2} \bmod 2\right). \end{aligned}$$

If $a \in Q$, then $aP = \{0\}$. Then by Lemma 8.2.3

$$\begin{aligned} S(\theta^a) &= \sum_{i\in P}\theta^{ai} + \sum_{i\in D_1}\theta^{ai} \\ &= [(q-1) \bmod 2] + \sum_{i\in D_1}\theta^{ai} \\ &= \sum_{i\in D_1}\theta^{ai} \\ &= \left(\frac{q-1}{2} \bmod 2\right). \end{aligned}$$

This completes the proof of the lemma. □

Note also that

$$S(1) = [q - 1 + (p-1)(q-1)/2] \bmod 2 = 0. \tag{8.7}$$

Lemma 8.2.5 $2 \in D_0$ *if and only if* $S(\theta) \in \{0, 1\}$.

Proof: By (8.5) and definition

$$\begin{aligned} S(\theta) &= \sum_{i \in P} \theta^i + \sum_{i \in D_1} \theta^i \\ &= 1 + \sum_{i \in D_1} \theta^i. \end{aligned}$$

If $2 \in D_0$ then by Lemma 8.2.2 $2D_i = D_i$. Thus by Lemma 8.2.4

$$S(\theta)^2 = S(\theta^2) = S(\theta).$$

Hence, $S(\theta) = 0$ or 1.

If $2 \in D_1$, then we have similarly by Lemma 8.2.4

$$S(\theta)^2 = S(\theta^2) = 1 + S(\theta).$$

Hence, $S(\theta) \notin \{0, 1\}$.

Since $2 \in D_0 \cup D_1$, we have completed the proof. □

In the sequel, we need the following Generalized Chinese Remainder Theorem [134].

Lemma 8.2.6 *Let m be the least common multiple of two positive integers m_1 and m_2. The system of congruences*

$$x \equiv a_1 \pmod{m_1}, \quad x \equiv a_2 \pmod{m_2} \tag{8.8}$$

has solutions if and only if

$$\gcd(m_1, m_2) | a_1 - a_2, \tag{8.9}$$

where $a|b$ means that a divides b. When the condition (8.9) holds, the system of congruences of (8.8) has only one solution modulo m.

Proof: It is left as an exercise. □

Lemma 8.2.7 *We have* $2 \in D_0$ *if and only if* $p \equiv \pm 1 \pmod 8$ *and* $q \equiv \pm 1 \pmod 8$ *or* $p \equiv \pm 3 \pmod 8$ *and* $q \equiv \pm 3 \pmod 8$.

Proof: Assume that $2 \in D_0$. By definition there is an integer s with $0 \le s \le e-1$ such that $2 \equiv g^s \bmod pq$. It follows that

$$g^s \equiv 2 \pmod p, \quad g^s \equiv 2 \pmod q.$$

If s is even, then 2 is a quadratic residue modulo both p and q. Hence $p \equiv \pm 1 \pmod 8$ and $q \equiv \pm 1 \pmod 8$.

If s is odd, then 2 is a quadratic nonresidue modulo both p and q. Hence $p \equiv \pm 3 \pmod 8$ and $q \equiv \pm 3 \pmod 8$. This proves the necessity.

If $p \equiv \pm 1 \pmod 8$ and $q \equiv \pm 1 \pmod 8$, then 2 is a quadratic residue modulo both p and q. Thus, there are even s_1 and even s_2 with $0 \le s_1 \le p-1$ and $0 \le s_2 \le q-1$ such that

$$g^{s_1} \equiv 2 \pmod p, \quad g^{s_2} \equiv 2 \pmod q. \tag{8.10}$$

Note that $\gcd(p-1, q-1) = 2$ and s_1 and s_2 both are even. By the Generalized Chinese Remainder Theorem described in Lemma 8.2.6, there is an integer s with $0 \le s \le e-1$ such that

$$s \equiv s_1 \pmod{p-1}, \quad s \equiv s_2 \pmod{q-1}.$$

Hence, $g^s \equiv 2 \pmod{pq}$, and $2 \in D_0$.

If $p \equiv \pm 3 \pmod 8$ and $q \equiv \pm 3 \pmod 8$, we can similarly prove that $2 \in D_0$. □

Let θ be a pqth root of unity. Among the pq pqth roots of unity θ^i, where $0 \le i \le pq-1$, the q elements θ^i, $i \in P \cup R$, are qth roots of unity, the p elements θ^i, $i \in Q \cup R$, are pth roots of unity. Hence,

$$x^p - 1 = \prod_{i \in Q \cup R} (x - \theta^i), \quad x^q - 1 = \prod_{i \in P \cup R} (x - \theta^i).$$

Let

$$d(x) = \prod_{i \in D_0 \cup D_1} (x - \theta^i).$$

It follows that

$$x^{pq} - 1 = \prod_{i=0}^{pq-1} (x - \theta^i) = \frac{(x^p - 1)(x^q - 1)}{x - 1} d(x),$$

where $d(x) \in GF(2)[x]$.

In the sequel let L and $m(x)$ denote the linear complexity and minimal polynomial of our generalized cyclotomic sequence of order 2 with respect to the two primes p and q.

Theorem 8.2.8 [124, 125]

(I) *If* $p \equiv 1 \pmod 8$ *and* $q \equiv 3 \pmod 8$ *or* $p \equiv -3 \pmod 8$ *and* $q \equiv -1 \pmod 8$, *then*

$$L = pq - 1, \quad m(x) = \frac{x^{pq} - 1}{x - 1}.$$

(II) *If* $p \equiv -1 \pmod 8$ *and* $q \equiv 3 \pmod 8$ *or* $p \equiv 3 \pmod 8$ *and* $q \equiv -1 \pmod 8$, *then*

$$L = (p - 1)q, \quad m(x) = \frac{x^{pq} - 1}{x^q - 1}.$$

(III) *If* $p \equiv -1 \pmod 8$ *and* $q \equiv -3 \pmod 8$ *or* $p \equiv 3 \pmod 8$ *and* $q \equiv 1 \pmod 8$, *then*

$$L = pq - p - q + 1, \quad m(x) = \frac{(x^{pq} - 1)(x - 1)}{(x^p - 1)(x^q - 1)}.$$

Proof: We have $\gcd(p - 1, q - 1) = 2$, so by Lemma 8.2.7 the six cases described in this theorem are the only ones such that $2 \notin D_0$.

In the two cases of (I), by Lemma 8.2.4

$$S(\theta^a) = \begin{cases} 0, & a = 0 \quad \text{(by (8.7))}, \\ \neq 0, & a \in D_0 \cup D_1 \quad \text{(by Lemma 8.2.7)}, \\ 1, & a \in P \cup Q. \end{cases}$$

Hence, $\gcd(x^{pq} - 1, S(x)) = x - 1$. It follows that

$$\begin{aligned} m(x) &= \frac{x^{pq} - 1}{\gcd(x^{pq} - 1, S(x))} = \frac{x^{pq} - 1}{x - 1}, \\ L &= \deg(m(x)) = pq - 1. \end{aligned}$$

In the two cases of (II), by Lemma 8.2.4

$$S(\theta^a) = \begin{cases} 0, & a = 0 \quad \text{(By (8.7))}, \\ \neq 0, & a \in D_0 \cup D_1 \quad \text{(By Lemma 8.2.7)}, \\ 0, & a \in P, \\ 1, & a \in Q. \end{cases}$$

Hence, $\gcd(x^{pq}-1, S(x)) = x^q - 1$. It follows that

$$\begin{aligned} m(x) &= \frac{x^{pq}-1}{\gcd(x^{pq}-1, S(x))} = \frac{x^{pq}-1}{x^q-1}, \\ L &= \deg(m(x)) = pq - q = (p-1)q. \end{aligned}$$

In the two cases of (III), by Lemma 8.2.4

$$S(\theta^a) = \begin{cases} 0, & a = 0 \quad \text{(by (8.7))}, \\ \neq 0, & a \in D_0 \cup D_1 \quad \text{(by Lemma 8.2.7)}, \\ 0, & a \in P \cup Q, \end{cases}$$

Hence, $\gcd(x^{pq}-1, S(x)) = (x^p-1)(x^q-1)/(x-1)$. It follows that

$$\begin{aligned} m(x) &= \frac{x^{pq}-1}{\gcd(x^{pq}-1, S(x))} = \frac{(x^{pq}-1)(x-1)}{(x^p-1)(x^q-1)}, \\ L &= \deg(m(x)) = pq - p - q + 1. \end{aligned}$$

□

Define

$$d_a(x) = \prod_{i \in D_a} (x - \theta^i), \quad a = 0, 1.$$

In case $2 \in D_0$, by Lemma 8.2.2 we have $2D_0 = D_0$ and

$$\begin{aligned} d_0(x)^2 &= \prod_{i \in D_0} (x^2 - \theta^{2i}) \\ &= \prod_{j \in 2D_0} (x^2 - \theta^j) \\ &= \prod_{j \in D_0} (x^2 - \theta^j) \\ &= d_0(x^2). \end{aligned}$$

Hence, $d_0(x) \in GF(2)[x]$. Similarly, we can prove that $d_1(x) \in GF(2)[x]$.

By definition

$$d(x) = d_0(x)d_1(x).$$

Thus, in the case $2 \in D_0$ we get

$$x^{pq} - 1 = \frac{(x^p-1)(x^q-1)d_0(x)d_1(x)}{x-1}. \tag{8.11}$$

Note that $d_0(x)$ and $d_1(x)$ depend on the choice of θ. However, by Lemmas 8.2.4 and 8.2.5, exactly one of $S(\theta)$ and $S(\theta^a)$ is zero, where a is an element of D_1. Thus, we can choose our θ such that $S(\theta) = 0$. With this choice the polynomials $d_0(x)$ and $d_1(x)$ are uniquely determined.

Theorem 8.2.9 [124, 125]

(I) *If* $p \equiv 1 \pmod 8$ *and* $q \equiv -1 \pmod 8$ *or* $p \equiv -3 \pmod 8$ *and* $q \equiv 3 \pmod 8$, *then*

$$L = \frac{pq+p+q-3}{2}, \quad m(x) = \frac{x^{pq}-1}{(x-1)d_0(x)}.$$

(II) *If* $p \equiv -1 \pmod 8$ *and* $q \equiv 1 \pmod 8$ *or* $p \equiv 3 \pmod 8$ *and* $q \equiv -3 \pmod 8$, *then*

$$L = \frac{(p-1)(q-1)}{2}, \quad m(x) = d_1(x).$$

(III) *If* $p \equiv -1 \pmod 8$ *and* $q \equiv -1 \pmod 8$ *or* $p \equiv 3 \pmod 8$ *and* $q \equiv 3 \pmod 8$, *then*

$$L = \frac{(p-1)(q+1)}{2}, \quad m(x) = \frac{(x^p-1)d_1(x)}{x-1}.$$

Proof: By Lemma 8.2.7 there are eight cases such that $2 \in D_0$, but two of them do not satisfy $\gcd(p-1, q-1) = 2$. It is easy to check that the six cases described in this theorem are the only ones such that $2 \in D_0$ and $\gcd(p-1, q-1) = 2$.

In the two cases of (I), by Lemma 8.2.4

$$S(\theta^a) = \begin{cases} 0, & a = 0 & \text{(by (8.7))}, \\ 0, & a \in D_0 & \text{(by the choice of } \theta), \\ 1, & a \subset D_1 & \text{(by the choice of } \theta), \\ 1, & a \subset P \cup Q & \text{(by Lemma 8.2.4)}. \end{cases}$$

Hence,

$$\begin{aligned} &\gcd(x^{pq}-1, S(x)) = (x-1)d_0(x), \\ &m(x) = \frac{x^{pq}-1}{\gcd(x^{pq}-1, S(x))} = \frac{x^{pq}-1}{(x-1)d_0(x)}, \\ &L = \deg(m(x)) = pq - 1 - (p-1)(q-1)/2 = (pq+p+q-3)/2. \end{aligned}$$

In the two cases of (II), by Lemma 8.2.4

$$S(\theta^a) = \begin{cases} 0, & a = 0 & \text{(by (8.7))}, \\ 0, & a \in D_0 & \text{(by the choice of } \theta), \\ 1, & a \in D_1 & \text{(by the choice of } \theta), \\ 0, & a \in P \cup Q & \text{(by Lemma 8.2.4)}. \end{cases}$$

Hence,

$$\gcd(x^{pq} - 1, S(x)) = \frac{(x^p - 1)(x^q - 1)d_0(x)}{x - 1},$$
$$m(x) = \frac{x^{pq} - 1}{\gcd(x^{pq} - 1, S(x))} = d_1(x),$$
$$L = \deg(m(x)) = \frac{(p-1)(q-1)}{2}.$$

In the two cases of (III), by Lemma 8.2.4

$$S(\theta^a) = \begin{cases} 0, & a = 0 & \text{(by (8.7))}, \\ 0, & a \in D_0 & \text{(by the choice of } \theta), \\ 1, & a \in D_1 & \text{(by the choice of } \theta), \\ 0, & a \in P & \text{(by Lemma 8.2.4)}, \\ 1, & a \in Q & \text{(by Lemma 8.2.4)}, \end{cases}$$

Hence,

$$\gcd(x^{pq} - 1, S(x)) = (x^q - 1)d_0(x),$$
$$m(x) = \frac{x^{pq} - 1}{\gcd(x^{pq} - 1, S(x))} = \frac{(x^p - 1)d_1(x)}{x - 1},$$
$$L = \deg(m(x)) = \frac{(p-1)(q-1)}{2} + p - 1 = \frac{(p-1)(q+1)}{2}.$$

□

Autocorrelation Values

The autocorrelation values of the output sequence s^∞ of the two-prime generator are described in the following theorem.

Theorem 8.2.10 [121]

1. Let $(p-1)(q-1)/4$ be even. Then

$$\mathrm{AC}_s(w) = \begin{cases} \frac{q-p-3}{pq}, & \textit{if } w \in P; \\ \frac{p+1-q}{pq}, & \textit{if } w \in Q; \\ \frac{-1}{pq}, & \textit{if } w \in Z_N^*. \end{cases}$$

2. *Let $(p-1)(q-1)/4$ be odd. Then*

$$\mathrm{AC}_s(w) = \begin{cases} \frac{q-p-3}{pq}, & \text{if } w \in P; \\ \frac{p+1-q}{pq}, & \text{if } w \in Q; \\ \frac{-3}{pq}, & \text{if } w \in D_0; \\ \frac{1}{pq}, & \text{if } w \in D_1. \end{cases}$$

Proof: It is left as an exercise. □

This theorem shows that the autocorrelation values of this generalized cyclotomic sequence of order two are quite flat when $|p-q|$ is very small.

The best case is when $q-p=2$, i.e., they are twin primes. In this case, if $(p-1)(q-1)/4$ is even, the $\mathrm{AC}_s(w)$ is two-valued, i.e., the sequence has the best autocorrelation property. In this càse, if $(p-1)(q-1)/4$ is odd, $\mathrm{AC}_s(w)$ is four-valued.

Another interesting case is when $q-p=4$. In this case $\mathrm{AC}_s(w)$ is four-valued when $(p-1)(q-1)/4$ is even, and three-valued when $(p-1)(q-1)/4$ is odd. In the case $q-p=4$ this sequence has also good autocorrelation property.

Hardware Implementation

By the Chinese Remainder Theorem the two-prime generator of order 2 can be implemented in hardware as in Figure 8.1, where CC1 and CC2 denote two cyclic counters that count the numbers $\{0, 1, 2, \cdots, p-1\}$ and $\{0, 1, 2, \cdots, q-1\}$ cyclically, respectively, and within CC1 and CC2 there are registers R1 and R2 that store the current counted number. The initial contents k_1 and k_2 of the two registers form the key of this generator, i.e., $k = (k_1, k_2)$, where $0 \le k_1 \le p-1$ and $0 \le k_2 \le q-1$. Cyclic counters are very efficient and frequently seen in modern electronic devices. In Figure 8.1 MEC1 and MEC2 are two special chips for modular exponentiation with respect to p and q respectively. They are similar to RSA chips, and can also be made relatively efficient as the two primes here are much smaller than those for an RSA public-key cryptosystem. Here we use primes having about 46 bits, while in RSA at least 512-bit primes are needed. MEC1 and MEC2 compute $x^{(p-1)/2} \bmod p$ and $y^{(q-1)/2} \bmod q$ respectively. The u_0 and v_0 denote the least significant bits of the output numbers of MEC1 and MEC2 respectively, and u_1 and v_1 the next bits of the output numbers of MEC1 and MEC2, respectively. The symbol below u_0 and u_1 denotes bit complementation. Finally, $\otimes$ and $\oplus$ denote the binary multiplier and adder that realize the multiplication and addition of $GF(2) = \{0, 1\}$.

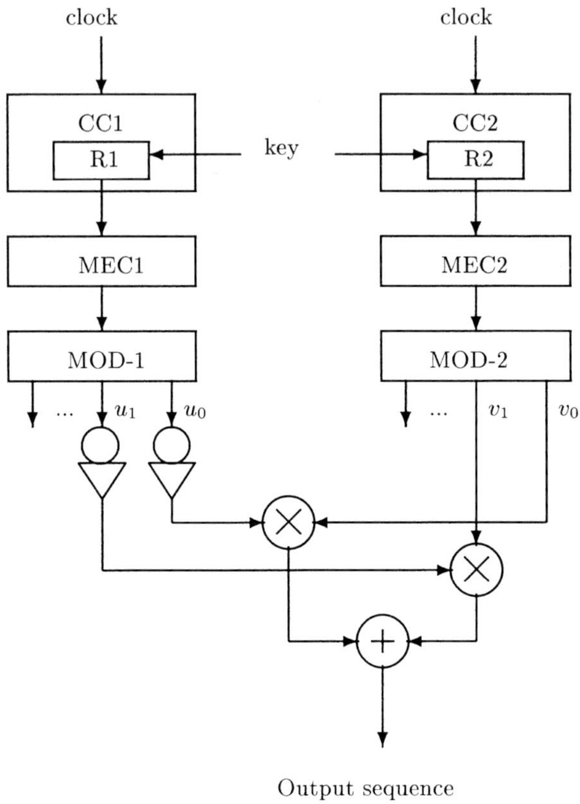

Figure 8.1: Hardware implementation of the two-prime generator of order 2.

Table 8.1: The relations.

(u, v)	(u_1, u_0, v_1, v_0)	s_j
$(1, 1)$	$(0, 1, 0, 1)$	0
$(1, q-1)$	$(0, 1, 1, 0)$	1
$(p-1, 1)$	$(1, 0, 0, 1)$	1
$(p-1, q-1)$	$(1, 0, 1, 0)$	0
$(0, 1)$	$(0, 0, 0, 1)$	1
$(0, q-1)$	$(0, 0, 1, 0)$	1
$(1, 0)$	$(0, 1, 0, 0)$	0
$(p-1, 0)$	$(1, 0, 0, 0)$	0
$(0, 0)$	$(0, 0, 0, 0)$	0

The correctness of this implementation is proved as follows. For each j, $0 \leq j \leq pq - 1$, let

$$\begin{aligned} u &= j^{(p-1)/2} \bmod p \in \{0, 1, p-1\}, \\ v &= j^{(q-1)/2} \bmod q \in \{0, 1, q-1\}. \end{aligned}$$

By the definitions of u_0, u_1, v_0, v_1, and s_j, we have the following correspondence depicted by Table 8.1. It is easily seen that

$$s_j = (u_1 \oplus 1) \otimes v_1 \oplus (u_0 \oplus 1) \otimes v_0.$$

Then the correctness of this implementation follows from the Chinese Remainder Theorem.

We define an additive synchronous stream cipher based on this generator as usual. For this purpose we suggest using two 48-bit primes, then the keysize is 96 bits which should be large enough as far as brute-force attack is concerned. Twin primes might be better than others. With current chips for modular exponentiation with respect to such primes, this specific cipher should be able to encrypt and decrypt at least at 30 Kbytes per second. Note that one page of English text (A4 size, ASCII) is about 3 Kbytes. Thus, such a cipher could encrypt 10 pages of English text per second. This performance may be slow for multimedia applications, but is certainly reasonable in military and diplomatic communications, where the size of a communication is usually not very large, say less than 600 pages of English text in each communication. Note that ciphering a 600-page document takes only one minute. It can also be used to encrypt and decrypt a classified large data base where performance is not so important, but secrecy is the primary concern.

When p and q are twin primes, the two-prime generator is referred to as the *twin-prime generator*. It could be among the best two-prime generators, as $|p-q|=2$.

8.3 Two-Prime Generator of Order 4

Two-prime generators based on the generalized cyclotomic numbers of order 2 were analyzed in Section 8.2. These generators are ideal keystream generators if the two primes are chosen properly. In Section 4.4.3 we saw that for two primes p and q of the form $4t+1$, the generalized cyclotomic numbers of order 4 are roughly stable. Due to these facts we now construct a generator based on the generalized cyclotomy of order 4.

Let $p = 4f+1$ and $q = 4f'+1$ with $\gcd(f, f') = 1$. Then as in Section 4.4.3 we have $d = \gcd(p-1, q-1) = 4$ and $e = 4ff'$. First, we define a cryptographic function from Z_{pq} to Z_2 by

$$F(j) = \begin{cases} 1, & j \in \{0, q, 2q, \cdots, (p-1)q\}; \\ 0, & j \in \{p, 2p, \cdots, (q-1)p\}; \\ ((j^{(q-1)/4} \bmod q) \bmod 2) \oplus & \\ ((j^{(p-1)/4} \bmod p) \bmod 2) \oplus 1, & \text{otherwise.} \end{cases}$$

It is easily seen that $F(x)$ has characteristic set $C_1 = D_i \cup D_j \cup Q$, where $Q = \{0, q, 2q, \cdots, (p-1)q\}$, D_i and D_j are two of the four cyclotomic classes defined in Section 4.4.3. Thus, the stability of the generalized cyclotomic numbers of order 4 ensures ideal nonlinearity of the above $F(x)$ with respect to the additions of Z_{pq} and Z_2.

With this $F(x)$ we build a generator based on the generalized cyclotomy of order 4, and defined by

$$s(i_0)_i = F((i_0 + i) \bmod N) \text{ for all } i \geq 0,$$

where $N = pq$ and $0 \leq i_0 \leq N-1$ is the key of this generator. The linear complexity of the output sequences can be computed by a method similar to that in Section 8.2, and the generator can be implemented in hardware in a way similar to that in Figure 8.1.

Summarizing our analysis, we conclude that the parameters of this generator should be chosen as follows:

1. p and q both are large enough with $\gcd(p-1, q-1) = 4$;
2. $|p-q|$ is very small, compared with pq;
3. both of $\text{ord}_p(2)$ and $\text{ord}_q(2)$ are large enough (see Theorem 3.8.2).

Problem 8.3.1 *Calculate the linear complexity of the output sequences of the two-prime generator of order four with the approach of Section 8.2.*

Problem 8.3.2 *Give a hardware implementation similar to Figure 8.1 for the two-prime generator of order four.*

8.4 Prime-Square Generator

It was shown in Section 3.7 that some sequences with period equal to the square of an odd prime are cryptographically attractive, since their linear and sphere complexity are easy to control. Let p be a large prime such that 2 is a primitive root of both p and p^2. Then Corollary 3.7.1 shows that for any sequence of period $N = p^2$ over Z_2, we have

1. $\mathrm{L}(s^\infty) \geq p - 1$;
2. $\mathrm{SC}_k(s^\infty) \geq p - 1$, if $k < \min\{\mathrm{WH}(s^N), N - \mathrm{WH}(s^\infty)\}$.

The two functions constructed in Section 4.5 have ideal nonlinearity with respect to the additions of Z_{p^2} and Z_2. Using these facts, we now construct a prime square generator based on the second function in Section 4.5:

$$F_C(x) = \begin{cases} 1, & x \in R; \\ (x^{p(p-1)/2} \bmod p^2) \bmod 2, & \text{otherwise}, \end{cases}$$

where $R = \{0, p, 2p, \cdots, (p-1)p\}$. With this function the prime-square generator is described by

$$s(i_0)_i = F_C(i_0 + i \bmod p^2) = ((i_0 + i)^{p(p-1)/2} \bmod p^2) \bmod 2, \quad i \geq 0,$$

where $0 \leq i_0 \leq p^2 - 1$ is the key of this generator. Other slight modifications of the values of the function $F_C(x)$ are also possible. In this way we get a slightly different generator with about the same cryptographic properties.

Computation of the Linear Complexity

In general the linear and sphere complexity and the nonlinearity of the cryptographic function can be also controlled, provided that a large prime p is chosen such that $\mathrm{ord}_p(2)$ is large enough, as shown by Basic Theorem 3.3.1. In this case $\mathrm{ord}_p(2)$ is the lower bound for both the linear and sphere complexity.

In fact, we do not need to use special properties of primes to control the linear complexity, as it can be computed exactly. The computation of the linear complexity given in [127] has technical errors, and was recomputed by Park, Hong and Chun [340].

Let g be a primitive root of p^2. Then g is also a primitive root of p. Recall that the generalized cyclotomic classes of order 2 with respect to p^2 are defined by

$$D_0 = (g^2), \quad D_1 = gD_0,$$

where (g^2) denotes the subgroup of Z_{p^2} generated by g^2. By definition the order of g modulo p^2 is $p(p-1)$. Hence,

$$D_0 \cap D_1 = \emptyset, \quad D_0 \cup D_1 = Z_{p^2}^*.$$

The generalized cyclotomic binary sequence s^∞ of order 2 with respect to p^2 is defined by

$$s_i = \left\{ \begin{array}{ll} 0, & \text{if } (i \bmod p^2) \in D_0; \\ 1, & \text{if } (i \bmod p^2) \in D_1 \cup R; \end{array} \right\}, \quad i \geq 0,$$

where R is defined to be $R = \{0, p, 2p, \cdots, (p-1)p\}$. It is not hard to see that the output sequences of the prime-square generator are shifted versions of the above sequence s^∞. So we need only to compute the linear complexity of this sequence.

To compute the linear complexity of the sequence, we need a number of lemmas.

Lemma 8.4.1 *1.* $\mathrm{ord}_{p^2}(g) = p(p-1)$.

2. D_0 is a subgroup of $Z_{p^2}^*$ with $|D_0| = p(p-1)/2$.

3. $aD_0 = \left\{ \begin{array}{ll} D_0, & a \in D_0; \\ D_1, & a \in D_1, \end{array} \right. \quad aD_1 = \left\{ \begin{array}{ll} D_1, & a \in D_0; \\ D_0, & a \in D_1. \end{array} \right.$

Proof: The first two parts follow from the definitions of g and D_0. We only need to prove the third part.

If $a \in D_0$, by definition there is an integer s such that $a = g^{2s}$. It follows that

$$\begin{aligned} aD_0 &= \{g^{2s+2t} : t = 0, 1, \cdots, p(p-1)-1\} = D_0; \\ aD_1 &= \{g^{2s+2t+1} : t = 0, 1, \cdots, p(p-1)-1\} = D_1. \end{aligned}$$

The remaining part can be similarly proved. □

Lemma 8.4.2 *Let b be any integer. Then we have*

$$D_i + bp = D_i \text{ for } i = 0, 1.$$

Proof: We first prove the lemma for the case $i = 0$. Since $|D_0 + bp| = |D_0|$, it suffices to show that $(D_0 + bp) \cap D_1 = \emptyset$. Suppose for contradiction that $(D_0 + bp) \cap D_1 \neq \emptyset$. Then there exist two integers r, s such that $g^{2r} + bp = g^{2s+1}$. So, we obtain

$$g^{2r} \equiv g^{2s+1} \pmod{p}.$$

This means that $g^{2(s-t)+1} \equiv 1 \pmod{p}$. This contradicts to the fact that the order of g modulo p is $p-1$, which is even. Therefore, $D_0 + bp = D_0$. In the case $i = 1$, the proof is similar. □

Let m be the order of 2 modulo p^2, where p is an odd prime, then there is a primitive p^2th root of unity over $GF(2^m)$, say θ.

Lemma 8.4.3 $\sum_{i \in R} \theta^i = 0$.

Proof: Note that

$$(1 - \theta^p) \sum_{i \in R} \theta^i = 1 - \theta^{p^2} = 0.$$

Since θ is a p^2th primitive root of unity, $\theta^p \neq 1$. The conclusion then follows. □

To compute the linear complexity and minimal polynomial for our generalized cyclotomic sequence, we define

$$S(x) = \left(\sum_{i \in R} + \sum_{i \in D_1} \right) x^i \in GF(2)[x].$$

To finish our work, we need to calculate $\gcd(x^{p^2} - 1, S(x))$.

Since

$$\left(\sum_{i \in R} + \sum_{i \in D_1} + \sum_{i \in D_0} \right) \theta^i = \sum_{i=0}^{p^2-1} \theta^i = 0,$$

by Lemma 8.4.3 we obtain

$$S(\theta) = \sum_{i \in D_1} \theta^i = \sum_{i \in D_0} \theta^i. \tag{8.12}$$

Lemma 8.4.4 *$S(\theta) = 0$ or 1.*

Proof: By (8.12) and Lemma 8.4.1

$$\begin{aligned} S(\theta)^2 &= \sum_{i\in D_1} \theta^{2i} = \sum_{j\in 2D_1} \theta^j \\ &= \begin{cases} \sum_{j\in D_1} \theta^j = S(\theta), & \text{if } 2 \in D_0 \\ \sum_{j\in D_0} \theta^j = S(\theta), & \text{if } 2 \in D_1. \end{cases} \end{aligned}$$

Thus, $S(\theta)(S(\theta)-1) = 0$. □

We now calculate $S(\theta)$. To this end, we use generalized cyclotomic numbers of order 2 with respect to p^2, which are defined to be

$$(i,j) = |(D_i + 1) \cap D_j|, \quad i,j = 0,1.$$

Thus, there are four cyclotomic numbers and some of them may be equal. The values for the four constants are given in Section 4.5, but we shall only use some facts about these cyclotomic numbers without using their exact values.

Lemma 8.4.5 *$-1 \in D_0$ if and only if $p \equiv 1 \pmod 4$.*

Proof: If $p = 4t+1$ for some integer t, then $g^{p(p-1)} = g^{4tp} = 1$. It follows that $(g^{2tp}-1)(g^{2tp}+1) = g^{4tp} - 1 = 0$. Because g is a primitive root of p^2, $-1 = g^{2tp} \in D_0$.

If $p = 4t+3$ for some integer t, then $g^{p(p-1)} = g^{(4t+2)p} = 1$. It follows that $(g^{(2t+1)p}-1)(g^{(2t+1)p}+1) = g^{(4t+2)p} - 1 = 0$. Because g is a primitive root of p^2, $-1 = g^{(2t+1)p} \in D_1$. □

Lemma 8.4.6 *Let p be an odd prime and $p \equiv 3 \pmod 4$ Then,*

$$|D_0 \cap (D_0 + 1)| = |D_1 \cap (D_1 + 1)|.$$

Proof: By Lemma 8.4.5, $-1 \in D_1$. Hence

$$|D_0 \cap (D_0+1)| = |-D_0 \cap (-D_0 - 1)| = |D_1 \cap (D_1 - 1)| = |(D_1+1) \cap D_1)|.$$

□

Lemma 8.4.7 *If p is an odd prime, then $S(\theta) = 0$.*

Proof: Assume that $p \equiv 1 \pmod 4$. By Lemma 8.4.5 we have $-1 \in D_0$. Then by Lemma 8.4.1 we have $-D_i = D_i$ for $i = 0, 1$ and so

$$\begin{aligned}(j,i) &= |D_i \cap (D_j + 1)| \\ &= |(D_i - 1) \cap D_j| \\ &= |D_j \cap (D_i - 1)| \\ &= |(-D_j) \cap (-D_i + 1)| \\ &= |D_j \cap (D_i + 1)| \\ &= (i,j).\end{aligned}$$

Let $u = S(\theta)$. By Lemma 8.4.5, $u = S(\theta) \in \{0,1\}$ and

$$\begin{aligned}u = u^2 &= \left(\sum_{i\in D_1} \theta^i\right)\left(\sum_{j\in D_0} \theta^j\right) \\ &= \left(\sum_{i\in D_1} \theta^{-i}\right)\left(\sum_{j\in D_0} \theta^j\right) \\ &= \sum_{i\in D_1}\sum_{j\in D_0} \theta^{j-i} \\ &= \sum_{a=1}^{p^2-1} |D_1 \cap (D_0 + a)|\theta^a \\ &= \sum_{b=1}^{p-1} |D_1 \cap (D_0 + bp)|\theta^{bp} + \\ &\quad \sum_{a\in D_0} |D_1 \cap (D_0 + a)|\theta^a \\ &\quad \sum_{a\in D_1} |D_1 \cap (D_0 + a)|\theta^a \\ &= 0 \text{ (because } |D_1 \cap (D_0 + bp)| = |D_1 \cap D_0| = 0 \text{ by Lemma 8.4.2)} + \\ &\quad \sum_{a\in D_0} |a^{-1}D_1 \cap (a^{-1}D_0 + 1)|\theta^a + \\ &\quad \sum_{a\in D_1} |a^{-1}D_1 \cap (a^{-1}D_0 + 1)|\theta^a \\ &= \sum_{a\in D_0} |D_1 \cap (D_0 + 1)|\theta^a + \\ &\quad \sum_{a\in D_1} |D_0 \cap (D_1 + 1)|\theta^a\end{aligned}$$

$$\begin{aligned} &= (0,1)\sum_{a\in D_0}\theta^a + (1,0)\sum_{a\in D_1}\theta^a \\ &= [(1,0)+(0,1)]u. \end{aligned}$$

Since we have proved that $(0,1)=(1,0)$, we have $u=[(1,0)+(0,1)]u=0$. This completes the proof of the first part.

Assume now that $p\equiv 3 \pmod 4$, then $-1\in D_1$. We have similarly

$$\begin{aligned} u=u^2 &= \left(\sum_{i\in D_1}\theta^i\right)\left(\sum_{j\in D_0}\theta^j\right) \\ &= \left(\sum_{i\in D_1}\theta^i\right)\left(\sum_{j\in D_1}\theta^{-j}\right) \quad \text{(by the fact } -D_1=D_0) \\ &= \sum_{i\in D_1}\sum_{j\in D_1}\theta^{i-j} \\ &= |D_1| + \sum_{a=1}^{p^2-1}|D_1\cap(D_1+a)|\theta^a \\ &= |D_1| + \sum_{b=1}^{p-1}|D_1\cap(D_1+bp)|\theta^{bp} + \\ &\quad \sum_{a\in D_0}|D_1\cap(D_1+a)|\theta^a + \\ &\quad \sum_{a\in D_1}|D_1\cap(D_1+a)|\theta^a \\ &= |D_1|-|D_1|+ \\ &\quad \text{(Because } |D_1\cap(D_1+bp)|=|D_1| \text{ by Lemma 8.4.2 and} \\ &\quad \sum_{b=1}^{p-1}\theta^{bp}=-1) \\ &\quad (1,1)\sum_{a\in D_0}\theta^a + (0,0)\sum_{a\in D_1}\theta^a \\ &= [(1,1)+(0,0)]u. \end{aligned}$$

Therefore, $u=0$ by Lemma 8.4.6. □

In what follows we need cyclotomic classes of order 2 with respect to p. Let $g_1=(g \bmod p)$. Then g_1 is a primitive root of p. The cyclotomic classes D_0' and D_1' of order 2 with respect to p are defined by

$$D_0' = (g_1^2), \quad D_1' = g_1 D_0',$$

where (g_1^2) denotes the subgroup generated by g_1^2 of Z_p^*. It is easily seen that

$$D_0' \cup D_1' = Z_p^*, \quad D_0' \cap D_1' = \emptyset.$$

Let $\theta_1 = \theta^p$. Then θ_1 is a pth primitive root of unity. Define

$$S'(x) = \sum_{i \in D_1'} x^i.$$

Lemma 8.4.8 *$S'(\theta_1) \in \{0,1\}$ if and only if $2 \in D_0'$ if and only if $p \equiv \pm 1 \pmod 8$.*

Proof: It is well-known that 2 is a quadratic residue if and only if $p \equiv \pm 1 \pmod 8$.

Similar to Lemma 8.4.1, we can easily prove the following:

$$aD_0' = \begin{cases} D_0', & a \in D_0'; \\ D_1', & a \in D_1', \end{cases} \quad aD_1' = \begin{cases} D_1', & a \in D_0'; \\ D_0', & a \in D_1'. \end{cases} \tag{8.13}$$

Note that

$$\sum_{j \in D_1'} \theta_1^j + \sum_{j \in D_0'} \theta_1^j + 1 = \sum_{i=0}^{p-1} \theta_1^i = 0. \tag{8.14}$$

If $2 \in D_0'$, then

$$S'(\theta_1)^2 = \sum_{i \in D_1'} \theta_1^{2i} = \sum_{i \in 2D_1'} \theta_1^i = \sum_{i \in D_1'} \theta_1^i = S'(\theta_1).$$

Hence, $S'(\theta_1) \in \{0,1\}$.

If $2 \in D_1'$, then by (8.13) and (8.14)

$$S'(\theta_1)^2 = \sum_{i \in D_1'} \theta_1^{2i} = \sum_{i \in 2D_1'} \theta_1^i = \sum_{i \in D_0'} \theta_1^i = S'(\theta_1) + 1.$$

Hence, $S'(\theta_1) \notin \{0,1\}$. □

Lemma 8.4.9

$$S(\theta^a) = \begin{cases} \frac{p+1}{2} \bmod 2, & \textit{if } a = 0; \\ S'(\theta_1) + 1, & \textit{if } a = a_1 p, \ a_1 \in D_0, \ 1 \le a_1 \le p-1; \\ S'(\theta_1), & \textit{if } a = a_1 p, \ a_1 \in D_1, \ 1 \le a_1 \le p-1; \\ S(\theta), & \textit{if } a \in Z_{p^2}^*. \end{cases}$$

Proof: By definition $R = \{0, p, 2p, \cdots, (p-1)p\}$ and

$$S(1) = p + p(p-1)/2 \bmod 2 = \frac{p+1}{2} \bmod 2.$$

Since every $a \in R \setminus \{0\}$ is of the form $a_1 p$ for some a_1, where $1 \le a_1 \le p-1$, we have

$$\begin{aligned} S(\theta^a) &= \sum_{i \in R} \theta^{a_1 p i} + \sum_{i \in D_1} \theta^{a_1 p i} \\ &= p + \sum_{i \in a_1 D_1} \theta^{p i} \\ &= \begin{cases} 1 + \sum_{i \in D_1} \theta^{p i}, & a_1 \in D_0; \\ 1 + \sum_{i \in D_0} \theta^{p i}, & a_1 \in D_1 \end{cases} \\ &= \begin{cases} 1 + p \sum_{i \in D_1'} \theta_1^i, & a_1 \in D_0; \\ 1 + p \sum_{i \in D_0'} \theta_1^i, & a_1 \in D_1 \end{cases} \\ &= \begin{cases} 1 + S'(\theta_1), & a_1 \in D_0; \\ S'(\theta_1), & a_1 \in D_1, \end{cases} \end{aligned}$$

where we have made use of (8.14) and the following facts:

$$D_1 \bmod p = D_1', \quad D_0 \bmod p = D_0'$$

and each element of D_i' is obtained p times.

If $a \in Z_{p^2}^*$, we have $aR = R$ and

$$\begin{aligned} S(\theta^a) &= \sum_{i \in R} \theta^{a i} + \sum_{i \in D_1} \theta^{a i} \\ &= \sum_{i \in aR} \theta^i + \sum_{i \in a D_1} \theta^i \\ &= \sum_{i \in R} \theta^i + \sum_{i \in a D_1} \theta^i \\ &= \begin{cases} \sum_{j \in D_1} \theta^j, & a \in D_0; \\ \sum_{j \in D_0} \theta^j, & a \in D_1 \end{cases} \\ &= S(\theta), \end{aligned}$$

where we have made use of Lemma 8.4.3 and (8.12). □

Lemma 8.4.10 *We have* $2 \in D_i$ *if and only if* $2 \in D'_i$.

Proof: Obviously, $2 \in D_i$ implies that $2 \in D'_i$. Suppose that $2 \in D'_0$ and $2 \notin D_0$. Then $2 \in D_1$. It follows that $2 \in D'_1$. Hence

$$2 \in D'_0 \cap D'_1 = \emptyset.$$

This is impossible. Thus, $2 \in D'_0$ implies that $2 \in D_0$. Similarly, $2 \in D'_1$ implies that $2 \in D_1$. □

For $i = 0$ and 1 define

$$\begin{aligned} d'_i(x) &= \prod_{a \in D'_i} (x - \theta_1^a), \\ d_i(x) &= \prod_{a \in D_i} (x - \theta^a). \end{aligned}$$

It is easily seen that

$$x^p - 1 = (x-1)d'_0(x)d'_1(x). \tag{8.15}$$

and

$$x^{p^2} - 1 = (x-1)d'_0(x)d'_1(x)d_0(x)d_1(x). \tag{8.16}$$

Lemma 8.4.11 $d'_i(x) \in GF(2)[x]$ *or* $d_i(x) \in GF(2)[x]$ *if and only if* $p \equiv \pm 1 \pmod 8$.

Proof: By Lemmas 8.1.8 and 8.4.10, $2 \in D'_0$ if and only if $2 \in D_0$ if and only if $p \equiv \pm 1 \pmod 8$. If $p \equiv \pm 1 \pmod 8$, then $2 \in D_0 \cap D'_0$. Hence, $2D_a = D_a$ and

$$\begin{aligned} d_a(x)^2 &= \prod_{j \in D_a} (x^2 - \theta^{2j}) \\ &= \prod_{j \in 2D_a} (x^2 - \theta^j) \\ &= \prod_{i \in D_a} (x^2 - \theta^j) \\ &= d_a(x^2). \end{aligned}$$

It follows that $d_a(x) \in GF(2)[x]$.

If $p \equiv \pm 3 \pmod 8$, then $2 \in D_1 \cap D_1'$. Hence, $2D_a = D_{(a+1) \bmod 2}$ and

$$\begin{aligned} d_a(x)^2 &= \prod_{j \in D_a} (x^2 - \theta^{2j}) \\ &= \prod_{j \in 2D_a} (x^2 - \theta^{j}) \\ &= \prod_{j \in D_{a+1 \bmod 2}} (x^2 - \theta^{j}) \\ &= d_{(a+1) \bmod 2}(x^2) \\ &\neq d_a(x^2). \end{aligned}$$

It follows that $d_a(x) \notin GF(2)[x]$. □

Note that $d_i'(x)$ and $d_i(x)$ depend on the choice of θ, but any new choice leads to at most an exchange between $d_0'(x)$ and $d_1'(x)$ and between $d_0(x)$ and $d_1(x)$.

Theorem 8.4.12 [340] *Let L and $m(x)$ denote the linear complexity and the minimal polynomial of our generalized cyclotomic sequence.*

1. *If $p \equiv 1 \pmod 8$, then $L = \frac{p+1}{2}$ and*

$$m(x) = \begin{cases} (x-1)d_0'(x), & S'(\theta_1) = 0; \\ (x-1)d_1'(x), & S'(\theta_1) = 1. \end{cases}$$

2. *If $p \equiv -1 \pmod 8$, then $L = \frac{p-1}{2}$ and*

$$m(x) = \begin{cases} d_0'(x), & S'(\theta_1) = 0; \\ d_1'(x), & S'(\theta_1) = 1. \end{cases}$$

3. *If $p \equiv 3 \pmod 8$, then $L = p - 1$ and*

$$m(x) = d_0'(x)d_1'(x).$$

4. *If $p \equiv -3 \pmod 8$, ten $L = p$ and*

$$m(x) = x^p - 1.$$

Proof: Similar to the proof of Lemma 8.4.10, we can prove that

$$a_1 \in D_i' \text{ if and only if } a_1 \in D_i,\ i = 0, 1, \tag{8.17}$$

where $1 \le a_1 \le p-1$.If $p \equiv 1 \pmod 8$, then $2 \in D_0 \cap D_0'$ by Lemmas 8.4.8 and 8.4.10. By Lemma 8.4.11, $d_0'(x) \in GF(2)[x]$ and $d_1'(x) \in GF(2)[x]$. By Lemma 8.4.8, $S'(\theta_1)$ or $S'(\theta_1)+1$ is zero. Then by Lemma 8.4.9 and Lemma 8.4.7 as well as (8.17), we obtain

$$S(\theta^a) = \begin{cases} 1, & a = 0; \\ S'(\theta_1)+1, & a = a_1 p, a_1 \in D_0'; \\ S'(\theta_1), & a = a_1 p, a_1 \in D_1'; \\ 0, & a \in Z_{p^2}^*. \end{cases}$$

It follows that

$$\gcd(x^{p^2-1}-1, S(x)) = \begin{cases} d_1'(x)d_0(x)d_1(x), & S'(\theta_1)=0; \\ d_0'(x)d_0(x)d_1(x), & S'(\theta_1)=1. \end{cases}$$

Hence,

$$\begin{aligned} m(x) &= \frac{x^{p^2}-1}{\gcd(x^{p^2}-1, S(x))} \\ &= \begin{cases} (x-1)d_0'(x), & S'(\theta_1)=0; \\ (x-1)d_1'(x), & S'(\theta_1)=1 \end{cases} \end{aligned}$$

and

$$L = \deg(m(x)) = \frac{p+1}{2}.$$

If $p \equiv -1 \pmod 8$, then $2 \in D_0 \cap D_0'$ by Lemmas 8.4.8 and 8.4.10. By Lemma 8.4.11, $d_0'(x) \in GF(2)[x]$ and $d_1'(x) \in GF(2)[x]$. By Lemma 8.4.8, $S'(\theta_1)$ or $S'(\theta_1)+1$ is zero. Then by Lemma 8.4.9 and Lemma 8.4.7 as well as (8.17), we obtain

$$S(\theta^a) = \begin{cases} 0, & a = 0; \\ S'(\theta_1)+1, & a = a_1 p, a_1 \in D_0'; \\ S'(\theta_1), & a = a_1 p, a_1 \in D_1'; \\ 0, & a \in Z_{p^2}^*. \end{cases}$$

It follows that

$$\gcd(x^{p^2-1}-1, S(x)) = \begin{cases} (x-1)d_1'(x)d_0(x)d_1(x), & S'(\theta_1)=0; \\ (x-1)d_0'(x)d_0(x)d_1(x), & S'(\theta_1)=1. \end{cases}$$

Hence,

$$\begin{aligned} m(x) &= \frac{x^{p^2}-1}{\gcd(x^{p^2}-1, S(x))} \\ &= \begin{cases} d_0'(x), & S'(\theta_1)=0; \\ d_1'(x), & S'(\theta_1)=1 \end{cases} \end{aligned}$$

and

$$L = \deg(m(x)) = \frac{p-1}{2}.$$

If $p \equiv 3 \pmod 8$, then $2 \notin D_0' \cup D_0$ by Lemmas 8.4.8 and 8.4.10. Again by Lemma 8.4.8, both $S'(\theta_1)$ and $S'(\theta_1)+1$ are nonzero. Then byLemma 8.4.9 and Lemma 8.4.7, we obtain

$$S(\theta^a) = \begin{cases} 0, & a = 0; \\ \neq 0, & a = a_1 p, a_1 \in D_0' \cup D_1'; \\ 0, & a \in Z_{p^2}^*. \end{cases}$$

It follows that

$$\gcd(x^{p^2} - 1, S(x)) = (x-1)d_0(x)d_1(x).$$

Hence,

$$m(x) = \frac{x^{p^2} - 1}{\gcd(x^{p^2} - 1, S(x))} = d_0'(x)d_1'(x)$$

and

$$L = \deg(m(x)) = p - 1.$$

If $p \equiv -3 \pmod 8$, then $2 \notin D_0 \cup D_0'$ by Lemmas 8.4.8 and 8.4.10. By Lemma 8.4.11,$d_0'(x) \notin GF(2)[x]$ and $d_1'(x) \notin GF(2)[x]$. By Lemma 8.4.8, both $S'(\theta_1)$ and $S'(\theta_1)+1$ are non-zero. Then by Lemma 8.4.9 and Lemma 8.4.7 as well as (8.17), we obtain

$$S(\theta^a) = \begin{cases} 1, & a = 0; \\ \neq 0, & a = a_1 p, a_1 \in D_0' \cup D_1'; \\ 0, & a \in Z_{p^2}^*. \end{cases}$$

It follows that

$$\gcd(x^{p^2} - 1, S(x)) = d_0(x)d_1(x).$$

Hence,

$$m(x) = \frac{x^{p^2} - 1}{\gcd(x^{p^2} - 1, S(x))} = x^p - 1$$

and

$$L = \deg(m(x)) = p.$$

□

8.5 Implementation and Performance

For application we are concerned with the implementation and performance of the cyclotomic generators described in this chapter. With chips for modular exponentiation (MEC), all of them can be implemented in hardware and software. The cyclotomic generator of order $2k$ and the prime-square generator can be implemented straightforward in hardware with a ring cyclic counter and MEC. When the cyclotomic generator of order $2k$ is used for additive syncronous stream ciphering, the prime for this generator should have about 120 bits for the time being. The prime for the latter generator should have about 60 bits. The two-prime generator of order four can be implemented similar to the two-prime generator.

With modern chips for modular exponentiation the additive syncronous stream ciphers with these generators as their keystream generators should encrypt and decrypt at least 30 Kbytes per second. Thus, they should encrypt and decrypt at least 10 pages (A4, ASCII) of English text per second. As made clear in Section 8.2, this is clearly a reasonable performance in military and diplomatic communications, password encryption, database encryption, and applications where the data in communication is of small amount.

These generators are clearly slow in multimedia applications. But in many applications security is the primary concern, while performance is not so important. These generators are suitable for such applications.

8.6 A Summary of Binary Cyclotomic Generators

Since we have controlled the difference property of the cryptographic functions and the linear and sphere complexity of the output sequences of the binary cyclotomic generators, the formulae in Section 2.4 and theorems and corollaries regarding the linear and sphere complexity in Chapters 3 and 4 show that these generators have the following properties:

1. the cryptographic function $f(x)$ has ideal difference property;
2. the cryptographic function $f(x)$ has ideal nonlinearity with respect to the additions of Z_N and Z_2;
3. the cryptographic function $f(x)$ has ideal autocorrelation property;
4. the affine approximation of $f(x)$ with respect to $(Z_N, +)$ and $(Z_2, +)$ makes no sense, since there are only two trivial affine functions from Z_N to Z_2 for odd N;
5. the output sequence has ideal autocorrelation property;

6. the output sequence has ideal two-bit pattern distribution property;

7. the output sequence has ideal linear and sphere complexity;

8. the mutual information $I(i;\ z_i z_{i+t-1})$ has ideal stability, here z^∞ denotes the output sequence of the NSG; and

9. the additive stream cipher system with this NSG as the keystream generator has ideal density of encryption (resp. decryption) transformations, i.e., the probability of agreement between two encryption (resp. decryption) transformations specified by two keys is approximately 1/2.

In fact we can calculate exact values of measures (such as autocorrelation values, the mutual information) for the above aspects based on the formulae in Section 2.4 if we have formulae for the difference parameters. For example measures for the above aspects for the cyclotomic generator of order 2 can be expressed exactly in terms of N, the modulus for the modulo N ring counter. The exact cryptographic properties formulated in Chapter 9 can illustrate this. If we have bounds for the difference parameters, then using the formulae in Section 2.4 gives bounds for measures on the above aspects.

In addition, the Weil Theorem (see Section A.5) and the formulae for cyclotomic numbers may indicate that the output sequences of these cyclotomic generators have a roughly ideal distribution property for any pattern with length $1 \le l \le \lfloor \log_2 N \rfloor$.

Chapter 9

Analysis of Cyclotomic Generators of Order 2

The cyclotomic generator of order 2 described in Section 8.1 is based on the function

$$f(x) = (x^{(N-1)/2} \bmod N) \bmod 2,$$

where N is a prime. If $N \equiv 3 \pmod 4$, then the corresponding generator is the DSC generator, otherwise it is the ADSC generator. We make such a distinction since the characteristic set of $f(x)$ is a difference set of Z_N if $N \equiv 3 \pmod 4$, and an almost difference set otherwise. It was pointed out in Section 8.6 that the cyclotomic generators of order 2 have many ideal cryptographic attributes. The nonlinearity of the above cryptographic function and the autocorrelation analysis of the output sequences of DSC and ADSC generators were analyzed in Chapter 6. We have also seen that the linear and affine approximation of the cryptographic function with respect to the additions of Z_N and Z_2 make no sense. It is also clear that the differential attack (see Section 4.2.3) does not work computationally for the cyclotomic generators of order 2, for the above cryptographic function $f(x)$ has optimal difference property with respect to the addition of Z_N. In this chapter we describe some other attributes precisely and carry out a security analysis for the cyclotomic generators of order 2. Our analysis for the cyclotomic generators of order 2 is roughly applicable to other cyclotomic generators. For some related randomness properties of Legendre sequences we refer to [95]. It is interesting that Legendre sums are related to the weight distribution of some circulant codes (see Helleseth [198]).

9.1 Crosscorrelation Property

The output sequences of both DSC and ADSC generators are the 0-1 version of modified Legendre sequences. Since Legendre sequences are based on either difference sets or almost difference sets, their autocorrelation property is optimal. This means that it is impossible to approximate a Legendre sequence with its delayed versions. Now a cryptographically interesting question is: What is the extent of correlation between two Legendre sequences of different prime periods?

Let s^∞ be a Legendre sequence[1] of prime period p defined by the modified Legendre symbol

$$s_i = \left(\frac{i}{p}\right)' = \begin{cases} 1, & \text{if } p|i, \\ \left(\frac{i}{p}\right), & \text{otherwise,} \end{cases}$$

where $\left(\frac{i}{p}\right)$ is the usual Legendre symbol. Let t^∞ be another Legendre sequence of prime period $q \neq p$ defined by the same modified Legendre symbol. Then the crosscorrelation of the two sequences with respect to period pq is measured by

$$\begin{aligned} \mathrm{CC}_{s,t} &= \left[\sum_{i=0}^{pq-1} \left(\frac{i}{p}\right)' \left(\frac{i}{q}\right)'\right] /pq \\ &= \left[\sum_{i=0}^{pq-1} \left(\frac{i}{p}\right) \left(\frac{i}{q}\right) + \sum_{i=1}^{q-1} \left(\frac{p}{q}\right) \left(\frac{i}{q}\right) + \sum_{i=1}^{p-1} \left(\frac{q}{p}\right) \left(\frac{i}{p}\right) + 1\right] /pq. \end{aligned}$$

It is easy to prove that

$$\sum_{x=0}^{q-1} \left(\frac{ux+v}{q}\right) = q\left(\frac{v}{q}\right)\left[1 - \left(\frac{u}{q}\right)^2\right].$$

It follows that

$$\sum_{i=1}^{q-1} \left(\frac{i}{q}\right) = \sum_{i=1}^{p-1} \left(\frac{i}{p}\right) = 0,$$

and

$$\sum_{x=0}^{pq-1} \left(\frac{x}{p}\right) \left(\frac{x}{q}\right) = \sum_{j=0}^{p-1} \sum_{i=0}^{q-1} \left(\frac{ip+j}{p}\right) \left(\frac{ip+j}{q}\right)$$

[1] Here we use the $\{1, -1\}$-version of Legendre sequence. But sometimes we refer to the $\{0, 1\}$-version.

$$\begin{aligned}
&= \sum_{j=0}^{p-1}\left(\frac{j}{p}\right)\sum_{i=0}^{q-1}\left(\frac{ip+j}{q}\right) \\
&= \sum_{j=0}^{p-1}\left(\frac{j}{p}\right)q\left(\frac{j}{q}\right)\left[1-\left(\frac{p}{q}\right)^2\right] \\
&= 0.
\end{aligned}$$

Hence, we have $\mathrm{CC}_{s,t} = 1/pq$. This result means that the crosscorrelation between two Legendre sequences defined by the modified Legendre symbol is almost minimal. Thus it is also impossible to approximate a Legendre sequence with another one.

9.2 Decimation Property

The decimation of sequences is also cryptographically interesting. Let s^∞ be a Legendre sequence defined by the modified Legendre symbol. If a is a quadratic residue modulo the prime p then it is easy to see that $s_{ai}^\infty = s_i^\infty$; otherwise we have

$$s_{ai}^\infty + s_i^\infty = \underbrace{200\cdots0}_{p}\underbrace{200\cdots0}_{p}\cdots.$$

Thus each decimation of a Legendre sequence is either the original one or almost the complement sequence of the original one. It follows that the multipliers of Legendre sequences are the quadratic residues. This fact also shows that most Legendre sequences based on Mersenne primes cannot be maximum-length sequences.

9.3 Linear Complexity

The results about the linear and sphere complexity of sequences presented in Chapters 3 and 4 are naturally true for Legendre sequences (we refer to the 0-1 version of Legendre sequences). But those results only depend on the special properties of the period. For Legendre sequences we have an exact result, as described below.

Let p be a prime. The 0-1 version of the Legendre sequence s^∞ with respect to the prime p is defined by

$$s_i = \begin{cases} 1, & \text{if } i \bmod p \text{ is a quadratic residue;} \\ 0, & \text{if } i \bmod p \text{ is a quadratic nonresidue;} \\ 0, & \text{if } i \bmod p \text{ is } 0, \end{cases}$$

for each $i \geq 0$. Here and hereafter Legendre sequences are viewed as binary sequences over the finite field $GF(2)$. The 0-1 version of a Legendre sequence could also be defined as the complement of the sequence defined above. This alternative definition would make little difference.

Let s^∞ be the Legendre sequence of period p over $GF(2)$, and define

$$S^p(x) = s_0 + s_1 x + \cdots + s_{p-1} x^{p-1}.$$

We shall need the basic result of Lemma 8.2.1.

Let Q denote the set of quadratic residues q with $0 < q \leq p-1$ and N the set of quadratic nonresidues modulo p. Before calculating the actual linear complexity, we have to mention the following basic facts:

Lemma 9.3.1 *Let β be a pth root of unity over the field $GF(2^m)$ that is the splitting field of $x^p - 1$.*

B1: *$(Q, \cdot)$ is a group with $|Q| = (p-1)/2$ and $q \cdot N = N$ for any $q \in Q$, where "$\cdot$" denotes integer multiplication modulo p.*

B2: *$S^p(\beta^q) = S^p(\beta)$ for any $q \in Q$ and $S^p(\beta^n) = 1 + S^p(\beta)$ for any $n \in N$.*

B3: *$S^p(\beta) \in \{0, 1\}$ iff $(S^p(\beta))^2 = S^p(\beta)$ iff $2 \in Q$.*

B4: *$2 \in Q$ if and only if $p = 8t \pm 1$ for some t.*

Proof: Basic fact B1 is straightforward. Since $(Q, \cdot)$ is a group, we have $qQ = Q$ and $q^{-1} \in Q$ for any $q \in Q$. Hence,

$$S^p(\beta^q) = \sum_{x \in Q} \beta^{qx} = \sum_{y \in Q} \beta^y = S^p(\beta).$$

Because $n^{-1} \in N$ for any $n \in N$ and $nQ = N$, we have

$$S^p(\beta^n) = \sum_{x \in Q} \beta^{nx} = \sum_{y \in N} \beta^y = 1 + S^p(\beta).$$

This completes the proof of B2. Since the characteristic of the field $GF(2^m)$ is 2, the first part of B3 is obvious. Since $(S^p(\beta))^2 = S^p(\beta^2)$, it then follows from B2 that $(S^p(\beta))^2 = S^p(\beta)$ if and only if $2 \in Q$. The proof of B4 can be found in many books about number theory. □

Parts 2 and 4 of the following theorem are due to Ding, Helleseth, and Shan [132] and Parts 1 and 3 due to Jungnickel [223].

Theorem 9.3.2 *Let s^∞ be the Legendre sequence of period p as before. Then*

1. *if* $p = 8t - 1$ *for some* t, *then* $L(s^\infty) = (p+1)/2$;

2. *if* $p = 8t + 1$ *for some* t, *then* $L(s^\infty) = (p-1)/2$;

3. *if* $p = 8t + 3$ *for some* t, *then* $L(s^\infty) = p$; *and*

4. *if* $p = 8t + 5$ *for some* t, *then* $L(s^\infty) = p - 1$.

Proof: For simplicity we use L_p to denote the linear complexity of the Legendre sequence with period p. Define

$$S^p(x) = \sum_{i \in Q} x^i$$

and let β be a primitive pth root of unity over the field $GF(2^m)$ that is the splitting field of $x^p - 1$. Then by (8.4) we have

$$\begin{aligned} L_p &= \deg[(x^p - 1)/\gcd(x^p - 1, S^p(x))] \\ &= p - |\{j : S^p(\beta^j) = 0,\ 0 \le j \le p-1\}|. \end{aligned} \tag{9.1}$$

The proof of the theorem is then completed by considering two cases depending on whether 2 is a quadratic residue.

We first consider the case that $2 \in Q$, which happens if and only if $p \equiv 1$ or $7 \pmod 8$ by basic fact B4 above. It follows from basic fact B3 that in this case we have $S^p(\beta) \in \{0, 1\}$ and either $S^p(\beta^q) = 0$ for all $q \in Q$ or $S^p(\beta^n) = 0$ for all $n \in N$. Since $S^p(1) = (p-1)/2 \bmod 2$, it follows that $S^p(1) = 0$ if $p \equiv 1 \pmod 8$ and $S^p(1) = 1 \ne 0$ if $p \equiv 7 \pmod 8$. Hence if $p \equiv 1 \pmod 8$ then by (9.1)

$$\begin{aligned} L_p &= p - |\{j : S^p(\beta^j) = 0,\ 0 \le j \le p-1\}| \\ &= p - (p-1)/2 - 1 = (p-1)/2 \end{aligned}$$

and if $p \equiv 7 \pmod 8$ then by (9.1)

$$L_p = p - (p-1)/2 = (p+1)/2.$$

Finally, we consider the case $2 \notin Q$, which happens if and only if $p \equiv 3$ or $5 \pmod 8$ by basic fact B4. By basic fact B3 we have $S^p(\beta) \notin \{0, 1\}$. Since $-1 = 1$, it follows from basic fact B2 that $S^p(\beta^j) \ne 0$ for all j with $0 < j \le p - 1$. Since $S^p(1) = (p-1)/2 \bmod 2$, it follows that $S(1) = 1$ if $p \equiv 3 \pmod 8$ and $S^p(1) = 0$ if $p \equiv 5 \pmod 8$. Thus, if $p \equiv 3 \pmod 8$ then by (9.1)

$$L_p = p - |\{j : S^p(\beta^j) = 0,\ 0 \le j \le p-1\}| = p$$

and if $p \equiv 5 \pmod 8$ then by (9.1)

$$L_p = p - |\{j : S^p(\beta^j) = 0,\ 0 \le j \le p-1\}| = p - 1.$$

Hence, we have completed the proof of this theorem. □

We now determine the minimal polynomial (also called the feedback polynomial) of Legendre sequences.

In the case that $2 \in Q$, let β be a primitive pth root over $GF(2^m)$ as before. Since $S^p(\beta)^2 = S^p(\beta)$, we have $S^p(\beta) = 0$ or 1. By basic fact B3 we can choose the primitive root β such that $S^p(\beta) = 1$. Because $2 \in Q$, we have $Q = 2Q$. Thus by basic fact B3

$$q(x) = \prod_{q \in Q} (x - \beta^q) \text{ and } n(x) = \prod_{n \in N} (x - \beta^n)$$

have coefficients from $GF(2)$. The polynomials $q(x)$ and $p(x)$ depend on the choice of β, but there are only two possibilities. In the sequel, we shall fix β as above, i.e. we choose β so that $S^p(\beta) = 1$.

Theorem 9.3.3 *Let s^∞ be the Legendre sequence of period p as before and $m(x)$ its minimal (feedback) polynomial. Then*

1. *if $p = 8t - 1$ for some t, then $m(x) = (x-1)q(x)$;*
2. *if $p = 8t + 1$ for some t, then $m(x) = q(x)$;*
3. *if $p = 8t + 3$ for some t, then $m(x) = x^p - 1$; and*
4. *if $p = 8t + 5$ for some t, then $m(x) = (x^p - 1)/(x - 1)$.*

Proof: We use some facts from the proof of Theorem 9.3.2 above. We consider first the case $2 \in Q$, which is equivalent to $p = 8t \pm 1$ for some t. As pointed out before, the polynomials $q(x)$ and $n(x)$ have coefficients from $GF(2)$. Obviously, we have

$$x^p - 1 = (x-1)q(x)n(x).$$

Recall that $2 \in Q$ is equivalent to $p \equiv 1$ or $7 \pmod 8$. In the case that $p \equiv 7 \bmod 8$, we have $S^p(1) = 1$ from the proof of Theorem 9.3.2. By our choice of β and the proof of Theorem 9.3.2, $S^p(\beta^q) = S(\beta) = 1 \neq 0$ for all $q \in Q$ and therefore $S^p(\beta^n) = 0$ for all $n \in N$. We obtain that

$$\gcd(x^p - 1, S^p(x)) = n(x).$$

Hence, by (8.3) we have

$$m(x) = \frac{x^p - 1}{\gcd(x^p - 1, S^p(x))} = (x-1)q(x).$$

Similarly, if $p \equiv 1 \pmod 8$, we have $S^p(1) = 0$ from the proof of Theorem 9.3.2 . By our choice of β we get $S^p(\beta^q) = 1$ for all $q \in Q$ and $S^p(\beta^n) = 0$ for all $n \in N$. It follows that

$$\gcd(x^p - 1, S^p(x)) = (x-1)n(x).$$

Hence, by (8.3) we have

$$m(x) = \frac{x^p - 1}{\gcd(x^p - 1, S^p(x))} = q(x).$$

Now we consider the case $2 \notin Q$. The proof of Theorem 9.3.2 has shown that in this case we have

$$S^p(\beta^j) \neq 0 \text{ for all } 0 < j \leq p-1$$

and also $S^p(1) = 1$ if $p \equiv 3 \pmod 8$, and $S^p(1) = 0$ if $p \equiv 5 \pmod 8$. It follows that

$$\gcd(x^p - 1, S^p(x)) = \begin{cases} 1, & p \equiv 3 \pmod 8 \\ x-1, & p \equiv 5 \pmod 8. \end{cases}$$

Thus, by (8.3)

$$m(x) = \frac{x^p - 1}{\gcd(x^p - 1, S^p(x))} = \begin{cases} x^p - 1, & p \equiv 3 \pmod 8 \\ \frac{x^p-1}{x-1}, & p \equiv 5 \pmod 8. \end{cases}$$

Hence, we have completed the proof of this theorem. □

9.4 Security against a Decision Tree Attack

In this section we analyze the security of the DSC generator with respect to a decision-tree based key-recovering attack. With a little modification the same analysis can be carried out for the ADSC generator.

Assume that a cryptanalyst knows the algorithm, and a number of successive plaintext-ciphertext bit pairs, then she has got a number of successive keystream bits $z_i z_{i+1} \cdots z_{i+v-1} = z^v$, say. The purpose of key-determining attacks is to recover the key or an equivalent key by making use of the information about the key, which is implied in the obtained ciphertext or known plaintext-ciphertext pairs. Under the assumption that the cryptographic algorithm is known to the enemy, it is cryptographically beneficial to require that the mutual information

$$I(k = a; c_1 c_2 \cdots c_n = b_1 b_2 \cdots b_n)$$

be approximately the same for as many a's in the key space K as possible, where $b_1 \cdots b_n$ is a given piece of ciphertext. At least the average mutual information

$$I(k = a; c_1 c_2 \cdots c_n)$$

for fixed n should be about the same for as many a's in K as possible. Similar requirements for the mutual information between key and known plaintext-ciphertext pairs are also imperative. We now introduce a decision-tree based attack which is based on the analysis of the mutual information between the key and the keystream sequence.

Decision-tree based attacks determine the time-varying key i, i.e., the content of the register of the DSC generator which is used to produce the first bit of the known piece of keystream z^v, by making use of the information about i contained in the keystream sequence z^v. For the DSC generator all the keys in Z_N are equally likely.

The stability analysis of the mutual information $I(i; z_i z_{i+1} \cdots z_{i+v-1} = i_0 i_1 \cdots i_{v-1})$ is not easy when $v \geq 3$. To carry out the stability analysis, we need to develop some bounds on the number of occurrences of patterns of quadratic residues and nonresidues.

Let s^∞ be the Legendre sequence of period N output by the DSC generator. Define the set

$$C_i = \{j \in Z_N : s_j = i\}$$

for each $i = 0, 1$. By the definition of Legendre sequences, C_1 is the set of quadratic residues modulo N, and $C_0 = Z_N \setminus C_1$.

Let $i_1, i_2, \cdots, i_{v-1} \in \{0, 1\}$ and $r_0, r_1, \cdots, r_{v-1}$ be v pairwise distinct elements of Z_N. Define

$$D_{i_0 \cdots i_{v-1}}(r_0, \cdots, r_{v-1}) = \cap_{k=0}^{v-1} (C_{i_k} - r_k),$$

$$d_{i_0 \cdots i_{v-1}}(r_0, \cdots, r_{v-1}) = |\cap_{k=0}^{v-1} (C_{i_k} - r_k)|.$$

Then for a fixed v and a set of fixed $r_0, \cdots, r_{v-1}$ the set

$$\{D_{i_0 \cdots i_{v-1}}(r_0, \cdots, r_{v-1}) : i_0, \cdots, i_{v-1} \in \{0, 1\}\} \tag{9.2}$$

forms a partition of Z_N. The above parameters $d_{i_0 \cdots i_{v-1}}(r_0, \cdots, r_{v-1})$ measure the number of patterns distributed in a cycle of the sequence.

Generally, the mutual information $I(i; z_i z_{i+1} \cdots z_{i+v-1} = i_0 i_1 \cdots i_{v-1})$ depends on $| \cap_{j=0}^{v-1} (D_{i_j} - j)|$ (see Section 6.4). Thus, the analysis of the mutual information is equivalent to that of the parameters $d_{i_0 \cdots i_{v-1}}(r_0, \cdots, r_{v-1})$. This is also equivalent to the distribution analysis of patterns of length v in the DSC sequences.

To ensure ideal pattern distributions of length v in the DSC sequences and a stable mutual information $I(i; z_i z_{i+r_1} \cdots z_{i+r_{v-1}} = i_0 i_1 \cdots i_{v-1})$, it is necessary and sufficient to require

$$d_{i_0 \cdots i_{v-1}}(r_0, \cdots, r_{v-1}) \approx \frac{N}{2^v}$$

for all pairwise distinct nonzero elements $r_0, r_1, \cdots, r_{v-1}$ of Z_N.

The analysis for the cases $v = 1$ and 2 shows that this condition is satisfied in these two cases. The calculation of $d_{i_0 \cdots i_{v-1}}(r_0, \cdots, r_{v-1})$ becomes quite difficult for $v \geq 3$. We now derive some bounds on the parameters. Before doing so, let us observe some relations among the parameters.

There are many relations among these parameters which can be described by

$$\sum_{i_{j_1}, \cdots, i_{j_u}} d_{i_0 \cdots i_{v-1}}(r_0, \cdots, r_{v-1}) = d_{i_{j'_1} \cdots i_{j'_{v-u}}}(r_{i_{j'_1}}, \cdots, r_{i_{j'_{v-u}}}),$$

where

$$\{j'_1 \cdots j'_{v-u}\} = \{0, 1, \cdots, v-1\} \setminus \{j_1, \cdots, j_u\}.$$

These equations give some conservation relations among these parameters.

For the distribution of patterns of length two in Legendre sequences we have the following exact result, which shows that Legendre sequences have the best possible distribution of patterns of length two. Similar results for the ordinary Legendre symbol are given by Hasse [194, pp. 149-158].

Proposition 9.4.1 *If* $N \equiv 3 \pmod 4$, *then*

$$d_{ij}(r_0, r_1) = \begin{cases} (N-3)/4, & \text{for } (i,j) = (1,1), \\ (N+1)/4, & \text{for } (i,j) \neq (1,1), \end{cases}$$

where $r = r_0 - r_1 \neq 0$.

If $N \equiv 1 \pmod 4$, *then*

$$d_{11}(r_0, r_1) = \begin{cases} (N-5)/4, & \text{for } (N-1)/2 \text{ nonzero } r \text{ of } Z_N, \\ (N-1)/4, & \text{for the remaining nonzero elements,} \end{cases}$$

$$d_{10}(r_0, r_1) = \begin{cases} (N+3)/4, & \textit{for } (N-1)/2 \textit{ nonzero } r \textit{ of } Z_N, \\ (N-1)/4, & \textit{for the remaining nonzero elements,} \end{cases}$$

$$d_{01}(r_0, r_1) = \begin{cases} (N+3)/4, & \textit{for } (N-1)/2 \textit{ nonzero } r \textit{ of } Z_N, \\ (N-1)/4, & \textit{for the remaining nonzero elements,} \end{cases}$$

$$d_{00}(r_0, r_1) = \begin{cases} (N+3)/4, & \textit{for } (N-1)/2 \textit{ nonzero } r \textit{ of } Z_N, \\ (N-1)/4, & \textit{for the remaining nonzero elements,} \end{cases}$$

where $r = r_0 - r_1 \neq 0$.

Proof: Let D_0 and D_1 denote the sets of quadratic residues and nonresidues modulo N respectively. Here 0 is neither a quadratic residue nor a quadratic nonresidue.

By definition $C_1 = D_0$ and $C_0 = D_1 \cup \{0\}$. Note that D_0 is a multiplicative group. We have

$$\begin{aligned} d_{11}(r_0, r_1) &= |(D_0 + r_0) \cap (D_0 + r_1)| \\ &= |(D_0 + r) \cap D_0| \\ &= |(D_j + 1) \cap D_j| \\ &= (j, j), \end{aligned}$$

where $r = r_0 - r_1 \neq 0$, $r^{-1} \in D_j$ for some $j \in \{0, 1\}$ and (j, j) is a cyclotomic number of order 2.

Similarly,

$$\begin{aligned} d_{01}(r_0, r_1) &= |(D_1 \cup \{0\} + r_0) \cap (D_0 + r_1)| \\ &= |(D_1 \cup \{0\} + r) \cap D_0| \\ &= |(D_{(j+1) \bmod 2} + 1) \cap D_j| + |\{r\} \cap D_0| \\ &= ((1 + j) \bmod 2, j) + |\{r\} \cap D_0|, \end{aligned}$$

where $r = r_0 - r_1 \neq 0$ and $r^{-1} \in D_j$ for some j.

With a similar argument, we have

$$d_{10}(r_0, r_1) = (j, (1 + j) \bmod 2) + |\{0\} \cap (D_0 + r)|$$

and

$$\begin{aligned} &d_{00}(r_0, r_1) \\ = \ &((j + 1) \bmod 2, (j + 1) \bmod 2) + |\{0\} \cap (D_1 + r)| + |\{r\} \cap D_1|, \end{aligned}$$

where $r = r_0 - r_1 \neq 0$ and $r^{-1} \in D_j$ for some j.

From Proposition 4.3.2, we get the following formulae: If $N \equiv 1 \pmod 4$, the cyclotomic numbers of order two are given by

$$(0, 0) = \frac{N - 5}{4}, \ (0, 1) = (1, 0) = (1, 1) = \frac{N - 1}{4}.$$

If $N \equiv 3 \pmod 4$, they are given by

$$(0,0) = (1,0) = (1,1) = \frac{N-3}{4}, \ (0,1) = \frac{N+1}{4}.$$

Note that $-1 \in D_0$ if and only if $N \equiv 1 \pmod 4$ and $r \in D_i$ if and only if $r^{-1} \in D_i$. The conclusions of this proposition then follow from the cyclotomic numbers of order 2 and the above four formulas for $d_{ij}(r_0, r_1)$. □

We note that the above proposition can also be proved with the following formulae of Jacobsthal [96]:

$$\sum_{x=0}^{N-1} \left(\frac{ax+b}{N}\right) = N\left(\frac{b}{N}\right)\left[1 - \left(\frac{a}{N}\right)^2\right],$$

$$\sum_{x=0}^{N-1} \left(\frac{ax^2+bx+c}{N}\right) = \begin{cases} (N-1)\left(\frac{a}{N}\right), & \text{if } b^2 - 4ac \equiv 0 \pmod N \\ -\left(\frac{a}{N}\right), & \text{if } b^2 - 4ac \not\equiv 0 \pmod N. \end{cases}$$

It is easy to see that the study of the parameters $d_{i_0 \cdots i_{v-1}}(r_0, \cdots, r_{v-1})$ of the quadratic residue partition is actually that of the number of occurrences of patterns of quadratic residues and nonresidues. The cases for $v = 1, 2$ have already been solved, as described above. However, they are very difficult to calculate for $v \geq 3$. According to Davenport [96], in the case $N \equiv -1 \pmod 4$, Jacobsthal proved

$$d_{000}(r, r+1, r+2), \ d_{111}(r, r+1, r+2) = \frac{N}{2^3} + O(1),$$

as $N \to \infty$; and in the case $N \equiv 1 \pmod 4$, we have

$$d_{000}(r, r+1, r+2), \ d_{111}(r, r+1, r+2) = \frac{N}{2^3} + O(\sqrt{N})$$

Hasse also gave tight bounds on $d_{i_0 i_1 i_2}(r_0, r_1, r_2)$ [194, pp. 165-167]. Some bounds on $d_{0000}(r, r+1, r+2, r+3)$ and $d_{1111}(r, r+1, r+2, r+3)$ were given by Dörg [143] and Hopf [208], which are of little value for our cryptographic purpose since they are not tight at all. Davenport [96] developed upper bounds of the form $N/2^v + N^{c_v}$ on both $d_{00\cdots 0}(r, r+1, \cdots, r+v-1)$ and $d_{11\cdots 1}(r, r+1, \cdots, r+v-1)$ for v with $4 \leq v \leq 9$. For each of these bounds, $c_v \geq \frac{2}{3}$. These bounds are also too loose to have cryptographic value. The case $v = 4$ was also discussed by Hudson [214]. Certain patterns of values of totally multiplicative functions were investigated by Walum [431]. An important development in the topic is a general lower and upper bound

$\frac{N}{2^v} \pm v(3+\sqrt{N})$ on $d_{i_0 \cdots i_{v-1}}(r_0, \cdots, r_{v-1})$ developed by Peralta [344]. These will be referred to as Peralta bounds. Peralta treated zero as a quadratic residue, while we take it as a nonresidue in order to make it easy to calculate the DSC sequence. These two kinds of treatment make little difference. The Peralta bounds can be used to estimate the pattern distribution of the DSC sequences and the above mutual information to some extent. However, they are still too loose. We now derive a bound on $d_{i_0 \cdots i_{v-1}}(r_0, r_1, \cdots, r_{v-1})$, which is much superior to Peralta's.

To this end, we need the calculation and estimation of the character sum

$$\phi_r(a_1, \cdots, a_r) = \sum_{x=0}^{N-1} \left(\frac{(x+a_1)\cdots(x+a_r)}{N} \right),$$

where $a_1, \cdots, a_r$ are pairwise distinct elements of Z_N.

The Peralta bounds are based mainly on the following estimate:

Lemma 9.4.2 *Let $r \geq 2$. Then*

$$|\phi_r(a_1, \cdots, a_r)| < (r-1)\sqrt{N},$$

where the a_i's are pairwise distinct elements of Z_N.

This result is a special case of a more general result about multiplicative characters due to Weil (see Schmidt [382, Theorem 2C, p. 43]). Following Peralta [344], we call this inequality the *Weil bound.*

We know $\phi_2(a_1, a_2)$ can be evaluated exactly, since it can be reduced to the calculation of $\phi_1(a)$. This can be generalized into the fact that $\phi_r(a_1, \cdots, a_r)$ reduces to $\phi_{r-1}(0, 1, b_1, \cdots, b_{r-3})$ if r is even, to $\phi_r(0, 1, b_1, \cdots, b_{r-2})$ if r is odd. The proof of the case $r = 4$ was given by Davenport [96]. We now generalize Davenport's proof for the case $r = 4$ to the general case of r even.

Lemma 9.4.3 *Let $r \geq 2$ be even, and let $a_1, \cdots, a_r$ be r pairwise distinct elements of Z_N. Then*

$$\phi_r(a_1, a_2, \cdots, a_r) = -1 + \left(\frac{c}{N}\right) \phi_{r-1}(d_2, \cdots, d_r),$$

where these d_i are pairwise distinct and

$$d_i = \frac{(a_i - a_2)(a_1 - a_3)}{(a_i - a_1)(a_2 - a_3)}, \quad i \geq 2,$$

$$c = (a_1 - a_2)(a_1 - a_3) \prod_{i=2}^{r} [(a_i - a_1)(a_2 - a_3)].$$

Proof: The key to such a reduction is the use of the transformation

$$x = \frac{uy+v}{gy+h},$$

with $uh - gv \not\equiv 0 \pmod{N}$, which gives

$$\sum_{x \neq u/g} F(x) = \sum_{y \neq -h/g} F\left(\frac{uy+v}{gy+h}\right),$$

where $F(x)$ is a polynomial in $Z_N[x]$. We choose now

$$u = -a_1(a_2 - a_3), \qquad v = -a_2(a_1 - a_3),$$
$$g = a_2 - a_3, \qquad h = a_1 - a_3.$$

In our case, define

$$F(x) = \left(\frac{(x+a_1)\cdots(x+a_r)}{N}\right).$$

Since $\left(\frac{x^2}{N}\right) = 1$ for all $x \not\equiv 0 \pmod{N}$, we have (using the hypothesis that r is even)

$$\left(\frac{c}{N}\right)\left(\frac{\prod_{i=2}^{r}(-h/g + d_i)}{N}\right) = \left(\frac{1}{N}\right) = 1.$$

Then applying the above transformation yields

$$\begin{aligned}
\phi_r(a_1, \cdots, a_r) &= \sum_{x \neq -a_1} F(x) \\
&= \sum_{y \neq -h/g} F\left(\frac{uy+v}{gy+h}\right) \\
&= \sum_{y \neq -h/g} \left(\frac{(gy+h)^{-r} \prod_{i=1}^{r}[(a_i g + u)y + a_i h + v]}{N}\right) \\
&- \sum_{y \neq -h/g} \left(\frac{\prod_{i=1}^{r}[(a_i g + u)y + a_i h + v]}{N}\right) \\
&= \left(\frac{c}{N}\right) \sum_{y \neq -h/g} \left(\frac{\prod_{i=2}^{r}(y + d_i)}{N}\right) \\
&= \left(\frac{c}{N}\right) \sum_{y=0}^{N-1} \left(\frac{\prod_{i=2}^{r}(y + d_i)}{N}\right) - 1 \\
&= -1 + \left(\frac{c}{N}\right) \phi_{r-1}(d_2, \cdots, d_r).
\end{aligned}$$

It can be easily proven that $d_2, \cdots, d_r$ are pairwise distinct. □

We will refer to this result as the *Davenport reduction theorem*, which will play an important role in developing the new bounds.

The following combinatorial results are needed in the sequel.

Lemma 9.4.4 *Let $n \geq 2$. Then*

1. $\sum_{i \text{ even}} \binom{n}{i} = \sum_{i \text{ odd}} \binom{n}{i} = 2^{n-1}$.
2. $\sum_{i=1}^{n} \binom{n}{i}(i-1) = 2^{n-1}(n-2)+1$.

Proof: The first part is a standard formula that is obtained from

$$(x+1)^n = \sum_{i=0}^{n} \binom{n}{i} x^i.$$

Since

$$n(x+1)^{n-1} = \sum_{i=1}^{n} i \binom{n}{i} x^{i-1},$$

we have

$$n2^{n-1} = \sum_{i=1}^{n} i \binom{n}{i}.$$

It follows that

$$\begin{aligned} \sum_{i=1}^{n}(i-1)\binom{n}{i} &= \sum_{i=1}^{n} i \binom{n}{i} - \sum_{i=1}^{n} \binom{n}{i} \\ &= n2^{n-1} - 2^n + 1. \end{aligned}$$

□

In the sequel we assume $v \geq 3$. To derive the bounds, we make use of the function

$$G(x) = \frac{1}{2^v} \prod_{k=0}^{v-1} \left[1 + (-1)^{i_k+1} \left(\frac{x - r_k}{N} \right) \right].$$

The following lemma is needed later.

Lemma 9.4.5 *Define*

$$B(t, v, i_0, \cdots, i_{v-1}) = \sum_{0 \le k_1 < \cdots < k_t \le v-1} (-1)^{\sum_{j=1}^{t}(i_{k_j}+1)},$$

where t *and* v *are positive integers with* $t \le v$ *and* $i_0, \cdots, i_{v-1} \in \{0, 1\}$. *Then*

$$|B(t, v, i_0, \cdots, i_{v-1})| \le \binom{v}{t}.$$

Since $\sum_x \left(\frac{x}{N}\right) = 0$, the following lemma is clear.

Lemma 9.4.6

$$\sum_{x=0}^{N-1} \sum_{k=0}^{v-1} (-1)^{i_k} \left(\frac{x - r_k}{N}\right) = 0.$$

As we assume that $r_0, r_1, \cdots, r_{v-1}$ are pairwise distinct, the following conclusion follows from the second Jacobsthal formula described before.

Lemma 9.4.7

$$\begin{aligned} &\sum_{x=0}^{N-1} \sum_{0 \le k_1 < k_2 \le v-1} (-1)^{i_{k_1}+i_{k_2}} \left(\frac{x - r_{k_1}}{N}\right) \left(\frac{x - r_{k_2}}{N}\right) \\ = \;& -B(2, v, i_0, \cdots, i_{v-1}). \end{aligned}$$

The following lemma derives directly from the Davenport reduction theorem.

Lemma 9.4.8 *Let* w *be even. Then*

$$\begin{aligned} &\sum_{\substack{t=4 \\ t \text{ even}}}^{w} \sum_{0 \le k_1 < \cdots < k_t \le v-1} (-1)^{\sum_{j=1}^{t}(i_{k_j}+1)} \phi_t(-r_{k_1}, \cdots, -r_{k_t}) \\ = \;& \sum_{\substack{t=4 \\ t \text{ even}}}^{w} \sum_{0 \le k_1 < \cdots < k_t \le v-1} (\;\; 1)^{\sum_{j=1}^{t}(i_{k_j}+1)} \left(\frac{c(r_{k_1}, \cdots, r_{k_t})}{N}\right) \times \\ & \phi_{t-1}(-r'_{k_1}, \cdots, -r'_{k_{t-1}}) \\ & - \sum_{\substack{t=4 \\ t \text{ even}}}^{w} B(t, v, i_0, \cdots, i_{v-1}), \end{aligned}$$

where $c(r_{k_1}, \cdots, r_{k_t}) \not\equiv 0 \pmod{N}$ *and* $r'_{k_1}, \cdots, r'_{k_{t-1}}$ *are pairwise distinct and are determined by the formula in the Davenport reduction theorem.*

By Lemmas 9.4.6 and 9.4.7 we obtain that

$$
\begin{aligned}
2^v \sum_x G(x) &= N + \sum_{t=1}^{v} \sum_{0 \le k_1 < \cdots < k_t \le v-1} (-1)^{\sum_{j=1}^{t}(i_{k_j}+1)} \phi_t(-r_{k_1}, \cdots, -r_{k_t}) \\
&= N - B(2, v, i_0, \cdots, i_{v-1}) + \\
&\quad \sum_{t=3}^{v} \sum_{0 \le k_1 < \cdots < k_t \le v-1} (-1)^{\sum_{j=1}^{t}(i_{k_j}+1)} \phi_t(-r_{k_1}, \cdots, -r_{k_t}) \\
&= \sum_{\substack{t \ge 3 \\ t \text{ odd}}} \sum_{0 \le k_1 < \cdots < k_t \le v-1} (-1)^{\sum_{j=1}^{t}(i_{k_j}+1)} \phi_t(-r_{k_1}, \cdots, -r_{k_t}) + \\
&\quad \sum_{\substack{t \ge 4 \\ t \text{ even}}} \sum_{0 \le k_1 < \cdots < k_t \le v-1} (-1)^{\sum_{j=1}^{t}(i_{k_j}+1)} \phi_t(-r_{k_1}, \cdots, -r_{k_t}) + \\
&\quad N - B(2, v, i_0, \cdots, i_{v-1}).
\end{aligned}
$$

Applying Lemma 9.4.8 and the Weil bound to the formula above yields

$$
\begin{aligned}
&\left| 2^v \sum_x G(x) - N + \sum_{\substack{t \ge 2 \\ t \text{ even}}} B(t, v, i_0, \cdots, i_{v-1}) \right| \\
\le\ & \sqrt{N} \left(\sum_{\substack{t \ge 3 \\ t \text{ odd}}} \binom{v}{t} (t-1) + \sum_{\substack{t \ge 4 \\ t \text{ even}}} \binom{v}{t} (t-2) \right) \\
=\ & \sqrt{N} \left(\sum_{t=3}^{v} \binom{v}{t} (t-1) - \sum_{\substack{t \ge 4 \\ t \text{ even}}} \binom{v}{t} \right) \\
=\ & \sqrt{N} \left(\sum_{t=1}^{v} \binom{v}{t} (t-1) - \sum_{\substack{t \ge 1 \\ t \text{ even}}} \binom{v}{t} \right) \\
=\ & \sqrt{N} \left(\sum_{t=1}^{v} \binom{v}{t} (t-1) - \sum_{\substack{t \ge 0 \\ t \text{ even}}} \binom{v}{t} + 1 \right) \\
=\ & \sqrt{N} \left(2^{v-1}(v-2) + 1 - 2^{v-1} + 1 \right) \\
=\ & \sqrt{N} \left(2^{v-1}(v-3) + 2 \right).
\end{aligned}
$$

By Lemma 9.4.5

$$\begin{aligned}\left|\sum_{\substack{t\geq 2\\ t \text{ even}}} B(t, v, i_0, \cdots, i_{v-1})\right| &\leq -1 + \sum_{\substack{0\leq t\leq v\\ t \text{ even}}} \binom{v}{t}\\ &= 2^{v-1} - 1.\end{aligned}$$

Combining the above two formulae yields

$$\frac{N}{2^v} - T \leq \sum_x G(x) \leq \frac{N}{2^v} + T, \tag{9.3}$$

where

$$T = \frac{\sqrt{N}(2^{v-1}(v-3)+2) + 2^{v-1} - 1}{2^v}.$$

Now let $\mathrm{WH}(i_0 \cdots i_{v-1})$ denote the Hamming weight of vector $(i_0 \cdots i_{v-1})$, i.e., the number of 1s in the binary vector. Assume that $\mathrm{WH}(i_0 \cdots i_{v-1}) = v - t$. Note that $r_j \in C_{i_j} + r_j$ if and only if $i_j = 0$, so there are at most t elements in $\{r_0, r_1, \cdots, r_{v-1}\}$ which belong to the set $D_{i_0\cdots i_{v-1}}(r_0, \cdots, r_{v-1})$, where $t \leq v$. We now assume that there are exactly u such elements, say $r_0, r_1, \cdots, r_u$, where $u \leq t \leq v$. Let $B = \{r_0, r_1, \cdots, r_u\}$ and $A = D_{i_0\cdots i_{v-1}}(r_0, \cdots, r_{v-1}) \setminus B$. Then we have

$$\sum_{x\in A} G(x) = d_{i_0\cdots i_{v-1}}(r_0, \cdots, r_{v-1}) - u,$$

$$\sum_{x\in B} G(x) = u/2.$$

It follows that

$$d_{i_0\cdots i_{v-1}}(r_0, \cdots, r_{v-1}) - \frac{t}{2} \leq \sum_{x\in D_{i_0\cdots i_{v-1}}} G(x) \leq d_{i_0\cdots i_{v-1}}(r_0, \cdots, r_{v-1}),$$

where $D_{i_0\cdots i_{v-1}}$ denotes $D_{i_0\cdots i_{v-1}}(r_0, \cdots, r_{v-1})$. Hence, for each $(i_0 \cdots i_{v-1})$ we have

$$d_{i_0\cdots i_{v-1}}(r_0, \cdots, r_{v-1}) - \frac{v}{2} \leq \sum_{x\in D_{i_0\cdots i_{v-1}}} G(x) \leq d_{i_0\cdots i_{v-1}}(r_0, \cdots, r_{v-1}).$$

If the Hamming distance between the vectors $(j_0 \cdots j_{v-1})$ and $(i_0 \cdots i_{v-1})$ is greater than or equal to two, it is not difficult to see that

$$\sum_{x\in D_{j_0\cdots j_{v-1}}(r_0,\cdots,r_{v-1})} G(x) = 0.$$

Furthermore, if the Hamming distance between the two vectors $(j_0 \cdots j_{v-1})$ and $(i_0 \cdots i_{v-1})$ is one, it is clear that

$$0 \leq \sum_{x \in D_{j_0 \cdots j_{v-1}}(r_0, \cdots, r_{v-1})} G(x) \leq \frac{1}{2}.$$

Note that

$$\cup_{i_0, \cdots, i_{v-1} \in \{0,1\}} D_{i_0 \cdots i_{v-1}}(r_0, \cdots, r_{v-1}) = Z_N. \tag{9.4}$$

It follows that

$$d_{i_0 \cdots i_{v-1}}(r_0, \cdots, r_{v-1}) - \frac{v}{2} \leq \sum_{x \in Z_N} G(x) \leq d_{i_0 \cdots i_{v-1}}(r_0, \cdots, r_{v-1}) + \frac{v}{2}. \tag{9.5}$$

Combining (9.3) and (9.5) proves the following result.

Proposition 9.4.9 [120] *Let the symbols and assumptions be as before and* $v \geq 3$. *We have*

$$-A \leq d_{i_0 \cdots i_{v-1}}(r_0, r_1, \cdots, r_{v-1}) - \frac{N}{2^v} \leq A,$$

where $r_0, r_1, \cdots, r_{v-1}$ *are pairwise distinct elements of* Z_N *and*

$$A = \frac{\sqrt{N}(2^{v-1}(v-3)+2) + 2^{v-1}(v+1) - 1}{2^v}.$$

Note that A is essentially $\sqrt{N}\frac{v-3}{2} + \frac{v+1}{2}$. Our new bounds are much superior to the Peralta bounds.

For the case $v = 3$, the above proof shows that

$$\left| d_{i_0 i_1 i_2}(r_0, r_1, r_2) - \frac{N}{2^3} \right| \leq \frac{2\sqrt{N} + 15}{2^3}.$$

Numerical computation shows that these lower and upper bound for $v = 3$ are quite tight.

It can be seen from the development of the bounds that the new bounds are usually tight for small v. Proposition 9.4.9 shows that Legendre sequences have an ideal distribution of patterns of length v when v is small.

For cryptographic purposes, we need better bounds. Thus, solving the following open problem is of cryptographic interest.

Research Problem 9.4.10 *Develop better bounds on* $d_{i_0 \cdots i_{v-1}}$ *for* v *with* $3 \leq v \leq \lfloor \log_2 N \rfloor$.

Now we are ready to discuss the stability of the mutual information

$$I(i; z_{r_0} z_{r_1} \cdots z_{r_{v-1}} = i_0 i_1 \cdots i_{v-1}),$$

where $r_0 = 0$ and $r_1, \cdots, r_{v-1}$ are arbitrary distinct elements of Z_N. Owing to the bounds of Proposition 9.4.9 on the parameters $d_{i_0 \cdots i_{v-1}}(r_0, \cdots, r_{v-1})$ we can conclude that the stability of this mutual information is ideal for small v. This means also that the pattern distribution of length v in the DSC sequences is ideal for small v. Further deductions about the stability of this mutual information and about the pattern distribution need tight bounds on the parameters $d_{i_0 \cdots i_{v-1}}(r_0, \cdots, r_{v-1})$ when v is large.

We now introduce a decision-tree based attack on the DSC generator. The decision is based on mutual information analysis. To illustrate this, let us take the example of $N = 4t - 1 = 19$ with $t = 5$. The difference set and its complement are respectively

$$D = \{1, 4, 5, 6, 7, 9, 11, 16, 17\},$$
$$D^* = \{0, 2, 3, 8, 10, 12, 13, 14, 15, 18\}.$$

We show now how to determine the key i by the following procedure. Let $s_j = z_{i+j}$ for each $j \geq 0$. The procedure can be described as follows:

Step 1:

$$s_0 = 1 \Rightarrow i \in D_1 = D \ (s_0 = 0 \Rightarrow i \in D_1^* = D^*)$$

Step 2:

$$s_0 s_1 = \left\{ \begin{array}{l} 10 \Rightarrow i \in \{1, 7, 9, 11, 17\} = D_{10} \\ 11 \Rightarrow i \in \{4, 5, 6, 16\} = D_{11} \end{array} \right\} D_1$$

Step 3:

$$s_0 s_1 s_2 = \left\{ \begin{array}{l} \left. \begin{array}{l} 100 \Rightarrow i \in \{1, 11, 17\} = D_{100} \\ 101 \Rightarrow i \in \{7, 9\} = D_{101} \end{array} \right\} D_{10} \\ \left. \begin{array}{l} 110 \Rightarrow i \in \{6, 16\} = D_{110} \\ 111 \Rightarrow i \in \{4, 5\} = D_{111} \end{array} \right\} D_{11} \end{array} \right\} D_1$$

Step 4:

$$s_0 s_1 s_2 s_3 = \left\{ \begin{array}{l} \left. \begin{array}{l} \left. \begin{array}{l} 1000 \Rightarrow i \in \{11\} = D_{1000} \\ 1001 \Rightarrow i \in \{1, 17\} = D_{1001} \end{array} \right\} D_{100} \\ \left. \begin{array}{l} 1010 \Rightarrow i \in \{7, 9\} = D_{1010} \\ 1011 \Rightarrow i \in \emptyset = D_{1011} \end{array} \right\} D_{101} \end{array} \right\} D_{10} \\ \left. \begin{array}{l} \left. \begin{array}{l} 1100 \Rightarrow i \in \{16\} = D_{1100} \\ 1101 \Rightarrow i \in \{6\} = D_{1101} \end{array} \right\} D_{110} \\ \left. \begin{array}{l} 1110 \Rightarrow i \in \{5\} = D_{1110} \\ 1111 \Rightarrow i \in \{4\} = D_{1111} \end{array} \right\} D_{111} \end{array} \right\} D_{11} \end{array} \right\} D_1$$

Step 5:

$$s_0s_1s_2s_3s_4 = \left\{ \begin{array}{l} 10100 \Rightarrow i \in \{9\} = D_{10100} \\ 10101 \Rightarrow i \in \{7\} = D_{10101} \end{array} \right\} D_{1010}$$

Thus, we obtain the shortest consecutive bits of key stream

$$z^u = z_i z_{i+1} \cdots z_{i+u+1},$$

which specifically determine the key $i \in D$. For each key $i \in D^*$, we can similarly determine the sequence. The shortest sequences which determine the keys in D are

$$\begin{array}{rclrclrcl} 1 &=& 1001, & 4 &=& 1111, & 5 &=& 1110, \\ 6 &=& 1101, & 11 &=& 1000, & 9 &=& 10100, \\ 16 &=& 1100, & 17 &=& 1011, & 7 &=& 10101. \end{array}$$

The foregoing discussion shows that the DSC generator of Section 8.1 is theoretically breakable if a cryptanalyst knows enough consecutive bits of plaintext-ciphertext-bit pairs. We analyze now whether the generator is *computationally breakable.* Before doing so, we observe first the lower bound in Proposition 9.4.9. For stream ciphering purposes the modulus N is usually very large. So the lower bound in Proposition 9.4.9 is approximately

$$\frac{N}{2^v} - \frac{(v-3)\sqrt{N} + v + 1}{2}.$$

Thus determining each key requires at least $M = \lceil B \rceil$ consecutive bits of key stream, where B is the solution of the following equation

$$\frac{N}{2^x} - \frac{(x-3)\sqrt{N} + x + 1}{2} = 1.$$

Suppose that a cryptanalyst gets the keystream segment

$$z_i \cdots z_{i+u} = s_0 \cdots s_u,$$

where $u > \lfloor \log_2 N \rfloor$. To determine the present key i with the above method, she has to use an algorithm to compute $f(i) = (x^{(N-1)/2} \bmod N$ for arbitrary $i \in Z_N$. Assume that the cryptanalyst uses the fast exponentiation algorithm to determine each $D_{s_0 \cdots s_v}$ for each v with $0 \le v \le \lfloor \log_2 N \rfloor$. If we take each integer multiplication, each integer addition, each modulo-2 and each modulo-N as one unit of computation, then the number of operations needed to determine the key i is at least

$$\begin{aligned}
& 2M \sum_{i=0}^{M-1} N/2^i \quad \text{(from the part of exponentiation mod 2)} \\
+ \quad & 2 \sum_{i=0}^{M-1} N/2^i \quad \text{(from the part of of the counter)} \\
+ \quad & \sum_{i=0}^{M-1} N/2^i \quad \text{(from the part of } y \text{ mod 2)} \\
= \quad & 2N(M+3)(1 - \frac{1}{2^M}) \\
= \quad & \Theta(NM),
\end{aligned}$$

where M is given above. If N is large enough, it is clearly impossible to determine the key computationally.

Generally, let $U_A(N)$ be the minimal number of operations to compute $f(i)$ for each $i \in Z_N$ by an algorithm A. Then we can similarly prove that the minimal number of operations for this kind of key-determining attack based on the algorithm A is at least

$$2U_A(N) \sum_{i=0}^{M-1} N/2^i = 4NU_A(N) \left(1 - \frac{1}{2^M}\right).$$

Since for any algorithm A we have $U_A(N) \geq 1$, we see that for any such a key-determining attack, the minimum number of operations needed is at least

$$4N \left(1 - \frac{1}{2^M}\right).$$

If N is chosen to be large enough, for example, say about 2^{100}, any attack of this kind is computationally infeasible at the present time.

On the other hand, it seems that the storage space needed is at least $O(N/2)$ with this procedure. This may also be infeasible for large N. Computational complexity is one source of deterministic randomness, and some of its cryptographic uses may be found in [123, 422, 468, 469, 256, 162, 299].

9.5 Sums of DSC Sequences

Since DSC (difference-set characterized) sequences are cryptographically attractive in many aspects, we analyze the bitwise-XOR of two DSC sequences. For ADSC sequences the analysis is almost the same.

Let $N_1 = 4t_1 - 1$ and $N_2 = 4t_2 - 1$ be two distinct large primes with t_1 and t_2 odd, and let z_1^∞ and z_2^∞ be the corresponding DSC sequences of Section 8.1.

9.5.1 Linear Complexity Analysis

The generating functions of the two sequences can be written as

$$\begin{aligned} z_1^\infty &= z_1^{N_1}(x)/(x^{N_1}+1), \\ z_2^\infty &= z_2^{N_2}(x)/(x^{N_2}+1). \end{aligned}$$

Hence

$$z^\infty(x) = (z_1 + z_2)^\infty(x) = \frac{z_1^{N_1}(x)(x^{N_2}+1) + z_2^{N_2}(x)(x^{N_1}+1)}{(x^{N_1}+1)(x^{N_2}+1)}.$$

Since both N_1 and N_2 are primes and $N_1 \neq N_2$, $\gcd(x^{N_1}+1,\ x^{N_2}+1) = x+1$. On the other hand, it follows from $\mathrm{L}(z_1^\infty) = N_1$ and $\mathrm{L}(z_2^\infty) = N_2$ that

$$\gcd(z_1^{N_1}(x),\ x^{N_1}+1) = \gcd(z_2^{N_2}(x),\ x^{N_2}+1) = 1.$$

This ensures that

$$\gcd((x^{N_1}+1)(x^{N_2}+1),\ z_1^{N_1}(x)(x^{N_2}+1) + z^{N_2}(x)(x^{N_1}+1)) = 1.$$

Thus, the minimal polynomial of the sequence z^∞ is $(x^{N_1}+1)(x^{N_2}+1)$, and therefore

$$\mathrm{L}(z_1^\infty + z_2^\infty) = N_1 + N_2.$$

9.5.2 Balance Analysis

It is easily seen that

$$\begin{aligned} &\Pr(z_{1i} = 0) = (N_1+1)/2N_1, \quad \Pr(z_{1i} = 1) = (N_1-1)/2N_1, \\ &\Pr(z_{2i} = 0) = (N_2+1)/2N_2, \quad \Pr(z_{2i} = 1) = (N_2-1)/2N_2. \end{aligned}$$

Since z_{1i} and z_{2i} are statistically independent, we have

$$\begin{aligned} \Pr(z_i = 1) &= \Pr(z_{1i} + z_{2i} = 1) \\ &= \Pr(z_{1i} = 0,\ z_{2i} = 1) + \Pr(z_{1i} = 1,\ z_{2i} = 0) \\ &= \frac{N_1+1}{2N_1}\frac{N_2-1}{2N_2} + \frac{N_1-1}{2N_1}\frac{N_2+1}{2N_2} \\ &= \frac{N_1N_2-1}{2N_1N_2}. \end{aligned}$$

This means that the sequence z^∞ is almost balanced, with almost the same balance property as the sequences z_1^∞ and z_2^∞.

9.5.3 Correlation Analysis

It is easily verified that

$$\Pr(z_i = z_{1i}) = \Pr(z_{2i} = 0) = (N_2 + 1)/2N_2,$$
$$\Pr(z_i = z_{2i}) = \Pr(z_{1i} = 0) = (N_1 + 1)/2N_1.$$

This means that any correlation attack by making use of the correlation between z_i and z_{1i}, as well as z_i and z_{2i} is impossible.

9.5.4 Differential Analysis

The equivalence between the autocorrelation and the differential analysis of binary sequences has been proved in Section 2.4. For the correlation property of the sequence z^∞, we have the following result.

Theorem 9.5.1 *The autocorrelation function of the sequence z^∞ is four-valued, i.e.,*

$$\mathrm{AC}_z(j) = \begin{cases} 1, & j = 0; \\ -\frac{1}{N_2}, & j = kN_1,\ k \neq 0; \\ -\frac{1}{N_1}, & j = kN_2,\ k \neq 0; \\ \frac{1}{N_1 N_2}, & \gcd(j,\ N_1 N_2) = 1. \end{cases}$$

Proof: If $j = kN_1,\ k \neq 0$, then

$$\begin{aligned} \mathrm{AC}_z(j) &= \sum_{i=0}^{N_1 N_2 - 1} (-1)^{z_{2,i} + z_{2,i+j}} / N_1 N_2 \\ &= AC_{z_2}(j \bmod N_2) = -1/N_2. \end{aligned}$$

Similarly, we can prove

$$\mathrm{AC}_z(j) = -1/N_1, \quad \text{for } j = kN_2,\ k \neq 0.$$

If $\gcd(j,\ N_1 N_2) = 1$, then by definition we have

$$\begin{aligned} \mathrm{AC}_z(j) &= [2|\{z_{1,i} + z_{1,i+j} = 0\} \cap \{z_{2,i} + z_{2,i+j} = 0\}| \\ &\quad +2|\{z_{1,i} + z_{1,i+j} = 1\} \cap \{z_{2,i} + z_{2,i+j} = 1\}| - N_1 N_2]/N_1 N_2 \\ &= 2\Pr(z_{1,i} + z_{1,i+j} = 0)\Pr(z_{2,i} + z_{2,i+j} = 0) \\ &\quad +2\Pr(z_{1,i} + z_{1,i+j} = 1)\Pr(z_{2,i} + z_{2,i+j} = 1) - 1 \\ &= (AC_{z_1}(j \bmod N_1) + 1)(AC_{z_2}(j \bmod N_2) + 1)/2 \\ &\quad +(1 - AC_{z_1}(j \bmod N_1))(1 - AC_{z_2}(j \bmod N_2))/2 - 1 \\ &= 1/N_1 N_2. \end{aligned}$$

This completes the proof. □

Now we calculate the difference parameters $d_z(g, g'; j)$ defined in Section 4.2. Let D be the characteristic set of z^∞, then $|D| = (N_1N_2 - 1)/2$. On the other hand, we have

$$|(D + j) \cap D| + |(D + j) \cap D^*| = (N_1N_2 - 1)/2,$$
$$|(D^* + j) \cap D^*| + |(D^* + j) \cap D| = (N_1N_2 + 1)/2,$$
$$|(D + j) \cap D| + |(D^* + j) \cap D| = (N_1N_2 - 1)/2,$$

where $D^* = Z_N \setminus D$. Consequently, we obtain

$$\begin{aligned} \mathrm{AC}_z(j) &= [|(D + j) \cap D| + |(D^* + j) \cap D^*| \\ &\quad -|(D + j) \cap D^*| - |(D^* + j) \cap D|]/N_1N_2 \\ &= [4|(D + j) \cap D| + 2 - N_1N_2]/N_1N_2, \end{aligned}$$

whence for $j \neq 0$,

$$|(D + j) \cap D| = \begin{cases} (N_1N_2 - N_1 - 2)/4, & j = kN_1, \ k \neq 0; \\ (N_1N_2 - N_2 - 2)/4, & j = kN_2, \ k \neq 0; \\ (N_1N_2 - 1)/4, & \gcd(j;\ N_1N_2) = 1 \end{cases}$$

and

$$|(D + j) \cap D^*| = \begin{cases} (N_1N_2 + N_1)/4, & j = kN_1, \ k \neq 0; \\ (N_1N_2 + N_2)/4, & j = kN_2, \ k \neq 0; \\ (N_1N_2 - 1)/4, & \gcd(j;\ N_1N_2) = 1. \end{cases}$$

These results show that D is not a difference set of $Z_{N_1N_2}$, but has a relatively good difference property. Speaking specifically, for all $j \in Z_N$ with $\gcd(j,\ N_1N_2) = 1$, the equations

$$j = d_1 - d_2, \quad d_1,\ d_2 \in D$$

have the same number of solutions. Only for those j's with $j \bmod N_1 = 0$ and $j \bmod N_2 = 0$, the above equation has a different number of solutions. It is necessary to choose two primes N_1 and N_2 such that $|N_1 - N_2|$ is small, in order to get a better sum sequence concerning the autocorrelation and difference property.

Chapter 10

Nonbinary Cyclotomic Generators

In the foregoing chapter we constructed a number of binary generators. In some applications nonbinary sequences may be needed. In this chapter we describe the rth-order cyclotomic generator and analyze its properties. In Section 10.5 we summarize some cryptographic ideas behind binary and nonbinary cyclotomic generators. Sections 10.1, 10.2, 10.3 and 10.4 are based on Ding and Helleseth [128].

10.1 The rth-Order Cyclotomic Generator

Let $p = rt + 1$ where r and p are both primes. Let β be a generator of the multiplicative group of $GF(p)$ (i.e. β has order $p-1$). The cyclotomic classes of order r give a partition of $GF(p)^* = GF(p) \setminus \{0\}$ defined by

$$D_0 = (\beta^r), \quad D_1 = \beta D_0, \quad \cdots, \quad D_{r-1} = \beta^{r-1} D_0,$$

where D_0 is the multiplicative subgroup generated by β^r.

The rth-order cyclotomic generator is defined by

$$s(k)_i = \begin{cases} j, & \text{if } [i+k \bmod p] \in D_j, \quad j = 0, 1, \ldots, r-1; \\ 0, & \text{if } i+k \bmod p = 0, \end{cases}$$

for each $i \geq 0$, where $0 \leq k \leq p-1$ is the initial state of the generator. Thus, $s(k)^\infty$ is a semi-infinite sequence of period p over $GF(r)$, and is a shift of $s(0)^\infty$.

We call $s(0)^\infty$ the cyclotomic sequence of order r over $GF(r)$ with respect to the prime p, and denote it by s^∞. Thus, s^∞ is a semi-infinite sequence of period p over $GF(r)$. The distribution of elements of $GF(r)$ over a cycle of s^∞ is the best possible, i.e., 0 appears $t+1$ times, and each other element t times. When $r = 2$ a cyclotomic sequence of order 2 is simply a Legendre sequence.

For small r, the rth-order cyclotomic generator can be implemented easily. As an example, we consider the ternary cyclotomic generator. Let $p = 3t + 1$ be a prime. To implement the ternary generator, we need the cryptographic function $F(x)$ defined in Section 4.3.

In Section 4.3.2 it was proved that the $F(x)$ can be expressed as

$$F(x) = a(x^t \bmod p) \bmod 3,$$

with $a(x) = (2t + 1)[3 + (u - 1)x - (u + 2)x^2] \bmod p$. With this function the ternary cyclotomic generator based on cyclotomic numbers of order 3 is described by

$$s(k)_i = (a(i + k)^t \bmod p) \bmod 3, \quad i \geq 0, \tag{10.1}$$

where $0 \leq k \leq p - 1$ is the key of this generator. This gives an easy implementation of the ternary cyclotomic generator

Remark: The r-th order cyclotomic generator defined above is based on the cyclotomy of $GF(p)$, where p is a prime. It can naturally be generalized using cyclotomy of $GF(p^m)$. This has been considered by Dai, Yang, Gong and Wang [93].

10.2 Linear Complexity

We are concerned with the linear complexity of the rth-order cyclotomic sequence over $GF(r)$. By a proper choice of the prime p, we can control the linear complexity with the results of Chapters 3 and 4. However, we can actually compute the linear complexity of these sequences, as done below.

Let s^∞ be a sequence of period n over a field F, and define

$$S^n(x) = s_0 + s_1 x + \cdots + s_{n-1} x^{n-1}.$$

Recall the following basic results (see Lemma 8.2.1):

1. the minimal polynomial of s^∞ is given by $(x^n - 1)/\gcd(x^n - 1, S^n(x))$; and

2. the linear complexity of s^∞ is given by $n - \deg(\gcd(x^n - 1, S^n(x)))$.

Let θ be a pth root of unity over $GF(r^m)$ and

$$S(x) = \sum_{i=1}^{r-1} i \sum_{u \in D_i} x^u \in GF(r)[x].$$

Define

$$U_i = \sum_{u \in D_i} \theta^u, \quad i = 0, 1, \ldots, r-1.$$

Lemma 10.2.1 *Let $d \in D_j$, then $S(\theta^d) = S(\theta) + j$.*

Proof: By definition it follows that

$$dD_i = D_{i+j}$$

for $i = 0, 1, \ldots, r-1$, where the indices are computed modulo r. Therefore,

$$\begin{aligned} S(\theta^d) &= \sum_{i=1}^{r-1} i \sum_{u \in D_i} \theta^{ud} \\ &= \sum_{i=1}^{r-1} i \sum_{u \in dD_i} \theta^u \\ &= \sum_{i=1}^{r-1} iU_{i+j} \\ &= \sum_{i=1}^{r-1} (i-j)U_i. \end{aligned}$$

Since $1 + \sum_{i=0}^{r-1} U_i = 0$, it follows that

$$\begin{aligned} S(\theta^d) - S(\theta) &= \sum_{i=0}^{r-1} ((i-j) - i)U_i \\ &= -j \sum_{i=0}^{r-1} U_i \\ &= j. \end{aligned}$$

□

The following result is due to Ding and Helleseth [128].

Theorem 10.2.2 *Let L be the linear complexity of the cyclotomic sequence s^∞ of order $r \geq 3$. Then*

$$L = \begin{cases} p-1, & \text{if } r \notin D_0; \\ \frac{(r-1)(p-1)}{r}, & \text{if } r \in D_0. \end{cases}$$

Proof: Since $(S(\theta))^r = S(\theta^r)$, it follows from Lemma 10.2.1 that $S(\theta) \in GF(r)$ if and only if $r \in D_0$. Observe that by definition $S(1) = (p-1)(r-1)/2 = rt(r-1)/2$, and therefore $S(1) = 0$ for $r \geq 3$ since $r-1$ is even. The proof is divided into two cases depending on whether $r \in D_0$ or $r \notin D_0$.

Case 1: *($r \notin D_0$).* In this case $S(\theta) \notin GF(r)$ and Lemma 10.2.1 implies that $S(\theta^d) \neq 0$ for all $d \in GF(r)^*$. Therefore, for $r \geq 3$,

$$\gcd(x^p - 1, S(x)) = x - 1.$$

This proves the first part of the theorem.

Case 2: *($r \in D_0$).* In this case we have $S(\theta) \in GF(r)$ and Lemma 10.2.1 implies that $S(\theta^d) = 0$ for d in exactly one cyclotomic class. Hence,

$$\deg(\gcd(x^p - 1, S(x))) = \frac{p-1}{r} + 1.$$

This proves the second part of the theorem. □

We now compute the minimal polynomial of the rth-order cyclotomic sequence over $GF(r)$.

In the case that $r \in D_0$, let

$$d_i(x) = \sum_{u \in D_i} (x - \theta^u), \quad i = 0, 1, \ldots, r-1.$$

Since $(d_i(x))^r = d_i(x^r)$, the coefficients of the polynomials $d_i(x)$ belong to $GF(r)$. Obviously, we have

$$x^p - 1 = (x-1) \prod_{i=0}^{r-1} d_i(x).$$

The polynomials $d_i(x)$ depend on the choice of the primitive root θ. However, this only results in a permutation of the subscripts i of the $d_i(x)$.

Since $S(\theta^d)$ takes on all elements of $GF(r)$ when d ranges over D_0, D_1, $\cdots$, D_{r-1}, we can fix our θ above such that $S(\theta) = 0$. From the proof of Theorem 10.2.2 we obtain the following result due to Ding and Helleseth [128].

Theorem 10.2.3 *Let $m(x)$ be the minimal polynomial of a cyclotomic sequence of order $r \geq 3$. Then*

$$m(x) = \begin{cases} \frac{x^p - 1}{x - 1}, & \text{when } r \notin D_0; \\ \frac{x^p - 1}{(x-1)d_0(x)}, & \text{when } r \in D_0. \end{cases}$$

□

10.3 Autocorrelation Property

Lemma 10.3.1 *Let* $a_h(\tau) = |\{j : s_{j+\tau} - s_j = h,\ 0 \le j \le p-1\}|$ *for* $h = 0, 1, \ldots, r-1$, *then*

$$a_h(\tau) = |D_h \cap \{\tau\}| + |D_{-h} \cap \{-\tau\}| + \sum_{a=0}^{r-1}(a, a+h).$$

Proof: For $\tau \not\equiv 0 \pmod{p}$ and $h \neq 0$ we have

$$\begin{aligned}
a_h(\tau) &= \sum_{a=0}^{r-1} |\{j : s_{j+\tau} = a+h, s_j = a,\ 0 \le j \le p-1\}| \\
&= |(D_h - \tau) \cap (D_0 \cup \{0\})| + |(D_0 \cup \{0\} - \tau) \cap D_{-h}| + \\
&\qquad \sum_{a \in GF(r)\setminus\{0,-h\}} |(D_{a+h} - \tau) \cap D_a| \\
&= |(D_h - \tau) \cap \{0\}| + |(\{0\} - \tau) \cap D_{-h}| + \sum_{a=0}^{r-1} |(D_{a+h} - \tau) \cap D_a| \\
&= |D_h \cap \{\tau\}| + |D_{-h} \cap \{-\tau\}| + \sum_{a=0}^{r-1} |-\tau^{-1} D_a \cap (-\tau^{-1} D_{a+h} + 1)| \\
&= |D_h \cap \{\tau\}| + |D_{-h} \cap \{-\tau\}| + \sum_{a=0}^{r-1}(a, a+h).
\end{aligned}$$

For $\tau \not\equiv 0 \pmod{p}$ and $h = 0$ we have

$$\begin{aligned}
a_0(\tau) &= \sum_{a=0}^{r-1} |\{j : s_{j+\tau} = a, s_j = a,\ 0 \le j \le p-1\}| \\
&= |((D_0 - \tau) \cup \{-\tau\}) \cap (D_0 \cup \{0\})| + \sum_{a \in GF(r)\setminus\{0\}} |(D_a - \tau) \cap D_a| \\
&= |D_0 \cap \{\tau\}| + |D_0 \cap \{-\tau\}| + \sum_{a=0}^{r-1} |-\tau^{-1} D_a \cap (-\tau^{-1} D_a + 1)| \\
&= |D_0 \cap \{\tau\}| + |D_0 \cap \{-\tau\}| + \sum_{a=0}^{r-1}(a, a).
\end{aligned}$$

□

Let $\epsilon = e^{2\pi\sqrt{-1}/r}$ be a rth primitive root of unity. The periodic autocorrelation of s^∞ is defined to be

$$\mathrm{AC}_s(\tau) = \sum_{j=0}^{p-1} \epsilon^{s_{j+\tau}-s_j}.$$

In terms of the cyclotomic numbers the autocorrelation is

$$\mathrm{AC}_s(\tau) = \sum_{h=0}^{r-1} a_h(\tau)\epsilon^h. \tag{10.2}$$

For the case $r = 3$, with the cyclotomic numbers of order 3 we can easily prove the following conclusion.

Theorem 10.3.2 [128] *Let $-1 \in D_j$ and $\tau \in D_l$. The autocorrelation function of the ternary cyclotomic sequence of order 3 is given by*

$$\mathrm{AC}_s(\tau) = -1 + (\epsilon^l + \epsilon^{-l-j})$$

where $\epsilon = e^{2\pi\sqrt{-1}/3}$.

Recently, the autocorrelation function of the r-th order cyclotomic sequence has been determined, as stated in the following theorem.

Theorem 10.3.3 [199] *If t is even,*

$$\mathrm{AC}_s(\tau) = -1 + \epsilon^l + \epsilon^{-l},$$

where $\tau \in D_l$.

If t is odd,

$$\mathrm{AC}_s(\tau) = -1 + \epsilon^l - \epsilon^{-l},$$

where $\tau \in D_l$.

10.4 Decimation Property

Let d be an integer with $1 \le d \le p-1$. The d-step decimation sequence t^∞ of s^∞ is defined to be

$$t_i = s_{di} \text{ for all } i \ge 0.$$

It follows by definitions that for $d \in D_a$, we have

$$t_i = \begin{cases} s_i + a, & \text{if } i \not\equiv 0 \pmod{p}; \\ s_i, & \text{if } i \equiv 0 \pmod{p}. \end{cases}$$

In this case the t^∞ is a linear translation of s^∞ except the entries i corresponding to $i \equiv 0 \pmod{p}$.

Thus, decimation of a cyclotomic sequence gives essentially the same sequence. This is rather different from m-sequences.

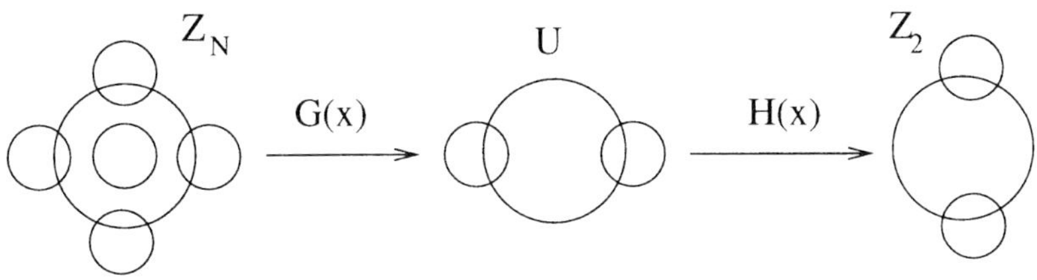

Figure 10.1: A description of the cryptographic idea behind cyclotomic generators.

10.5 Ideas Behind the Cyclotomic Generators

There are several cryptographic ideas behind the construction of the cyclotomic generators. The *first one* is the order of choosing the design parameters for the generator. Contrary to the traditional approach, we first control the period of the output sequence. This automatically ensures the linear complexity and its stability. Then we choose the cryptographic function for other purposes.

The *second cryptographic idea* behind the design and analysis of cyclotomic generators is the idea of introducing good "partners", in order to get a stable system. In particular, we search for pairs consisting of a period and finite field so that it is easy to control the linear complexity and its stability for those sequences over those fields with corresponding partner periods. We say that such pairs work in harmony with respect to the aspects of linear complexity and its stability. For example, some Mersenne and Fermat primes are not good partners of the field $GF(2)$, since it is difficult to control the linear complexity and its stability for binary sequences with period equal to some Fermat and Mersenne primes. Our analysis in Chapters 3 and 5 shows that it is sensible to use $\text{ord}_p(q)$ as a measure of the partnership between a prime p and an integer q when designing sequences of period p over $GF(q)$. We call them the *best partners* with respect to $GF(q)$ when q is a primitive root modulo p.

Another kind of partnership is to find an integer r which is a power of prime such that $\min\{\text{ord}_{p_1}(r), \cdots, \text{ord}_{p_h}(r)\}$ is large enough when designing sequences of period $N = p_1 \cdots p_h$ over $GF(r)$, where $p_1, \cdots, p_h$ are distinct primes. We say that r is a *best common partner* of $p_1, \cdots, p_h$ if r is a common primitive root of these primes.

The *third cryptographic idea* is to use some techniques for ensuring "good + bad = good". With a simple argument each cryptographic function employed in the generators described in Chapter 8 and this chapter can be expressed as

$$F(x) = H(G(x)),$$

where $G(x)$ is a mapping from Z_N to U which is a subgroup of the group $(Z_N^*, \cdot)$ with order d, and $H(x)$ a mapping from U to Z_d. The nonlinearity of $G(x)$ with respect to $(Z_N, +)$ and $(U, \cdot)$ is determined mainly by the (generalized) cyclotomic numbers of order d, which usually have ideal stability; while the function $H(x)$ is almost linear (or with a good linearity) with respect to $(U, \cdot)$ and $(Z_d, +)$. Thus, it is clear that one cryptographic idea behind the cyclotomic generators is

"GOOD + BAD = GOOD".

The *fourth cryptographic idea* is to make use of the relativity about nonlinearity and linearity. It is well-known that nonlinearity and linearity are relative to the operations considered, and that both linear components and nonlinear components should be employed in many cipher systems. To find some cryptographic functions with good nonlinearity with respect to some operations, one can try to find some linear cryptographic function with respect to some other operations and use them in the context of the former operations. This is to say that bad things in one sense may be good ones in another sense, and one way to get goodness is to use badness in a proper way and proper context. To illustrate this philosophy, we first take the corresponding function $G(x) = x^{(p-1)/d} \bmod p$ used to construct the cyclotomic generator of order $2k$. Then $G(x)$ is linear with respect to $(Z_p^*, \cdot)$ and $(U, \cdot)$, where U is the multiplicative subgroup of Z_p^* with order d. But $G(x)$ has ideal nonlinearity with respect to $(Z_p, +)$ and $(U, +)$ if we define $G(0)$ to be any fixed element of U. And we use $G(x)$ in the context of the latter pair of operations exactly. The same idea has been used for other generators.

The *fifth cryptographic idea* is to choose the design parameter of the NSG of Figure 2.5(b) such that $|G|$ does not divide N. This ensures that there are only trivial affine functions from Z_N to the Abelian group G over which the sequence is constructed. Thus, affine approximation makes no sense.

Chapter 11

Generators Based on Permutations

In this chapter permutations of finite fields $GF(q)$ and of residue class rings Z_m with good nonlinearity and with a simple implementation are discussed. Then some generators based on those permutations are constructed. There are a number of promising generators in this class including the RSA bit generator. "Good plus bad equals good" is still the cryptographic idea of this chapter, but the technique for ensuring this is quite different from that for cyclotomic generators.

11.1 The Cryptographic Idea

For a prime $p = df + 1$, it is not difficult to see that the cryptographic function $f(x) = x \bmod d$ from Z_p to Z_d has very bad difference property and nonlinearity with respect to the additions of Z_p and Z_d, though it is nonlinear. Taking $d = 2$ as an example, we have

$$f(x+y) = \begin{cases} f(x) + f(y), & x + y < p \\ f(x) + f(y) + 1, & \text{otherwise.} \end{cases}$$

This function $f(x)$ clearly has bad difference property and bad nonlinearity. Now one question arises: Can we find a permutation π of Z_p such that the function

$$f(x) = \pi(x) \bmod d$$

has optimum nonlinearity with respect to the additions of the two rings? To answer the question, we first observe an example for the case $d = 2$. Let $\pi(x)$ be a permutation of Z_7 as follows:

x	0	1	2	3	4	5	6
$\pi(x)$	0	1	3	4	5	2	6

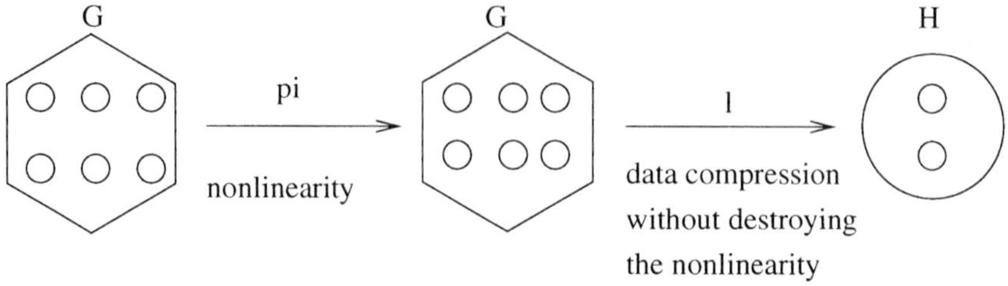

Figure 11.1: An intuitive description of the cryptographic idea.

Then the characteristic set of this $f(x) = \pi(x) \bmod 2$ is a difference set. This means that $f(x)$ has the best nonlinearity with respect to the additions of Z_7 and Z_2. Apparently, there should be many such permutations.

The cryptographic idea of this chapter can be described as follows. To find good cryptographic functions from an Abelian group $(G, +)$ to another Abelian group $(H, +)$, we first find one permutation π of G with good nonlinearity. Then we choose some linear (affine) function or some function $l(x)$ from G to H which is close to linear with respect to the additions of the two groups. Finally, by combining the two functions we get a function $f(x) = l(\pi(x))$ with ideal nonlinearity. The function π is responsible for the nonlinearity of the function $f(x)$; and $l(x)$ is responsible for the data compression without destroying the nonlinearity of π. The linearity of $l(x)$ is necessary for keeping the nonlinearity of $\pi(x)$. The permutation π and the linear function $l(x)$ are respectively responsible for key confusion and diffusion when the function is properly used in a cipher system. The idea can be described with Figure 11.1. This is a realization of the equation

$$\text{GOOD} + \text{BAD} = \text{GOOD}.$$

The idea of this chapter is very different from that of Chapters 7 and 8. In Chapter 7 we employ exponential functions x^d with $d|(p-1)$. These functions are not permutations. They have already produced data compression in some sense because they are functions from a large set Z_p to a small one. In this chapter we use exponential functions x^d with $\gcd(d, p-1) = 1$. These functions are only responsible for nonlinearity, not for data compression.

11.2 Permutations on Finite Fields

It is an elementary fact that every function $f(x)$ from $GF(q)$ into $GF(q)$ can be expressed as a polynomial

$$g(x) = \sum_{c \in GF(q)} f(c)(1 - (x - c)^{q-1}).$$

If a polynomial is a permutation of $GF(q)$, it is called a *permutation polynomial.*

One criterion for whether a polynomial is a permutation is the following theorem of Hermite.

Theorem 11.2.1 [276] *Let $GF(q)$ be a finite field of characteristic p. Then $f \in GF(q)[x]$ is a permutation polynomial of $GF(q)$ if and only if the following two conditions hold:*

1. *f has exactly one root in $GF(q)$;*
2. *for each integer t with $1 \leq t \leq q-2$ and $t \not\equiv 0 \pmod p$, the reduction of $f(x)^t \bmod (x^q - x)$ has degree $\leq q - 2$.*

It follows from this theorem that, if $d > 1$ is a divisor of $q-1$, then there is no permutation polynomial of $GF(q)$ of degree d. Another criterion is the following.

Theorem 11.2.2 *Let $GF(q)$ be of characteristic p. Then $f \in GF(q)[x]$ is a permutation polynomial of $GF(q)$ if and only if the following two conditions hold:*

1. *the reduction of $f(x)^{q-1} \bmod (x^q - x)$ has degree $q - 1$;*
2. *for each integer t with $1 \leq t \leq q-2$ and $t \not\equiv 0 \pmod p$, the reduction of $f(x)^t \bmod (x^q - x)$ has degree $\leq q - 2$.*

There are also other criteria [276]. But they are all not practical.

For the natural sequence generator, we need permutation polynomials of Z_p that have ideal nonlinearity and can be realized efficiently. Permutations of $GF(q)$ with high nonlinearity are also needed in constructing certain keystream generators. Hence permutation polynomials of specific forms are interesting to us. Below we briefly describe the known classes of permutation polynomials. It is of cryptographic importance to analyze the nonlinearity of these permutation polynomials.

11.2.1 Dickson Permutation Polynomials

A class of Dickson polynomials $D_n(x, a)$ of degree n over $GF(q)$ is defined by

$$D_n(x, a) = \sum_{j=0}^{\lfloor n/2 \rfloor} \frac{n}{n-j} \binom{n-j}{j} (-a)^j x^{n-2j}.$$

For $a \neq 0$, $D_n(x, a)$ is a permutation polynomial of $GF(q)$ if and only if $\gcd(n, q^2 - 1) = 1$ [276, Theorems 7.8, 7.16]. Note that whether or not a Dickson polynomial $D_n(x, a)$ permutes $GF(q)$ depends only on its degree and is independent of a. The Dickson polynomials can be viewed as generalizations of the power polynomials since $D_n(x, 0) = x^n$. Note that x^n is a permutation polynomial if and only if $\gcd(n, q - 1) = 1$.

11.2.2 Linearized Permutation Polynomials

Clearly, every linear polynomial over $GF(q)$ is a permutation polynomial of $GF(q)$. Let $GF(q)$ be of characteristic p. Then the p-polynomial

$$L(x) = \sum_{i=0}^{m} a_i x^{p^i} \in GF(q)[x]$$

is a permutation polynomial of $GF(q)$ if and only if $L(x)$ only has the root 0 in $GF(q)$. These linear functions could be of much cryptographic value if they are used properly.

11.2.3 Permutation Polynomials of the Form $x^{(q+m-1)/m} + ax$

For odd q if m divides $q - 1$, then there are permutation polynomials of the form $x^{(q+m-1)/m} + ax$ [314]. When $m = 2$ this gives a class of cryptographically interesting permutation polynomials

$$x^{(q+1)/2} + ax \in GF(q)[x]$$

with q odd and $a = (c^2 + 1)(c^2 - 1)^{-1}$ for some $c \in GF(q)^*$ with $c^2 \neq 1$.

11.2.4 Permutation Polynomials of the Form $x^r(g(x^s))^{(q-1)/s}$

Polynomials of the form $x^r(g(x^s))^{(q-1)/s}$ are permutations of $GF(q)$ if $\gcd(r, q - 1) = 1$, $s|(q - 1)$ and $g(x^s)$ has no nonzero roots in $GF(q)$ [276, Theorem 7.10].

11.2.5 Cohen Permutation Polynomials

Let $L(x)$ be a linearized polynomial of the form

$$L(x) = \sum_{i=0}^{k} a_i x^{p^i}$$

with the property that for some $s \geq 1$, $a_i = 0$ unless $s|i$. Such an $L(x)$ is called a p^s-polynomial. Let d divide $p^s - 1$ with p not dividing d. Then $L(x) = xM(x^d)$ and $S(x) = xM^d(x)$ is called a (p^s, d)-polynomial. If M has no roots in $GF(q)$, then S is a permutation polynomial [81, 314].

11.3 A Generator Based on Inverse Permutations

Let p be a prime. Then the mapping $\pi(x) = x^{p-2} = x^{-1}$ is a permutation on Z_p. It has already been shown that the permutation π has ideal nonlinearity with respect to the addition of Z_p [333, 19]. Actually, for each $a \neq 0$ it has been proven that the difference function

$$D_a(x) = \pi(x + a) - \pi(x)$$

takes on the value a^{p-2} at most four times, and other possible values of Z_p at most two times [333].

Assume that $f(x)$ is a permutation on a finite Abelian group $(A, +)$, and $g(x)$ is an affine function from $(A, +)$ to another Abelian group $(B, +)$ such that it takes on each element of B equally often and $g(-x) = -g(x) + l$ for some fixed $l \in B$. Setting

$$h(x) = g(f(x)),$$

we give now an intuitive analysis of the nonlinearity of this function with respect to the additions of A and B.

By assumption the characteristic class of the affine function is some set $\{D_0, D_0 + a_1, \cdots, D_0 + a_{d-1}\}$, where D_0 is a subgroup of $(A, +)$. If $f(x)$ has good nonlinearity with respect to the addition of A, then for each $a \neq 0$, the difference function defined by a and f, i.e.,

$$D_a(x) = f(x + a) - f(x)$$

should take on as many elements of A as possible and also have good nonlinearity. This means that the image $D_a(A)$ should have a roughly equally likely distribution among the characteristic classes, i.e., the elements of

$D_a(A) \cap D_i$ should be approximately the same as i ranges from 0 to $d-1$. It follows that the function

$$\begin{aligned} D_{h,a}(x) &= h(x+a) - h(x) = g(f(x+a)) - g(f(x)) \\ &= g(f(x+a) - f(x)) - u, \end{aligned}$$

should take on as many elements of B as possible and each of those possible elements approximately the same number of times, where u is some fixed element of B. This shows that $h(x)$ has good nonlinearity if $f(x)$ does. A detailed proof of the "good + bad = good" will be given in the following section.

Now we analyze the specific permutation $\pi(x) = x^{p-2}$ on Z_p, which is the inverse of x if $x \neq 0$. Let

$$g(x) = x \bmod 2.$$

At the beginning of this chapter we have seen that the linearity of $g(x)$ is similar to that of nontrivial affine functions. On the other hand, because p is odd, we have

$$g(-x) = g(p-x) = 1 + g(x) = -g(x) + 1.$$

It follows that the function $F(x) = (x^{p-2} \bmod p) \bmod 2$ has good nonlinearity with respect to the additions of Z_p and Z_2. Using this $F(x)$, we have a binary natural generator described by

$$s_i = [(i_0 + i)^{p-2} \bmod p] \bmod 2, \quad i \geq 0,$$

where $0 \leq i_0 \leq p-1$ is the key of the generator.

To control the linear complexity and its stability of the output sequence of the generator, we may choose a Stern prime or Sophie Germain prime. Such a prime ensures the best linear and sphere complexity as described in Corollaries 3.4.10 and 3.4.11. Generally, it suffices to choose a prime p such that $\mathrm{ord}_p(2)$ is large enough due to Basic Theorem 3.3.1.

If one wishes to have a software implementation of this generator, one may use the Extended Euclidean Algorithm to compute the value of x^{p-2}, which is the inverse of x with respect to the multiplication of Z_p. Generally speaking, the calculation of x^{-1} based on the Extended Euclidean Algorithm may be faster than that of x^{p-2} based on fast exponentiation algorithms.

11.4 Binary Generators and Permutations of $GF(2^n)$

In the above section we have intuitively shown a technique for ensuring "good + bad = good". By further making use of this approach, we shall

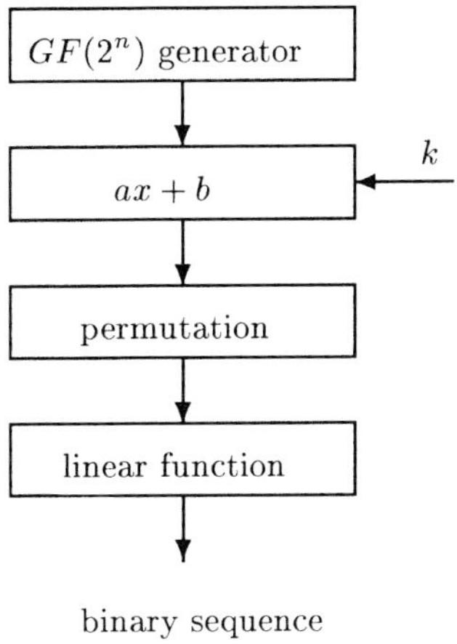

Figure 11.2: A binary generator based on permutations.

develop cryptographic functions from $GF(2^n)$ to $GF(2)$ or from $GF(2)^n$ to $GF(2)$ for some binary generators. We first prove that the technique of last section can ensure the holding of the equation "good + bad = good".

Let $(G, +)$, $(H, +)$ and $(K, +)$ be finite Abelian groups with $|G| \geq |H| \geq |K|$, f a function from G onto H and g a function from H to K. Then we have the composition function $h = g \circ f$, which is a function from G to K. Suppose that g is a nontrivial linear surjection such that $g(-x) = -g(x)$; then it follows that $|K|$ divides $|H|$. Let $K = \{k_0 = 0, k_1, \cdots, k_{d-1}\}$ and

$$V_i = \{x : \; g(x) = k_i\}.$$

Then V_0 is a subgroup of $(G, +)$ and for each i the V_i can be expressed as $V_i = V_0 + h_i$ for some $h_i \in H$. It follows that for each $a \in G$ and $k_i \in K$

$$\begin{aligned} \Pr(h(x+a) - h(x) = k_i) &= \Pr(g(f(x+a) - f(x)) = k_i) \\ &= \Pr(f(x+a) - f(x) \in V_i) \\ &= \sum_{b \in V_i} \Pr(f(x+a) - f(x) = b). \end{aligned}$$

Thus if $f(x)$ has good nonlinearity with respect to the additions of G and H, then the function h must have good nonlinearity with respect to the additions of G and K. This provides the theoretical foundation of this chapter.

With the above result we can build the binary generators of Figure 11.2, where the $GF(2^n)$ generator produces each of the element of $GF(2^n)$

with equal probability. In this generator the key is the pair (a, b) with $0 \neq a, b \in GF(2^n)$. If we choose a permutation of $GF(2^n)$ with good nonlinearity with respect to the addition of $GF(2^n)$ and a linear function from $GF(2^n)$ to $GF(2)$, then the composite cryptographic function has good nonlinearity with respect to the additions of $GF(2^n)$ and $GF(2)$. What remains to be investigated is the control of the linear complexity of the output sequence. However, one should keep in mind that it is sometimes unnecessary to require large linear complexity of sequences for some nonadditive stream ciphers (see Section 2.1.3).

To design such a generator, we should first choose an easily implementable permutation of $GF(2^n)$ with good nonlinearity with respect to the addition of the finite field. The best candidates are the power permutations of $GF(2^n)$. The nonlinearity of such permutations has been investigated in [19, 333]. In the following subsections some of the power permutations with ideal nonlinearity with respect to the addition of $GF(2^n)$ or $GF(2)^n$ are introduced.

11.4.1 APN Permutations and their Properties

Recall that in Section 2.4 and in Chapter 6 the nonlinearity measure of a permutation of $GF(q)$ is defined by

$$P_g = \max_{0 \neq a \in G} \max_{b \in H} \Pr(g(x+a) - g(x) = b),$$

where $\Pr(A)$ denotes the probability of the occurrence of event A.

For a permutation f of $GF(2)^n$, the minimum value for P_f is 2^{1-n}. Permutations of $GF(2)^n$ with $P_f = 2^{1-n}$ are said to be almost perfect nonlinear (APN) [334]. From the definition of APN permutations, it is clear that the following Lemma 11.4.1 holds:

Lemma 11.4.1 *Let $f(x)$ be a permutation of $GF(2^n)$ (resp. $GF(2)^n$) and $g(x, a) = f(x)+f(x+a)$. Then $f(x)$ is APN iff $g(x, a)$ takes on exactly 2^{n-1} different nonzero elements of $GF(2^n)$ (resp. $GF(2)^n$) and each of them two times when x ranges over $GF(2^n)$ (resp. $GF(2)^n$) for each $a \neq 0$.*

It may be cryptographically beneficial to require that $g(x, a)$ takes on each nonzero element of $GF(2^n)$ (resp. $GF(2)^n$) equally often, i.e., $g(x, a)$ takes on each element of $GF(2^n)$ (resp. $GF(2)^n$) 2^n times when x ranges over $GF(2^n)$ (resp. $GF(2)^n$) and a over $GF(2^n) \setminus \{0\}$ (resp. $GF(2)^n \setminus \{0\}$). Such functions are called *difference uniformly distributed (DUD)*. The $f(x)$ in the following Example 11.4.2 is APN, but not DUD, while the one in Example 11.4.3 is both APN and DUD.

Example 11.4.2 *Let* $f(x) = (f_1, f_2, f_3)$ *in* $GF(2)[x_1, x_2, x_3]^3$, *where* $f_1(x) = x_1 + x_2 + 1 + x_2x_3$, $f_2(x) = x_1 + x_3 + x_1(x_2 + x_3)$, $f_3(x) = x_2 + x_1x_3$.

Example 11.4.3 *Let* $f(x) = (f_1, f_2, f_3)$ *in* $GF(2)[x_1, x_2, x_3]^3$, *where* $f_1(x) = x_1x_2 + x_1x_3 + x_1 + x_2$, $f_2(x) = x_1 + x_2x_3 + x_3$, $f_3(x) = 1 + x_1 + x_1x_2 + x_2x_3$.

If $f(x_1, \cdots, x_n) = (f_1(x), \cdots, f_n(x))$ is a permutation of $GF(2)^n$, let $B = \{\alpha_1, \cdots, \alpha_n\}$ be any basis of $GF(2^n)$ over $GF(2)$, then

$$F(X) = \sum_{i=1}^{n} f_i(x_1, \cdots, x_n)\alpha_i \tag{11.1}$$

is a permutation of $GF(2^n)$, and vice versa, where $X = \sum x_i\alpha_i \in GF(2^n)$. So there is a one-to-one correspondence between the permutations of $GF(2)^n$ and those of $GF(2^n)$ under a chosen basis of $GF(2^n)$ over $GF(2)$. We denote here and hereafter the permutation $f(x) = (f_1(x), \cdots, f_n(x))$ in (11.1) by $[F(X)]_B$.

For an odd n, let $\{\alpha_1^*, \cdots, \alpha_n^*\}$ be the dual basis of B, then each component of $f(x)$ can be expressed as

$$f_i(x) = Tr(F(X)\alpha_i^*), \tag{11.2}$$

where $X = \sum x_i\alpha_i$.

The following result about the nonlinearity of the function $F(X)$ and $f(x)$ in (11.1) is obviously true, which is the theoretical foundation for constructing permutations of $GF(2)^n$ with good nonlinearity from those of $GF(2^n)$.

Proposition 11.4.4 *Let* $B = \{\alpha_1, \cdots, \alpha_n\}$ *be a basis of* $GF(2^n)$ *over* $GF(2)$, $x = (x_1, \cdots, x_n)$, $y = (y_1, \cdots, y_n)$, $a = (a_1, \cdots, a_n)$, $b = (b_1, \cdots, b_n) \in GF(2)^n$, *and* $X = \sum x_i\alpha_i$, $Y = \sum y_i\alpha_i$, $A = \sum a_i\alpha_i$, $B = \sum b_i\alpha_i \in GF(2^n)$, *then*
1) $\Pr(F(X) + F(Y) = A \mid X + Y = B) = \Pr(f(x) + f(y) = a \mid x + y = b)$;
2) $P_F = P_f$;
3) $P_F = P_{F^{2^i}}$ *for each integer* i.

This proposition shows that the nonlinearity of $F(X)$ and that of $f(x)$ are the same.

Proposition 11.4.5 *Let* $f(x) = (f_1(x), \cdots, f_n(x))$ *be an APN (DUD) permutation of* $GF(2)^n$, *then for each nonsingular* $n \times n$ *matrix* A *over* $GF(2)$, $g(x) = (f_1(x), \cdots, f_n(x))A$ *is also APN (DUD).*

The above Proposition 11.4.5 is useful in constructing APN permutations. Two permutations $f(x)$ and $g(x)$ of $GF(2)^n$ are said to be linearly equivalent if there are a nonsingular $n \times n$ matrix A over $GF(2)$ and a vector b of $GF(2)^n$ such that $f(x) = g(Ax + b)$.

Let $f(x) = [F(X)]_B$. For the changing of the basis, let $B' = \{\beta_1, \cdots, \beta_n\}$ be another basis of $GF(2^n)$ over $GF(2)$, $f'(x) = [F(X)]_{B'}$ and

$$(\beta_1, \cdots, \beta_n) = (\alpha_1, \cdots, \alpha_n)A^t; \tag{11.3}$$

then A is nonsingular and

$$f'(x) = (f_1(xA), \cdots, f_n(xA))A^{-1}. \tag{11.4}$$

This result shows that permutations obtained from a permutation of $GF(2^n)$ by changing the basis are usually not linearly equivalent.

We now consider the *conjugacy classes* of $Z^*_{2^n-1}$ with respect to the modulus $(2^n - 1)$. A conjugacy class C_k is the set $\{k2^i \bmod (2^n - 1), i = 0, 1, \cdots, \}$. Proposition 11.4.4 shows that $P_F = P_{F^{2^i}}$ for any permutations of $GF(2^n)$, so we can construct a class of permutations with good nonlinearity, provided that we have one.

We need the notion of the nonlinear order of a permutation $f(x) = (f_1(x), \cdots, f_n(x))$, which is defined as

$$\mathrm{ord}(f) = \max_{1 \le i \le n} \mathrm{ord}(f_i),$$

where $\mathrm{ord}(f_i)$ is the nonlinear order (or degree) of $f_i(x)$. It is trivial to see that the maximum nonlinear order of an APN permutation in $GF(2)^n$ is $n - 1$. This upper bound is achievable (see Examples 11.4.2 and 11.4.3).

It is well known that X^d is a permutation of $GF(2^n)$ iff $\gcd(d, 2^n - 1) = 1$. In the following subsections we mainly introduce the permutations X^d in $GF(2^n)$ with good nonlinearity. The following result is cryptographically useful. It was obtained in [55] according to [333].

Proposition 11.4.6 *Let B be a basis of $GF(2^n)$ over $GF(2)$, and let d be an integer; then* $\mathrm{ord}([X^d]_B) = \mathrm{WH}(d)$, *where* $\mathrm{WH}(d)$ *is the Hamming weight of the binary representation of the integer d.*

Proof: Let $B = \{\alpha_1, \cdots, \alpha_n\}$ and

$$d = 2^{k_t} + 2^{k_{t-1}} + \cdots + 2^{k_1}, \; k_t > k_{t-1} > \cdots > k_1,$$

then for $X = \sum x_i \alpha_i$, we have

$$\begin{aligned} X^d &= \prod_{i=1}^{t} \sum_{l=1}^{n} x_l \alpha_l^{2^{k_i}} \\ &= \sum_{1 \le j_1, \cdots, \le j_t \le n} x_{j_1} \cdots x_{j_t} \alpha_{j_1}^{2^{k_1}} \cdots \alpha_{j_t}^{2^{k_t}}, \end{aligned}$$

whence $\text{ord}([X^d]_B) = t = \text{WH}(d)$. □

The following propositions are cryptographically important, because they show some cryptographic properties of the component functions of an APN permutation.

Proposition 11.4.7 *If $f(x) = (f_1(x), \cdots, f_n(x))$ is an APN permutation of $GF(2)^n$, then none of $f_1, \cdots, f_n$ is affine.*

Proof: Suppose that $f_1(x) = b_{1n}x_n + \cdots + b_{11}x_1 + b_0$, then

$$f_1(x) + f_1(x+c) = \sum_{i=1}^{n} b_{1i}c_i,$$

so we can find a vector $c \neq 0$ such that $f_1(x) + f_1(x+c) = 0$. Hence

$$f(x) + f(x+c) = (0, f_2(x) + f_2(x+c), \cdots, f_n(x) + f_n(x+c)).$$

To ensure that $f(x) + f(x+c)$ takes on 2^{n-1} distinct vectors of $GF(2)^n$, there must exist a vector x such that

$$f(x) + f(x+c) = (0, \cdots, 0).$$

This contradicts the one-to-one property of $f(x)$, and therefore completes the proof of the proposition. □

This proposition demonstrates that each component function of an APN permutation is not affine. We now discuss the nonlinear terms $x_i x_j$ with $i \neq j$ of APN permutations.

Proposition 11.4.8 *If $f(x) = (f_1(x), \cdots, f_n(x))$ is an APN permutation of $GF(2)^n$, then every quadratic term $x_i x_j (i \neq j)$ must appear in at least one of the component functions $f_1, \cdots, f_n$.*

Proof: For $c, x \in GF(2)^n$, let $x^c = 0$ when $x \neq c$, and $x^c = 1$ otherwise. Therefore $f(x)$ can be expressed as

$$\begin{aligned} f(x) &= \sum_{c \in GF(2)^n} x^c f(c) = \prod_{i=1}^{n} x_i \sum_{c \in GF(2)^n} f(c) \\ &+ \sum_{i=1}^{n-1} \sum_{1 \le k_1 \le \cdots \le k_t \le n} \Big(\prod_{j \neq k_1, \cdots, k_t} x_j \Big) \sum_{c} c'_{k_1} \cdots c'_{k_t} f(c) \end{aligned}$$

where $c_i^{'} = 1 + c_i$. Without loss of generality, we consider the coefficient of the term $x_{n-1}x_n$, which is

$$f(0\cdots001) + f(0\cdots000) + f(0\cdots010) + f(0\cdots011),$$

not equal to the zero vector by the definition of APN permutations. This proves the proposition. □

Proposition 11.4.8 tells us that any APN permutation must be dependent on all the quadratic terms, which may show the importance of the quadratic terms of an APN permutation.

11.4.2 Quadratic Permutations with Controllable Nonlinearity

Nyberg and Knudsen have studied the permutations f in $GF(2^m) = GF(2^{dn})$ which satisfy the property that every nonzero linear combination of the components of f is a balanced quadratic form x^tCx in n indeterminates over $GF(2^d)$ with $\text{rank}(C + C^t) = n - 1$ [334]. General results about the quadratic APN permutations were obtained in [19, 88, 333].

Proposition 11.4.9 *Let $f(x) = (f_1, \cdots, f_n)$ be a permutation in $GF(2)^n$, where*

$$f_l(x) = \sum_{1 \le i < j \le n} a_{ij}^{(l)} x_i x_j + \sum_{i=1}^{n} b_i^{(l)} x_i + b_0^{(l)}, \quad 1 \le l \le n.$$

If the entries $a_{ij}^{(l)}$ of the matrix A_l are 0 when $i = j$, and are $a_{\min\{i,j\}\max\{i,j\}}^{(l)}$ otherwise, then $f(x)$ is APN iff $\text{rank}(A_1w^t, \cdots, A_nw^t) = n - 1$ for each $w \neq 0$.

Proof: Let

$$\begin{aligned} g_l(x, w) &= f_l(x) + f_l(x + w) = xA_lw^t + \sum_{1 \le i < j \le n} a_{ij}^{(l)} w_i w_j + \sum_{i=1}^{n} b_i^{(l)} w_i \\ &= xA_lw^t + f_l(w) + f_l(0). \end{aligned}$$

For each $w \neq 0$, the set of linear equations

$$(g_1(x, w), \cdots, g_n(x, w)) = (d_1, \cdots, d_n) \neq 0 \tag{11.5}$$

has no solution or only two solutions iff $\text{rank}(A_1w^t, \cdots, A_nw^t) = n - 1$ for each $w \neq 0$. This proves the theorem. □

From the foregoing proof it follows that the following Corollary 11.4.10 holds:

Corollary 11.4.10 *Let the symbols and notations be the same as in Proposition 11.4.9. If*

$$\max_{w\neq 0} \operatorname{rank}(A_1 w^t, \cdots, A_n w^t) = k,$$

then $P_f \leq 2^{-k}$.

Proposition 11.4.11 *If* $d = 2^l(2^k+1)$, $\gcd(d, 2^n-1) = 1$ *(i.e.,* $n/\gcd(k,n)$ *is odd) and* $m = \gcd(2^n - 1, 2^k - 1)$, B *is any basis of* $GF(2^n)$ *over* $GF(2)$, $f(x) = [X^d]_B, x \in GF(2)^n$ *and* $F(X) = X^d, X \in GF(2^n)$, *then* $P_f = P_F \leq (m+1)/2^n$.

Proof: Because of Proposition 11.4.4 it suffices to prove the case $d = 2^k+1$. Let

$$G(X,\beta) = X^d + (X+\beta)^d = X^{2^k}\beta + X\beta^{2^k} + \beta^{2^k+1} = \alpha \tag{11.6}$$

Since $G(X,\beta)$ is a linearized function of X, we only need to consider the number of solutions of the equation

$$\beta X^{2^k} + \beta^{2^k} X = 0, \tag{11.7}$$

which is equivalent to $X = 0$ or $(X\beta^{-1})^{2^k-1} = 1$.

Set $H = \{x : x^u = 1, x \in GF(2^n)\}$. Clearly, H is a subgroup of the cyclic group $GF(2^n)^*$. So it is also cyclic, say $H = (h)$, then $h^m = 1$ for some integer m. Hence $\operatorname{ord}(h)$ divides m. It follows that the number of solutions of (11.7) is at most $m+1$, so is that of (11.6). This proves the theorem. □

Corollary 11.4.12 *If* $\gcd(2^k + 1, 2^n - 1) = 1$, *then the permutations* $X^{2^l(2^k+1)}$ *and* $[X^{2^l(2^k+1)}]_B$ *are APN iff* $\gcd(k,n) = 1$.

Proof: The permutation $[X^{2^l(2^k+1)}]_B$ is APN iff $m = \gcd(2^k-1, 2^n-1) = 1$, which is equivalent to $\gcd(k,n) = 1$. Of course the conditions $\gcd(k,n) = 1$ and $\gcd(2^k + 1, 2^n - 1) = 1$ together imply that n is odd. □

The result of Corollary 11.4.12 has been proved in [19] and [334]. Note that $\operatorname{ord}([X^{2^l(2^k+1)}]_B) = 2$ by Proposition 11.4.6.

11.4.3 Permutations of Order 3

Linear structures could be fatal for the security of some block ciphers [152]. For a quadratic APN permutation $f = (f_1, \cdots, f_n)$ in $GF(2)^n$, it is not difficult to see that each $f_i(x)$ has a linear structure, i.e., there is a nonzero vector w such that $f_i(x) + f_i(x+w) = f_i(w) + f_i(0)$. This may be a cryptographic fault. In this sense it is important to construct permutations which have good nonlinearity and high nonlinear order.

Proposition 11.4.13 *Let n be odd, $d = 2^{i+2} + 2^{i+1} + 2^i$, and $i \geq 0$. If B is a basis of $GF(2^n)$ over $GF(2)$, $f(x) = [X^d]_B$ and $F(x) = X^d$, then* $\text{ord}(f) = 3$ *and $P_f = P_F = 2^{1-n}$ or $3 \times 2^{1-n}$.*

Proof: Because of Proposition 11.4.4 and $d = 7 \times 2^i$, it suffices to prove the case $d = 7$. Let

$$G(X, \beta) = X^d + (X+\beta)^d, \quad \beta \neq 0, \tag{11.8}$$

then $G(X,\beta) = \alpha$ is equivalent to

$$Y^d + (Y+1)^d = r, \tag{11.9}$$

where $Y = X/\beta$, $r = \alpha\beta^{-d}$. If $r = 1$ and $d = 7$, then (11.9) is equivalent to $Y(Y^6 - 1) = 0$. Since $\gcd(6, 2^n - 1) = \gcd(3, 2^n - 1) = 1$, (11.9) has only two solutions.

If $r \neq 0$, assume that (11.9) has two solutions in $GF(2^n)$, say Y_1, $1 + Y_1$. Suppose it has another two solutions Y_2 and $1 + Y_2$ in $GF(2^n)$, let Y_3 and $1 + Y_3$ be the other two solutions of (11.9) in an extension field of $GF(2^n)$. By making use of the relationships between the coefficients and roots of (11.9), we get

$$Y_1 + Y_2 + Y_3 = 0 \text{ or } 1.$$

This means that $Y_3 \in GF(2^n)$. Thus $G(X,\beta) = \alpha$ has either no solution or two solutions or six solutions in $GF(2^n)$. This proves the first part of the theorem. Finally, it follows from Proposition 11.4.6 that $\text{ord}(f) = 3$. □

In the following subsections we will see that permutation $[X^7]_B$ of $GF(2)^5$ is APN. We now discuss when the $f(x)$ in Proposition 11.4.13 is APN. If (11.9) has more than two solutions in $GF(2^n)$, then it follows from the above proof that it has six solutions, say, Y_1, $1 + Y_1$, Y_2, $1 + Y_2$, Y_3, $1 + Y_3$. By making use of the relations between the coefficients and roots of (11.9), we get

$$\begin{cases} (Y_1^2 + Y_1)^2 + (Y_2^2 + Y_2)^2 + (Y_1^2 + Y_1)(Y_2^2 + Y_2) = 1 \\ (Y_1^2 + Y_1)(Y_2^2 + Y_2)(Y_2^2 + Y_2 + Y_1^2 + Y_1) = r + 1 \end{cases}$$

Let $Y_1^2 + Y_1 = a$, $Y_2^2 + Y_2 = b$, then a, $b \in GF(2^n)$. Thus we obtain

$$\begin{cases} a^2 + b^2 + ab = 1 \\ ab(a+b) = r + 1, \end{cases}$$

which is equivalent to

$$\begin{cases} b^3 + b + r + 1 = 0 \\ a^3 + a + r + 1 = 0 \\ (a+b)^3 + a + b + r + 1 = 0, \end{cases}$$

because a, $b \neq 0$. Hence the equation

$$X^3 + X + r + 1 = 0 \tag{11.10}$$

has three solutions in $GF(2^n)$. On the other hand, let

$$X^3 + X + r + 1 = (X + a)(X^2 + aX + c),$$

then we have

$$\begin{cases} a^2 + c = 1 \\ ac = r + 1. \end{cases}$$

It is easy to prove that $Y^2 + Y + e$ is irreducible in $GF(2^n)$ if and only if $\mathrm{Tr}(e) = 1$. Thus

$$X^2 + aX + a^2 + 1 = a^2[(X/a)^2 + (X/a) + (a^2+1)/a^2]$$

is irreducible in $GF(2^n)$ if and only if

$$\begin{aligned} \mathrm{Tr}((a^2+1)/a^2) &= \mathrm{Tr}(1 + a^{-2}) \\ &= \mathrm{Tr}(1) + \mathrm{Tr}(a^{-2}) \\ &= \mathrm{Tr}(1) + \mathrm{Tr}(a^{-1}) \\ &= 1. \end{aligned}$$

Since n is odd, we have $\mathrm{Tr}(1) = 1$. Therefore $X^3 + X + r + 1$ has only one solution if and only if $\mathrm{Tr}(a^{-1}) = 0$. Thus, if we can find a condition that ensures $\mathrm{Tr}((Y^2+Y)^{-1}) = 0$ for every solution Y of (11.9), the permutations f and $F(X)$ in Proposition 11.4.13 must be APN, where n is odd.

11.4.4 APN Permutations of Order $n - 1$

It has already been mentioned that constructing higher order permutations with ideal nonlinearity is cryptographically desirable. This subsection presents a class of maximum order APN permutations of $GF(2)^n$.

Proposition 11.4.14 *Let n be odd and $d = 2^n - 2^i - 1$, $0 \le i \le n-1$. If B is a basis of $GF(2^n)$ over $GF(2)$, then $f(x) = [X^d]_B$ is a maximum order APN permutation of $GF(2)^n$ and $F(X) = X^d$ is an APN permutation of $GF(2^n)$.*

Proof: We first consider the case $i = 0$. Then $F(X) = X^d = 0$ when $X = 0$, $F(X) = X^{-1}$ otherwise. Now we analyze the number of solutions of the equation

$$X^d + (X + \beta)^d = \alpha \tag{11.11}$$

If $\alpha = \beta^d$, then 0 and β are two solutions of (11.11) in $GF(2^n)$. Suppose that $X \neq 0$, β, is another solution of (11.11) in $GF(2^n)$, then from (11.11) we get

$$X^2 + \beta X + \beta^2 = 0. \tag{11.12}$$

It follows that $X^3 = \beta^3$, which gives $X = \beta$, because $\gcd(3, 2^n - 1) = 1$. This is a contradiction. Hence, in this case (11.11) has only two solutions.

If $\alpha \neq \beta^d$, then 0 and β are not solutions of (11.11), whence (11.11) can be written as

$$G(X, \beta) = X^{-1} + (X + \beta)^{-1} = \beta / X(X + \beta) = \alpha, \tag{11.13}$$

which is equivalent to

$$X^2 + \beta X + \alpha^{-1}\beta = 0. \tag{11.14}$$

Obviously, (11.14) has at most two solutions for each $\alpha \neq \beta^d$, hence so does (11.13).

Summarizing the above results, we see that $[X^d]_B$ and X^d are APN. Since $d = 2^n - 2$, we get $\mathrm{WH}(d) = n - 1$, whence $\mathrm{ord}(f) = n - 1$. Finally, it follows from Proposition 11.4.4 that for each $d = 2^n - 1 - 2^i$ the permutation is APN. Thus, the conclusion of the proposition is true. □

11.4.5 Permutations of Order $n - 2$

This section presents a class of permutations of order $n - 2$ in $GF(2)^n$ and in $GF(2^n)$ with good nonlinearity.

Theorem 11.4.15 *Let n be odd, $\gcd(3, n) = 1$ and $d = 2^n - 2^{i+1} - 2^i - 1$, $0 \le i \le n - 2$. Then the permutation $f(x) = [X^d]_B$ has order $n - 2$ and nonlinearity $P_f = 2^{1-n}$ or $3 \times 2^{1-n}$.*

Proof: Because of Proposition 11.4.4 it suffices to prove the case $d = 2^n - 1 - 3$. Consider now the equation

$$G(X, \beta) = X^d + (X + \beta)^d = \alpha, \ \alpha \neq \beta^d, 0. \tag{11.15}$$

Clearly, 0 and β are not solutions of (11.15). Therefore (11.15) is equivalent to

$$X^6 + X^5\beta + X^4\beta^2 + X^3\beta^3 + X^2\alpha^{-1}\beta + X\alpha^{-1}\beta^2 + \alpha^{-1}\beta^3 = 0. \tag{11.16}$$

Similar to the proof of Proposition 11.4.13, we can prove that (11.15) has either no solution or two or six solutions in $GF(2^n)$.

What remains to be considered, is the equation

$$X^d + (X + \beta)^d = \beta^d. \tag{11.17}$$

Let $Y = X/\beta$, then (11.17) is equivalent to

$$Y^d + (1 + Y)^d = 1. \tag{11.18}$$

We claim that (11.18) has only two solutions 0 and 1 in $GF(2^n)$. If not, say that $Y_1 \neq 0, 1$, is another one in $GF(2^n)$. Then we get

$$1 + Y_1 + Y_1^2 + Y_1^3 + Y_1^4 + Y_1^5 + Y_1^6 = 0, \tag{11.19}$$

whence $Y_1^7 = 1$. Since $\gcd(3, n) = 1$, it follows that $Y_1 = 1$, a contradiction. Hence (11.18) has only two solutions in $GF(2^n)$.

Summarizing the above results, we see that $P_f = 2^{1-n}$ or $3 \times 2^{1-n}$. It can be easily seen that $\text{ord}(f)=n-2$. □

11.4.6 Permutations X^d with $d = 2^m - 1$

For $d = 2^m + 1$ with $\gcd(m, n) = 1$, we have seen that X^d is APN in $GF(2^n)$. It is natural to ask whether the permutation X^{2^m-1} is APN. A simple example is x^7 which is APN in $GF(2^5)$, but not APN in $GF(2^4)$. Therefore X^{2^m-1} may be APN or not in $GF(2^n)$, depending on the structure of the field $GF(2^n)$. To investigate the problem, we may need the following lemma:

Lemma 11.4.16 *Assume that $2^n - 1$ is prime, then each nonzero conjugacy class of $Z^*_{2^n-1}$ mod $(2^n - 1)$ has n elements, and there are $(2^n - 2)/n$ such conjugacy classes.*

Since $d = 2^m - 1$, we get

$$G(X, \beta) = X^d + (X + \beta)^d = \beta^d(Y^d + 1)/(Y + 1),\ X \neq \beta,$$

where $Y = X/\beta$. Therefore we need only to discuss the number of solutions of the equation

$$Y^{2^m-1} + 1 = r(Y + 1),\ r \neq 0,\ 1. \tag{11.20}$$

Counting the solutions of this equation remains an open problem.

11.4.7 APN Permutations via Crosscorrelation Function

By definition APN permutations are related to the autocorrelation function of permutations. A crosscorrelation function approach to permutations of $GF(2^n)$ was given by Cusick [89]. With this approach it is possible to get some new classes of permutations with ideal nonlinearity which might be difficult to get with the direct approach in the foregoing subsections. In this section we introduce some APN permutations via the crosscorrelation approach [89].

It is well-known that all maximum-length sequences of period $q^m - 1$ over $GF(q)$ can be obtained by decimating one maximum-length sequence of period $q^m - 1$ over $GF(q)$ (see [276] or [169]). Since we are concerned with permutations of $GF(2^n)$, our discussion will be restricted to $GF(2)$. By choosing an additive character of $GF(q)$, it is possible to generalize the following discussion.

Let a^∞ and b^∞ be two maximum-length sequences of period $2^n - 1$ over $GF(2)$, then there exist an integer d with $\gcd(d, 2^n - 1) = 1$ and an integer t such that $b_{j+t} = a_{dj}$ for all j. Thus, the periodic crosscorrelation function of the two sequences defined in Section 2.3.3 becomes

$$\mathrm{CC}(d, t) = \sum_{j=0}^{2^n-2} (-1)^{a_{j+t}+a_{dj}} \qquad (0 \leq t \leq 2^n - 2).$$

Since the crosscorrelation function depends only on d and t, but not on the particular maximum-length sequence chosen, by the trace representation theorem (see [276] or [169]) we can take

$$a_j = \mathrm{Tr}(\alpha^j) \quad \text{for all } j,$$

where α is a primitive element of $GF(2^n)$. Hence the crosscorrelation function is converted into

$$\mathrm{CC}(d, t) = \sum_{j=0}^{2^n-2} (-1)^{\mathrm{Tr}(\alpha^{j+t}+\alpha^{dj})} = \sum_{x \in GF(2^n), x \neq 0} (-1)^{\mathrm{Tr}(x\alpha^t + x^d)} \tag{11.21}$$

To make things simple, consider the function $A_d(y)$ defined by

$$A_d(y) = \mathrm{CC}(d,t) + 1 = \sum_{x \in GF(2^n)} (-1)^{\mathrm{Tr}(xy+x^d)} \qquad (y = \alpha^t), \qquad (11.22)$$

where the second equality follows from (11.21).

One of the main results obtained by Cusick [89] with the crosscorrelation function approach is the class of APN permutations described by the following theorem.

Theorem 11.4.17 *Suppose n is odd, $d = 2^{2k} - 2^k + 1$ and $\gcd(k,n) = 1$. Then the permutation of $GF(2^n)$ given by $f(x) = x^d$ is APN.*

To prove the theorem, we need a number of lemmas as described below.

Lemma 11.4.18 *Let symbols be the same as before. Then*

1. $\sum_{y \in GF(2^n)} (A_d(y))^2 = 2^{2n}$.
2. $\sum_{y \in GF(2^n)} (A_d(y))^3 = 2^{2n} S_d$,

where S_d is the number of pairs x_1, x_2 of elements of $GF(2^n)$ such that $x_1 + x_2 + 1 = 0$ and $x_1^d + x_2^d + 1 = 0$ simultaneously.

The proof of the first part is easy (see Niho [327, p. 25] or Helleseth [196, p. 214]). The proof of the second part is given by Helleseth in [196, pp. 214-215].

Define the polynomial

$$f_d(x) = x^d + (x+1)^d + 1.$$

It is seen that the S_d defined in Lemma 11.4.18 is equal to the number of roots of $f_d(x)$ in $GF(2^n)$.

Let $\{\delta_i : 1 \le i \le G(n)\}$ denote the set of all possible distinct nonzero values of $f_d(x)$ for $x \in GF(2^n)$ and $\sharp(\delta_i)$ denote the number of values of x in $GF(2^n)$ such that $f_d(x) = \delta_i$. Then it follows that

$$\sum_{i=1}^{G(n)} \sharp(\delta_i) = 2^n - S_d. \qquad (11.23)$$

Further we define

$$M_n = \sum_{i=1}^{G(n)} \sharp(\delta_i)^2.$$

The key idea of Cusick's proof of Theorem 11.4.17 is the evaluation of the fourth power sum in the following lemma.

Lemma 11.4.19 *Suppose* $\gcd(d, 2^n - 1) = 1$. *Then*

$$\sum_{y \in GF(2^n)} (A_d(y))^4 = 2^{2n}(S_d^2 + M_n).$$

Proof: By (11.22) we have

$$\sum_{y \in GF(2^n)} (A_d(y))^4 = \sum_{x_1+x_2+x_3+x_4=0} (-1)^{\mathrm{Tr}(x_1^d+\cdots+x_4^d)} =: 2^n R. \qquad (11.24)$$

Let h denote the number of solutions $(x_1, x_2, x_3) \in GF(2^n)^3$ of the following equations

$$\begin{cases} x_1 + x_2 + x_3 + 1 = 0, \\ x_1^d + x_2^d + x_3^d + 1 = 0. \end{cases} \qquad (11.25)$$

For any solution $\alpha \neq 0$ in $GF(2^n)$ of (11.25) the $2^n - 1$ associated quadruples $\alpha(x_1, x_2, x_3, 1)$ together contribute $2^n - 1$ to the sum R in (11.24). If a triple (x_1, x_2, x_3) satisfies

$$\begin{cases} x_1 + x_2 + x_3 + 1 = 0, \\ x_1^d + x_2^d + x_3^d + 1 \neq 0, \end{cases} \qquad (11.26)$$

then the $2^n - 1$ associated quadruples $\alpha(x_1, x_2, x_3, 1)$ together contribute -1 to the sum R in (11.24). The only quadruples (x_1, x_2, x_3, x_4) which are not associated with a solution of either (11.25) or (11.26) are those of the form $(x_1, x_2, x_3, 0)$. By the second part of Lemma 11.4.18 the contribution of those quadruples of the sum R is $2^n S_d$. Combining the above results together, we have

$$R = (2^n - 1)h + (-1)(2^{2n} - h) + 2^n S_d = 2^n(h - 2^n + S_d). \qquad (11.27)$$

Now it remains to evaluate h. In (11.25) we first suppose that $x_1 + x_2 = \beta \neq 0$ for some $\beta \in GF(2^n)$ and define γ by $\gamma^{-1} = \beta$. Now $x_3 = x_1 + x_2 + 1 = \beta + 1$, so the simultaneous equations (11.25) are equivalent to

$$x^d + (x + \beta)^d + (\beta + 1)^d + 1 = 0, \qquad (11.28)$$

where we put $x_1 = x$. If we replace x by βx, then (11.28) becomes

$$\beta^d(x^d + (x+1)^d + 1) + \beta^d + (\beta+1)^d + 1 = 0$$

or

$$f_d(x) = f_d(\gamma). \qquad (11.29)$$

Thus solutions of (11.25) with $x_1 + x_2 = \gamma^{-1} \neq 0$ and $x_3 = \gamma^{-1} + 1$ are in one-to-one correspondence with solutions x of (11.29).

There are $S_d - 1$ nonzero values of γ such that $f_d(\gamma) = 0$; each of these gives S_d values of x satisfying (11.29). Thus, we obtain $S_d(S_d - 1)$ solutions of (11.25). The solutions of $f_d(x) = \delta_i$, where $1 \leq i \leq G(n)$, give M_n solutions of (11.24). The only remaining solutions of (11.25) are the 2^n solutions with $x_1 + x_2 = 0, x_3 = 1$. Hence we have

$$h = S_d(S_d - 1) + M_n + 2^n. \tag{11.30}$$

Combining (11.24), (11.27) and (11.30) completes the proof of Lemma 11.4.19. □

The crosscorrelation spectrum is defined to be the set of values taken on by the crosscorrelation function together with a count of the number of times each value occurs. The next lemma [228, 327] gives the crosscorrelation spectrum for the value of d in Theorem 11.4.17.

Lemma 11.4.20 *Suppose $n = 2^{2k} - 2^k + 1, g = \gcd(k, n)$ and n/g is odd. Then the crosscorrelation spectrum for $A_d(y)$ is given by Table 11.1.*

Table 11.1: The spectrum.

Value of $A_d(y)$	Number of times given value occurs
$2^{(n+g)/2}$	$2^{n-g-1} + 2^{(n-g-2)/2}$
0	2^{n-1}
$-2^{(n+g)/2}$	$2^{n-g-1} - 2^{(n-g-2)/2}$

Proof of Theorem 11.4.17: We have $d = 2^{2k} - 2^k + 1$. By Lemma 11.4.20 we obtain

$$\sum_{y \in GF(2^n)} (A_d(y))^3 = 2^{2n+1}, \qquad \sum_{y \in GF(2^n)} (A_d(y))^4 = 2^{3n+1}.$$

It then follows from Lemmas 11.4.18 and 11.4.19 that

$$S_d = 2 \text{ and } M_n = 2^{n+1} - 4.$$

Now (11.23) and the definition of M_n give

$$2 \sum_{i=1}^{G(n)} \sharp(\delta_i) = \sum_{i=1}^{G(n)} \sharp(\delta_i)^2 = 2^{n+1} - 4. \tag{11.31}$$

Since $\sharp(\delta_i)$ is always an even positive integer (because $f_d(x) = f_d(x+1)$), the first equality in (11.31) implies

$$\sharp(\delta_i) = 2 \text{ for all } i. \tag{11.32}$$

The second equality in (11.31) gives $G(n) = 2^{n-1} - 1$ and we have from (11.32)

$$|\{x : x^d + (x+1)^d = \beta\}| = 2 \text{ for all } \beta \in GF(2^n). \tag{11.33}$$

Since for any $\alpha \in GF(2^n)$ we have

$$x^d + (x+1)^d = \beta \text{ implies } y^d + (y+\alpha)^d = \alpha^d \beta,$$

where $y = \alpha x$, it follows from (11.33) that $\delta(x^d) = 2$ and this proves Theorem 11.4.17. □

Another possible proof for Theorem 11.4.17 can be described as follows: If n is odd, $d = 2^{2k} - 2^k + 1$ and $\gcd(k, n) = 1$, then Lemma 11.4.20 implies that the function $f(x) = x^n$ is "almost bent" in the sense of Chabaud and Vaudenay [67]. Now Theorem 2 of [67] says that the almost bent function $f(x)$ must be APN.

11.4.8 Other Power Functions with Good Nonlinearity

Recently new power functions x^d over $GF(2^n)$ with good nonlinearity have been found. Let $n = 2m - 1$, $d = 2^m - 1$, where $m \geq 2$. Helleseth and Sandberg have proved that x^d is APN [202]. They have also found two families of other power mappings with good nonlinearity [202]. For other advances in this direction we refer to Dobbertin [141].

11.4.9 Choosing the Linear Functions

To choose the linear functions for the generator of Figure 11.2, we may use the trace functions $\mathrm{T}_a(x) = \mathrm{Tr}_{GF(2^n):GF(2)}(ax)$, where $a \neq 0$ is an element of $GF(2^n)$. The choice of the parameter a may be used to control the linear complexity of the output sequences. Other linear functions from $GF(2^n)$ to $GF(2)$ may also be suitable. This depends on the $GF(2^n)$ generator, which can be designed to generate the elements of $GF(2^n)$ in some prescribed order so that it together with the cryptographic functions ensure large linear and sphere complexity of the output sequence.

For the design of cryptographic functions from $GF(2)^n$ to $GF(2)$ for similar generators, linear functions $l(x) = \sum_{i=0}^{n-1} l_i x_i$ can be used. But to ensure ideal diffusion, the function $x_0 + x_1 + \cdots + x_{n-1}$ may be the best one from this point of view.

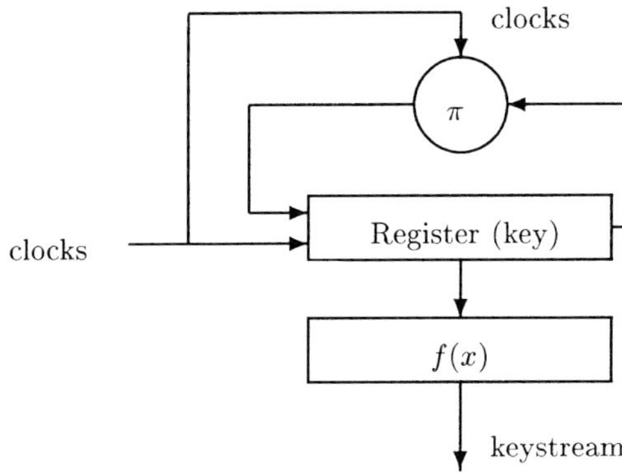

Figure 11.3: A general description of cyclic-key generators.

11.5 Cyclic-Key Generators and their Problems

Some generators based on permutations have been discussed in the foregoing two sections. In this section we will give an overview of some generators from a high level. The kind of cyclic-key generators described in this section includes various known generators, such as counter generators, the natural sequence generator, the nonlinear filter, and the nonlinear combiner. Furthermore, new cyclic-key generators could be designed.

11.5.1 Cyclic-Key Generators

Cyclic-key generators refer to those whose keys are initial states of generators with internal memory which changes cyclically according to the clock of the generator. Thus the key of the generators is time-varying. Typically, many cyclic-key generators can be depicted by Figure 11.3, where π is a permutation of the key space K, and $f(x)$ is a cryptographic function which applies to the content of the register. Thus we regard the permutation π as one part of the cryptographic algorithms, not as one part of the key. Once the key space and the permutation are chosen, the key at time i produces the key for the time $i+1$, i.e., $k_{i+1} = \pi(k_i), i = 0, 1, ...$ Thus, the cyclic-key sequence must repeat. If the key space has n elements, then for any permutation π we have $k_{i+n} = k_i$ for all $i \geq 0$. To describe the cyclic-key generators better, we need some notions in group theory.

Let S be a set and G any group. By a *group action* of G or a *G-action* on S we mean a mapping $\mu : S \times G \to S$, i.e., a binary operation associating with any $s \in S$, $g \in G$ an element $\mu(s, g)$, such that

1. $\mu(s, gh) = \mu(\mu(s, g), h)$ for all $s \in S, g, h \in G$;

2. $\mu(s, 1) = s$ for all $s \in S$.

We express this fact also by saying that G *acts* on S and call S a G-*set.* Usually we write $g(s)$ instead of $\mu(s, g)$. When we have a G-action on S, each $g \in G$ defines a mapping $\varphi_g(s) = g(s)$ of S into itself. In terms of these mappings the rules (1) and (2) above are expressed by the equations:

$$\varphi_{gh} = \varphi_g \varphi_h, \ \varphi_1 = 1. \tag{11.34}$$

Thus the mapping

$$g \mapsto \varphi_g \tag{11.35}$$

is a homomorphism (of monoids) from G to the monoid of all mappings of S into itself. This means that each φ is actually a permutation of S.

Let S be a set, then any group acting on S can be used to define an equivalence relation on S by putting

$$x \sim y \text{ iff } y = g(x) \text{ for some } g \in G. \tag{11.36}$$

The equivalence classes are called the *orbits* of the action and the orbit containing x is written Gx. If S consists of a single orbit, G is said to act *transitively.*

For an arbitrary G-set S we call the set

$$G_s = \{g \in G | g(s) = s\}$$

the *stabilizer* of s under the action of G. Concerning the number of elements of an orbit we have the following result:

Proposition 11.5.1 *If a group G acts on a set S and $s \in S$ is a point lying in a finite orbit Gs, then the number of elements of Gs equals the index of the stabilizer G_s of s:*

$$|Gs| = [G : G_s]. \tag{11.37}$$

For a key space K of n elements, all the permutations on K form a group, i.e., the symmetric group on n symbols. Clearly, not all the permutations of K can be used for the cyclic-key generator of Figure 11.3. As the cyclic key has a period, it is cryptographically necessary to require that the cyclic key has a large period under the permutation π. Let Sym_K denote all the permutations on K, and $\Pi = (\pi)$ the cyclic subgroup of Sym_K generated by the permutation π. If Π acts on K transitively, we call π a *primitive permutation* of K. Under a primitive permutation all the cyclic keys have the maximum period $|K|$. The primitive permutations give the maximum least period of the output sequences provided that the cryptographic function $f(x)$ is chosen properly. For a primitive permutation π there is only one orbit when the group Π acts on the set K.

To design the cyclic-key generator of Figure 11.3 we choose

- a key space K, which has an algebraic structure (for example, a ring, a field, or an Abelian group).
- a permutation π on K;
- a ring $(G, +, \cdot)$;
- a mapping $f(x)$ from K to G

such that

1. the action of Π on K results only in orbits with large numbers of elements for the purpose of ensuring a large period of the output sequences;
2. the permutation π can be realized algebraically and efficiently;
3. $|K|$ has special forms with respect to $(G, +, \cdot)$ for the control of the linear and sphere complexity of the output sequences;
4. the function $f(x)$ has good difference property and good nonlinearity with respect to the two binary operations of K and G respectively. The choice of the binary operation of $(G, +)$ depends on that of the realization operation of the permutation π.

Before discussing specific designs of the generator, we turn to some basic facts about orbits. Proposition 11.5.1 gives the number of elements in an orbit. Concerning the number of orbits when a finite group acts on a G-set S, we have the following well known proposition:

Proposition 11.5.2 *Let G be a finite group acting on a finite G-set S. For each $g \in G$ let c_g be the number of points fixed by g, then the number of orbits is*

$$t = \frac{1}{|G|} \sum_{g \in G} c_g. \tag{11.38}$$

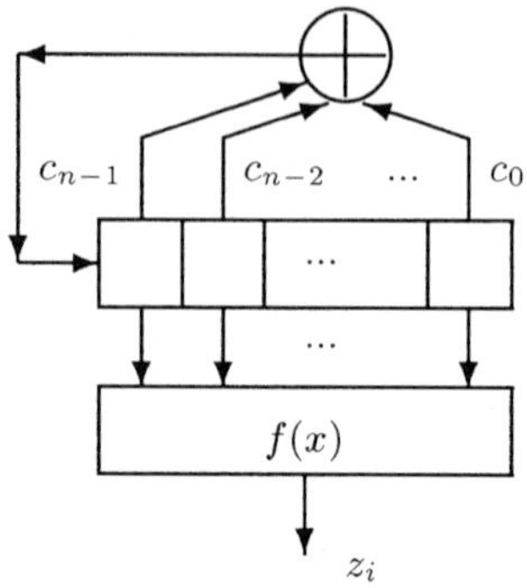

Figure 11.4: The nonlinear filter.

Thus t is the "average" number of points fixed by a permutation. This is a fundamental result. To prove (11.38) we count the number of pairs $(x, g) \in S \times G$ such that $g(x) = x$ in two ways: on the one hand, for each $g \in G$, the number of pairs occurring is c_g; on the other hand, for each orbit, of k points say, each x is fixed by the elements of its stabilizer, which by the orbit formula has $|G|/k$ elements. Thus each orbit contributes $|G|$ pairs in all and so $\sum c_g = |G| \cdot t$, where t is the number of orbits. Now (11.38) follows on dividing by $|G|$.

11.5.2 Several Specific Forms: An Overview

Now we show that some generators can be considered as special forms of the generator of Figure 11.3. Consider the nonlinear filter $f(x)$ in Figure 11.4, where $c(x) = c_0 + c_1 x + \cdots + c_{n-1}x^{n-1}$ is the feedback polynomial of the LFSR and $f(x)$ is the filter function. Usually, $c(x)$ is taken over a finite field $GF(q)$ with $c_0, c_{n-1} \neq 0$ and $f(x)$ is a mapping from $GF(q)$ to $GF(r)$, where $GF(r)$ is a subfield of $GF(q)$. The key of the generator may consist of the state vector of the LFSR and the feedback polynomial $c(x)$. But it is usually assumed that the key consists only of the state vector of the LFSR. In other words, it is usually assumed that the feedback polynomial $c(x)$ is known. Usually, the function $f(x)$ is only applied to some taps of the LFSR, say, in positions n_1, ..., n_m, where $0 \leq n_i \leq n-1$. By extending the function $f(x)$, we can generally assume that the cryptographic function $f(x)$ applies to all the positions.

Let us consider the generator over the finite field $GF(q)$, and take $(GF(q)^n)^* := GF(q)^n \setminus \{0^n\}$ as the key space. Suppose a feedback polynomial $c(x) = c_0 + c_1 x + \cdots + c_n x^n$ with $c_0 = 1, c_1, c_n \neq 0$ is applied to the

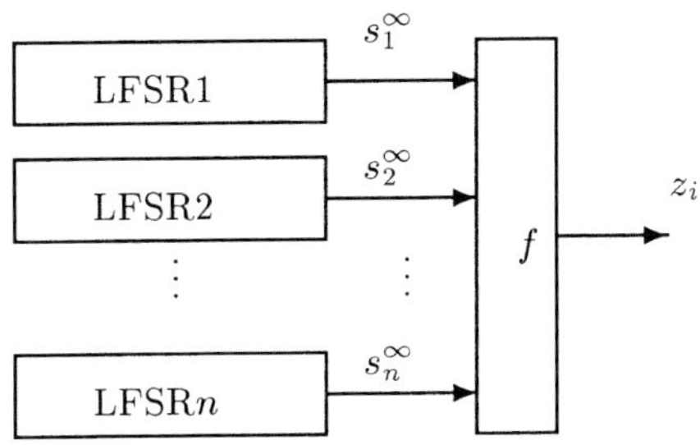

Figure 11.5: The nonlinear combiner.

LFSR. Let the matrix A be defined by

$$A = \begin{pmatrix} 0 & 0 & 0 & \cdots & 0 & 0 & -c_n \\ 1 & 0 & 0 & \cdots & 0 & 0 & -c_{n-1} \\ 0 & 1 & 0 & \cdots & 0 & 0 & -c_{n-2} \\ \vdots & \vdots & \vdots & \vdots & \vdots & \vdots & \vdots \\ 0 & 0 & 0 & \cdots & 1 & 0 & -c_2 \\ 0 & 0 & 0 & \cdots & 0 & 1 & -c_1 \end{pmatrix} \tag{11.39}$$

and the permutation π be

$$\pi(S) = SA,$$

where S is the state vector. Then the nonlinear filter of Figure 11.4 is a special case of the generator of Figure 11.3. Obviously, $c(x)$ is primitive iff the permutation π is transitive.

Another intensively studied generator is the nonlinear combiner depicted in Figure 11.5, where the output sequences of n LFSRs are combined by a cryptographic function $f(x)$. This generator can also be viewed as a special case of the generator of Figure 11.3. For this, let $l_1, \cdots, l_n$ be the lengths of the LFSRs of Figure 11.5, and let $c_i(x)$ be the feedback polynomial of LFSRi, where $i = 1, \cdots, n$. Denote the corresponding matrix by A_i and take the set

$$K = (GF(q)^{l_1})^* \times \cdots \times (GF(q)^{l_n})^*$$

as the key space of the generator of Figure 11.5. If we define the permutation

$$\pi(S_1, \cdots, S_n) = (S_1 A_1, \cdots, S_n A_n),$$

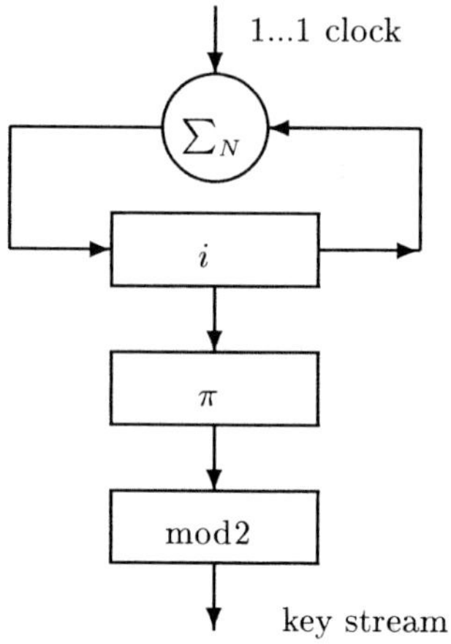

Figure 11.6: A NSG based on permutations of Z_N.

where $S_i \in (GF(q)^{l_i})^*$ for each i, we see that the nonlinear combiner of Figure 11.5 is also a special realization of the cyclic-key generator of Figure 11.3.

It is also easily seen that the generator of Figure 2.5 based on a ring counter is a special realization of the cyclic-key generator. Since every periodic generator can be realized by the NSGs of Figure 2.5, it follows that the NSGs are equivalent to the generators of Figure 11.3. Consequently, each periodic generator has a cyclic-key generator realization.

11.6 A Generator Based on Permutations of Z_m

To design the NSG of Figure 11.6, we wish to find a permutation π of Z_N such that its nonlinearity with respect to the addition of Z_N is good enough. As already made clear, our cryptographic idea here is still "good + bad = good". To design generators based on permutations of Z_N, the first thing we want to do is to find permutations of Z_N which have an efficient realization.

Our first consideration is the permutations of the form x^n on Z_m. If m is a prime, then it is well known that x^n is a permutation of Z_m iff $\gcd(n, m-1) = 1$. For m composite, there are also such permutations. If $m = p_1 \cdots p_r$ where p_i, $1 \le i \le r$, are distinct primes, we have the following conclusion due to Cordes [84].

Proposition 11.6.1 *Let $m = p_1 \cdots p_r$, where the p_i, $1 \le i \le r$, are distinct primes, and let n be a positive integer such that* $\gcd(n, (p_1-1)\cdots(p_r-1)) =$

1. *Then x^n yields a permutation mod m.*

Proof: The proof is by induction on r. The case $r = 1$ is well-known. Assume it holds for $r \leq k-1$ and consider $m = p_1 \cdots p_r$. Let $\{a_1, \cdots, a_{\phi(m)}\}$ be a complete reduced residue system mod m (i.e., $\gcd(a_i, m) = 1$). If $a_i^n \equiv a_j^n \pmod{m}$, then $(a_i/a_j)^n \equiv x^n \equiv 1 \pmod{m}$ and, hence $x^n \equiv 1 \pmod{p_i}, 1 \leq i \leq r$. So $\gcd(n, p_i - 1) = 1$ and it follows from the conclusion of the case $r = 1$ that $x \equiv 1 \pmod{p_i}$. But then $x \equiv 1 \pmod{m}$. Thus $\{a_1^n, \cdots, a_{\phi(m)}^n\}$ are distinct mod m.

Now suppose $a^n p_i^n \equiv b^n p_i^n \pmod{m}$ where $1 \leq a, b \leq m/p_i$. Then $a^n p_i^{n-1} \equiv b^n p_i^{n-1} \pmod{m/p_i}$ and so $a^n \equiv b^n \pmod{m/p_i}$ since p_i is invertible mod m/p_i. By the induction hypothesis, $a = b$. Consequently, $\{a^n p_i^n : a = 1, ..., m/p_i\}$ are distinct mod m. Moreover, for any i, the sets $\{a^n p_i^n : a = 1, ..., m/p_i\}$ and $\{a_1^n, \cdots, a_{\phi(m)}^n\}$ are disjoint since $a_j^n \not\equiv 0 \pmod{p_i}$, $1 \leq j \leq \phi(m)$. If $a^n p_i^n \equiv b^n p_j^n \pmod{m}$ for $i \neq j$ with $1 \leq a \leq m/p_i, 1 \leq b \leq m/p_i$, then clearly $p_i | b, p_j | a$. Let $a = p_j a', b = p_i b'$. It follows that

$$a'^n (p_i p_j)^{n-1} \equiv b'^n (p_i p_j)^{n-1} \pmod{\frac{m}{p_i p_j}},$$

and

$$p_i p_j \text{ invertible mod } \frac{m}{p_i p_j} \text{ implies } a'^n \equiv b'^n \pmod{\frac{m}{p_i p_j}}.$$

By the induction hypothesis and by $1 \leq a', b' \leq m/p_i p_j$, $a' = b'$, and so $ap_i = bp_j$. The only intersection the sets $\{ap_k^n : a = 1, ..., m/p_k\}$, $k = i, j$, have then is at the common multiples of $p_i p_j$. Suppose now that $x^n \equiv y^n \pmod{m}$ with $x \not\equiv y \pmod{m}$. From the beginning of this paragraph neither x nor y can be relatively prime to m. But then $x = ap_i, y = bp_j$ for some i, j with $1 \leq i, j \leq r$. Again by the above argument this implies $x \equiv y \pmod{m}$. This results in a contradiction. Thus $\{x^n : x = 1, ..., m\}$ are distinct mod m. This completes the induction and the proof of the proposition. □

A simple alternative proof of this proposition is to show that x^n is a permutation of Z_{p_i}. Then the conclusion follows from the Chinese remainder theorem.

Conversely suppose $\gcd(n, (p_1 - 1) \cdots (p_r - 1)) \neq 1$. Then there is at least one i such that $\gcd(n, p_i - 1) \neq 1$ and by the conclusion of the case $r = 1$, x^n does not yield a permutation mod p_i. If $x \not\equiv y \pmod{p_i}$ and $x^n \equiv y^n \pmod{p_i}$, then for $z = m/p_i, xz \not\equiv yz \pmod{m}$ and $(xz)^n \equiv (yz)^n \pmod{m}$. So x^n is not a permutation mod m. This gives the following proposition due to Cordes [84].

Proposition 11.6.2 *Let m, n be positive integers. Then $\{x^n : x = 1, ..., m\}$ are distinct mod m if and only if m is the product $p_1 \cdots p_r$ of distinct primes p_i, $1 \le i \le r$, and $\gcd(n, (p_1-1)\cdots(p_r-1)) = 1$.*

The smallest n that works then is the smallest prime q for which $\gcd(q, (p_1-1)\cdots(p_r-1)) = 1$. In [84] all the l such that x^{kl+1} yields a permutation mod m for all $k \ge 1$ has been found. By Proposition 11.6.2, $\gcd(kl+1, (p_1-1)\cdots(p_r-1)) = 1$ for all $k \ge 1$ is true if and only if $\gcd(kl+1, q) = 1$, $k \ge 1$, for all primes q dividing $\mathrm{lcm}(p_1-1, \cdots, p_r-1)$. And $\gcd(kl+1, q) = 1$ for all $k \ge 1$ is equivalent to $q|l$. Combining the above, Cordes arrived at the following conclusion:

Proposition 11.6.3 *Let $m = p_1 \cdots p_r$ be a product of the distinct primes $p_i, 1 \le i \le r$. Suppose $\{q_i : i = 1, ..., s\}$ are distinct prime factors of $lcm(p_1-1, \cdots, p_r-1)$. Then x^{kl+1} yields a permutation mod m for all $k \ge 1$ if and only if l is a multiple of $q_1 \cdots q_s$.*

Now we have all the possible permutations of the form x^n on Z_N when N is the product of distinct primes. For our generator, we need further to know whether there are permutations in this class which have good nonlinearity with respect to the addition of Z_N. If there are some, which ones are they? In what follows we present some permutations of Z_m and analyze their nonlinearity. The following theorem plays an important role in the nonlinearity analysis.

Theorem 11.6.4 *Let $f(x) = f_0 + f_1 x + \cdots + f_l x^l \in Z_m[x]$, where $m = m_1 m_2 \cdots m_t$ and m_i are pairwise relatively prime. Define*

$$f_i(x) = f_{i,0} + f_{i,1}x + \cdots + f_{i,l}x^l \in Z_{m_i}, \tag{11.40}$$

where $f_{i,j} = f_j \bmod m_i$ and $1 \le i \le t, 0 \le j \le l$. Then

$$\Pr(f(x+a) - f(x) = b) = \prod_{i=1}^{t} \Pr(f_i(x+a_i) - f_i(x) = b_i), \tag{11.41}$$

where $a_i = a \bmod m_i$, $b_i = b \bmod m_i$, variable x is random and takes on each possible element of Z_m (respectively, Z_{m_i}) equally often.

Proof: Let α be any random variable on Z_m and

$$\phi(x) = (x \bmod m_1, \cdots, x \bmod m_t).$$

Since ϕ is an isomorphism between Z_m and $Z_{m_1} \times \cdots \times Z_{m_t}$, the random variables $\alpha_i = \alpha \bmod m_i$ must be independent. It follows that the events

$(f_i(x+a_i)-f_i(x)=b_i)$ must be independent. By definition and the Chinese remainder theorem $f(x+a)-f(x)=b$ if and only if $f_i(x+a_i)-f_i(x)=b_i$ for all $i=1,2,...,t$. Thus, (11.41) follows. □

An alternative proof of the theorem is to prove the following equality

$$\begin{aligned} & |\{x \in Z_m : f(x+a)-f(x)=b\}| \\ = & \prod_{i=1}^{t} |\{y \in Z_{m_i} : f_i(y+a_i)-f_i(y)=b_i\}|. \end{aligned}$$

which is easy to derive from the Chinese remainder theorem.

We now analyze the nonlinearity of some permutations of the form x^e over $Z_{p_1p_2\cdots p_t}$.

Theorem 11.6.5 *Let $m=p_1p_2\cdots p_t$, where p_i are pairwise distinct primes no less than 5, and let $e=m-2$ and $f(x)=x^e$. For any pair of $a\neq 0, b\in Z_m$, define $a_i=a \bmod p_i$ and $b_i=b \bmod p_i$ for $i=1,...,t$.*

1. *If all $a_i\neq 0$, then*

$$\begin{aligned} & \Pr(f(x+a)-f(x)=b)=0 \text{ or} \\ & 1/m \le \Pr(f(x+a)-f(x)=b) \le 3^t/m. \end{aligned}$$

2. *If $a_{i_j}=0$ for $j=1,2,...,s$ and otherwise $a_i\neq 0$, where $1\le s<t$, and $b_{i_j}=0$ for $j=1,2,...,s$, then $\Pr(f(x+a)-f(x)=b)=0$ or*

$$\frac{1}{\prod_{v\neq i_1,\ldots,i_s} p_v} \le \Pr(f(x+a)-f(x)=b) \le \frac{4^{t-s}}{\prod_{v\neq i_1,\ldots,i_s} p_v}.$$

3. *If $a_{i_j}=0$ for $j=1,2,...,s$ and otherwise $a_i\neq 0$, where $1\le s<t$, and $b_{i_j}\neq 0$ for some j with $1\le j\le s$, then $\Pr(f(x+a)-f(x)=b)=0$.*

Proof: Define $e_i=p_i-2$ so $e_i\equiv e \bmod p_i$. Let $f_i(x)=x^{e_i}$ be the corresponding permutation polynomial of Z_{p_i}. Then

$$f_i(x+a_i)-f_i(x)=(x+a_i)^{p_i-2}-x^{p_i-2}.$$

If $a_i=0, b_i=0$, then $\Pr(f_i(x+a_i)-f_i(x)=b_i)=1$. If $a_i=0, b_i\neq 0$, then $\Pr(f_i(x+a_i)-f_i(x)=b_i)=0$. If $a_i\neq 0$ and $b_i=a_i^{-1}\neq 0$, then $x=0$ and $x=-a_i$ are obviously solutions of $f_i(x+a_i)-f_i(x)=b_i$, which may have other solutions. Suppose that x is such a solution. Thus, we have

$$\frac{1}{x+a_i}-\frac{1}{x}=b_i$$

which is equivalent to $x(x+a_i) = -b_i^{-1}a_i$ and has either no solution or two solutions other than $x = -a_i$. Hence in the case $a_i \neq 0$ and $b_i = a_i^{-1}$,

$$\Pr(f_i(x+a_i) - f_i(x) = b_i) = \frac{2}{p_i} \text{ or } \frac{4}{p_i} \tag{11.42}$$

If $a_i \neq 0$ and $b_i = a_i^{-1} \neq 0$, (11.42) also holds. If $a_i \neq 0$ and $b_i \neq a_i^{-1}$, we can similarly prove that

$$\Pr(f_i(x+a_i) - f_i(x) = b_i) = 0 \text{ or } \frac{2}{p_i}$$

If $a_i \neq 0$ and $b_i = 0$, then $\Pr(f_i(x+a_i) - f_i(x) = b_i) = 0$ since $f_i(x)$ is a permutation of Z_{p_i}. Combining the above results gives

$$\Pr(f_i(x+a_i) - f_i(x) = b_i) = \begin{cases} 1, & \text{if } a_i = 0, b_i = 0; \\ 0, & \text{if } a_i = 0, b_i \neq 0; \\ 0, & \text{if } a_i \neq 0, b_i = 0; \\ 2/p_i \text{ or } 4/p_i, & \text{if } a_i \neq 0, b_i = a_i^{-1}, \\ 0 \text{ or } 2/p_i, & \text{if } a_i \neq 0, b_i \neq a_i^{-1}. \end{cases} \tag{11.43}$$

With these formulae we are ready to prove the theorem. If all $a_i \neq 0$, then by (11.43) $\Pr(f_i(x+a_i) - f_i(x) = b_i)$ must take on one of $0, 2/p_i, 3/p_i, 4/p_i$. The conclusion of part one then follows from Theorem 11.6.4. If the conditions of part two hold, by (11.43) $\Pr(f_{i_j}(x+a_{i_j}) - f_{i_j}(x) = b_{i_j}) = 1$ for $j = 1, ..., s$. Then the conclusion of part two follows from that of part one. If the conditions of part three hold, then there must exist an integer i_j such that $b_{i_j} \neq 0$ and

$$\Pr(f_{i_j}(x+a_{i_j}) - f_{i_j}(x) = b_{i_j}) = \Pr(0 = b_{i_j}) = 0.$$

This proves part three. □

This theorem shows that the permutation polynomial x^{m-2} of Z_m has good nonlinearity when $|p_i - p_j|$ is small for each pair of (i, j) with $i \neq j$ and t is small.

If $\gcd(3, p_i - 1) = 1$ for $i = 1, 2, ..., t$, then x^3 is a permutation polynomial of $Z_{p_1 p_2 \cdots p_t}$. This permutation is cryptographically interesting since only two multiplications of $Z_{p_1 p_2 \cdots p_t}$ are needed to compute x^3. Thus, it could be an ideal cryptographic function if it has good nonlinearity.

Theorem 11.6.6 *Let $m = p_1 p_2 \cdots p_t$, where p_i are pairwise distinct primes with $\gcd(3, p_i - 1) = 1$, and let $\pi(x) = x^3$. For any pair of $0 \neq a, b \in Z_m$, define $a_i = a \bmod p_i$ and $b_i = b \bmod p_i$ for $i = 1, ..., t$.*

1. *If all $a_i \neq 0$, then*

$$\Pr(f(x+a) - f(x) = b) = 0 \text{ or } 2^t/m.$$

2. *If $a_{i_j} = 0$ for $j = 1, 2, ..., s$ and otherwise $a_i \neq 0$, where $1 \leq s < t$, and $b_{i_j} = 0$ for $j = 1, 2, ..., s$, then* $\Pr(f(x+a) - f(x) = b) = 0$ *or* $2^{t-s}/m$.

3. *If $a_{i_j} = 0$ for $j = 1, 2, ..., s$ and otherwise $a_i \neq 0$, where $1 \leq s < t$, and $b_{i_j} \neq 0$ for some j with $1 \leq j \leq s$, then* $\Pr(f(x+a) - f(x) = b) = 0$.

Proof: Consider first the permutation $\pi_i(x) = x^3$ of Z_{p_i}. Since $\pi_i(x+a_i) - \pi_i(x) = 3a_ix^2 + 3a_i^2x + a_i^3$, we have

$$\Pr(\pi_i(x+a_i) - \pi_i(x) = b_i) = \begin{cases} 1, & \text{if } a_i = 0, b_i = 0; \\ 0, & \text{if } a_i = 0, b_i \neq 0; \\ 0 \text{ or } 2/p_i, & \text{if } a_i \neq 0. \end{cases} \quad (11.44)$$

Thus, if all $a_i \neq 0$, by (11.44) $\Pr(\pi_i(x+a_i) - \pi_i(x) = b_i) = 0$ or $2/p$ for all i. If one of them is zero, by Theorem 11.6.4 $\Pr(\pi(x+a) - \pi(x) = b) = 0$; otherwise each $\Pr(\pi_i(x+a_i) - \pi_i(x) = b_i) = 2/p_i$. Again by Theorem 11.6.4 $\Pr(\pi(x+a) - \pi(x) = b) = 2^t/m$. This proves part one.

If the conditions of part two hold, then $\Pr(\pi_{i_j}(x+a_i) - \pi_{i_j}(x) = b_{i_j}) = 1$ for $i = 1, 2, ..., s$. Then the conclusion of part two follows from that of part one. If the conditions of part three hold, there must exist a $b_{i_j} \neq 0$, and thus $\Pr(\pi_{i_j}(x + a_{i_j}) - \pi_{i_j} = b_{i_j}) = 0$. The conclusion of part three thus follows from Theorem 11.6.4. □

This theorem shows that the nonlinearity of the permutation x^3 is also good when $|p_i - p_j|$ is small for each pair (p_i, p_j) with $i \neq j$ and t is small. Another interesting permutation polynomial of $Z_{p_1 \cdots p_t}$ is x^5, where p_i are pairwise distinct primes with $\gcd(5, p_i - 1) = 1$ for $i = 1, ..., t$. Similar to Theorem 11.6.6, one can prove the following conclusion about the nonlinearity of the permutation x^5.

Theorem 11.6.7 *Let $m = p_1p_2 \cdots p_t$, where p_i are pairwise distinct primes with* $\gcd(5, p_i - 1) = 1$, *and let $\pi(x) = x^5$. For any pair of $0 \neq a, b \in Z_m$, define $a_i = a \bmod p_i$ and $b_i = b \bmod p_i$ for $i = 1, ..., t$.*

1. *If all $a_i \neq 0$, then* $0 \leq \Pr(f(x+a) - f(x) = b) \leq 4^t/m$.

2. *If $a_{i_j} = 0$ for $j = 1, 2, ..., s$ and otherwise $a_i \neq 0$, where $1 \leq s < t$, and $b_{i_j} = 0$ for $j = 1, 2, ..., s$, then* $0 \leq \Pr(f(x+a) - f(x) = b) \leq 4^{t-s}/m$.

3. *If* $a_{i_j} = 0$ *for* $j = 1, 2, ..., s$ *and otherwise* $a_i \neq 0$, *where* $1 \leq s < t$, *and* $b_{i_j} \neq 0$ *for some* j *with* $1 \leq j \leq s$, *then* $\Pr(f(x+a) - f(x) = b) = 0$.

By the Chinese remainder theorem, one can easily prove that

$$\pi(x) = \sum_{i=1}^{t} (m/p_i) w_i x^{e_i} \tag{11.45}$$

are permutations of $Z_{p_1 \cdots p_t}$, where $\gcd(e_i, p_i - 1) = 1$ for all i, $m = p_1 p_2 \cdots p_t$, and where w_i are integers with $\gcd(w_i, p_i) = 1$. Thus, if $e_i = p_i - 2$ or 3, the permutation of (11.45) has good nonlinearity. For more about permutations on Z_m we refer to [134].

In what precedes the nonlinearity of a number of permutation polynomials of $Z_{p_1 \ldots p_t}$ has been analyzed. These permutations have good nonlinearity. The interesting case is $t = 2$ and the two primes are chosen to be approximately as equal as possible, and the most interesting case is when the two primes are twins. These permutation polynomials in the case that $t = 2$ and $|p_1 - p_2|$ is small could be ideal cryptographic functions for the sequence generator of Figure 11.6.

Finally, we consider the nonlinearity of the permutation $\pi(x) = (x^{p(p-1)} - 1)x + x^e$ of Z_{p^2}, where $e = p(p-1) - 1$. Note that

$$\pi(x) = \begin{cases} -x, & x \in R = \{0, p, 2p, ..., (p-1)p\}, \\ x^e, & x \in Z_{p^2}^*. \end{cases}$$

Consider first the case $a \in R$. Then $x \in R$ if and only if $x + a \in R$, and $x \in Z_{p^2}^*$ if and only if $x + a \in Z_{p^2}^*$. Hence

$$\begin{aligned} |\{x \in R : \pi(x+a) - \pi(x) = b\}| &= |\{x \in R : -a = b\}| \\ &= \begin{cases} p, & \text{if } b = -a, \\ 0, & \text{otherwise,} \end{cases} \end{aligned}$$

and

$$\begin{aligned} &|\{x \in Z_{p^2}^* : \pi(x+a) - \pi(x) = b\}| \\ = \; &|\{x \in Z_{p^2}^* : (x+a)^e - x^e = b\}| = 0 \text{ or } 2. \end{aligned}$$

Thus, $|\{x \in Z_{p^2} : \pi(x+a) - \pi(x) = b\}| \leq p + 2$ and

$$\Pr(\pi(x+a) - \pi(x) = b) \leq (p+2)/p^2.$$

Then we consider the case $a \in Z_{p^2}^*$. Let

$$h_1 = |\{x \in R : \pi(x+a) - \pi(x) = b\}| = |\{x \in R : (x+a)^e + x = b\}|.$$

The number of roots of $(x+a)^e + x = b$ can be estimated as follows. Since $a \in Z_{p^2}^*, x + a \in Z_{p^2}$. So $x \in R$ and $(x+a)^e + x = b$ if and only if $(x+a)^{e+1} + x(x+a) = b(x+a)$ if and only if $x^2 + (a-b)x + 1 - ab = 0$. Thus, $0 \le h_1 \le 2$.

Let

$$\begin{aligned} h_2 &= |\{x \in Z_{p^2}^* : \pi(x+a) - \pi(x) = b\}| \\ &= |\{x \in (R-a) : -(x+a) - x^e = b\}| + \\ &+ |\{x \in Z_{p^2}^* \setminus (R-a) : (x+a)^e - x^e = b\}| \\ &\le 2 + 2 = 4. \end{aligned}$$

Hence, if $a \in Z_{p^2}^*$, then

$$0 \le \Pr(\pi(x+a) - \pi(x) = b) \le (h_1 + h_2)/p^2 = 6/p^2.$$

Combining the above results proves the following theorem.

Theorem 11.6.8 *Let p be a prime, $e = (p-1)p-1$, and $\pi(x) = x^{p(p-1)+1} + x^{p(p-1)-1} - x$. Then*

$$\Pr(\pi(x+a) - \pi(x) = b) \le \begin{cases} (p+2)/p^2, & \text{if } a \in R, \\ 6/p^2, & \text{if } a \in Z_{p^2}^*, \end{cases}$$

where $\pi(x)$ is considered as a permutation polynomial of Z_{p^2}.

Thus, this permutation polynomial has good nonlinearity. On the other hand, it can be realized as $\pi(x) = x^{p(p-1)-1}(x^2+1) - x$. Thus, at most $\lceil \log_2(p(p-1)-1) \rceil + 2$ multiplications and two additions are needed to compute $\pi(x)$.

It is possible that some permutations of $Z_{p_1 p_2 \cdots p_t}$ have even better nonlinearity.

Research Problem 11.6.9 *Let $N = p_1 \cdots p_r$ be a product of r distinct primes. Find permutations of the form x^n of Z_N which have better nonlinearity with respect to the addition of Z_N.*

As shown in Chapter 3, the linear complexity of sequences of period $N = pq$, where p and q are distinct primes, is easy to control. So we are much more interested in permutations of Z_{pq} which have good nonlinearity with respect to the addition of Z_{pq}, where p and q are distinct primes with special forms.

There are two special cases for the generator of Figure 11.6 which are cryptographically interesting. If we choose N to be a large prime, and

$\pi(x) = x^e$ to be a permutation of Z_N with good nonlinearity with respect to the addition of Z_N, we get a generator of Figure 11.6 which includes the generator of Section 11.3 as a special case.

If we choose $N = pq$ and e a positive integer such that $\gcd(e, (p-1)(q-1)) = 1$, where p and q are primes, then x^e is the same as the RSA permutation [367]. The generator of Figure 11.6 based on this permutation is the RSA bit generator. Here we require that x^e has good nonlinearity with respect to the addition of Z_N. Thus we are only interested in special RSA bit generators. This generator is naturally different from the two-prime generator of Section 8.2 which is based on generalized cyclotomy.

If we choose $N = p(p+2)$ and e a positive integer such that $\gcd(e, (p-1)(p+1)) = 1$, where p and $p+2$ are twin primes, the generator of Figure 11.6 based on the permutation x^e could also be cryptographically interesting if the integer e is chosen such that x^e has good nonlinearity. This twin-prime based generator is different from the twin-prime generator of Section 8.2 based on cyclotomy.

Another interesting special case for the NSG of Figure 11.6 is when $N = p^2$, where p is a large prime. To design this generator, we need permutations of Z_{p^2} with good nonlinearity and a simple implementation. There might exist permutations of Z_{p^2} having better nonlinearity than the one described by Theorem 11.6.8. The control of the linear and sphere complexity of these generators is the same as that of the cyclotomic generators, as shown clearly by the results of Chapters 3 and 4.

Chapter 12

Quadratic Partitions and Cryptography

The quadratic partition problem of solving the Diophantine equation $p = x^2 + ny^2$ for a given integer n and prime p has been attacked by many mathematicians. Indeed, Dickson [110] lists results of over 100 mathematicians who made contributions to this problem. Among them are Lagrange, Legendre, Gauss, Goldbach and many others. However, they have investigated this problem only mathematically. Our main cryptographic interest in this problem comes from cyclotomic numbers. As we have seen in some of the preceding chapters, cyclotomic numbers are very useful in designing some keystream generators. The relation between cyclotomic numbers and the quadratic partition, already known to Gauss, has led us to the theory of quadratic partitions. Another cryptographic application of the quadratic partition $p = x^2 + ny^2$ is the search for primes with large norms in some integer domains other than Z. We need those primes to construct generators based on the arithmetic of those integer domains. This chapter is not intended to present all of the mathematical theories concerning quadratic partitions, but to mention those theories associated with some cryptographic quadratic partitions and to propose some problems from our cryptographic point of view. There may be some other applications of this mathematical problem to cryptography which remain to be investigated. It is interesting to note that an elementary result about the quadratic partition $p = x^2 + y^2$ has already been used to crack the Ong-Schnorr-Shamir digital signature scheme [151] successfully. This is discussed further in the last section of this chapter.

12.1 Quadratic Partition and Cryptography

The previous chapters have made it clear that the determination and stability analysis of (generalized) cyclotomic numbers are of considerable importance in the design and analysis of some stream ciphers. As we saw in Chapter 4, the determination and stability of cyclotomic numbers are completely determined by the quadratic partition of a prime p or ap, where a is an integer.

By the results of Chapter 4 and Appendix A, the cyclotomic numbers of order 3 are determined by the partition $4p = L^2+27M^2 = L^2+3(3M)^2$ with $L \equiv 1 \pmod 3$; the cyclotomic numbers of order 9 by $4p = L^2 + 27M^2 = L^2 + 3(3M)^2$ with $L \equiv 7 \pmod 9$ and a factorization of p in the field of 9th roots of unity [14]; the cyclotomic numbers of order 4 by $p = x^2 + 4y^2$ with $x \equiv 1 \pmod 4$; the cyclotomic numbers of orders 5 and 10 by

$$16p = x^2 + 50u^2 + 50v^2 + 125w^2$$

with $x \equiv 1 \pmod 5$ and $v^2 - 4uv - u^2 = xw$; the cyclotomic numbers of order 6 by $p = A^2 + 3B^2$; the cyclotomic numbers of 7 by $p = T^2 + 7U^2$ with $t \equiv 1 \pmod 7$; the cyclotomic numbers of orders 8 and 16 by

$$p = x^2 + 4y^2 = a^2 + 2b^2 \quad (x \equiv a \equiv 1 \pmod 4);$$

the cyclotomic numbers of order 12 by

$$p = x^2 + 4y^2 = A^2 + 3B^2 \quad (x \equiv 1 \pmod 4),\ A \equiv 1 \pmod 6));$$

and the cyclotomic numbers of order 15 by

$$\begin{aligned} p &= a^2 + 3b^2,\ a \equiv -1 \pmod 3 \\ &= c^2 + 15d^2,\ c \equiv -1 \pmod 3 \\ &= x^2 + 5u^2 + 5v^2 + 5w^2,\ xw = v^2 - uv - u^2,\ x \equiv -1 \pmod 5. \end{aligned}$$

The cyclotomic numbers of order 11 depend partially on the representation

$$4p = a^2 + 11b^2,$$

and those of order 24 on

$$\begin{aligned} p &= X^2 + 4Y^2,\ X \equiv 1 \pmod 4 \\ &= A^2 + 3B^2,\ A \equiv 1 \pmod 6 \\ &= C^2 + 2D^2,\ c \equiv 1 \pmod 4 \\ &= U^2 + 24V^2,\ U \equiv -c \pmod 3. \end{aligned}$$

Thus, most of the cyclotomic numbers depend completely or partially on the representation of primes in the form $p = x^2 + ny^2$, which will be the main topic of the following sections.

As seen in some of the foregoing chapters, residue difference sets are cryptographically attractive. Some of them are also related to the quadratic partition of primes. The biquadratic residues modulo a prime p form a difference set if and only if $p = 3^2 + (2t)^2$. If $p = 6f + 1 = x^2 + 3 \times 3^2$, then $D_0 + D_1 + D_3$ form a difference set, where the D_i's are defined in Chapter 4. If $p = 8f + 1$, then $D_0 \cup \{0\}$ is a difference set if and only if the following quadratic partitions hold at the same time:

$$\begin{aligned} p &= (21)^2 + (8y)^2 \\ &= 7^2 + 8b^2. \end{aligned}$$

These facts again show the cryptographic importance of the quadratic partition of primes.

12.2 $p = x^2 + y^2$ and $p = x^2 + 4y^2$

The two-square problem in number theory has two parts: (a) characterize the set of integers, for which the diophantine equation

$$x^2 + y^2 = n \tag{12.1}$$

has solutions in integers x and y; (b) determine the solutions of (12.1) for given n if it has some. These two parts are both cryptographically meaningful. For our application we need only one specific solution of (12.1) for evaluating the nonlinearity and difference property of some cryptographic functions. Cryptographically we are mostly interested in (12.1) when n is a prime. In this case we can assume the partition is of the form $p = x^2 + 4y^2$, since p is prime.

There are quite a number of mathematicians who have attacked and contributed to this two-square problem, including Mohamed Ben Alhocain, Leonard da Pisa (better known as Fibonacci), Vieta, Xylander, Bachet, Girard, Fermat. For details about the historical development of this topic, one may consult [179, 86, 450]. Fermat's contribution to this problem can be described as follows.

Proposition 12.2.1 *We have $p = x^2 + y^2$, $x, y \in Z$ if and only if $p \equiv 1 \pmod 4$.*

To determine the cyclotomic numbers of order 4, 12 and 24, we need to know the values of x and y in the decomposition $p = x^2 + 4y^2$ with $x \equiv 1$

(mod 4). Consider the following specific examples:

$$257 = 1^2 + 4 \times 8^2;\ 61 = 5^2 + 4 \times 3^2$$
$$101 = 1^2 + 4 \times 5^2;\ 313 = 13^2 + 4 \times 6^2.$$

These examples show that in such a decomposition the ratio $|x/y|$ varies to a large extent. Then one question arises:

Question 12.2.2 *Can we find an explicit expression for the x and y in terms of p in such a quadratic partition?*

To answer this question, we first have a look at a proof of Proposition 12.2.1, which was given by Euler. To introduce the proof, we need the following lemma.

Lemma 12.2.3 *Suppose that N is a sum of two relatively prime squares, and that $q = x^2 + y^2$ is a prime divisor of N. Then N/q is also a sum of two relatively prime squares.*

The classical proof of this lemma can be described as follows [86]. Write $N = a^2 + b^2$, where a and b are relatively prime. By assumption we have $q = x^2 + y^2$, and thus q divides

$$\begin{aligned} x^2N - a^2q &= x^2(a^2 + b^2) - a^2(x^2 + y^2) \\ &= x^2b^2 - a^2y^2 = (xb - ay)(xb + ay). \end{aligned}$$

Because q is prime, it divides one of these two factors. By changing the sign of a if necessary, we can assume that $q|xb - ay$. It follows that $xb - ay = dq$ for some integer d.

We claim $x|a + dy$. Since x and y are relatively prime, this is equivalent to $x|(a + dy)y$. However,

$$\begin{aligned} (a + dy)y &= ay + dy^2 = xb - dq + dy^2 \\ &= xb - d(x^2 + y^2) + dy^2 = xb - dx^2, \end{aligned}$$

which is obviously divisible by x. Furthermore, if we set $a + dy = cx$, then the above equation implies $b = dx + cy$. Thus, we have

$$\begin{aligned} a &= cx - dy \\ b &= dx + cy \end{aligned} \tag{12.2}$$

Then employing the classical identity

$$(x^2 + y^2)(z^2 + w^2) = (xz \pm yw)^2 + (xw \mp yz)^2, \tag{12.3}$$

we obtain

$$\begin{aligned} N &= a^2 + b^2 = (cx - dy)^2 + (dx + cy)^2 \\ &= (x^2 + y^2)(c^2 + d^2) = q(c^2 + d^2). \end{aligned}$$

Thus $N/q = c^2 + d^2$ is a sum of squares, and (12.2) shows that c and d must be relatively prime since a and b are. This proves the lemma. □

If $p = x^2 + y^2$, then congruences modulo 4 easily imply that $p \equiv 1 \pmod 4$. Proving the converse is not easy. The modern version of Euler's proof consists of two steps. Given an odd prime p, the two steps are:

Descent Step: If $p|a^2 + b^2$, $\gcd(a, b) = 1$, then p can be written as $x^2 + y^2$.

Reciprocity Step: If $p \equiv 1 \pmod 4$, then $p|a^2 + b^2$, $\gcd(a, b) = 1$.

To complete the proof of the Descent Step, let p be an odd prime dividing $N = a^2 + b^2$, where a and b are relatively prime. If a and b are changed by multiples of p, we still have $p|a^2 + b^2$. We may thus assume that $|a| < p/2$ and $|b| < p/2$, which in turn implies that $N < p^2/2$. The new a and b may have a greatest common divisor $d > 1$, but p doesn't divide d, so dividing a and b by d, we may assume that $p|N$, $N < p^2/2$, and $N = a^2 + b^2$ where $\gcd(a, b) = 1$. Then all prime divisors $q \neq p$ of N are less than p. If q were a sum of two squares, then Lemma 12.2.3 would show N/q would be a multiple of p, which is also a sum of two squares. If all such q's were sums of two squares, then repeatedly applying Lemma 12.2.3 would imply that p itself was of the same form. So if p is not a sum of two squares, there must be a smaller prime q with the same property. Repeating this process indefinitely, we get an infinite decreasing sequence of prime numbers. This contradiction completes the Descent Step.

This is a classical descent argument, and as Weil [450, pp. 68-69] argues, it is probably similar to what Fermat did. There is also another approach to the Descent Step which is based on the reduction theory of positive definite quadratic forms.

The Reciprocity Step is simple. Since $p \equiv 1 \pmod 4$, we can write $p = 4k + 1$. Then Fermat's Little Theorem implies that

$$(x^{2k} - 1)(x^{2k} + 1) = x^{4k} - 1 \equiv 0 \pmod p$$

for all $x \not\equiv 0 \pmod p$. If $x^{2k} - 1 \not\equiv 0 \pmod p$ for some x, then $p|x^{2k} + 1$, so that p divides a sum of relatively prime squares, as desired. It is easy to see that the required x exists, since $x^{2k} - 1$ is a polynomial over the field Z_p and hence has at most $2k < p - 1$ roots. Euler's first proof that x exists was quite different, for it used the calculus of finite differences [86, p. 69]. So Proposition 12.2.1 has been proved. □

For our application we are concerned with whether it is possible to have an efficient algorithm for finding the solutions of the two square partition $p = x^2 + y^2$ for given primes p. Concerning the Reciprocity Step we can usually find an x with $0 < x < p$ such that $p|x^{2k} + 1$ with ease, because any quadratic nonresidue of p is such an x, where $p = 4k + 1$. Thus, if ξ is a quadratic nonresidue of p, then $p|N'$, where $N' = (\xi^k)^2 + 1$. To get an $N = a^2 + b^2$ with $\gcd(a, b) = 1$ and $N < p^2/2$ such that $p|N$, we need only to calculate u with $\xi^k \bmod p = \pm u$ where $0 < u < p/2$.

Let $N = u^2 + 1$. It was already known to Fermat that a positive integer M is the sum of two squares if and only if the quotient of M by its largest square factor is the product of primes congruent to 1 modulo 4 [179, 86]. It follows that $N_1 = N/p$ must be a sum of two squares. Assume

$$N_1 = m^2 \prod_i p_i, \quad p_i \equiv 1 \pmod 4,$$

where m^2 is the largest square factor of N_1, and $p_i < p$ for each p_i. Then each p_i is the sum of two squares. If we can find the square partitions of each p_i, then by repeatedly employing the classical identity (12.3) we can get many two square partitions of N_1.

With one obtained two square partition $N_1 = z^2 + w^2$ we can try to solve the equation

$$\begin{aligned} pN_1 &= (x^2 + y^2)(z^2 + w^2) = (xz \pm yw)^2 + (xw \mp yz)^2 \\ &= u^2 + 1^2, \end{aligned}$$

which results in the equations

$$\begin{cases} xz \pm yw = a \\ xw \mp yz = b, \end{cases} \tag{12.4}$$

where $(a, b) = (\pm 1, \pm u)$, and $(\pm u, \pm 1)$. The solutions (x, y) are some of the two-square partitions of the given p. It can be seen that all of the two square partitions of p can be obtained in this way if those of N_1 are given.

Thus, one possible approach to the two square partition of a given prime $p = 4k + 1$ may be summarized as follows:

Step 1: Choose a quadratic nonresidue ξ of p.

Step 2: Calculate $\xi^k \bmod p = u$. If $u > p/2$, then set $u = p - u$.

Step 3: Let $N = u^2 + 1$, $N_1 = N/p$. Find the decomposition

$$N_1 = m^2 \sum_i p_i, \ p_i \equiv 1 \pmod 4,$$

where m^2 is the largest square factor of N_1, and $p_i < p$ for each p_i.

Step 4: Find the two square partition of $p_i's$.

Step 5: Use the classical identity (12.3) to find the two-square partitions of N_1.

Step 6: For each partition of N_1, solve (12.4) to get the two square partitions of p.

To illustrate the above approach, we take the prime $p = 149$. Note that 2 is a quadratic nonresidue (in fact, a primitive root) modulo p. By calculation we have $u = 44$. It follows that $N = 44^2 + 1$. Thus, $N_1 = N/p = 13$. It is easy to check that there are only four two-square partitions $13 = (\pm 3)^2 + (\pm 2)^2$. Solving (12.4) by choosing eight possibilities of (a, b), we get only four solutions $p = (\pm 10)^2 + (\pm 7)^2$. Actually, these are all the two-square partitions of 149.

Now we turn to the complexity of the above approach. For cryptographic purposes the primes should usually be quite large. The N in Step 2 is larger than the prime p. Thus finding prime factors of N_1 is usually difficult. Step 4 is the Descent Step. Step 5 and Step 6 are relatively easy. Steps 1 and 2 are very easy. Thus it is in general very difficult to get the two-square partitions of given large primes of the form $p = 4k + 1$ with the above method. But for special primes of this form the above approach may be simple.

A concise exposition of four different constructions for x and y in the partition $p = x^2 + y2$ is given in [97, pp. 120-123]. Here we give a detailed discussion of the most efficient of these methods.

According to Lehmer [265], in a one-page note Hermite [204] published the following efficient method for representing a given prime $p \equiv 1 \pmod 4$ as a sum of squares:

1. Find the solution x_0 of $x^2 \equiv -1 \pmod p$, where $0 < x_0 < p/2$.

2. Expand x_0/p into a simple continued fraction to the point where the denominators of its convergents A'_n/B'_n satisfy the inequality $B'_{k+1} < \sqrt{p} < B'_{k+2}$. Then

$$p = (x_0 B'_{k+1} - pA'_{k+1})^2 + (B'_{k+1})^2.$$

This method, which was the best method known before 1967 (see Shanks [393]) for computing x and y in $p = x^2 + y^2$, appeared simultaneously with a paper of Serret [391] on the same subject. Hermite's method, however, is superior, in that it contains a criterion for ending the algorithm at the right place, while Serret's does not (see Brillhart [37]). In 1972 Brillhart gave an improvement of the algorithm, basing on the fact that the calculation of the convergents in Step 2 can be dispensed with, since the values needed for the representation are already at hand in the continued fraction expansion itself. The shortened algorithm by Brillhart is the following:

1. The same as in the Hermite's.

2. Carry out the Euclidean algorithm on p/x_0 (not x_0/p), producing the sequence of remainders $R_1, R_2, \cdots$, to the point where R_i is first less than $\sqrt{p}$, and

$$\begin{array}{ll} p = R_k^2 + R_{k+1}^2, & \text{if} \quad R_1 > 1, \\ p = x_0^2 + 1, & \text{if} \quad R_1 = 1. \end{array}$$

Brillhart's proof of the shortened algorithm is the following. Assume $R_1 > 1$. Since $0 < x_0 < p/2$ and $p|(x_0 + 1)$, then from Perron [342] we see that the following properties hold:

(i) The continued fraction expansion of p/x_0 has an even number of partial quotients and is palindromic, i.e.,

$$p/x_0 = [q_0, q_1, \cdots, q_k, q_k, \cdots, q_1, q_0] = A_{2k+1}/B_{2k+1},$$

$k \geq 0$. (Observe that the convergents A'_{n+1}/B'_{n+1} for the expansion of x_0/p are the reciprocals of the convergents A_n/B_n for p/x_0.)

(ii) $A_{2k+1} = p$ and $A_{2k} = x_0$.

(iii) $p = A_k^2 + A_{k-1}^2$.

(iv) From (ii), the recursion formula for the numerators A_n gives the following set of equations:

$$p = q_0 x_0 + A_{2k-1}, \ x_0 = q_1 A_{2k-1} + A_{2k-2}, \cdots.$$

The equations in (iv) are clearly identical with those in the Euclidean algorithm for p/x_0. Hence, $A_{2k-1} = R_1, A_{2k-2} = R_2, \cdots, A_{k+1} = R_{k-1}, A_k = R_k, A_{k-1} = R_{k+1}, \cdots$. Using these equations with (iii), gives $p = R_k^2 + R_{k+1}^2$. Certainly, then $R_k < \sqrt{p}$. If $k = 1$, then R_k is the first $R_k < \sqrt{p}$. If $k > 1$, then from the observation in (i), $R_{k-1} = A_{k+1} = B'_{K+2}$. But, from Hermite's development, $B'_{k+2} > \sqrt{p}$, so R_k is the first remainder less than $\sqrt{p}$.

If $R_1 = 1$, then $p = q_0 x_0 + 1$ and $p/x_0 = [q_0, q_0]$. Together, these imply $q_0 = x_0$, so $p = x_0^2 + 1$. This completes the proof.

As already made clear in our first approach, the solution x_0 of $x^2 \equiv -1 \pmod{p}$ can be obtained by computing $x_0 = c^{(p-1)/4} \bmod p$, where c is a quadratic nonresidue of p. Brillhart pointed out that $c = 2$ and $c = 3$ can be used when $p \equiv 5 \pmod{8}$ and $p \equiv 17 \bmod 24$, respectively. In the remaining case, $p \equiv 1 \pmod{24}$, c can be found by using the quadratic reciprocity law.

To illustrate the shortened algorithm, we take the example given by Brillhart. Let $p = 10006721 \equiv 17 \pmod{24}$. Then $c = 3$ and $x_0 = 3^{2501680} \equiv 2555926 \pmod{p}$. Then

$$\begin{array}{lcllr}
10006721 & = & 3 \cdot 2555926 & + & 2338943 \\
2555926 & = & 1 \cdot 2338943 & + & 216983 \\
2338943 & = & 10 \cdot 216983 & + & 169113 \\
216983 & = & 1 \cdot 169113 & + & 47870 \\
169113 & = & 3 \cdot 47870 & + & 25503 \\
47870 & = & 1 \cdot 25503 & + & 22367 \\
\cdots\cdots & . & \cdots\cdots & . & \cdots\cdots \\
25503 & = & 1 \cdot 22367 & + & 3136 \\
22367 & = & 7 \cdot 3136 & + & 415
\end{array}$$

Hence, since $22357^2 > p$ and $3136^2 < p$, we have

$$p = 3136^2 + 415^2.$$

Clearly, some primes of special form can be expressed as a sum of two squares without much calculation. For example, the prime number $N = (2^{691} - 2^{346} + 1)/5$, discovered by Brillhart and Selfridge [37], can be easily written as $N = [(3 \cdot 2^{345} - 1)/5]^2 + [(2^{345} - 2)/5]^2$. Also, the identity $U_{2k+1} = U_k^2 + U_{k+1}^2$, where U_n is the nth Fibonacci number, provides such a representation for Fibonacci primes in terms of the Fibonacci numbers themselves.

The above shortened algorithm works very efficiently, since we have a fast exponentiation modulo p algorithm for finding an x_0, and the step (2) of the algorithm is based on the Euclidean algorithm which is efficient.

If we need such an algorithm for the purpose of getting some Gaussian primes, one or several solutions of the quadratic partition may be enough. However, for the purpose of analyzing the stability of some cyclotomic numbers we need some special solutions as described in Chapter 4. So the problem now is how many distinct quadratic partitions $p = x^2 + y^2$ a prime

$p = 4t + 1$ has. The following Proposition 12.2.5 shows that there are only four distinct integer solutions (x, y), that is every prime $p = 4t + 1$ is in one and only in one way a sum of two squares of positive integers. This was already known to Fermat in 1640. Euler proved the converse of the above conclusion in 1742, which led also to a primality test.

Now we turn to (12.1). For the solvability of (12.1) we have the following result, which was known to Fermat and was first proved by Euler [179].

Proposition 12.2.4 *The Diophantine equation (12.1) is solvable if and only if all prime divisors q of n with $q \equiv 3 \pmod 4$ occur in n to an even power.*

Concerning the Diophantine equation (12.1), the following more general result was proved by Gauss with the help of quadratic forms [159], and by Jacobi [217] with the help of elliptic functions.

Proposition 12.2.5 *Denote the number of divisors of n by $d(n)$, and write $d_a(n)$ for the number of those divisors with $d \equiv a \pmod 4$. Let $n = 2^f n_1 n_2$, where $n_1 = \prod_{p\equiv 1 \pmod 4} p^r$, $n_2 = \prod_{q\equiv 3 \pmod 4} q^s$, and let $r(n)$ be the number of solutions of (12.1); then $r(n) = 0$ if any of the exponents s is odd. If all s are even, then $r(n) = 4d(n_1) = 4(d_1(n) - d_3(n))$.*

12.3 $p = x^2 + 2y^2$ and $p = x^2 + 3y^2$

The cyclotomic numbers of orders 6, 12 and 24 depend on or partially on the quadratic partition

$$p = x^2 + 3y^2, \; x \equiv 1 \pmod 3, \tag{12.5}$$

and that of order 8 partially on the partition

$$p = x^2 + 2y^2, \; x \equiv 1 \pmod 4. \tag{12.6}$$

It has been proven by Euler that the following conclusion about the partition holds:

Proposition 12.3.1 *An odd prime p can be represented as $p = x^2 + 3y^2$ if and only if $p = 3$ or $p \equiv 1 \pmod 3$, $p = x^2 + 2y^2$ if and only if $p \equiv 1$ or $3 \pmod 8$.*

Euler used the same two-step strategy in his proofs for $x^2 + 2y^2$ and $x^2 + 3y^2$ [86]. The Descent Steps are:

If $p|x^2 + 2y^2, \gcd(x, y) = 1$, then p is of the form $x^2 + 2y^2$.
If $p|x^2 + 3y^2, \gcd(x, y) = 1$, then p is of the form $x^2 + 3y^2$.

The Reciprocity Steps are:

If $p \equiv 1, 3 \pmod 8$, then $p|x^2 + 2y^2, \gcd(x, y) = 1$.
If $p \equiv 1 \pmod 3$, then $p|x^2 + 3y^2, \gcd(x, y) = 1$,

where p is always an odd prime. The proof of Proposition 12.3.1 can be found, for example, in [450, 86]. We can give a similar approach to the determination of the solutions of (12.6) and (12.5), which is analogous to the first approach in the foregoing section. But the complexity of the approach is large. It seems unknown whether Hermite's algorithm can be modified into one for this kind of quadratic partition. Thus, an efficient algorithm for finding the solutions of (12.5) and (12.6) should be developed.

12.4 $p = x^2 + ny^2$ and Quadratic Reciprocity

Before going further into the cryptographic aspects of the quadratic partition $p = x^2 + ny^2$, we need to study the relation between the partition and quadratic reciprocity. The well known law of quadratic reciprocity is described as follows.

Proposition 12.4.1 (Quadratic Reciprocity) *If p and q are distinct odd primes, then*

$$\left(\frac{p}{q}\right)\left(\frac{q}{p}\right) = (-1)^{(p-1)(q-1)/4},$$

where $(\frac{\cdot}{\cdot})$ is the Legendre symbol.

This theorem is not only theoretically beautiful, but also computationally very useful. It is easy to prove that the above theorem of quadratic reciprocity is equivalent to the following proposition [86, p. 15].

Proposition 12.4.2 *If p and q are distinct odd primes, then $(\frac{q}{p}) = 1$ if and only if $p \equiv \pm\alpha^2 \pmod{4q}$ for some odd integer α.*

The Reciprocity Step in treating the quadratic partition is closely connected to quadratic residues, as described by the following proposition. As pointed out in [86, p. 13], the Reciprocity Step was one of the main things that led Euler to discover quadratic reciprocity. The definition of quadratic residue immediately gives:

Proposition 12.4.3 *Let n be a nonzero integer, and p an odd prime not dividing n. Then*

$$p|x^2 + ny^2, \quad \gcd(x, y) = 1 \Longleftrightarrow \left(\frac{-n}{p}\right) = 1.$$

12.5 $p = x^2 + 7y^2$ and Quadratic Forms

Integral quadratic forms in two variables

$$f(x, y) = ax^2 + bxy + cy^2, \quad a, b, c \in Z \tag{12.7}$$

were studied by Lagrange, Gauss and many others. Our cryptographic partition $p = x^2 + ny^2$ is obviously a special form of (12.7). In this section we introduce only some basic facts about quadratic forms in two variables.

A form $ax^2 + bxy + cy^2$ is said to be *primitive* if its coefficients a, b and c are relatively prime. It follows from this definition that any form is an integer multiple of a primitive form. An integer m is *represented* by a form $f(x, y)$ if the equation

$$m = f(x, y) \tag{12.8}$$

has an integer solution in x and y. If the x and y in (12.8) are relatively prime, we say that m is *properly represented* by $f(x, y)$.

Two forms $f(x, y)$ and $g(x, y)$ are called *equivalent* if there are integers p, q, r and s such that

$$f(x, y) = g(px + qy, rx + sy) \quad \text{with} \quad ps - qr = \pm 1. \tag{12.9}$$

Writing

$$M = \begin{pmatrix} p & q \\ r & s \end{pmatrix},$$

we have $\det(M) = ps - qr = \pm 1$. This means that M is in the group of 2×2 invertible integer matrices $GL(2, Z)$. It can be easily proven that the equivalence of forms is an equivalence relation which can be further divided into two kinds: the proper and improper equivalence. An equivalence is said to be a *proper equivalence* if $ps - qr = 1$, i.e., $M \in SL(2, Z)$, and it is an *improper equivalence* if $ps - qr = -1$ [86]. The following elementary facts can be easily proved.

Proposition 12.5.1 *Elementary Facts:*

1. *Proper equivalence is an equivalence relation, but improper equivalence is not.*

2. *Equivalent forms represent the same numbers, and the same holds for proper representations.*

3. *Any form equivalent to a primitive form is itself primitive.*

A very nice relation between proper representation and proper equivalence is the following:

Proposition 12.5.2 *A form $f(x, y)$ properly represents an integer m if and only if $f(x, y)$ is properly equivalent to the form $mx^2 + bxy + cy^2$ for some $b, c \in Z$.*

To prove the proposition, we first suppose that $f(p, q) = m$, where p and q are relatively prime. By the Extended Euclidean Algorithm we can find integers r and s so that $ps - qr = 1$, and then

$$\begin{aligned} f(px + ry, qx + sy) &= f(p, q)x^2 + (f(p, s) + f(r, q))xy + f(r, s)y^2 \\ &= mx^2 + bxy + cy^2 \end{aligned}$$

is of the desired form. To prove the converse, note that $mx^2 + bxy + cy^2$ represents m properly by taking $(x, y) = (1, 0)$, and this completes the proof of the proposition.

To study the equivalence, we need the notion of *discriminant*, which is defined to be $D = b^2 - 4ac$ for the form $ax^2 + bxy + cy^2$. Concerning the discriminant we have the following elementary facts:

1. Suppose two forms $f(x, y)$ and $g(x, y)$ have discriminants D and D' respectively, and that

   $$f(x, y) = g(px + qy, rx + sy), \quad p, q, r, s \in Z.$$

   Then

   $$D = (ps - qr)^2 D'.$$

2. Equivalent forms have the same discriminant.

3. If $D > 0$, then $f(x, y)$ represents both positive and negative integers. In this case the form is called *indefinite*.

4. If $D < 0$, then the form represents only positive integers or only negative ones, depending on the sign of a, and $f(x, y)$ is accordingly called *positive* or *negative definite*.

5. b is even (resp. odd) if and only if $D \equiv 0$ (resp. 1) $\pmod 4$.

These facts can be easily proved. For example, for $f(x, y) = ax^2 + bxy + cy^2$, we can use the identity

$$4af(x, y) = (2ax + by)^2 - Dy^2 \tag{12.10}$$

to prove Facts 3) and 4).

A necessary and sufficient condition for a number m to be represented by a form of discriminant D is the following:

Proposition 12.5.3 *Let $D \equiv 0, 1 \pmod 4$ be an integer and m be an odd integer relatively prime to D. Then m is properly represented by a primitive form of discriminant D if and only if D is a quadratic residue modulo m.*

Proof: If $f(x, y)$ properly represents m, then by Proposition 12.5.2, we may assume $f(x, y) = mx^2 + 2bxy + cy^2$. Thus $D = b^2 - 4mc$, and $D \equiv b^2 \pmod m$ follows immediately.

Conversely, suppose that $D \equiv b^2 \pmod m$. Since m is odd, we can assume that D and b have the same parity (replace b by $b + m$ if necessary), and then $D \equiv 0, 1 \pmod 4$ implies that $D \equiv b^2 \pmod{4m}$. This means that $D = b^2 - 4mc$ for some c. Then $mx^2 + bxy + cy^2$ represents m properly and has discriminant D, and the coefficients are relatively prime since m is relatively prime to D. This completes the proof. □

Because $-4n$ is a quadratic residue modulo p if and only if $(\frac{-4n}{p}) = (\frac{-n}{p}) = 1$, we get immediately from Proposition 12.5.3 the most useful version of the above proposition:

Corollary 12.5.4 *Let n be an integer and let p be an odd prime not dividing n. Then $(\frac{-n}{p}) = 1$ if and only if p is represented by a primitive form of discriminant $-4n$.*

The importance of this corollary is that primes p which satisfy $(\frac{-n}{p}) = 1$ can be represented by forms of discriminant $-4n$. But there are usually many quadratic forms of a given discriminant. For our application only the simple forms $x^2 + my^2$ are interesting.

We now turn to the positive definite forms, which include the forms $x^2 + ny^2$ with $n > 0$. Their theories are simple and elegant. A primitive positive definite form $ax^2 + bxy + cy^2$ is said to be *reduced* if

$$|b| \le a \le c, \text{ and } b \ge 0 \text{ if either } |b| = a \text{ or } a = c. \tag{12.11}$$

Note that a and c are positive since the form is positive definite. The basic theorem is the following [86]:

Proposition 12.5.5 *Every primitive positive definite form is properly equivalent to a unique reduced form.*

Now we consider some examples. The forms $3x^2 \pm 2xy + 5y^2$ are clearly equivalent. However, since they are both reduced, Proposition 12.5.5 implies that they are not properly equivalent. On the other hand, $2x^2 + 2xy + 3y^2$ is reduced, but it is properly equivalent to $2x^2 - 2xy + 3y^2$, which is not reduced.

Now one question is whether there are only a finite number of reduced forms of a given discriminant D. To answer the question, we make some observations. For a given discriminant $D < 0$, suppose that $ax^2 + bxy + cy^2$ is a reduced form with discriminant D. Then $b^2 \leq a^2$ and $a \leq c$, so we have $-D = 4ac - b^2 \geq 3a^2$. It follows that

$$a \leq \sqrt{(-D)/3}. \tag{12.12}$$

This shows that the answer to the above question is "yes". Two forms are said to be *in the same class* if they are properly equivalent. Letting $h(D)$ denote the number of classes of primitive positive definite forms of discriminant D, which by Proposition 12.5.5 is just the number of reduced forms, we have thus proved the following proposition:

Proposition 12.5.6 *Let $D < 0$ be fixed. Then the number $h(D)$ of classes of primitive positive definite forms of discriminant D is finite, and furthermore $h(D)$ is equal to the number of reduced forms of discriminant D.*

According to [86, p. 29] there is an algorithm for computing reduced forms and class numbers which, for small discriminants, is easily implemented on a computer. Table 12.1 gives some examples.

Note that $x^2 + ny^2$ is always a reduced form of discriminant $D = -4n$. So if $h(-4n) = 1$ for some given n, by Proposition 12.5.5 any odd prime not dividing n with $\left(\frac{-n}{p}\right) = 1$ must be represented as $x^2 + ny^2$. Thus, the characterization of $p = x^2 + ny^2$ works when $h(-4n) = 1$. In fact it works only in this case. Thus, the analysis of $h(-4n)$ is important for this purpose. The following result was conjectured by Gauss and proved by Landau [86, 258]:

Proposition 12.5.7 *Let n be a positive integer. Then $h(-4n) = 1$ if and only if $n = 1, 2, 3, 4$ or 7.*

With this proposition and Proposition 12.5.5 we arrive easily at

$$p = x^2 + 7y^2 \iff p \equiv 1, 9, 11, 15, 23, 25 \pmod{28}$$

Table 12.1: Some examples of class numbers and reduced forms.

D	$h(D)$	Reduced forms of Discriminant D
-4	1	x^2+y^2
-8	1	x^2+2y^2
-12	1	x^2+3y^2
-20	2	$x^2+5y^2,\ 2x^2+2xy+3y^2$
-28	1	x^2+7y^2
-56	4	$x^2+14y^2,\ 2x^2+7y^2,\ 3x^2\pm 2xy+5y^2$
-108	3	$x^2+27y^2,\ 4x^2\pm 2xy+7y^2$
-256	2	$x^2+64y^2,\ 4x^2+4xy+17y^2,\ 5x^2\pm 2xy+13y^2$

for primes $p \neq 7$, since $(\frac{-7}{p}) = 1$ holds only for these primes. To determine the cyclotomic numbers of order 7, we need the quadratic partition $p = x^2 + 7y^2$ with $x \equiv 1 \pmod 7$ (see Appendix A or [273]). As in the cases $n = 1, 2, 3$ and 4, what we need is an efficient algorithm to find the actual partition of large primes of the form $7k + 1$.

12.6 $p = x^2 + 15y^2$ and Genus Theory

It can be shown that the calculation of cyclotomic numbers of order 15 needs the partition $p = x^2 + 15y^2$ with $x \equiv -1 \pmod 3$ [44]. Since $h(-60) > 1$ by Proposition 12.5.7, there is more than one reduced form of discriminant -60. Thus, new methods of separating reduced forms of the same discriminant are needed. Genus theory can fulfill this task. The basic idea of genus theory, due to Lagrange, is to consider the congruence classes represented in Z_D^* by a single form, and then group together forms that represent the same classes.

To show the idea, we first take one example. The theory of quadratic forms, quadratic reciprocity and Table 12.1 give $h(-20) = 2$ and

$$p = \left\{ \begin{array}{c} x^2 + 5y^2 \\ 2x^2 + 2xy + 3y^2 \end{array} \right\} \Longleftrightarrow p \equiv 1, 3, 7, 9 \pmod{20}.$$

To separate the forms in a set of reduced forms with the same discriminant, we simply consider all D^2 residues

$$\{f(x, y) \bmod |D| : \ (x, y) \in Z_{|D|} \times Z_{|D|}\} \subseteq Z_{|D|},$$

then reject the residues which are not relatively prime to $|D|$. For our example $D = -20$, by computation we easily get

$$\begin{array}{llll} x^2 + 5y^2 & \text{represents} & 1, 9 & \text{in} \quad (Z_{20})^* \\ x^2 + 2xy + 3y^2 & \text{represents} & 3, 7 & \text{in} \quad (Z_{20})^* \end{array}$$

While for $D = -56$ we have

$$\begin{array}{llll} x^2 + 14y^2,\ 2x^2 + 7y^2 & \text{represents} & 1, 9, 15, 23, 25, 29 & \text{in} \quad (Z_{56})^* \\ 3x^2 \pm 2xy + 5y^2 & \text{represents} & 3, 5, 13, 19, 27, 45 & \text{in} \quad (Z_{56})^*. \end{array}$$

Thus $p = x^2 + 5y^2$ if and only if $p \equiv 1, 9 \pmod{20}$, and $p = 2x^2 + 2xy + 3y^2$ if and only if $p \equiv 3, 7 \pmod{20}$.

Generally, two primitive positive definite forms of discriminant D are said to be in the same *genus* if they represent the same values in $Z_{|D|}$. Note that equivalent forms represent the same numbers and hence are in the same genus. Furthermore, each genus consists of a finite number of classes of forms. For the above example $D = -20$, there are two genera, each consisting of a single class; and for the example $D = -56$, there are again two genera, but this time each genus consists of two classes.

We do not intend to go further into the genus theory. For details we refer to [86]. Here we just want to mention the result about the partition $p = x^2 + 15y^2$ and the relationship between some quadratic partition and genus theory. With genus theory it has been proved that

$$p = x^2 + 15y^2 \Longleftrightarrow p \equiv 1, 19, 31, 49 \pmod{60}.$$

Again, we need an efficient algorithm for solving the Diophantine equation. Finally, it was noted above that genus theory cannot solve the partition $p = x^2 + ny^2$ for all n's. To treat this problem in general, the theory of class fields is needed.

12.7 $p = x^2 + ny^2$ and Class Field Theory

As mentioned at the beginning of this chapter, we need to know the quadratic partition $p = x^2 + ny^2$ for two reasons: the nonlinearity analysis of some cryptographic functions based on cyclotomic numbers and the search for primes in some integer domains other than Z. For our applications, there are two different approaches:

Approach 1: Find large primes of certain forms which can be represented as $p = x^2 + ny^2$, where n is designed for our cryptographic purposes. Then find efficient algorithms to get the partitions we need.

Approach 2: Given the cryptographic parameter n, find some large primes from the set

$$B(n) = \{x^2 + ny^2 : (x, y) \in Z \times Z\} \tag{12.13}$$

if it contains large primes.

Concerning approach 1, we have two questions as follows:

Question 12.7.1 *For a given n, which primes can be represented as $p = x^2 + ny^2$?*

Question 12.7.2 *For a given n, if a large prime can be represented as $p = x^2 + ny^2$, how many solutions (x, y) are there? And how can we develop algorithms for finding the solutions?*

Regarding approach 2, we have again two questions which need to be answered:

Question 12.7.3 *For which n are there infinitely many primes in the set $B(n)$ defined above?*

Question 12.7.4 *For a given n such that there are infinitely many primes in the set $B(n)$, how can we find large primes in the set $B(n)$?*

With reference to Question 12.7.1, the classical two-step strategy and genus theory have answered the question for many n's. However, these nice methods are limited and cannot solve the problem for arbitrary $n > 0$. To treat this problem generally, we need class field theory, which might be tentatively regarded as the search for those Abelian extension fields which make possible the solution of the problem of the representation of a prime by a quadratic form. We do not intend to go further into the class field theory here. For details about the theory we refer to [86, 80]. Here we shall only present a general answer to Question 12.7.1 developed with the help of class field theory. For a proof of the following result, one may see, for example, Cox [86, pp. 110-112].

Proposition 12.7.5 *Let $n > 0$ be a squarefree integer with $n \not\equiv 3 \pmod 4$. Then there is a monic irreducible polynomial $f_n(x) \in Z[x]$ of degree $h(-4n)$ such that if an odd prime p divides neither n nor the discriminant of $f_n(x)$, then*

$$p = x^2 + ny^2 \Longleftrightarrow \left\{ \begin{array}{l} (-n/p) = 1 \textit{ and } f_n(x) \equiv 0 \pmod p \\ \textit{has an integer solution.} \end{array} \right\}$$

Furthermore, $f_n(x)$ may be taken to be the minimal polynomial of a real algebraic integer α for which $L = K(\alpha)$ is the Hilbert class field $K = Q(\sqrt{-n})$.

So far we have not found efficient algorithms which enable us to answer Question 12.7.2 when $n \neq 2, 4$. This problem remains to be investigated. To answer Question 12.7.3, we need the theory of ring class fields together with Dirichlet density. The classical theorem that answers the question is that a primitive positive definite quadratic form $ax^2 + by^2 + cy^2$ represents infinitely many prime numbers. Generally, we have the following proposition [86, 35, 451]:

Proposition 12.7.6 *Let $ax^2 + bxy + cy^2$ be a primitive positive definite quadratic form of discriminant $D < 0$, and let $PB(a, b, c)$ be the set of primes represented by this form. Then the Dirichlet density $\delta(PB(a, b, c))$ exists and is given by the formula*

$$\delta(PB(a,b,c)) = \begin{cases} \frac{1}{h(D)} & \text{if this form is properly} \\ & \text{equivalent to its opposite} \\ \frac{1}{2h(D)} & \text{otherwise.} \end{cases}$$

In particular, $ax^2 + bxy + cy^2$ represents infinitely many prime numbers.

As an example of what this proposition tells us, we consider forms of discriminant -56. Table 12.1 shows that the class number is 4 and gives the reduced forms. It follows from this proposition

$$\begin{aligned} \delta(\{p \text{ prime } : \ p = x^2 + 14y^2\}) &= \frac{1}{8} \\ \delta(\{p \text{ prime } : \ p = 2x^2 + 7y^2\}) &= \frac{1}{8} \\ \delta(\{p \text{ prime } : \ p = 3x^2 \pm 2xy + 5y^2\}) &= \frac{1}{4}. \end{aligned}$$

Note that these densities sum to $1/2$, which is the density of primes for which $(-56/p) = 1$. Generally, for any given negative discriminant, the densities of primes represented by the reduced forms (counted properly) always sum to $1/2$ [86].

Owing to the difficulty of answering Question 12.7.2, Question 12.7.4 is especially important for our application. It is unknown how to find large primes in the set $B(n)$.

Research Problem 12.7.7 *Develop methods for finding large primes in the set $B(n)$.*

Since partitioning a prime p into $p = x^2 + ny^2$ is necessary for analyzing a number of cryptographic attributes of some cyclotomic generators, an investigation into the following problem is important.

Research Problem 12.7.8 *Develop an efficient algorithm for the partition of a prime into $p = x^2 + ny^2$ for $n \geq 3$.*

12.8 Other Cryptographic Quadratic Partitions

Quadratic partition $4p = x^2 + 27y^2$ with $x \equiv 1 \pmod 3$ are needed for analyzing the stability of cyclotomic numbers of order 3. In fact if we can find the partitions $p = x^2 + 27y^2$, then we get

$$4p = (2x)^2 + 27(2y)^2.$$

As mentioned in Section 12.1, there are some other quadratic partitions of primes or multiples of primes we need for analyzing the nonlinearity of some cryptographic functions. The determination of these partitions is much more complicated. Thus, some quadratic partition problems for cryptographic purposes remain to be investigated.

According to the literature only cyclotomic numbers of orders in the range $[2, 24]$ are known. To construct generators in Chapter 8, we may need cyclotomic numbers of order $2k$ with $k > 12$. Thus, partitions $p = x^2 + ny^2$ for more n's may be needed. It is not possible to develop here all the mathematical theories associated with this problem. But it might be worthwhile to point out some of them.

The partition problem $p = x^2 + ny^2$ is related to the following mathematical theories: the classification of quadratic forms, genus theory, Euler's convenient numbers, quadratic reciprocity, cubic reciprocity, biquadratic and higher reciprocity, the Hilbert class field, ring class fields, elliptic curves, Gauss and Jacobi sums. Details about the relations can be found, for example, in [86, 215].

As we saw in Chapter 4, the cyclotomic numbers of order 10 depend on the quadratic partition

$$16p = x^2 + 50u^2 + 50v^2 + 125w^2$$

with $x \equiv 1 \pmod 5$ and $v^2 - 4uv - u^2 = xw$. Similar complicated quadratic partitions are needed to calculate cyclotomic numbers of other orders. It seems to be an open problem how to compute the values of x, u, v, w efficiently, given p. Such a problem is of course important for the corresponding cyclotomic generators, since quite a number of cryptographic attributes of the generators depend on the cyclotomic constants.

To show the cryptographic importance of the quadratic partition $p = x^2 + y^2$, we mention the Ong-Schnorr-Shamir signature scheme. Here we will follow the description of the system by McCurley [300, p.152]. In 1984, Ong, Schnorr, and Shamir [336] proposed a very efficient digital signature scheme based on the difficulty of solving a polynomial congruence modulo a composite integer. The original scheme was the following. A trusted authority chooses an odd integer $n = pq$ that is presumed hard to factor and publishes the number n (alternatively, each user could choose his own modulus n). Each user who wishes to sign a message m chooses a secret random integer s, computes $k \equiv s^2 \pmod{n}$ and gives k to the trusted authority. The trusted authority publishes all the public keys k. In order to sign m, the user will then produce a solution x, y to the congruence $x^2 - ky^2 \equiv m \pmod{n}$. Anyone can easily verify the validity of the signature x, y. Moreover, the user who holds the secret key s can easily produce a solution by first choosing a random integer r and then applying the extended Euclidean algorithm to calculate

$$x = 2^{-1}(mr^{-1} + r) \bmod n,$$
$$y = (2s)^{-1}(mr^{-1} - r) \bmod n.$$

It has been pointed out in [300] that the security of the scheme depends on a forger's apparent inability to find a solution to the congruence $x^2 - ky^2 \equiv m \pmod{n}$ when k, m and n are given, but s is kept secret. Unfortunately, the system was cracked shortly afterwards by Pollard [300]. Pollard and Schnorr [351] later proved that the congruence could be solved in random polynomial time assuming the extended Riemann hypothesis. This result was later improved by Adleman, Estes and McCurley [1].

As made clear in Section 12.1, a prime p can be represented as $p = x^2+y^2$ if and only if $p \equiv 1 \pmod{4}$; also, there is an efficient algorithm for finding such a representation. The method of solving the congruence $x^2 \pm y^2 \equiv m \pmod{n}$ is closely related to the quadratic partition of primes into $p = x^2 + y^2$ [300]. First, note that a solution to $x^2 - y^2 \equiv m \pmod{n}$ can be constructed trivially by solving the linear congruences

$$x - y \equiv m \pmod{n}, \quad x + y \equiv 1 \pmod{n}.$$

The case $x^2 + y^2 \equiv m \pmod{n}$ can be done as follows: we can use a method to find a prime p satisfying $p \equiv m \pmod{n}$ and $p \equiv 1 \pmod{4}$ [300]. Then we use the algorithm in Section 12.1 to find one quadratic partition of the prime p, i.e., $p = x^2 + y^2$. Then we have a solution of $x^2 + y^2 = p \equiv m \pmod{n}$.

According to [300], Pollard's key idea for solving the congruence $x^2 - ky^2 \equiv m \pmod{n}$ is to reduce it to solving a congruence of the same form, but with k replaced by some k_1 with $|k_1| \le \sqrt{4|k|/3}$ (and a new m). After a small number of such reduction steps, we eventually reach the case of solving a congruence of the form $x^2 \pm y^2 \equiv m \pmod{n}$, which has been solved by the approach above. Here we see that Pollard's idea is similar to the classical descent approach to the quadratic partition $p = x^2 + y^2$.

After the original scheme of Ong, Schnorr and Shamir was broken with Pollard's method, a modification was proposed based on higher-degree congruences [351]. One concrete proposal had its security based on the presumed difficulty of solving for x, y, z in the quadratic congruence

$$(m_1 - 2kxy)^2 + 4z^2(dx^2 + k(y^2 + dz^2) - m_2) \equiv 0 \pmod{n},$$

where m_1, m_2, d, k and n are given. This scheme was also broken by Evertse, Adleman, Kompella, McCurley and Miller [151], using methods that are similar to Pollard's original idea.

Chapter 13

Group Characters and Cryptography

Group characters are connected with many mathematical problems: the morphism theorems of algebraic structures, the solution of equations over finite fields, Gauss sums, Jacobi sums, to mention only a few. Character sums are very powerful tools in calculating cyclotomic numbers and in treating difference sets as well as in solving many mathematical problems. Our cryptographic interest in group characters comes from the fact that there are many group characters which turn out to be good cryptographic functions for many keystream generators. This chapter is concerned with the following: the introduction of elementary facts about characters; an overview of the cryptographic functions based on group characters; and some further cryptographic properties of some group characters and their potential cryptographic applications.

13.1 Group Characters

Let G be a finite Abelian group (written multiplicatively) of order $|G|$ with identity element 1_G. A *character* χ of G is a homomorphism from G into the multiplicative group U of complex numbers of absolute value 1. This is to say that, χ is a mapping from G into U with

$$\chi(ab) = \chi(a)\chi(b) \quad \text{for all } a, b \in G.$$

It is trivial to prove the following elementary facts:

Proposition 13.1.1 *Let χ be a character of a finite Abelian group G as above, and $g \in G$. Then*
(a) $\chi(1_G) = 1$.
(b) $\chi(g)$ *is a $|G|$th root of unity.*

(c) $\chi(g^{-1}) = \chi(g)^{-1} = \overline{\chi(g)}$,
where the bar is complex conjugation.

For every G we have the *trivial* character χ_0 defined by $\chi_0 = 1$ for all $g \in G$. For each character χ the character $\overline{\chi}$ defined by $\overline{\chi} = \overline{\chi(g)}$, is called the *conjugate character* of χ. Given finitely many characters $\chi_1, \cdots, \chi_n$ of G, the product character $\chi_1 \cdots \chi_n$ is defined by

$$(\chi_1 \cdots \chi_n)(g) = \chi_1(g) \cdots \chi_n(g), \text{ for all } g \in G.$$

Thus, the meaning of χ^n is clear. Under this multiplication of characters, the set G^C of characters of G forms an Abelian group. It follows immediately from part (b) of the above proposition that G^C is Abelian.

Let G be a finite cyclic group of order n, and let g be a generator of G. For a fixed integer j, $0 \le j \le n-1$, the function

$$\chi_j(g^k) = e^{2\pi ijk/n}, \quad k = 0, 1, \cdots, n-1,$$

defines a character of G, and it is easy to see that G^C consists exactly of the characters $\chi_0, \chi_1, \cdots, \chi_{n-1}$.

Let H be a subgroup of the finite Abelian group G and let θ be a character of H. Then θ can be extended to a character of G. For any two distinct elements $g_1, g_2 \in G$ there exists a character χ of G with $\chi(g_1) \neq \chi(g_2)$. There are also several other important facts about characters as stated in the following proposition:

Proposition 13.1.2 *Let the symbols be the same as above.*

1. *If χ is nontrivial, then $\sum_{g \in G} \chi(g) = 0$.*
2. *If $g \in G$ with $g \neq 1_G$, then $\sum_{\chi \in G^C} \chi(g) = 0$.*
3. *The number of characters of a finite group G is equal to $|G|$.*

The proofs of the above facts are easy (see Lidl and Niederreiter [276, p. 189] or Schmidt [382, p. 40]). Using this proposition, we can derive the following *orthogonality relations for characters.* Let χ and ψ be characters of G. Then

$$\frac{1}{|G|} \sum_{g \in G} \chi(g)\overline{\psi(g)} = \begin{cases} 0 & \text{for } \chi \neq \psi, \\ 1 & \text{for } \chi = \psi. \end{cases} \tag{13.1}$$

Furthermore, if g and h are elements of G, then

$$\frac{1}{|G|} \sum_{\chi \in G^C} \chi(g)\overline{\chi(h)} = \begin{cases} 0 & \text{for } g \neq h, \\ 1 & \text{for } g = h. \end{cases} \tag{13.2}$$

Character theory is mathematically important in several aspects. It can be used to obtain expressions for the number of solutions of equations in a finite Abelian group G. Let f be an arbitrary map from the cartesian product $G^n = G \times \cdots \times G$ (n factors) into G. Then for a fixed $h \in G$, the number $N(h)$ of n-tuples $(g_1, \cdots, g_n) \in G^n$ with $f(g_1, \cdots, g_n) = h$ is given by

$$N(h) = \frac{1}{|G|} \sum_{(g_1,\cdots,g_n)\in G^n} \chi(f(g_1, \cdots, g_n))\overline{\chi(h)}, \tag{13.3}$$

on account of (13.2). On the other hand, the Gauss sums and Jacobi sums based on characters are very useful in solving some equations over finite fields [215].

13.2 Field Characters and Cryptography

There are two finite Abelian groups in a finite field $GF(q)$, i.e., the additive group and multiplicative group of the field. For our cryptographic applications, we have to make an important distinction between the corresponding two kinds of characters.

We first consider the additive group $(GF(q), +)$. Let p be the characteristic of $GF(q)$, and $q = p^m$. We identify the prime field of $GF(q)$ with Z_p. The *absolute trace function* $\mathrm{Tr}(a)$ from $GF(q)$ to $GF(p)$ is defined by

$$\mathrm{Tr}(a) = a + a^p + a^{p^2} + \cdots + a^{p^{m-1}}, \ a \in GF(q). \tag{13.4}$$

With the absolute trace function we can now define the function χ_1 by

$$\chi_1(c) = e^{2\pi i \ \mathrm{Tr}(c)/p} \text{ for all } c \in GF(q), \tag{13.5}$$

which is a character of the additive group $(GF(q), +)$. Following [276] and [382], we call the characters of the group $(GF(q), +)$ *additive characters*, and we call the above character χ_1 the *canonical additive character* of $GF(q)$. This character is important due to the following proposition:

Proposition 13.2.1 *For $b \in GF(q)$, the function χ_b with $\chi_b = \chi_1(bc)$ for all $c \in GF(q)$ is an additive character of $GF(q)$, and every additive character of $GF(q)$ is obtained in this way.*

In particular, for the finite field $GF(p)$, where p is prime, we see that the p characters

$$\chi_j(a) = e^{2\pi i j a/p}, \ j = 0, 1, \cdots, p-1 \tag{13.6}$$

are all the additive characters of $GF(p)$, since $(GF(p),+)$ is cyclic.

Characters of the multiplicative group $GF(q)^*$ are called *multiplicative characters* of $GF(q)$. Since $GF(q)^*$ is a cyclic group of order $q-1$, its characters can be easily determined.

Proposition 13.2.2 *Let g be a fixed primitive element of $GF(q)$. For each $j = 0, 1, \cdots, q-2$, the function ψ_j with*

$$\psi_j(g^k) = e^{2\pi ijk/(q-1)}, \quad k = 0, 1, \cdots, q-2$$

defines a multiplicative character of $GF(q)$, and every multiplicative character of $GF(q)$ is obtained in this way.

As a consequence of Proposition 13.2.2, we have the following result about G^C:

Corollary 13.2.3 *G^C is cyclic of order $q-1$ with identity element ψ_0.*

Applying the orthogonality relations (13.1) and (13.2) to additive characters of $GF(q)$, for additive characters χ_a and χ_b in Proposition 13.2.1, we have

$$\sum_{c \in GF(q)} \chi_a(c)\overline{\chi_b(c)} = \begin{cases} 0 & \text{for } a \neq b, \\ q & \text{for } a = b. \end{cases} \tag{13.7}$$

In particular,

$$\sum_{c \in GF(q)} \chi_a(c) = 0 \quad \text{for } a \neq 0. \tag{13.8}$$

Furthermore, if c and d are elements of $GF(q)$, then

$$\sum_{b \in GF(q)} \chi_b(c)\overline{\chi_b(d)} = \begin{cases} 0 & \text{for } c \neq d, \\ q & \text{for } c = d. \end{cases} \tag{13.9}$$

Similarly, by applying (13.1) and (13.2) to the multiplicative characters, we have, for multiplicative characters ψ and τ,

$$\sum_{c \in GF(q)^*} \psi(c)\overline{\tau(c)} = \begin{cases} 0 & \text{for } \psi \neq \tau, \\ q-1 & \text{for } \psi = \tau. \end{cases} \tag{13.10}$$

In particular,

$$\sum_{c \in GF(q)^*} \psi(c) = 0 \quad \text{for } \psi \neq \psi_0. \tag{13.11}$$

Furthermore, if c and d are elements of $GF(q)^*$, then

$$\sum_{\psi} \psi(c)\overline{\psi(d)} = \begin{cases} 0 & \text{for } c \neq d, \\ q-1 & \text{for } c = d, \end{cases} \tag{13.12}$$

where the sum is extended over all multiplicative characters ψ of $GF(q)$.

13.2.1 Field Multiplicative Characters: Most Used Ones

We now turn to the cryptographic aspects of the characters described above. To begin with the multiplicative characters, we observe that many of the cryptographic functions (or components of them) used in some of the preceding chapters are multiplicative characters of the fields $D/\pi D$, where the π is a prime defined in some PID with $\mathrm{N}(\pi)$ being a prime number. For example, the cryptographic function

$$\chi(a) = \left(\frac{a}{p}\right) \tag{13.13}$$

is a multiplicative character of Z_p, known as the *quadratic character*, where p is a prime of Z. The function $(\chi + 1)/2$ is exactly the cryptographic function we employed for the cyclotomic generator of order 2 with $N = p$, as described in Section 8.1. For a finite field $GF(q)$ with q odd, the function

$$\lambda(c) = \begin{cases} 1 & \text{if } c \text{ is a square root,} \\ -1 & \text{otherwise} \end{cases} \tag{13.14}$$

is the quadratic character of $GF(q)$.

For our cryptographic purposes, we need to generalize the notion of characters. Let G be a finite Abelian group (written multiplicatively). If χ is a mapping from G to $(V, \times)$ such that

$$\chi(ab) = \chi(a) \times \chi(b) \text{ for all } a, b \in G, \tag{13.15}$$

and $(V, \times) \simeq (U, \times)$, where U is the set of complex numbers with absolute value 1, then we say χ is a *character* of G.

Let $p = df + 1$ be a prime number of Z. According to the above definition the cryptographic function

$$f(x) = x^{(p-1)/d} \bmod p \tag{13.16}$$

used in some of the foregoing chapters, is also a multiplicative character of Z_p. Summarizing some of the cryptographic functions in some of the preceding chapters defined on Z_p, we generally have an expression for the

$$Z_p \xrightarrow{\quad f(x) \quad} V_d \xrightarrow{\quad g(x) \quad} Z_d$$

Figure 13.1: A general description of some cryptographic functions.

cryptographic functions as in Figure 13.1, where V_d is a set of dth roots of unity of Z_p, $f(x)$ is a slight modification of the function in (13.16), i.e.,

$$f(x) = \begin{cases} c, & \text{if } x = 0, \\ x^{(p-1)/d} \bmod p, & \text{otherwise,} \end{cases} \tag{13.17}$$

where c is a fixed element of V_d; and $g(x)$ is defined by $g(x) = \log_\theta x$, where θ is a primitive root of p. Denoting the multiplication of Z_p as $\times$, we can easily see that $(V_d, \times) \simeq (U_d, \cdot)$, where U_d is the set of dth roots of unity in the complex numbers and "$\cdot$" is complex number multiplication.

Now we analyze the nonlinearity of the functions $f(x)$, $g(x)$ and $h(x) = g(f(x))$. Let the partition $D_0, D_1, \cdots, D_{d-1}$ of Z_p^* be defined as in Section 4.1. By the definition of the $f(x)$ in (13.17), we can choose the constant c such that the characteristic class of $f(x)$ is $D_0', D_1', \cdots, D_{d-1}'$, where $D_0' = D_0 \cup \{0\}$, $D_i' = D_i$ for each i with $1 \le i \le d-1$. We first consider the nonlinearity of $f(x)$ with respect to $(Z_p, +)$ and $(V_d, \times)$.

For each $a \in Z_p^*$, $b \in V_d$, let

$$p(a,b) = \sum_{i \in V_d} |D_i \cap (D_{b^{-1}i} - a)|/p.$$

Then it is not difficult to arrive at the following inequalities

$$\begin{aligned} 0 &\le \Pr(f(x)/f(x+a) = b) - p(a,b) \le \\ &\le \left[\sum_{i \in V_d} (|D_i' \cap (D_{b^{-1}i}' - a)| - |D_i \cap (D_{b^{-1}i} - a)|) \right] /p \le \\ &\le 2/p, \end{aligned}$$

where $\Pr(f(x)/f(x+a) = b)$ denotes the probability of $f(x)/f(x+a)$ taking on b. This means that the nonlinearity of $f(x)$ with respect to the above two operations depends on the difference parameters defined in Section 4.2, which were proven in Section 4.2 to be determined by the cyclotomic numbers of order d.

Restricting $f(x)$ to Z_p^*, we see that it is linear with respect to $(Z_p^*, \times)$ and $(V_d, \times)$. Thus, we can say that $f(x)$ is almost linear with respect to $(Z_p, \times)$ and $(V_d, \times)$.

The linearity of $g(x)$ with respect to $(V_d, \times)$ and $(Z_d, +)$ is clear. Furthermore, we have

$$g(x^{-1}) = \log_\theta x^{-1} = -g(x).$$

It follows from the arguments of Section 4.3 that the nonlinearity of the function $h(x) = g(f(x))$ with respect to $(Z_p, +)$ and $(Z_d, +)$ is about the same as that of $f(x)$ with respect to $(Z_p, +)$ and $(V_d, \times)$ if the cyclotomic numbers of order d have good stability. Nevertheless, it is clear that $h(x)$ is almost linear with respect to $(Z_p, \times)$ and $(Z_d, +)$.

The above analysis of the nonlinearity of $f(x)$, $g(x)$ and $h(x)$ shows again that a main cryptographic technique of this book is

$$\text{GOOD} + \text{BAD} = \text{GOOD}.$$

It follows that "bad" could be exactly the "good" we search for, if we use the cryptographic "bad" ones in *another* way or in another context. This cryptographic philosophy will be further discussed in the following subsection.

Before ending this subsection, let us generalize the discussion of this subsection to $GF(q)$. Consider now the cryptographic function in Section 4.6. It is also a multiplicative character of $GF(q)$. The above conclusion about $h(x)$ also applies to this function. This means that it is almost linear with respect to $(GF(q), \times)$ and $(Z_d, +)$, but its nonlinearity with respect to $(GF(q), +)$ and $(Z_d, +)$ depends on the cyclotomic numbers of order d defined on $GF(q)$. For details about the cyclotomy on $GF(q)$, one may consult [414].

The *norm function* from $F = GF(q^m)$ to $K = GF(q)$ is defined by

$$N_{F/K} = x^{(q^m-1)/(q-1)},$$

which is also a multiplicative character of $GF(q^m)$ by our definition. Its cryptographic values have already been implied in the above discussions.

13.2.2 Field Additive Characters: Most Used Ones

By Proposition 13.2.1 all the additive characters of $GF(q)$ are determined by the absolute trace function of (13.4). It is well known that the absolute trace function is linear from $(GF(q), +)$ to $(GF(p), +)$. This linear property may be cryptographically fatal if it is not properly used. However, if the nonlinearity of the absolute trace function from $(GF(q)^*, \times)$ to $(GF(p), +)$ is good, some absolute trace and additive characters could be cryptographically quite attractive. To investigate the cryptographic values of the absolute trace functions and additive characters of $GF(q)$, we summarize some elementary properties of the absolute trace function. Details about the proofs of the following facts can be found in [276, pp. 54-56].

We begin with the general trace function. For $x \in F = GF(q^m)$ and $K = GF(q)$, the *trace function* $\mathrm{Tr}_{F/K}(x)$ from F onto K is defined by

$$\mathrm{Tr}_{F/K}(x) = x + x^q + \cdots + x^{q^{m-1}}. \qquad (13.18)$$

It is the absolute trace function of (13.4) if K is the prime subfield of F. Let $\alpha \in GF(q^m)$, then the elements $\alpha, \alpha^q, \alpha^{q^2}, \cdots, \alpha^{q^{m-1}}$ are called the *conjugates* of α with respect to $GF(q)$, and the set of these elements is said to be the *conjugate class* containing α with respect to $GF(q)$. Thus, the trace of α over K is the sum of the conjugates of α with respect to K. The following proposition summarizes some of the properties of the trace function.

Proposition 13.2.4 *Let $F = GF(q^m)$ and $K = GF(q)$. Then the trace function satisfies the following properties:*

1. $\mathrm{Tr}_{F/K}(\alpha + \beta) = \mathrm{Tr}_{F/K}(\alpha) + \mathrm{Tr}_{F/K}(\beta)$ *for all* $\alpha, \beta \in F$;
2. $\mathrm{Tr}_{F/K}(c\alpha) = c\mathrm{Tr}_{F/K}(\alpha)$ *for all* $c \in K, \alpha \in F$;
3. *It is a linear transformation from F onto K, where both F and K are viewed as vector spaces over K;*
4. $\mathrm{Tr}_{F/K}(a) = ma$ *for all* $a \in K$;
5. $\mathrm{Tr}_{F/K}(\alpha^q) = \mathrm{Tr}_{F/K}(\alpha)$ *for all* $\alpha \in F$.

One important feature of the trace function is that it is not only itself a linear transformation, but can also be used to describe all linear transformations from F into K as shown by the following proposition:

Proposition 13.2.5 *Let F be a finite extension of the finite field K, both considered as vector spaces over K. Then the linear transformations from F into K are exactly the mappings $L_\beta, \beta \in F$, where $L_\beta(\alpha) = \mathrm{Tr}_{F/K}(\beta\alpha)$ for all $\alpha \in F$. Furthermore, we have $L_\beta \neq L_\gamma$ whenever β and γ are distinct elements of F.*

This proposition is of cryptographic significance, since it shows that the nonlinearity of all nontrivial linear functions from F^* into K with respect to $(F^*, \times)$ and $(K, +)$ is the same as that of the trace function.

Now we turn to the nonlinearity of the trace function $\mathrm{Tr}_{F/K}$ from $(F^*, \times)$ to $(K, +)$, where $F = GF(q^m)$ and $K = GF(q)$. To analyze the nonlinearity, we need to introduce the *conjugate class* of Z_{q^m-1}, where q is a power of a prime. A conjugate class of Z_{q^m-1} is the set

$$C_k = \{kq^i \bmod (q^m - 1) : \ i = 0, 1, \cdots, m-1\}.$$

For example, the conjugate classes of Z_7 are

$$C_0 = \{0\}, \ C_1 = \{1,2,4\}, \ C_3 = \{3,6,5\}. \tag{13.19}$$

The conjugate classes of Z_{15} are

$$\begin{array}{l} C_0 = \{0\}, \ C_1 = \{1,2,4,8\}, \ C_3 = \{3,6,12,9\}, \\ C_7 = \{7,14,13,11\}, \ C_5 = \{5,10\}. \end{array} \tag{13.20}$$

The conjugate classes of Z_{31} are

$$\begin{array}{l} C_0 = \{0\}, \\ C_1 = \{1,2,4,8,16\}, \\ C_3 = \{3,6,12,24,17\}, \\ C_{15} = \{15,30,29,27,23\}, \\ C_7 = \{7,14,28,25,29\}, \\ C_5 = \{5,10,20,9,18\}, \\ C_{11} = \{11,22,13,26,21\}. \end{array} \tag{13.21}$$

Clearly, all the conjugate classes form a partition of Z_{q^m-1}, and all the conjugate classes except C_0 form a partition of $Z^*_{q^m-1}$.

Due to the linearity of the trace function it is an onto function and takes on each element of K exactly q^{m-1} times. This means that it is balanced. It follows from the elementary property (5) of Proposition 13.2.4 that each of the characteristic sets of the trace function

$$T_a = \{\alpha \in F^* : \ \mathrm{Tr}(\alpha) = a\}, \quad \text{for all } a \in K$$

consists of the set $\{\theta^i : \ i \in I_a\}$, where θ is a primitive element of F^* and I_a is some union of conjugate classes C_j. In other words, we have

$$T_a = \{\theta^i : \ i \in I_a\} \text{ for each } a \in K. \tag{13.22}$$

We call the I_a's the *index classes* of the trace function with respect to θ. The changing of θ leads to new index classes $(I_a + j)$ for some fixed j. With the above symbols and definition we can now describe the nonlinearity of the trace function from $(F^*, \times)$ to $(K, +)$.

For each $\alpha \neq 0$ of F^* and $b \in K$, write α as θ^j for some j. As x runs through F^*, it is easy to obtain

$$\begin{array}{rcl} \Pr(\mathrm{Tr}(x) - \mathrm{Tr}(x/\alpha) = b) & = & \sum_{a \in K} |T_a \cap \alpha T_{a-b}|/(q^m - 1) \\ & = & \sum_{a \in K} |I_a \cap (I_{a-b} + j)|/(q^m - 1). \end{array} \tag{13.23}$$

This result shows that the nonlinearity of the trace, and therefore of all nontrivial linear functions from F into K, is determined by the difference parameters

$$d(a,b;j) = |I_a \cap (I_b + j)|, \ (a,b) \in K \times K, \ j \in Z_{q^m-1}.$$

Of course, the sets I_a for all $a \in K$ form a partition of the residue class ring Z_{q^m-1}. Thus, our nonlinearity analysis is equivalent to the difference analysis of the index classes.

To analyze the difference property of the partition I_a's, we first consider some examples. Taking $q = 2^3$ and the primitive element θ of $GF(2^3)$ with minimal polynomial $x^3 + x + 1$ over $GF(2)$, we get

$$I_1 = C_0 \cup C_3, \ I_0 = C_1,$$

where C_0 and C_3 are defined in (13.19). Obviously, I_1 and I_0 are difference sets of Z_7. Taking $q = 2^4$ and the primitive element θ with minimal polynomial $x^4 + x + 1$ over $GF(2)$, we obtain

$$I_1 = C_3 \cup C_7, \ I_0 = C_0 \cup C_1 \cup C_5,$$

where C_0, C_1, C_3, C_5 and C_7 are defined in (13.20). Simple calculations show immediately that I_0 and I_1 are difference sets of Z_{15}. These two examples show that the trace functions in these two cases have the optimum nonlinearity with respect to $(F^*, \times)$ and $(K, +)$. This may lead us to guess that in general I_1 and I_0 are difference sets of Z_{2^m-1}. This is true, but we postpone the proof.

Since the trace function is only a special function from F into K, we now investigate the nonlinearity of all linear functions from F to K if they are considered from $(F^*, \times)$ to $(K, +)$. Let $L(x)$ be a nonzero linear function from F to K. Then for each $\alpha \in F$ with $\alpha \neq 1$, and each $b \in K$, we have

$$\begin{aligned} & \Pr(L(x) - L(x/\alpha) = b) \\ = \ & \Pr(L(x(1-\alpha^{-1})) = b) = \begin{cases} \frac{q^{m-1}-1}{q^m-1}, & \text{if } b = 0, \\ \frac{q^{m-1}}{q^m-1}, & \text{if } b \neq 0. \end{cases} \end{aligned} \tag{13.24}$$

This proves the following theorem.

Theorem 13.2.6 *For every nonzero linear function $L(x)$ from $F = GF(q^m)$ to $K = GF(q)$, its nonlinearity from $(F^*, \times)$ to $(K, +)$ is optimal as described by (13.24).*

Thus, the trace function has also the best nonlinearity with respect to $(F^*, \times)$ and $(K, +)$. Note that (13.23) holds for every linear function from

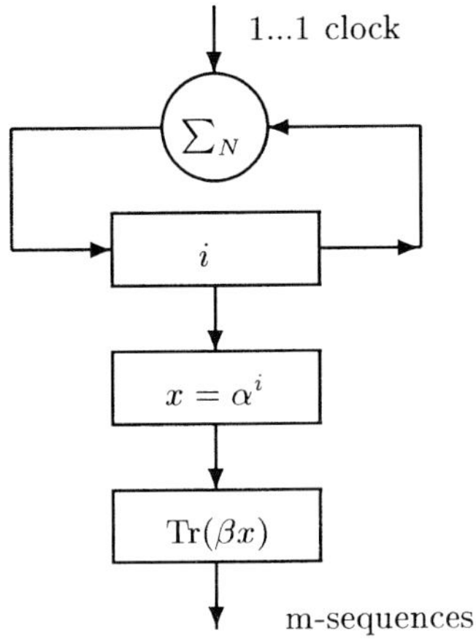

Figure 13.2: A NSG realization of m-sequences.

F to K. It follows from (13.23) and (13.24) that the index class I_0 is a difference set. This seems to have some relation with maximum-length sequences over K. In fact, we have the following general result:

Theorem 13.2.7 *Let θ be a primitive element of $F = GF(q^m)$ with minimal polynomial $m(x)$ over $K = GF(q)$ and $L(x)$ be any nonzero linear function from F to K. Define the sequence s^∞ over K by*

$$s_i = L(\theta^i) \ i = 0, 1, 2, \cdots.$$

Then the sequence s^∞ over K has least period $q^m - 1$ and minimal polynomial $m^*(x)$, where $m^*(x)$ is the reciprocal polynomial of $m(x)$.

Proof: We first prove that $L(ab) = 0$ for some $b \in F$ and for all $a \neq 0 \in F$ if and only if $b = 0$. If $b = 0$, then it is trivial. Conversely, suppose that $b \neq 0$; then ab ranges over F when a does. This means that $L(x) = 0$ for all $x \in F$, a contradiction. Let $a_i \in K, i = 0, 1, \cdots, l-1$ with $a_0, a_{l-1} \neq 0$. Then we have

$$\sum_{i=0}^{l-1} a_i s_{j-i} = L(\theta^j \sum_{i=0}^{l-1} a_i \beta^i), \tag{13.25}$$

where $\beta = \alpha^{-1}$ is another primitive element of F^*. It follows that, $\sum_{i=0}^{l-1} a_i s_{j-i} = 0$ if $m^*(x)$ divides $a_0 + a_1 x + \cdots + a_{l-1} x^{l-1} = a(x)$. On

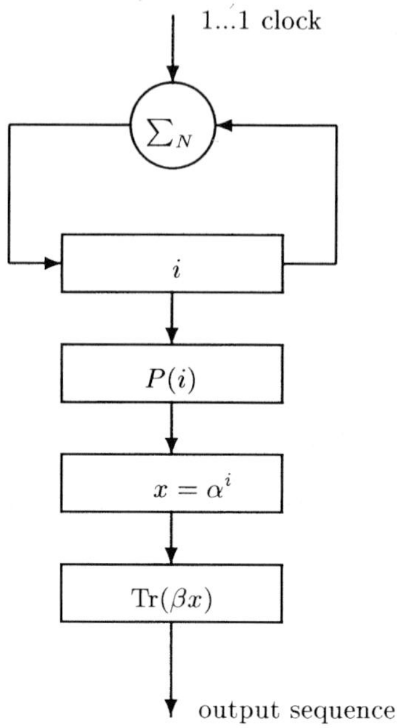

Figure 13.3: A modified generator of Figure 13.2.

the other hand, if $\sum_{i=0}^{l-1} a_i s_{j-i} = 0$ for all j with $j \geq l$, it follows from what we proved above that $a(\theta) = 0$. Thus, $m^*(x)$ must divide $a(x)$, since the minimal polynomial of β is $m^*(x)$. Due to the primitivity of θ the order of $m(x)$ and $m^*(x)$ must be $q^m - 1$. The least period of s^∞ follows. This completes the proof. $\square$

The above discussions, in particular Proposition 13.2.5 and Theorem 13.2.7, show that all of the maximum-length sequences over $GF(q)$ with periods $q^m - 1$ can be realized by the generator of Figure 13.2, where $N = q^m - 1$, $\sum_N$ denotes modulo N addition, $\mathrm{Tr}(x)$ denotes the trace function from $GF(q^m)$ to $GF(q)$, α is a chosen primitive element of $GF(q^m)$ and β is an element $\in GF(q^m)$.

By Theorem 13.2.6 every nonzero linear function from F to K has the best nonlinearity with respect to $(F^*, \times)$ and $(K, +)$. This rather remarkable result turns out to be easy to prove. However, the nonlinearity of the linear functions with respect to $(F^*, \times)$ and $(K, \times)$ seems more complicated. For some of them the nonlinearity depends on cyclotomic numbers. As shown in some of the preceding chapters, the nonlinearity of some of them is indeed good. Thus, it may be concluded that finite fields are cryptographically uesful in many ways. It may be difficult to access the cryptographic values of the balanced functions from one field to another one without putting them into specific cryptographic contexts. Sometimes in order to find "good" cryptographic building materials in some cryptographic context, we try to find some "bad" ones in some sense and to use them in proper ways.

We know that in the generator of Figure 13.2, its cryptographic function has the best nonlinearity with respect to $(F^*, \times)$ and $(K, +)$. This is one of its cryptographic advantages. However, the output sequences have only linear complexity m. To improve the generator, we can choose a permutation $P(x)$ of Z_{q^m-1} and produce the modified generator of Figure 13.3. Of course, affine permutations $P(x)$ of Z_{q^m-1} give only a decimation of the m-sequences. Thus, they give no improvement to the output sequences. It is not difficult to give examples to show there are some permutations which can improve the linear complexity of the sequences.

Naturally, we can choose the function $P(x)$ as a general mapping from Z_{q^m-1} to Z_{q^m-1}. That is to say, $P(x)$ need not be one to one. But in this case the balance property of m-sequences will be changed to some extent. The stop-and-go generator [21] can be realized by the generator of Figure 13.3 by choosing $P(x)$ properly (not one-to-one). How to choose an integer function $P(x)$ from Z_{q^m-1} to Z such that the output sequences of the generator of Figure 13.3 have some good cryptographic properties is an interesting problem.

Research Problem 13.2.8 *How can we choose an integer function $P(x)$ from Z_{q^m-1} to Z in Figure 13.3 to guarantee large linear complexity and ideal pattern distributions for the output sequence of the generator?*

13.3 Group Characters and Cyclotomic Numbers

Group characters are not only ideal cryptographic functions for certain applications, but also quite useful in calculating cyclotomic numbers, which determine a number of cryptographic attributes of cyclotomic generators. In fact, all known cyclotomic numbers are calculated based on some character sums, among which are Gauss sums, Jacobi sums and Dickson-Hurwitz sums. A connection between character sums and cyclotomic numbers is natural, since the number of solutions of many equations can be expressed as a kind of character sum. In this section we use group characters to calculate the cyclotomic numbers of order 2.

Let χ be a multiplicative character of $GF(q)$ with order d. Then d must divide $q-1$. As before, we let U_d denote the dth roots of unity in the complex numbers, say $U_d = \{u_0 = 1, ..., u_{d-1}\}$. For each i, where $0 \le i \le d-1$, let

$$D_i = \{x \in GF(q)^* | \chi(x) = u_i\}.$$

Clearly, $\{D_0, D_1, ..., D_{d-1}\}$ is a partition of $GF(q)^*$. Recall that the difference parameters of the partition are defined to be

$$d(i,j;a) = |D_i \cap (D_j - a)|,$$

where $0 \le i \le d-1$, $0 \le j \le d-1$ and $a \in GF(q)$. It is not hard to prove that the cyclotomic numbers of order d with respect to $GF(q)$ defined in Section 4.6 are given by

$$(i,j)_d = d(i,j;1).$$

By definition U_d is a subset of the set of complex numbers. Now we compute the difference parameters $d(i,j;a)$. To this end, we need to find the characteristic polynomial $F_i(x)$ of D_i, i.e.,

$$F_i(x) = \begin{cases} 1, & x \in D_i, \\ 0, & x \in GF(q)^* \setminus D_i. \end{cases}$$

By polynomial interpolation we obtain

$$F_j(x) = \frac{\prod_{i \ne j}(\chi(x) - u_i)}{\prod_{i \ne j}(u_j - u_i)}.$$

For simplicity we define $\chi(0) = 0$. Thus, $F_j(0) \neq 0$ and it is usually not an element of U_d. The following result plays an important role in calculating the difference parameters.

Theorem 13.3.1 *Let the symbols be the same as before. Then*

$$d(i,j;a) = \sum_{x \ne 0, a} F_i(x)F_j(x+a).$$

Proof: By definition $x \in D_i \cap (D_j - a)$ implies $x \neq 0, -a$, since 0 is not an element of any D_i. Hence $F_i(x)F_j(x+a) = 1$. Conversely, if $F_i(x)F_j(x+a) = 1$ and $x \neq 0, -a$, then by definition $x \in D_i \cap (D_j - a)$. This completes the proof. □

It is possible to calculate cyclotomic numbers of many orders with Theorem 13.3.1. Consider now cyclotomic numbers of order 2. Note that 2 divides $q-1$ for any finite field $GF(q)$ with characteristic $p \neq 2$. Then

$U_2 = \{1, -1\}$ and the two characteristic polynomials in this case are given by

$$F_0(x) = (\chi(x) + 1)/2, \quad F_1(x) = (1 - \chi(x))/2.$$

For the partition $\{D_0, D_1\}$ of $GF(q)^*$ the difference parameters are described by the following theorem.

Theorem 13.3.2 *Let the symbols be the same as before with q odd. Then for $0 \neq a \in GF(q)$*

$$\begin{aligned} 4d(0,0;a) &= q - 3 - \chi(a)[1 + \chi(-1)], \\ 4d(0,1;a) &= q - 1 - \chi(a)[1 - \chi(-1)], \\ 4d(1,0;a) &= q - 1 + \chi(a)[1 - \chi(-1)], \\ 4d(1,1;a) &= q - 3 + \chi(a)[1 + \chi(-1)], \end{aligned}$$

where

$$\chi(-1) = \begin{cases} -1, & \text{if } (q-1)/2 \text{ is odd,} \\ +1, & \text{if } (q-1)/2 \text{ is even.} \end{cases}$$

Proof: The proof is divided into several steps. By definition $\chi(0) = 0$. First we have the following basic fact about nontrivial characters (see Proposition 13.1.2).

$$\sum_x \chi(x) = 0. \tag{13.26}$$

Then we claim

$$\sum_x \chi(x(x+a)) = -1, \quad a \neq 0. \tag{13.27}$$

Since $\chi(0) = 0$, we have

$$\begin{aligned} \sum_{x \neq 0} \chi(x(x+a)) &= \sum_{y \neq 0} \chi\left(\frac{1+ay}{y^2}\right) \\ &= \sum_{y \neq 0} \chi(1 + ay) \\ &= \chi(a) \sum_{y \neq 0} \chi\left(\frac{1}{a} + y\right) \\ &= \chi(a) \left[\sum_{y \in GF(q)} \chi\left(\frac{1}{a} + y\right) - \chi\left(\frac{1}{a}\right) \right] \\ &= -\chi(1) = -1, \end{aligned}$$

where we have made use of (13.26) and (13.27) as well as the transformation $x = 1/y$. By Theorem 13.3.1, (13.26) and (13.27)

$$\begin{aligned} 4d(0,0;a) &= \sum_{x\neq 0,a} (\chi(x)+1)(\chi(x+a)+1) \\ &= \sum_{x} [\chi(x)\chi(x+a) + (\chi(x)+\chi(x+a))] \\ &\quad -\chi(a)[1+\chi(-1)] + q - 2 \\ &= q - 3 - \chi(a)[1+\chi(-1)]. \end{aligned}$$

This proves the first formula and the other three can be similarly proved. Finally we compute $\chi(-1)$.

Let α be a generating element of $GF(q)$. It follows from $\alpha^{q-1} = 1$ that $(\alpha^{(q-1)/2}+1)(\alpha^{(q-1)/2}-1) = 0$. Hence $\alpha^{(q-1)/2} = -1$ and $\chi(a)$ has order 2, so

$$\chi(-1) = \chi(\alpha)^{(q-1)/2} = (-1)^{(q-1)/2}.$$

If $(q-1)/2$ is even, it is clear that $\chi(-1) = 1$; otherwise $\chi(-1) = -1$. This completes the proof. □

Theorem 13.3.2 shows there are two sets of formulas for the difference parameters, depending on the parity of $(q-1)/2$. As a corollary of this theorem we have the following result.

Theorem 13.3.3 *Let $GF(q)$ be a finite field with characteristic $\neq 2$. If $(q-1)/2$ is even, then the cyclotomic numbers of order 2 are given by*

$$(0,0) = (q-5)/4, \quad (0,1) = (1,0) = (1,1) = (q-1)/4.$$

If $(q-1)/2$ is odd, they are given by

$$(1,0) = (q+1)/4, \quad (0,0) = (0,1) = (1,1) = (q-3)/4.$$

Cyclotomic numbers of other orders may be calculated in this way. For example, when $d = 3$, $U_3 = \{1, a, a^2\}$ with $a^2 + a + 1 = 0$. In this case the first characteristic polynomial is given by

$$F_0(x) = \frac{\chi(x)^2 + \chi(x) + 1}{(1-a)(1-a^2)}.$$

The other two characteristic polynomials are also easy to calculate.

13.4 The Nonlinearity of Characters

Let G be a finite Abelian group. Then the characters of G form an Abelian group under the product of characters. Thus every character χ will have $\chi^{|G|} = \chi_0$, where χ_0 is the trivial character (sometimes referred to as *principal character*). We say that χ is of *order d* if $\chi^d = \chi_0$, and if d is the smallest positive integer with this property. It is well known that d divides $|G|$.

As seen before, field characters play an important role in the design of some keystream generators. Our task in this section is to analyze the nonlinearity of field characters with respect to some operations. We will show that sometimes the nonlinearity is almost optimal. This fact indicates again that the linearity with respect to one pair of operations could indicate the best nonlinearity with respect to another pair of operations. It may follow that one way to get goodness is to make use of badness in a proper way.

13.4.1 The Nonlinearity of Multiplicative Characters

A multiplicative character χ is of course linear with respect to $(GF(q)^*, \times)$ and $(U, \times)$, where U is the set of complex numbers of absolute value 1. Let $\text{ord}(\chi) = d$, and let U_d denote the dth roots of unity in the complex numbers. Then χ is a mapping from $GF(q)^*$ to U_d. As before, we need to extend χ to $GF(q)$. This is done by defining

$$\chi(0) = c,$$

where 0 is the zero element of $GF(q)$, and c is any chosen element of U_d. We write χ^- for such an extended character of χ. Choosing a generator α of U_d, we could have a cryptographic function

$$F(x) = \log_\alpha \chi^-(x), \quad x \in GF(q),$$

which is a mapping from $GF(q)$ to Z_d.

Clearly, we have $F(xy) = F(x) + F(y)$ for each pair of nonzero x and y. Thus the nonlinearity of $F(x)$ with respect to $(GF(q), +)$ and $(Z_d, +)$ is the same as that of χ^- with respect to $(GF(q), +)$ and $(U_d, \times)$.

Lemma 13.4.1 [425] *Let $q-1 = dl$, and let q be an odd prime power. For the cyclotomic numbers of order d with respect to* GF(q) *we have*

$$\sum_{h=0}^{d-1}(h, h+k) = \begin{cases} l-1 & \text{if } k = 0, \\ l & \text{if } 1 \le k < d. \end{cases}$$

Theorem 13.4.2 [Carlet and Ding [60]] *Consider the nonlinearity of the extended multiplicative character χ^- of order d with respect to $(GF(q), +)$ and $(U_d, \times)$. Let q be odd and let $-1 \in D_s^{(d,q)}$ for some $0 \leq s \leq d-1$, where the $D_h^{(d,q)}$ are cyclotomic classes of order d.*

(1) *If $d - s \equiv 2k \pmod{d}$ has a solution k with $1 \leq k \leq d-1$, then*

$$P_{\chi^-} = \frac{l+2}{dl+1} = \frac{1}{d} + \frac{2d-1}{dq}.$$

(2) *Otherwise*

$$P_{\chi^-} = \frac{l+1}{dl+1} = \frac{1}{d} + \frac{d-1}{dq}.$$

In this case χ^- has optimal nonlinearity.

Proof: Since $\mathrm{ord}(\chi) = d$, $\chi = \psi_l$. Define $\beta = e^{2\pi i/d}$. Then β is a primitive d-th root of unity. Clearly,

$$\chi^-\left(D_0^{(d,q)} \cup \{0\}\right) = 1,$$
$$\chi^-\left(D_h^{(d,q)}\right) = \beta^h, \quad 1 \leq h < d.$$

For any $0 \neq a \in \mathrm{GF}(q)$ and $b = \beta^k \in U_d$, let $a^{-1} \in D_j^{(d,q)}$. By Lemma 13.4.1

$$\begin{aligned}
&|\{x \in \mathrm{GF}(q) | f(x+a)/f(x) = b\}| \\
= &\sum_{h=0}^{d-1} \left| D_h^{(d,q)} \cap \left(D_{k+h}^{(d,q)} - a\right)\right| + \left|\{a\} \cap D_k^{(d,q)}\right| + \left|\{-a\} \cap D_{d-k}^{(d,q)}\right| \\
= &\sum_{h=0}^{d-1} (h+j, h+j+k) + \left|\{a\} \cap D_k^{(d,q)}\right| + \left|\{-a\} \cap D_{d-k}^{(d,q)}\right| \\
= &\begin{cases} l - 1 + \left|\{a, -a\} \cap D_0^{(d,q)}\right|, & \text{if } k = 0 \\ l + \left|\{a\} \cap D_k^{(d,q)}\right| + \left|\{-a\} \cap D_{d-k}^{(d,q)}\right|, & \text{if } 1 \leq k < d. \end{cases}
\end{aligned}$$

If $d - s \equiv 2k \pmod{d}$ has a solution k with $1 \leq k \leq d-1$, then

$$\max_a \left|\{a\} \cap D_k^{(d,q)}\right| + \left|\{-a\} \cap D_{d-k}^{(d,q)}\right| = 2.$$

Otherwise the maximum value is 1. The conclusions of this theorem then follow. □

This theorem says that the nonlinearity of the extended multiplicative character χ^- with respect to $(GF(q), +)$ and $(U_d, \times)$ is either optimal or almost optimal.

13.4.2 The Nonlinearity of Additive Characters

Let ϕ be an additive character of $GF(q)$, and let d be its order. Then we have the trivial facts that $d > 1$ and $d|q$. By definition ϕ is linear with respect to $(GF(q), +)$ and $(U_d, \times)$. Writing ϕ_- for the restriction of ϕ to $GF(q)^*$, we consider now the nonlinearity of ϕ_- with respect to $(GF(q)^*, \times)$ and $(U_d, \times)$.

For each u in U_d we define the set D_u by $D_u = \{y : \phi(y) = u, y \in GF(q)\}$. Then it follows from the linearity of ϕ with respect to $(GF(q), +)$ and $(U_d, \times)$ that

$$|D_u| = q/d$$

for each u in U_d.

Combining the above facts and the fact that $\phi_-(xa)/\phi_-(x) = \phi_-(x(a-1))$, we have the following conclusion.

Theorem 13.4.3 *Let the symbols be the same as before. Then for each a of $GF(q)^*$ with $a \neq 1$ and each u of U_d with $u \neq 1$,*

$$\Pr(\phi_-(ax)/\phi_-(x) = u) = \frac{1}{d} + \frac{1}{qd}.$$

With this theorem we can now conclude that the nonlinearity of nontrivial additive characters of finite fields with respect to $(GF(q)^*, \times)$ and $(U_d, \times)$ is the best possible.

13.5 Ring Characters and Cryptography

Let $(R, +, \times)$ be a finite commutative ring with multiplicative identity 1_R. The additive characters of R are clear, since $(R, +)$ is an Abelian group. Let R^* be the set of all multiplicatively invertible elements of R. Then $(R^*, \times)$ is an Abelian group. The multiplicative characters of R are defined to be those of $(R^*, \times)$.

Ring characters could be cryptographically as attractive as field characters. In fact the twin-prime generator, the two-prime generator and the square generator in Chapter 8 employ the ring multiplicative characters of Z_{pq} and Z_{p^2}, where p and q are distinct prime numbers in Z. The nonlinearity of those cryptographic functions based on some of the ring multiplicative characters depends not only on the generalized cyclotomic numbers, but also on the assignment of the elements of $Z_{pq} \setminus Z_{pq}^*$ and $Z_{p^2} \setminus Z_{p^2}^*$. For these two kinds of rings, the assignment of those zero divisors does not contribute much to the nonlinearity of the cryptographic function due to the fact that

$|Z_{pq}^*|/|Z_{pq}|$ and $|Z_{p^2}^*|/|Z_{p^2}|$ are both approximately one. The Jacobi symbol is also a multiplicative character of the residue ring Z_n. But to extend it into a cryptographic function, the assignment of the zero divisors of Z_n will be of significance when n has many small factors.

Let a be a nonzero integer, b an odd integer, such that $\gcd(a, b) = 1$. The Jacobi symbol (a/b) is defined as an extension of Legendre symbol, in the following manner. Let $|b| = \prod_{p|b} p^{e_p}$ (with $e_p \geq 1$). Then

$$(\frac{a}{b}) = (\frac{a}{-b}) = \prod_{p|b} (\frac{a}{p}^{e_p}).$$

Therefore, (a/b) is equal to $+1$ or -1. Here are some of the properties of the Jacobi symbol:

1. $\left(\frac{a}{1}\right) = \left(\frac{a}{-1}\right) = 1.$
2. $\left(\frac{aa'}{b}\right) = \left(\frac{a}{b}\right)\left(\frac{a'}{b}\right).$
3. $\left(\frac{a}{bb'}\right) = \left(\frac{a}{b}\right)\left(\frac{a}{b'}\right).$
4. If a, b are relatively prime odd integers and $b \geq 3$, then we have the reciprocity law:

$$\left(\frac{a}{b}\right) = (-1)^{(a-1)(b-1)/4}\left(\frac{b}{a}\right).$$

5. If $b \geq 3$ and a is a square modulo b, then $\left(\frac{a}{b}\right) = 1$.

Apart from the moduli $n = pq$ and $n = p^2$, other cryptographically good moduli for employing the Jacobi character cryptographically, may be $2p$, $4p$ and $4pq$. The Jacobi character is also related to genus theory [86].

Let n be odd, then there are some nontrivial linear functions from Z_n to Z_2 with respect to $(Z_n, +)$ and $(Z_2, +)$; $f(x) = x \bmod 2$ is one example. Similar to the case of fields, we want to know the nonlinearity of the linear functions with respect to $(Z_n^*, \times)$ and $(Z_2, +)$. This problem seems complicated and remains open.

Chapter 14

P-Adic Numbers, Class Numbers and Sequences

The natural one-to-one correspondence between p-adic numbers and p-ary sequences is clear. The arithmetic of the field of p-adic numbers naturally gives a number of with-carry operations for p-ary sequences. The p-adic approach to the design and analysis of sequences turns out to be important for a number of reasons. In this chapter we will mainly concentrate on the 2-adic approach to the design and analysis of binary sequences. It is possible to extend many of the results for p-ary sequences.

Various class number problems are important topics in number theory. Among them is Gauss' class number problem for imaginary quadratic fields. It is interesting that some pseudorandom number sequences are related to the class numbers of imaginary quadratic fields.

Sections 14.4–14.7 are based on Klapper and Goresky [237, 240], and Section 14.8 on Cusick [88]. Other sections of this chapter are devoted to classical results about 2-adic numbers.

14.1 The 2-Adic Value and 2-Adic Expansion

Before defining the 2-adic value, we need to prove the following lemma [281, p. 6]:

Lemma 14.1.1 *Let $p \neq 0$ and $q \geq 1$ be integers such that* $\gcd(p, q) = 1$. *Then there exist a unique integer f and a pair of integers s and t such that*

$$2^f \frac{p}{q} = \frac{s}{t}, \ \gcd(s, t) = \gcd(2, s) = \gcd(2, t) = 1. \tag{14.1}$$

Proof: By assumption the integers p and q can be written as

$$p = 2^m s \quad \text{with } \gcd(2, s) = 1, \quad s \neq 0,$$
$$q = 2^n t \quad \text{with } \gcd(2, t) = 1, \quad t > 0,$$

where s and t are odd integers. It follows that

$$2^{n-m} \frac{p}{q} = \frac{s}{t}.$$

Since $\gcd(p, q) = 1$, at least one of the above m and n is zero and $\gcd(s, t) = 1$. The uniqueness of f follows from that of the above m and n. □

Let the notations be the same as in Lemma 14.1.1. If $\alpha = p/q \neq 0$, the 2-*adic value* of α is defined to be 2^f, and denoted by $|\alpha|_2$. The 2-adic value of zero is defined to be zero.

Rational numbers p/q with q odd are called 2-*adic integers* . It follows easily from Lemma 14.1.1 that p/q is a 2-adic integer if and only if $|p/q|_2 \leq 1$.

The following basic facts about the 2-adic value are fundamental and their proofs are trivial.

1. $|\alpha|_2 = 1$ if and only if $\alpha = p/q \neq 0$ with $\gcd(2, p) = \gcd(2, q) = 1$.
2. $|\alpha|_2 = 2^f$ if and only if $|2^f \alpha|_2 = 1$.
3. $|2^i \alpha|_2 = 2^{-i} |\alpha|_2$ for every integer i.
4. $|\alpha \pm \beta|_2 \leq \max\{|\alpha|_2, |\beta|_2\}$ (triangle inequality) and equality holds when $|\alpha|_2 \neq |\beta|_2$.
5. $|\alpha\beta|_2 = |\alpha|_2 |\beta|_2$.

There are close relations between the 2-adic value of rational numbers and the valuation for rings as well as the discrete valuation for algebraic function fields. We may come to some of these problems later.

The following lemma plays an important role in the 2-adic expansion of rational numbers.

Lemma 14.1.2 *For every rational number $\alpha = p/q \neq 0$, where $q \geq 1$ is odd, $|p| < q$ and $\gcd(p, q) = 1$, there exist two unique integers $u \in \{0, 1\}$ and p' with $0 < |p'| < q$ such that*

$$\frac{p}{q} = u + 2\frac{p'}{q}, \tag{14.2}$$

where $(u, p') = (0, p/2)$ if p is even, and $(u, p') = (1, (p - q)/2)$ if p is odd. Moreover, p' must be negative if $p < 0$.

Proof: It is easy to check that the (u, p') given in Lemma 14.1.2 for each case is a solution of (14.2). Assume that (u, p') and (u', p'') are two solutions. It follows from (14.2) that

$$p = uq + 2p', \quad p = u'q + 2p''.$$

Hence

$$0 = (u - u')q + 2(p' - p'').$$

It follows further from $\gcd(2, q) = 1$ and $u - u' \in \{-1, 0, 1\}$ that $u = u'$ and $p' = p''$.

If $p < 0$, the integer p' must be negative in both cases since $|p| < q$. □

A binary sequence $\{a_i\}_{i=-k}^{\infty}$ is the 2-*adic expansion* of a rational number α if

$$\lim_{n \to \infty} \left| \alpha - \sum_{i=-k}^{n-1} a_i 2^i \right|_2 = 0.$$

The 2-adic expansion of α is written as

$$\begin{aligned} \alpha &= a_{-k}a_{-k+1}...a_0a_1... \\ &= \textstyle\sum_{i=-k}^{\infty} a_i 2^i. \end{aligned} \tag{14.3}$$

If there exist two integers m and $N > 0$ such that

$$a_i = a_{i+N} \text{ for all } i \geq m,$$

the expression of (14.3) is said to be *eventually* or *ultimately periodic* with period N, and *periodic* if $m = -k$. For simplicity we sometimes write an ultimately periodic expansion as

$$\alpha = a_{-k}a_{-k+1}...a_{m-1}\overline{a_m...a_{m+N-1}},$$

where the bar represents the repeated part.

Proposition 14.1.3 *Let $\alpha = p/q \neq 0$ be a rational number with $q \geq 1$ being odd, $|p| < q$, and $\gcd(p, q) = 1$. And let $p = 2^m p_1$, where $\gcd(2, p_1) = 1$ and $m \geq 0$. Then α has the unique ultimately periodic 2-adic expansion*

$$\alpha = 0...01a_{m+2}...a_{m+h-1}\overline{a_{m+h}...a_{m+h+N-1}},$$

where at the beginning of the sequence there are exactly m zeros before the first 1, N is a positive integer with $1 \leq N \leq q-1$.

Proof: To get a 2-adic expansion for p/q, we repeat the procedure of finding the solution for (14.2) until a repeated rational number is found.

After repeating the procedure m times we get the first part of the 2-adic expansion 0...0 with m zeros and the rational number p_1/q. Since p_1 is odd and $\gcd(2, p_1) = 1$, repeating the procedure once more gives us a 1 after the zero sequence and a new rational number, denoted still by p_1/q, where $p_1 < 0$. Then all the following new p_1's remain negative when the procedure is further repeated. Because there are at most $q-1$ new rational numbers p_1/q with p_1 negative and $|p_1| < q$, after at most $q-1$ calls for the procedure we must get a rational number which had already appeared before. Then we get an ultimately period expansion for α as described in the proposition. The uniqueness of the expansion follows from that of the solution of (14.2). □

We now take an example to show how to get the 2-adic expansion for a rational number described in Proposition 14.1.3. Applying the constructive proof procedure for Proposition 14.1.3, we obtain

$$\begin{array}{rcrcrr} 4/9 & = & 0 & + & 2 & (2/9), \\ 2/9 & = & 0 & + & 2 & (1/9), \\ 1/9 & = & 1 & + & 2 & (-4/9), \\ -4/9 & = & 0 & + & 2 & (-2/9), \\ -2/9 & = & 0 & + & 2 & (-1/9), \\ -1/9 & = & 1 & + & 2 & (-5/9), \\ -5/9 & = & 1 & + & 2 & (-7/9), \\ -7/9 & = & 1 & + & 2 & (-8/9), \\ -8/9 & = & 0 & + & 2 & (-4/9). \end{array}$$

Therefore the expansion of 4/9 is

$$\frac{4}{9} = 001\overline{001110}.$$

The proof of Lemma 14.1.2 can be used to prove the following conclusion.

Lemma 14.1.4 *For every rational number $\alpha = p/q \neq 0$, where $q \geq 1$ is odd, $|p| > q$ and $\gcd(p, q) = 1$, there exist two unique integers $u \in \{0, 1\}$ and p' with $0 < |p'| < p$ such that*

$$\frac{p}{q} = u + 2\frac{p'}{q}, \tag{14.4}$$

where $(u, p') = (0, p/2)$ if p is even, and $(u, p') = (1, (p-q)/2)$ if p is odd.

Similar to Proposition 14.1.3, by repeating the procedure of finding the solution of (14.4) we can prove the following proposition.

Proposition 14.1.5 *Every rational number* $\alpha = p/q \neq 0$, *where* $q \geq 1$ *being odd,* $|p| > q$, *and* $\gcd(p, q) = 1$, *has the following unique expression*

$$\alpha = \sum_{i=0}^{h} a_i 2^i + 2^h \frac{p'}{q},$$

where $|p'| < q$, $\gcd(p', q) = 1$, *and* $a_i \in \{0, 1\}$ *for all* i.

Combining Lemma 14.1.1, Propositions 14.1.3 and 14.1.5, we obtain the following conclusion.

Proposition 14.1.6 *Every rational number has a unique ultimately periodic 2-adic expansion.*

The foregoing discussions show that the 2-adic expansion of a nonzero rational number $\alpha = p/q$ can be determined by the following procedure:

St1: Reduce p/q so that $\gcd(p, q) = 1$ and $q \geq 1$.

St2: With the proof procedure of Lemma 14.1.1 determine f and a pair of integers s and t such that

$$2^f \frac{p}{q} = \frac{s}{t}, \quad \gcd(s, t) = \gcd(2, t) = 1.$$

If $|s| < t$, then go to Step 4; otherwise respectively go to Step 3.

St3: With the procedure of Lemma 14.1.4, find the expression

$$\frac{s}{t} = \sum_{i=0}^{h} a_i 2^i + 2^h \frac{s'}{t},$$

where $|s'| < t$, $\gcd(s', t) = 1$, and $a_i \in \{0, 1\}$ for all i.

St4: Apply the proof procedure of Proposition 14.1.3 to s/t resp. s'/t to get the 2-adic expansion of s/t resp. s'/t, denoted by $\{b_i\}_{i=0}^{\infty}$.

St5: Output $\sum_{i=0}^{\infty} 2^{-f+i} b_i$ resp. $\sum_{i=0}^{h} 2^{-f+i} a_i + \sum_{j=0}^{\infty} 2^{h-f+j} b_j$ as the 2-adic expansion of the rational number.

The converse of Proposition 14.1.6 is the following conclusion.

Proposition 14.1.7 *For every ultimately periodic binary sequence a^∞ the associated 2-adic number $\sum_{i=0}^{\infty} a_i 2^i$ is the 2-adic expansion of a rational number.*

Proof: Because of the eventual periodicity let m and $N > 0$ be two integers such that

$$a_i = a_{i+N} \text{ for all } i \geq m.$$

First, we have

$$\alpha = \sum_{i=0}^{\infty} a_i 2^i = \left(\sum_{i=0}^{m-1} + \sum_{i=m}^{N+m-1} + \sum_{i=m+N}^{\infty} \right) a_i 2^i.$$

Then it follows that

$$\begin{aligned} 2^N \alpha &= 2^N \textstyle\sum_{i=0}^{m-1} a_i 2^i + \sum_{i=m}^{\infty} a_i 2^{i+N} \\ &= 2^N \textstyle\sum_{i=0}^{m-1} a_i 2^i + \sum_{i=m+N}^{\infty} a_i 2^i \\ &= (2^N - 1) \textstyle\sum_{i=0}^{m-1} a_i 2^i - \sum_{i=m}^{N+m-1} a_i 2^i + \alpha. \end{aligned}$$

Hence,

$$\alpha = \sum_{i=0}^{m-1} a_i 2^i - \frac{\sum_{i=m}^{N+m-1} a_i 2^i}{2^N - 1}, \tag{14.5}$$

which is a rational number. □

The above proof of Proposition 14.1.7, which parallels the classical proof of the rational expression $p(x)/q(x)$ for sequences over a field, follows the proof of the following conclusion [240].

Proposition 14.1.8 *Every periodic 2-adic integer $\sum_{i=0}^{\infty} a_i 2^i$ is the 2-adic expansion of a rational number $\alpha = p/q$ with q odd and $-q \leq p \leq 0$. Conversely, the 2-adic expansion of a rational number $\alpha = p/q$ with q odd and $-q \leq p \leq 0$ must be periodic.*

Proof: In the proof of Proposition 14.1.7, setting $m = 0$ proves the first part of this proposition. In particular, the 2-adic expansion of -1 is $\overline{1}$.

Conversely, suppose that $-q \leq p \leq 0$ and q is odd. Then $\text{ord}_q(2)$ exists. Let $N = \text{ord}_q(2)$, and set $s = (2^N - 1)/q$. Writing $s \cdot (-p) = \sum_{i=0}^{N-1} a_i 2^i$. Thus $\alpha = sp/(2^N - 1)$. The calculations leading to (14.5) may be run backwards to see that the segment $a_0, a_1, \cdots, a_{N-1}$ is a single period of a strictly periodic sequence. □

The proof of Proposition 14.1.7 may be used to prove the following old result of Gauss ([159], [26, Theorem 1]).

Corollary 14.1.9 *If p and q are relatively prime, $-q < p \leq 0$, and q is odd, then the period of the bit sequence for the 2-adic expansion of $\alpha = p/q$ is $T = \mathrm{ord}_q(2)$.*

14.2 A Fast Algorithm for the 2-Adic Expansion

Fast software-oriented algorithms for producing pseudorandom sequences are a necessity for some applications. We now describe a software-oriented algorithm for producing pseudorandom binary sequences, that is, the 2-adic expansion sequences of rational numbers.

For simplicity we consider the 2-adic expansion of rational numbers p/q such that $q \geq 1$, $\gcd(p, q) = 1$ and q is odd. Thus, the 2-adic expansion sequence of such a rational number is ultimately periodic. The analysis of Section 14.1 show that the following algorithm computes the 2-adic expansion sequence correctly.

An algorithm for computing the 2-adic expansion sequence:

begin: Input p and q.
repeat procedure:
If $|p|$ even, then output 0 and set $p \leftarrow p/2$; otherwise output 1 and set $p \leftarrow (p - q)/2$.
end

This algorithm is not only very simple, but also very efficient. To compute one bit output, one parity check and one even integer division by 2 plus at most one integer subtraction are needed. The memory requirement is less than $2\lceil \log_2 \max\{|p|, q\} \rceil$ bits. It is clearly a software-oriented algorithm.

It should be noted that the above algorithm works for all rational numbers p/q with q odd. This can be proved with arguments similar to those in Section 14.1.

14.3 The Arithmetic of $Q_{[2]}$ and $Z_{[2]}$

So far most investigations into the design and analysis of sequences are mainly based on the arithmetic of finite fields. Though few investigations into the cryptographic application of the 2-adic numbers have been done, it turns out that the arithmetic of the 2-adic numbers, which was introduced many years ago by mathematicians, is quite useful in constructing sequences for various applications.

In Section 14.1 it was proved that every rational number has a unique ultimately periodic expansion $\sum_{i=-f}^{\infty} a_i 2^i$, where $a_i \in \{0, 1\}$. Conversely, every ultimately period series $\sum_{i=-f}^{\infty} a_i 2^i$ is the 2-adic expansion of a unique rational number. A series $\sum_{i=-f}^{\infty} a_i 2^i$, where $a_i \in \{0,1\}$, is called a *2-adic number*, and a *2-adic integer* when $f = 0$. Let $Z_{[2]}$ denote the set of all 2-adic integers, and $Q_{[2]}$ the set of all 2-adic numbers. Then $Z_{[2]}$ is clearly a subset of $Q_{[2]}$. Let A denote the set of all rational numbers p/q, where $q \geq 1$ is odd. Then it follows from the discussion of Section 14.1 that there is a one-to-one correspondence between A and $Z_{[2]}$. This is why we refer to rational numbers of such a form also as 2-adic integers.

Before studying the structure of $Q_{[2]}$ and $Z_{[2]}$, we investigate the algebraic structure of the field Q and the ring A with respect to the usual addition and multiplication for rational numbers.

A number p/q is said to be a *reduced rational number* if $\gcd(p,q) = 1$ and $q \geq 1$ is an integer. The first result about the structure of A is the following conclusion.

Proposition 14.3.1 *A is a maximal proper ring of the field Q.*

Proof: By definition A is clearly a proper ring of Q. Assume that $B \supset A$ is a ring of Q. Let $\alpha \in B \setminus A$. Then α is not equal to zero and therefore can be expressed as $\alpha = v/u2^m$, where $m \geq 1$, $\gcd(2,v) = 1$ and v/u is reduced. Since $\gcd(2,v) = 1$, the reduced rational number $u/v \in A \subseteq B$. Thus, $\alpha u/v = 2^{-m} \in B$. On the other hand, $2 \in A \subseteq B$. Consequently, $2^{-1} = 2^{m-1}2^{-m} \in B$. It follows further from $B \supseteq A$ and the definition of A that $B = Q$. □

Proposition 14.3.2 *Every nonzero principal ideal I of A must be of the form $2^m A$.*

Proof: Let $p/q \in A$ be reduced, and $p = 2^m p_1/q$, where $m \geq 0$ and $\gcd(p_1, 2) = 1$. Consider the principal ideal $(p_1/q)A$. We first prove

$$\frac{p_1}{q} A = A. \tag{14.6}$$

The inclusion $(p_1/q)A \subseteq A$ is trivial. To prove the reverse inclusion, take any element $s/t \in A$ in reduced form. Then it is easily verified that

$$\gcd(sq, tp_1) = \gcd(s, p_1)\gcd(q, t),$$

which is odd. Set $x = sq/\gcd(sq, tp_1)$ and $y = tp_1/\gcd(sq, tp_1)$. Then y is odd, and $x/y \in A$ such that

$$\frac{s}{t} = \frac{p_1}{q}\frac{x}{y} \in \frac{p_1}{q} A.$$

This shows the reverse inclusion and proves (14.6).

Finally, we have

$$(p/q)A = 2^m((p_1/q)A) = 2^m A.$$

This completes the proof. □

Proposition 14.3.3 *A is a principal ideal domain.*

Proof: Let I be a nonzero ideal of A. Set

$$m = \min\{m : \text{ nonzero } 2^m p/q \in I, \text{ gcd}(2,p) = 1, \text{ and } p/q \text{ is reduced.}\}$$

Assume that $2^m p/q \in I$, where p/q is reduced and $\gcd(2,p) = 1$. Since $q/p \in A$, we have $2^m = 2^m(p/q)(q/p) \in I$. Thus, we have $I \supseteq 2^m A$.

On the other hand, each $0 \neq i \in I$ can be written as

$$i = 2^{m'}p/q,$$

where $m' \geq m$, the number p/q is reduced and $\gcd(2,p) = 1$. Thus, $i = 2^m 2^{m'-m} p/q \in 2^m A$. It follows $I \subseteq 2^m A$.

Finally, $\{0\}$ is clearly a principal ideal. Thus, every ideal of A is a principal ideal. □

A ring is called a *local ring* if it has only one maximal ideal. It follows from Propositions 14.3.2 and 14.3.3 that the first part of following proposition is true, while the second part is easily verified.

Proposition 14.3.4 *A is a local ring with the maximal ideal $2A$. Furthermore $A/2A \cong Z_2$.*

Let F be a field. A ring $O \subset F$ is said to be a *valuation ring* of F if $z \in O$ or $z^{-1} \in O$ for each $z \neq 0$. Let $O^* = \{z \in O : \text{there is a } w \in O \text{ with } zw = 1\}$. We can verify that the following claims are true:

1. Any valuation ring is a local ring.

2. Its unique maximal ideal, denoted by P and called a *place*, is a principal ideal.

3. If $P = tO$ then any nonzero $z \in F$ has a unique representation of the form $z = t^n u$ for some $n \in Z, u \in O^*$.

4. O is a principal ideal domain.

In the case of the rational number field, we have proved three of the four claims. Apparently, the element t, which is called a *prime element* , is equal to 2 with respect to A in this case. Set

$$v_P(z) = n,$$

where n is the integer in the unique expression $z = t^n u$ in the above Claim 3. Then it is easily verified that this function has the following properties:

1. $v_P(x) = \infty$ if and only if $x = 0$.
2. $v_P(xy) = v_P(x) + v_P(y)$ for any $x, y \in F$.
3. $v_P(x + y) \geq \min\{v_P(x), v_P(y)\}$ for any $x, y \in F$.
4. There is an element $z \in F$ with $v_P(z) = 1$.

Such a function is called a *discrete valuation* of the field. We can verify that there is a one-to-one correspondence between the valuation rings and the discrete valuations of a field. In the case of the rational number field Q the discrete valuation induced by the valuation ring A is $v_A : Q \to Z$ defined by

$$v_A(p/q) = m,$$

where p/q is reduced, and m is the unique integer such that $p/q = 2^m p_1/q$, where $\gcd(p_1, 2) = \gcd(q, 2) = 1$. The relation between this discrete valuation and the 2-adic value is

$$v_A(p/q) = -\log_2 |p/q|_2.$$

The above notions and results about rational numbers have already been extended into algebraic function fields which have applications in coding theory [413]. Now we consider the arithmetic of $Q_{[2]}$ and $Z_{[2]}$.

Suppose that α and β are two 2-adic numbers with

$$|\alpha|_2 \geq |\beta|_2, \ |\alpha|_2 = 2^f$$

and

$$\alpha = \sum_{i=-f}^{\infty} a_i 2^i, \quad \beta = \sum_{i=-f}^{\infty} b_i 2^i,$$

where $a_i, b_i \in \{0, 1\}$, and $a_{-f} \neq 0$, but one or more of the first digits $b_{-f}, b_{-f+1}, ...$ may be equal to zero.

The addition $\alpha + \beta$ is defined by the convergent series

$$\alpha + \beta = \sum_{i=-f}^{\infty} (a_i + b_i)2^i = \sum_{i=-f}^{\infty} r_i 2^i,$$

where each $r_i \in \{0, 1\}$ is calculated by

$$\begin{aligned} r_i &= (a_i + b_i + c_{i-1}) \bmod 2 \\ c_i &= (a_i + b_i + c_{i-1}) \operatorname{div} 2 \end{aligned}$$

(div 2 means remove the last binary digit) for each $i \geq -f$, where c_{-f-1} is defined to be 0, and the c_i's are carry bits.

As an example, let

$$\alpha = 1, \quad \beta = \sum_{i=0}^{\infty} 2^i,$$

then $\alpha + \beta = 0$.

Let $|\beta|_2 = 2^g$, so b_{-g} is the first digit distinct from 0, then

$$-\beta = 2^{-g} + \sum_{i=-g+1}^{\infty} (1 - b_i)2^i.$$

The subtraction $\alpha - \beta$ is defined to be $\alpha + (-\beta)$. It is obvious that $Q_{[2]}$ and $Z_{[2]}$ are Abelian groups with respect to the addition.

The multiplication of two 2-adic numbers is defined as follows. Let

$$\alpha = \sum_{i=-f}^{\infty} a_i 2^i, \quad \beta = \sum_{i=-g}^{\infty} b_i 2^i,$$

where $|\alpha|_2 = 2^f$ and $|\beta|_2 = 2^g$. After multiplying the series term by term and rearranging the terms, we obtain

$$\alpha\beta = \sum_{i=-(f+g)}^{\infty} u_i 2^i,$$

where

$$u_i = \sum_{k+j=i} a_k b_j,$$

for each $i \geq -(f+g)$. The u_i's could be much larger than 2. Then we use the same reduction procedure as before to get

$$\alpha\beta = \sum_{i=-(f+g)}^{\infty} r_i 2^i,$$

where each $r_i \in \{0, 1\}$ is calculated by

$$\begin{aligned} r_i &= (u_i + c_{i-1}) \bmod 2 \\ c_i &= (u_i + c_{i-1}) \operatorname{div} 2 \end{aligned}$$

for each $i \geq -f - g$, where $c_{-(f+g+1)}$ is defined to be 0, and the c_i's are carry digits.

Let

$$\alpha = \sum_{i=-f}^{\infty} a_i 2^i,$$

where $|\alpha|_2 = 2^f$ and $a_{-f} = 1$. One can easily prove that there is a unique 2-adic number

$$\beta = \sum_{i=f}^{\infty} b_i 2^i,$$

such that $\alpha\beta = 1$, where $b_f = 1$. This means that $Q_{[2]}$ forms a group with respect to the multiplication, but $Z_{[2]}$ does not.

Division for two 2-adic numbers is then defined to be $\alpha/\beta = \alpha\beta^{-1}$.

Let $Q_{<2>}$ be the set of ultimately periodic 2-adic numbers, and let $\phi : Q \to Q_{<2>}$ map each rational number to its unique 2-adic expansion. Then it is easily verified that ϕ is an isomorphism between $(Q, +, \cdot)$ and $(Q_{<2>}, +, \cdot)$. Thus, the structure of Q is the same as $Q_{<2>}$. Thus, we have the following conclusions.

Proposition 14.3.5 *The following conclusions regarding $Q_{<2>}$ and $Z_{[2]}$ are true:*

1. *$Z_{[2]}$ is a maximum proper ring of $Q_{<2>}$.*
2. *Every nonzero principal ideal of $Z_{[2]}$ must be of the form $2^m Z_{[2]}$.*
3. *$Z_{[2]}$ is a local ring.*
4. *the quotient field $Q_{<2>}/Z_{[2]}$ is isomorphic to Z_2.*

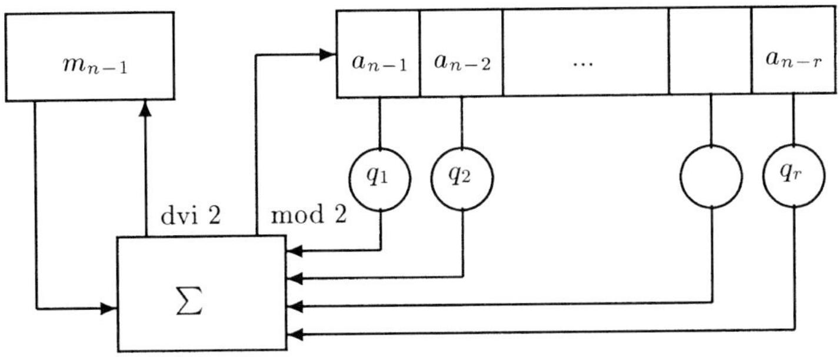

Figure 14.1: Feedback with carry shift register.

The one-to-one correspondence between the set of binary ultimately periodic sequences and the set of 2-adic integers defined by

$$\varphi : a^{\infty} \rightarrow \sum_{i=0}^{\infty} a_i 2^i$$

gives automatically the 2-*adic sum* and the 2-*adic product* of binary sequences.

14.4 Feedback Shift Registers with Carry

A kind of feedback shift register, feedback with carry shift registers (briefly FCSRs), was described by Klapper and Goresky [237, 240]. They can be thought of as LFSRs with ordinary addition in place of addition modulo 2, and auxiliary memory for storing the carry. The contents (0 or 1) of the tapped cells of the shift register are added as integers to the current contents of the memory to form a sum, $\sum$. The parity bit ($\sum \bmod 2$) of $\sum$ is fed back into the first cell, and the higher order bits ($\lfloor \sum /2 \rfloor$) are retained for the new value of the memory. The FCSR with connection integer q is depicted in Figure 14.1.

Note that $q_0 = -1$ does not correspond to a feedback tap, and that the coefficients of high powers of 2 are close to the output cell. In Figure 14.1, Σ denotes integer addition. The content of the register at any given time consists of r bits, denoted $a_{n-1}, a_{n-2}, \cdots, a_{n-r+1}, a_{n-r}$. The operation of the shift register is defined as follows:

A1. Form the integer sum $\sigma_n = \sum_{k=1}^{r} q_k a_{n-k} + m_{n-1}$.

A2. Shift the contents one step to the right, outputting the rightmost bit a_{n-r}.

A3. Place $a_n = \sigma_n \bmod 2$ into the leftmost cell of the shift register.

A4. Replace the memory integer m_{n-1} with $m_n = (\sigma_n - a_n)/2 = \lfloor \sigma_n/2 \rfloor$.

The integer q is referred to as the *connection integer* because its binary expansion gives the analog to the connection polynomial in the usual theory of linear feedback shift registers.

FCSRs were described to construct a feedback shift register whose output is the coefficient sequence of the 2-adic expansion

$$\alpha = \sum_{i=0}^{\infty} a_i 2^i = \frac{p}{q} \in Z_{[2]} \tag{14.7}$$

of a given rational number p/q with q odd and $0 \leq -p < q$ [240]. For the rest of this section, we fix an odd positive integer $q \in Z$ and let $r = \lfloor \log_2(q+1) \rfloor$. Write

$$q + 1 = q_1 2 + q_2 2^2 + \cdots + q_r 2^r \tag{14.8}$$

for the binary representation of the integer $q + 1$, where $q_r = 1$. The shift register uses r stages and no more than $\lfloor \log_2(r) \rfloor$ additional bits of memory. The feedback connections are given by the bits $\{q_1, q_2, \cdots, q_r\}$ appearing in (14.8).

The memory requirements can be easily seen as follows [240]. Let $w = \mathrm{WH}(q + 1)$ be the number of nonzero q_i, $i = 1, \cdots, r$, the Hamming weight of $q + 1$. If the memory needed for m_{n-1} is no more than w bits then the same will be true for all later m_i with $i \geq n$. This follows from (A1) and (A4) because $\sigma_n \leq w + m_{n-1} \leq 2w$ and $m_n \leq \sigma_n/2 \leq w$. Note that this is also true for the software algorithm in Section 14.2, due to Lemmas 14.1.1 and 14.1.4.

Moreover, if we initialize a FCSR with a larger memory of b bits for the initial carry, where $b > w$, then with each step, the memory will decrease at least by 1. After $b - w$ steps, the memory needed for later carries will be no more than w bits. This follows from (A1) and (A4) which give

$$m_n \leq \sigma_n/2 \leq (w + m_{n-1})/2 < m_{n-1}.$$

Thus the memory for the carries is never greater than the maximum of b and w. Allowing one additional bit for a sign if the initial carry is negative, the number of bits required for the memory of the carries is at most $\lceil \max(b, w) \rceil + 1$. For periodic sequences, the memory requirement is never greater than w bits.The same conclusion holds for the software algorithm in Section 14.2, due to Lemma 14.1.4.

FCSRs are clearly a hardware implementation of the computation of the 2-adic expansion sequence of some rational numbers. A software implementation was described in Section 14.2.

A variant of the feedback with carry shift register architecture was described by Klapper and Goresky [240]. It is based on arithmetic in the ring of π-adic integers, where addition and multiplication are just as in the ring of 2-adic integers. However, carried bits are advanced d steps. Feedback with carry shift registers over finite fields were described in [240].

14.5 Analysis and Synthesis of FCSRs

The analysis of a generator is concerned with properties of the output sequences. Among the important properties are periodicity and pattern distributions. Fundamental cryptographic problems concerning the FCSRs are the following:

1. The periodicity of the FCSR sequences.
2. The pattern distributions of the FCSR sequences.
3. The analysis of the linear and sphere complexity of the FCSR sequences.

Some of these problems have been solved to a certain extent, others remain open.

To determine the output sequence of a given FCSR, the arithmetic of 2-adic integers is needed. Suppose we fix an r-stage FCSR with connection integer $q = -1 + q_1 2 + q_2 2^2 + \cdots + q_r 2^r$, with initial memory m_{r-1}, and with initial loading $a_{r-1}, a_{r-2}, \cdots, a_1, a_0$, as depicted in Figure 14.1. The register will generate an infinite, ultimately periodic binary sequence $a^\infty = \{a_i\}_{i=0}^\infty$, to which the 2-adic integer $\alpha = a_0 + a_1 2 + a_2 2^2 + a_3 2^3 + \cdots \in Z_{[2]}$ is associated and is called the 2-*adic integer*[1] of the FCSR (with its given initial loading and initial memory). Note that we refer to both rational numbers p/q with q odd, and series $\sum_{i=0}^\infty a_i 2^i$ as 2-adic integers, where a^∞ is ultimately periodic due to the one-to-one correspondence between them. Define

$$p = \sum_{i=0}^{r-1} \sum_{j=0}^{i} q_j a_{i-j} 2^i - m_{r-1} 2^r, \tag{14.9}$$

[1]In [240] it is called the 2-adic value. To avoid a confusion with the 2-adic value in Section 14.1, we suggest this name.

where we have set $q_0 = -1$ so that $q = \sum_{i=0}^{r} q_i 2^i$.

One basic fact about FCSRs is the following result [289, 240, p.46].

Theorem 14.5.1 *The output a^∞ of a FCSR with connection integer q, initial memory value m_{r-1}, and initial loading $a_{r-1}, a_{r-2}, \cdots, a_1, a_0$, is the 2-adic expansion sequence of the rational number $\alpha = p/q$.*

Proof: Consider the transition from one state of the shift register to the next. Suppose that, for some given state, the value of the memory is m_{n-1} and that the content of the register is given by the r bits $a_{n-1}, a_{n-2}, \cdots, a_{n-r}$, with a_{n-1} the leftmost bit and a_{n-r} the rightmost bit, and where the register shifts towards the right. The next state is determined by calculating (A1)

$$\sigma_n = m_{n-1} + \sum_{i=1}^{r} q_i a_{n-i}, \tag{14.10}$$

writing the new memory contents as $m_n = \lfloor \sigma_n/2 \rfloor$, and writing the new content of the leftmost cell as $a_n = \sigma_n \bmod 2$ (see (A3) and (A4)). (The remaining bits are shifted once to the right.) These equations may be combined into the expression

$$\sigma_n = 2m_n + a_n.$$

It follows that

$$a_n = \sum_{i=1}^{r} q_i a_{n-i} + (m_{n-1} - 2m_n), \tag{14.11}$$

provided that $n \geq r$. Suppose the initial loading of the register consists of memory mem $= m_{r-1}$ and with register bit values $a_{r-1}, a_{r-2}, \cdots, a_1, a_0$. Now substituting (14.11) into the expression (14.7) for α gives

$$\begin{aligned} \alpha &= a_0 + a_1 2 + \cdots + a_{r-1} 2^{r-1} + \sum_{n=r}^{\infty} a_n 2^n \\ &= x + \sum_{n=r}^{\infty} \Big(\sum_{i=1}^{r} q_i a_{n-i}\Big) 2^n + \sum_{n=r}^{\infty} (m_{n-1} - 2m_n) 2^n, \end{aligned} \tag{14.12}$$

where

$$x = a_0 + a_1 2 + \cdots + a_{r-1} 2^{r-1}$$

is the integer represented by the initial loading of the register. The second summation in (14.12) cancels except for the first term, m_{r-1}, leaving

$$\begin{aligned}
\alpha &= x + m_{r-1}2^r + \sum_{n=r}^{\infty}\sum_{i=1}^{r} q_i 2^i a_{n-i} 2^{n-i} \\
&= x + m_{r-1}2^r + \sum_{i=1}^{r} q_i 2^i \Big(\sum_{n=r}^{\infty} a_{n-i} 2^{n-i}\Big) \\
&= x + m_{r-1}2^r + \sum_{i=1}^{r} q_i 2^i (\alpha - (a_0 2^0 + a_1 2^1 + \cdots + a_{r-i-1}2^{r-i-1})) \\
&= x + m_{r-1}2^r + \alpha\sum_{i=1}^{r} q_i 2^i - \sum_{i=1}^{r-1}\sum_{j=0}^{r-i-1} q_i 2^i a_j 2^j
\end{aligned}$$

(where the inner sum is empty, hence zero, when $i = r$ in the third line). These equations give

$$\alpha = \frac{x + m_{r-1}2^r - \sum_{i=1}^{r-1}\sum_{j=0}^{r-i-1} q_i 2^i a_j 2^j}{1 - \sum_{i=1}^{r} q_i 2^i} \tag{14.13}$$

$$= \frac{\sum_{i=0}^{r-1}\sum_{j=0}^{i} q_j a_{i-j} 2^i - m_{r-1}2^r}{q}. \tag{14.14}$$

This completes the proof. □

Combining Theorem 14.5.1 and Propositions 14.1.6, 14.1.7, and 14.1.8, we have the following conclusions [240].

Corollary 14.5.2 *If $a^\infty = \{a_i\}_{i=0}^{\infty}$ is an ultimately periodic binary sequence then the associated 2-adic number $\alpha = \sum a_i 2^i$ is a quotient of two integers, $\alpha = p/q$ and the denominator q is the connection integer of a FCSR which generates the sequence a^∞. The sequence a^∞ is periodic if and only if $-q < p \le 0$.*

If $-q < p < 0$ and p is relatively prime to q, then by Corollary 14.1.9, the sequence is periodic and the period is $T = \mathrm{ord}_q(2)$. If p and q have a common factor, then the period is a divisor of $\mathrm{ord}_q(2)$. This is shown clearly by the proof of Proposition 14.1.8.

If $p \ge 0$ or $p \le -q$ then the sequence has a transient prefix before it drops into a periodic state. If p is a multiple of q, then after the transient prefix the output consists of all 0's or all 1's, depending on whether p is positive or negative. The discussion in Section 14.1 has made this clear.

The synthesis problem for sequences over a field F with respect to a given generator with finite memory includes the following:

1. Is it possible for the generator to produce every ultimately periodic or periodic sequence over F by choosing a proper set of design parameters?

2. If possible, how can we determine the set of parameters of minimal sizes with which the generator can produce a given sequence?

3. If one has an algorithm to determine the set of parameters, how many consecutive characters of the given sequence are needed to determine the set of parameters with this algorithm? And what is the computational complexity of this algorithm?

For the FCSRs we have the same problems.

The first question with respect to the FCSR synthesis of binary sequences is: Given a 2-adic integer $\alpha = p/q$, how do we determine an initial loading, i.e., the set of design parameters, of the FCSR so that the output sequence coincides with the 2-adic expansion of α? This problem can be solved as follows [240].

Let q be an odd positive integer. Set $r = \lfloor \log_2(q+1) \rfloor$ and let p be an integer. Write $q = \sum_{i=0}^{r} q_i 2^i$ with $q_0 = -1$ and $q_i \in \{0, 1\}$ for $i > 0$. We want to determine the initial setting (including the extra memory) of the FCSR with connection integer q that outputs the 2-adic expansion of p/q. The number of nonzero taps in such a FCSR is $r = \mathrm{WH}(q+1)$, the Hamming weight of the binary expansion of $q+1$. The initial memory is related to p and q by (14.9).

For a given fraction p/q, the initial loading can be derived by the following steps.

C1. Compute $a_0, a_1, \cdots, a_{r-1}$ by the software algorithm for the 2-adic expansion described in Section 14.1, which is efficient.

C2. Compute $y = \sum_{i=0}^{r-1} \sum_{j=0}^{i} q_j a_{i-j} 2^i$, say by a polynomial evaluation algorithm.

C3. Compute $m = (y - p)/2^r$ in time $O(r)$.

We can then use $a_0, \cdots, a_{r-1}$ as the initial loading and m as the initial memory in a FCSR with connection integer q. This FCSR will output the 2-adic expansion of p/q. If the given 2-adic integer $\alpha = p/q$ is not reduced, by reduction we can find a shorter FCSR that produces the 2-adic expansion of p/q.

An initial loading is said to be degenerate if the 2-adic number $\alpha = p/q$ corresponding to the output sequence is an integer (in the usual sense, i.e. a "rational" integer). In this case, after a transient prefix, the FCSR outputs all 0's (if $\alpha > 0$) or all 1's (if $\alpha < 0$). The following theorem gives some properties of the prefix [240].

Theorem 14.5.3 *If the initial loading of an r-stage FCSR is degenerate, then the output will stabilize to all 0's or all 1's after no more than $max(\log_2(m), \log_2(\mathrm{WH}(q+1)+1))$ steps, where m denotes the initial memory value. If moreover the initial memory value is $m = 0$, then any degenerate initial loading will ultimately result in all 0's, which will occur in no more than $\log_2(\mathrm{WH}(q+1)+1)$ steps.*

Proof: Assume the value $\alpha = p/q$ of the FCSR is an integer. We consider the possibilities $\alpha \geq 0$ and $\alpha \leq 0$ separately. If $\alpha \geq 0$ then $p \geq 0$ and (14.13) gives

$$\begin{aligned} p &\leq \sum_{i=1}^{r-1} q_i 2^i \sum_{j=0}^{r-i-1} a_j 2^j \\ &\leq \sum_{i=1}^{r-1} q_i 2^i \sum_{j=0}^{r-i-1} 2^j \\ &\leq \sum_{i=1}^{r-1} q_i 2^i 2^{r-i} \\ &= \mathrm{WH}(q+1)2^r. \end{aligned}$$

Since $q > 2^r$, we have $\alpha < \mathrm{WH}(q+1)$. So in this case, the output sequence is the (reverse of the) binary expansion for α, which takes $\log_2(\mathrm{WH}(q+1))$ bits, after which we have all 0's.

If $\alpha < 0$ then $p < 0$ and (14.13) gives

$$|p| \leq \sum_{i=0}^{r-1} a_i 2^i + m2^r \leq (1+m)2^r$$

so $|\alpha| < 1+m$. In this case, the output sequence takes no more than $\log_2(1+m)$ steps before it stabilizes to all 1's. If the initial memory $m = 0$ then $|\alpha| < 1$ which contradicts the assumption that α is a negative integer. □

A specific FCSR with connection integer $q = 37 = 32+4+2-1$ was considered in [240], with the 5-stage shift register having feedback connections on the first, second, and fifth cells, counting from the left. The element $\gamma = 2^{-1} \in Z_{37}$ is $\gamma = 19$. Consider the initial loading such that the output sequence is given by

$$a_n = (\gamma^n \bmod 37) \bmod 2$$

Table 14.1: The states of a FCSR with $q = 37$.

mem	regis.	a_0	p	n
0	10011	1	1	0
1	01001	1	19	1
1	10100	0	28	2
1	01010	0	14	3
1	00101	1	7	4
1	00010	0	22	5
0	10001	1	11	6
1	01000	0	24	7
1	00100	0	12	8
0	10010	0	6	9
0	11001	1	3	10
1	11100	0	20	11
1	11110	0	10	12
1	11111	1	5	13
2	01111	1	21	14
2	00111	1	29	15
1	10011	1	33	16
1	11001	1	35	17

mem	regis.	a_0	p	n
2	01100	0	36	18
1	10110	0	18	19
1	01011	1	9	20
1	10101	1	23	21
1	11010	0	30	22
1	11101	1	15	23
2	01110	0	26	24
1	10111	1	13	25
1	11011	1	25	26
2	01101	1	31	27
2	00110	0	34	28
1	00011	1	17	29
1	00001	1	27	30
1	00000	0	32	31
0	10000	0	16	32
0	11000	0	8	33
1	01100	0	4	34
1	00110	0	2	35

for $n = 0, 1, 2, \cdots$. The index n is recorded as the last column in Table 14.1. The column "mem" indicates the integer value of the memory, and a_0 represents the output bit (i.e. the rightmost bit in the register). Each state S of the shift register corresponds to a rational number $f(S) = -p/37$ and the numerator p is recorded also in the table. The table therefore lists all the strictly periodic states of the FCSR.

14.6 The 2-Adic Span and 2-RA Algorithm

The 2-adic span of a binary, ultimately period sequence a^∞, denoted by $\lambda_2(a^\infty)$, is defined to be the smallest number of bits used by any FCSR whose output is the sequence a^∞ [240].

As in the case of linear span, the 2-adic span of a sequence, introduced by Klapper and Goresky [240], is intended to measure how large a FCSR is required to output the sequence. In the case of LFSRs, this is given by the number of bits in a register that outputs the sequence, and, when the sequence is periodic, this number coincides with the degree of the connection polynomial, i.e., the denominator of the reduced rational function giving the power series whose coefficients are the bits of the sequence.

The 2-adic span is more complex than the linear span. The number of bits in the connection number coincides with the size of the basic register, but additional space is required for the carry. For periodic sequences, this extra memory is small (at most log of the number of bits in the basic register), and if such sequences were our only concern we could ignore it. This would be convenient as the size of the basic register is well behaved under various algebraic operations on the sequence. However, non-purely periodic sequences arise naturally from with-carry algebraic operations on periodic sequences. Thus, in the definition of the 2-adic span the extra memory used for the carry was taken into account by Klapper and Goresky [240]. This is exactly the same as in the case of linear span. For uniformity we always assume there is at least one bit of extra memory.

The 2-adic span is cryptographically important only if there is an efficient algorithm for finding an initial loading of a FCSR with which this sequence can be produced. The cryptographic meaning of the 2-adic span will be shown by the rational approximation algorithm described later, where the function $\Phi(p, q) = \max(|p|, |q|)$ plays an important role.

Though the 2-adic span is more complicated than the linear span, we have the following result due to Klapper and Goresky [238, p. 269].

Proposition 14.6.1 *If* $\alpha = \sum_{i=0}^{\infty} a_i 2^i = p/q$ *is the rational number corresponding to* a^{∞}, *where* $\gcd(p, q) = 1$ *then the 2-adic span is bounded by*

$$\begin{aligned} & \lceil \log \Phi(p, q) \rceil - \lceil \log \log \Phi(p, q) \rceil \\ \leq\ & \lambda_2(a^{\infty}) \\ \leq\ & \lceil \log \Phi(p, q) \rceil + \lceil \log \log \Phi(p, q) \rceil . \end{aligned}$$

It follows that

$$\begin{aligned} & \lambda_2(a^{\infty}) - \lceil \log(\lambda_2(a^{\infty})) \rceil - 1 \\ \leq\ & \lceil \log \Phi(p, q) \rceil \\ \leq\ & \lambda_2(a^{\infty}) + \lceil \log(\lambda_2(a^{\infty})) \rceil + 1. \end{aligned}$$

Proof: Let $r = \lfloor \log(q + 1) \rfloor$ and write $p = p' + b2^r$ with $-2^r < p' \leq 0$. The absolute value of the memory for a FCSR with connection integer q that outputs p'/q is at most $\mathrm{WH}(q + 1) \leq \lfloor \log(q) \rfloor$. If we add b to the initial value, the resulting FCSR outputs a^{∞}. The absolute value of its memory never exceeds the maximum of the initial value and $\mathrm{WH}(q + 1)$. There are two cases to consider for the right hand inequality.

If $p < 0$, then $b < 0$ and $|b| < |p|/2^r$. The initial memory m for the FCSR that outputs p/q satisfies

$$0 \leq -b \leq m \leq \mathrm{WH}(q + 1) - b \leq \mathrm{WH}(q + 1) + |p|/2^r.$$

Hence

$$\begin{aligned}\lambda_2(a^\infty) &\leq \lceil \log(\mathrm{WH}(q+1) + |p|/2^r) \rceil + r \leq \\ &\leq \lceil \mathrm{WH}(q+1) \rceil + \lceil \log |p| \rceil \leq \\ &\leq \lceil \log\log \Phi(p,q) \rceil + \lceil \log \Phi(p,q) \rceil .\end{aligned}$$

If $p > 0$, then $0 < b < p/2^r + 1$. The initial memory m for the FCSR that outputs p/q satisfies

$$-p/2^r - 1 \leq -b \leq m \leq \mathrm{WH}(q+1) - b \leq \mathrm{WH}(q+1).$$

Therefore

$$|m| \leq \max(p/2^r + 1, \mathrm{WH}(q+1)).$$

In this case we need one extra bit for the sign, so

$$\begin{aligned}\lambda_2(a^\infty) &\leq \lceil \log(\max(|p|/2^r + 1, \mathrm{WH}(q+1))) \rceil + r + 1 \\ &\leq \lceil \log \Phi(p,q) \rceil + \lceil \log\log \Phi(p,q) \rceil\end{aligned}$$

(unless $q = 1$, in which case the result can be seen directly).

For the left hand inequality, note that the smallest basic register for a FCSR that outputs a^∞ is one with connection integer q, and this basic register has $\lfloor \log(q+1) \rfloor$ bits, therefore we have

$$\lfloor \log(q) \rfloor \leq \lambda_2(a^\infty).$$

Allowing at least one bit for the carry gives

$$\lceil \log(q) \rceil \leq \lambda_2(a^\infty).$$

Suppose $|p| > |q|$. First let $p < 0$. As seen above, the initial memory is at least $-b = |b| \lfloor |p|/2^r \rfloor$. Thus at least $\lceil \log \lfloor |p|/2^r \rfloor \rceil + r = \lceil \log |p| \rceil$ bits are required.

Finally, let $p > 0$. Then $b \geq p/2^r$. Let x be the initial memory for the FCSR with connection integer q that outputs p'/q. Then we have

$$\lceil \log |b - x| \rceil + r = \lceil \log(b) \rceil - \lceil \log(x) \rceil + r \geq \lceil \log(b2^r) \rceil - \lceil \log(x) \rceil$$

for integers in the range in question. The lower bound follows.

The last set of inequalities follows since the first set implies also

$$\lceil \log \Phi(p,q) \rceil \leq 2\lambda_2(a^\infty).$$

This completes the proof. □

For linear complexity we see that the linear span of the bitwise sum of two periodic sequences is less than or equal to the linear complexities of the two sequences. This is not true for the 2-adic span. But a similar result is the following conclusion [240, p. 130].

Theorem 14.6.2 *Suppose a^∞ and b^∞ are periodic binary sequences. Let c^∞ denote the 2-adic sum of the sequences a^∞ and b^∞. Then the 2-adic span of c^∞ is less than or equal to*

$$\lambda_2(a^\infty) + \lambda_2(b^\infty) + 2\lceil \log(\lambda_2(a^\infty))\rceil + 2\lceil \log(\lambda_2(b^\infty))\rceil + 2.$$

Proof: Suppose the binary sequences a^∞ and b^∞ correspond to 2-adic integers p_1/q_1 and p_2/q_2 respectively. The 2-adic sum sequence c^∞ corresponds to the 2-adic integer

$$\frac{p_1}{q_1} + \frac{p_2}{q_2} = \frac{p_1q_2 + p_2q_1}{q_1q_2}. \tag{14.15}$$

By Proposition 14.6.1,

$$\begin{aligned}
\lambda_2(c^\infty) &\le \lceil \log(\Phi(p_1q_2 + p_2q_1, q_1q_2))\rceil + \lceil \log\log(\Phi(p_1q_2 + p_2q_1, q_1q_2))\rceil \\
&\le \lceil \log(2\Phi(p_1, q_1)\Phi(p_2, q_2))\rceil + \lceil \log\log(2\Phi(p_1, q_1)\Phi(p_2, q_2))\rceil \\
&\le \lceil \log(\Phi(p_1, q_1))\rceil + \lceil \log(\Phi(p_2, q_2))\rceil + 1 + \\
&\quad \lceil \log(\log(\Phi(p_1, q_1)) + \log(\Phi(p_2, q_2)) + 1)\rceil \\
&\le \lambda_2(a^\infty) + \lambda_2(b^\infty) + \lceil \log(\lambda_2(a^\infty))\rceil + \lceil \log(\lambda_2(b^\infty))\rceil + 2 \\
&\quad + \lceil \log(\lambda_2(a^\infty) + \lambda_2(b^\infty) + \log(\lambda_2(a^\infty)) + \log(\lambda_2(b^\infty)) + 3)\rceil,
\end{aligned}$$

from which the result follows. □

The 2-adic span may be much less than this if the fraction (14.15) is not reduced. Although the relation between the linear span and 2-adic span remains unknown, we have the following conclusion, which shows there exist m-sequences of maximal 2-adic span [240].

Theorem 14.6.3 *Suppose a^∞ is a periodic sequence with period $N = 2^M - 1$. Suppose that $2^N - 1$ is prime. Then the 2-adic span of a^∞ is one greater than the period N.*

Proof: Consider a FCSR which generates the sequence a^∞, and let q denote the connection integer. Then $\mathrm{ord}_q(2) = N$. This says that $2^N - 1$ is divisible by q. However, by assumption, $2^N - 1$ is prime, hence $q = 2^N - 1$. The 2-adic span is then at least $\log_2(q + 1) + 1 = N + 1$. But any sequence of period N can be generated by a FCSR with N bits in the basic register and one bit of carry (which is always zero). □

More generally, the same proof shows that the 2-adic span of any periodic sequence with period N is greater than or equal to $\log_2(r + 1) + 1$, where r is the smallest prime divisor of $2^N - 1$.

Because of the efficient rational approximation algorithm described below, the 2-adic span becomes an important measure of the strength of keystream sequences for additive stream ciphering. Regarding the linear and 2-adic span investigations into the following problems are cryptographically necessary.

Research Problem 14.6.4 *Find the relation between the linear and 2-adic spans of binary sequences.*

Research Problem 14.6.5 *Find fast generators of binary sequences with both large linear and 2-adic span.*

It is possible for a binary sequence to have both large linear and 2-adic span. One example is the periodic sequence $\overline{0....01}$. But how to find such sequences with ideal pattern distributions remains open. For some cryptanalysis of some ciphers based on the 2-adic span we refer to Klapper and Goresky [240].

It is clear that the definition of linear span depends only on the shortest linear recurrence. It can be defined without reference to any special hardware—linear-feedback shift-registers. Similarly, the 2-adic span might also be defined without the special hardware model—FCSRs. We shall come to the hardware and software complexities in Chapter 16.

The 2-adic span is cryptographically meaningful only if there is an efficient algorithm to find an initial loading for a minimum FCSR which produces any given ultimately periodic sequence. An efficient algorithm for this purpose, which is derived from the procedure outlined by de Weger [452] and Mahler [280], was developed by Klapper and Goresky [240]. We shall refer to this algorithm as the *2-RA algorithm*. An important distinction between the 2-RA algorithm described by Klapper and Goresky and the original procedure outlined by de Weger and Mahler is that the 2-RA algorithm is adaptive.

The well-known LFSR synthesis problem for sequences over a finite field is: Find the smallest LFSR that generates a give sequence over a field F. There are several algorithms for the problem, such as the continued fraction algorithm [74, 94, 311, 453], and the Berlekamp-Massey algorithm [18, 291]. The Berlekamp-Massey algorithm is optimal in the following two senses:

1. It determines the shortest LFSR that generates a given sequence.

2. It does so with minimal information: Only the first $2L$ consecutive bits of the sequence are needed, where L denotes the linear span of the sequence.

Furthermore, the algorithm is iterative, and the complexity is $O(L^2)$, which is faster than the equation-solving approach which has the complexity $O(L^3)$.

According to [240] the continued fraction expansion in the field $Q_{[2]}$ of 2-adic numbers of the element $\alpha = \sum_{i=0}^{\infty} a_i 2^i$ does not exhibit similar optimality properties, and a number of decoding algorithms available in the context of Hensel and arithmetic codes [177, 250, 283] do not satisfy both of the optimality properties mentioned above.

However, the 2-RA algorithm, which is an analog of the Berlekamp-Massey algorithm, has both optimality properties: It constructs the smallest FCSR which generates the sequence a^∞, and it needs only a knowledge of the first $2M + 2\log M + 2$ bits, where M is the 2-adic span of a^∞.

The 2-RA algorithm is based on the p-adic approximation theory. The rate of convergence of the algorithm is controlled as described in Theorem 14.6.9. We now follow [240] to describe the 2-RA algorithm and its proof.

For any pair of integers p and q, as before define

$$\Phi(p,q) = \max(|p|, |q|).$$

Assume we have consecutive terms $a_0, a_1, \cdots$ of a binary sequence a^∞ which is the 2-adic expansion of a number α. We wish to determine a pair of integers (p,q) so that $\alpha = p/q$ and so that $\Phi(p,q)$ is minimal among all such pairs of integers. In the rational approximation algorithm, given in Figure 14.2, and in the rest of this section, the symbols $f = (f_1, f_2)$ and $g = (g_1, g_2)$ denote pairs of integers.

The congruence $\alpha g_2 - g_1 \equiv 0 \pmod{2^{k+1}}$ may be checked without performing the full multiplication at each stage, by saving and updating the previous values of $\alpha g_2 - g_1$ and $\alpha f_2 - f_1$. Inside the loop, in the second and third cases, the number d is chosen so as to minimize $\Phi(f+xg)$ (respectively, $\Phi(g + xf)$) among all possible odd integers x. It may be computed by division. For example, suppose we are in the second case: $\alpha g_2 - g_1 \not\equiv 0 \pmod{2^{k+1}}$ and $\Phi(g) < \Phi(f)$ If $g_1 \neq \pm g_2$, then d is among the odd integers immediately less than or greater than $(f_2-f_1)/(g_1-g_2)$ and $-(f_1+f_2)/(g_1+g_2)$. Thus it suffices to consider the value of $\Phi(f + dg)$ for these four values of d. When $g_1 = \pm g_2$, one or the other of these quotients is not considered. If $\Phi(g) > \Phi(f)$ then the roles of f and g are switched.

To have a better understanding of the 2-RA algorithm, we consider the following binary sequence $a = 00011$. Since a_3 is the first nonzero entry of the sequence a, by the algorithm of Figure 14.2 we have at the first step

$$\alpha = 2^3 = 8, \; f = (0,2), \; g = (2^3, 1).$$

So far the fraction $g_1/g_2 = 2^3$ has 0001 as the first four bits of its 2-adic expansion such that $\Phi(g)$ is minimal. Based on the above parameters, the

```
Rational_Approximation()
  begin
  input a_i s until the first nonzero a_{k-1} is found
  α = a_{k-1} · 2^{k-1}
  f = (0, 2)
  g = (2^{k-1}, 1)
  while there are more bits do
      input a new bit a_k
      α = α + a_k 2^k
      if α · g_2 − g_1 ≡ 0 (mod 2^{k+1}) then
          f = 2f
      else if Φ(g) < Φ(f) then
          Let d be odd and minimize Φ(f + dg)
          ⟨g, f⟩ = ⟨f + dg, 2g⟩
      else
          Let d be odd and minimize Φ(g + df)
          ⟨g, f⟩ = ⟨g + df, 2f⟩
      fi fi
      k = k + 1
  od
  return g
  end
```

Figure 14.2: The 2-RA algorithm.

2-RA algorithm will find a new fraction $g = (g_1, g_2)$ that has 00011 as the first five bits of its 2-adic expansion such that $\Phi(g)$ is minimal.

The next bit a_4 of a is 1, by the algorithm we put

$$\alpha = \alpha + a_4 2^4 = 8 + 16.$$

By the algorithm we should now check whether $\alpha g_2 - g_1 \equiv 0 \pmod{2^5}$ holds for the new α. It is easily seen that this is not satisfied. Then we see that $\Phi(g) = 8 > \Phi(f) = 2$. By the algorithm we compute

$$\frac{g_1 - g_2}{f_2 - f_1} = \frac{7}{2}, \quad -\frac{g_1 + g_2}{f_1 + f_2} = -\frac{9}{2}.$$

Since the odd integers closest to $7/2$ are 3 and 5 and those closest to $-9/2$ are also -3 and -5, we get four possible d. By simple computation we see that $d = -3$ minimizes $\Phi(g + df)$. Thus, we get new parameters

$$g = g - 3f = (8, -5), \ f = 2f = (0, 4).$$

Thus, the fraction g_1/g_2 that minimizes $\Phi(g)$ and has 00011 as the first five bits of its 2-adic expansion is $-8/5$.

We remark that the algorithm described in figure 14.2 is not given in full detail, since how to minimize $\Phi(f+dg)$ and $\Phi(g+df)$ is not shown. However, this has been explained in detail in the paragraph just before our example. It should be noted that the 2-RA algorithm minimizes $\Phi(g) = \max(|p|, |q|)$, and there could be two or more such fractions.

To prove the optimality of the approximation algorithm, We need some lemmas. Consider the kth *approximation lattice* for the 2-adic number α,

$$L_k = \{h \in Z \times Z : \alpha \cdot h_2 - h_1 \equiv 0 \pmod{2^k}\}.$$

It is a free module of rank 2 over the ring of integers Z and hence admits a Z-basis. The following lemma is a key observation of [452] and its proof is straightforward:

Lemma 14.6.6 *Two pairs of integers $f, g \in L_k$ form a basis for L_k if and only if $|f_1g_2 - f_2g_1| = 2^k$.*

The proofs of the following lemma and the following Theorems 14.6.9 and 14.6.8 presented in [240] utilize the methods of [280, 452].

Lemma 14.6.7 *For each k, at the top of the loop the following conditions hold:*

1. *f and g are in L_k;*
2. *$\langle f, g\rangle$ is a basis for L_k;*
3. *$f \subset 2(Z \times Z) - L_{k+1}$;*
4. *g minimizes $\Phi(h)$ over all elements $h \in L_k$ with h_2 odd.*

Proof: The proof is by induction. It is straightforward to check that the conditions hold initially. Let us suppose that the conditions hold at stage k. If $g \in L_{k+1}$, then it is again straightforward to check the conditions. Therefore, assume $g \notin L_{k+1}$. We treat the case when $\Phi(g) < \Phi(f)$. The other case is similar. Let f' and g' be the new values after updating.

1. We have

$$\begin{aligned}\alpha \cdot g_2' - g_1' &= \alpha \cdot (g_2 + df_2) - (g_1 + df_1) \\ &= (\alpha \cdot g_2 - g_1) + (\alpha \cdot f_2 - f_1) \\ &\equiv 2^k + d2^k \pmod{2^{k+1}} \\ &\equiv 0 \pmod{2^{k+1}},\end{aligned}$$

since f and g are in $L_k - L_{k+1}$ and d is odd. Therefore $g' \in L_{k+1}$. Also, g is in L_k, so $f' = 2g$ is in L_{k+1}.

2. By Lemma 14.6.6, we have $f_1 g_2 - f_2 g_1 = 2^k$. Therefore

$$f_1' g_2' - f_2' g_1' = 2g_1(f_2 + dg_2) - 2g_2(f_1 + dg_1) = 2(f_1 g_2 - f_2 g_1) = 2^{k+1}.$$

Again by Lemma 14.6.6, $\langle g', f' \rangle$ is a basis for L_{k+1}.

3. We have $g \in Z \times Z - L_{k+1}$, so $f' = 2g \in 2(Z \times Z) - L_{k+2}$.

4. Suppose that minimality fails. Since $\langle f', g' \rangle$ form a basis for L_{k+1}, there are integers a and b so that

$$\Phi(ag' + bf') < \Phi(g') \tag{14.16}$$

and $ag_2' + bf_2'$ is odd. The latter condition is equivalent to a being odd since f_2' is even and g_2' is odd. By possibly negating both a and b, we can assume a is nonnegative. Further, if $a = 1$, then $ag' + bf' = f + (d + 2b)g$ and this contradicts the choice of d in the algorithm. Thus we can assume that $a > 1$. Equation (14.16) can be rewritten as

$$\Phi(af + (ad + 2b)g) < \Phi(f + dg).$$

Let c be the odd integer closest to $d + 2b/a$. Then $|c - (d + 2b/a)| \le (a-1)/a$. It follows that

$$\begin{aligned}\Phi(f + cg) &\le \tfrac{1}{a}\Phi\left(f + \left(d + \tfrac{2b}{a}\right) g\right) + \tfrac{a-1}{a}\Phi(g) \\ &< \Phi(f + dg),\end{aligned}$$

which contradicts the choice of d. □

It follows immediately from Lemma 14.6.7 that the following theorem holds.

Theorem 14.6.8 *Let $g = (g_1, g_2)$ denote the output of the preceding algorithm when T bits a_i are used. Then g_2 is odd,*

$$\alpha \cdot g_2 - g_1 \equiv 0 \pmod{2^T},$$

and any other pair $g' = (g_1', g_2')$ which satisfies these two conditions has $\Phi(g') \ge \Phi(g)$.

The correctness of the 2-RA algorithm follows from the following result due to Klapper and Goresky [240].

Theorem 14.6.9 *Suppose a^∞ is an ultimately periodic sequence with associated 2-adic number $\alpha = \sum a_i 2^i = p/q$, with $p, q \in Z$, and* $\gcd(p,q) = 1$. *If $T \geq \lceil 2\log\Phi(p,q)\rceil + 2$ bits a_i are used, then the 2-RA algorithm outputs $g = (p,q)$. (Hence also if $T \geq 2\lambda_2(a^\infty) + 2\lceil\log(\lambda_2(a^\infty))\rceil + 3$.)*

Proof: By assumption, $\alpha = p/q$ so q is odd and $(p,q) \in L_k$ for all k. The output from the algorithm is a pair $g = (g_1, g_2) \in L_T$ which is Φ-minimal, so $\Phi(g_1, g_2) \leq \Phi(p,q)$. Hence

$$|g_1 q| \leq |g_1||q| \leq \Phi(g_1,g_2)\cdot\Phi(p,q) \leq \Phi(p,q)^2 \leq 2^{T-2},$$

since by assumption $T \geq 2\log_2\Phi(p,q) + 2$. Similarly, $|pg_2| \leq 2^{T-2}$. However, $\alpha g_2 - g_1 \equiv 0 \pmod{2^T}$ so $g_1 q \equiv pg_2 \pmod{2^T}$, which implies that $g_1 q = pg_2$. Therefore (g_1, g_2) is some odd integer multiple of (p,q). By Φ-minimality, this integer must be ± 1 which gives $g_1 = p$ and $g_2 = q$ (or else $g_1 = -p$ and $g_2 = -q$). □

The computational complexity of the 2-RA algorithm is similar to that of the Berlekamp-Massey algorithm. Suppose the rational approximation algorithm is executed with a sequence a^∞ which is ultimately periodic, with rational associated 2-adic number $\alpha = p/q$. Then the rational approximation algorithm takes

$$T = 2\log(\Phi(p,q)) + 2 \leq 2\lambda_2(a^\infty) + 2\lceil\log(\lambda_2(a^\infty))\rceil + 2$$

steps to converge.

Consider the kth step. If $\alpha g_2 - g_1 \not\equiv 0 \pmod{2^{k+1}}$, then we say that a discrepancy has occurred. The complexity of the algorithm depends on the number of discrepancies. To simplify the computation of αg_2, we maintain αf_2 as well. When no discrepancy occurs, these values and the value of f can be updated with k bit operations.

Suppose a discrepancy occurs. The minimization step can be done with two divisions of k bit integers. The remaining steps take time $O(k)$. Then αg_2 and αf_2 can be updated with $O(k)$ bit operations and two multiplications of k bit integers by d.

Let D be the number of discrepancies, and let M be the maximum time taken by a multiplication or division of T bit integers. The Schönhage-Strassen algorithm [386], gives $M = O(T\log T\log\log T)$. This can be improved to $M \sim T\log T$ using Pollard's nonasymptotic algorithm and Newton interpolation for $T < 2^{37}$ on a 32-bit machine or $T < 2^{70}$ on a 64-bit machine [350]. These are ranges that are typical in current usage.

The complexity of the algorithm is thus $4DM+O(T^2)$. Strictly in terms of T, this is $O(T^2 \log T \log\log T)$. However, if the sequence is chosen so the number of discrepancies is small, the complexity is lower.

It is important to note that FCSRs cannot be used as keystream generators directly; they can only be used as building blocks of keystream generators since we have the efficient 2-RA algorithm. This is exactly the same as LFSRs. Thus, how to combine several small FCSRs to get a generator that produces binary sequences of large linear and 2-adic span is an important cryptographic issue.

An arithmetic or with-carry analog of Blahut's theorem is developed by Goresky, Klapper and Washington [174]. This relates the length of the smallest feedback with carry shift register to the number of nonzero classical Fourier coefficients of a periodic sequence.

14.7 Some Properties of FCSR Sequences

To be employed as keystream sequences for some stream ciphers, the FCSR sequences should have not only large 2-adic and linear spans, but also other properties such as ideal pattern distributions. This section is concerned with some properties of FCSR sequences, which are based on [240].

The trace representation of sequences over finite fields has played an important role in the analysis of sequences with respect to the arithmetic of finite fields. As described in Chapter 7, under certain conditions periodic sequences over finite fields have the so-called trace representation (see Section 7.1). For binary periodic FCSR sequences we have the following similar representation [240].

Theorem 14.7.1 *Suppose a periodic sequence a^∞ is generated by a FCSR with connection integer q. Let $\gamma = 2^{-1} \in Z_q$ be the inverse of 2 in the cyclic group of integers modulo q. Then there exists $A \in Z_q$ such that for all $i = 0, 1, 2, \cdots$,*

$$a_i = (A\gamma^i \bmod q) \bmod 2.$$

Proof: First, recall the definition of $(x \bmod q) \bmod 2$ fixed at the beginning of Chapter 4. Suppose the FCSR is in a state S, meaning the memory has some value m and the register is loaded with bits $a_0, a_1, \cdots, a_{r-1}$. We also suppose that the FCSR is in periodic mode, i.e. that the output sequence a^∞ is periodic with no transient prefix. Let $T = \mathrm{ord}_q(2)$ denote the period of this sequence. To such a state S we associate its 2-adic integer, $f(S)$. By

Theorem 14.5.1, $f(S)$ is a 2-adic integer of the form

$$f(S) = -\frac{p}{q} = \sum_{i=0}^{\infty} a_i 2^i,$$

with $0 \leq p \leq q-1$. Now let S' denote the next state of the FCSR, so

$$f(S') = -\frac{p'}{q} = \sum_{i=0}^{\infty} a_{i+1} 2^i.$$

Thus, $0 \leq p' \leq q-1$ and

$$-2\frac{p'}{q} + a_0 = -\frac{p}{q},$$

or $p = 2p' - a_0 q \in Z$. If we read this equation modulo 2, we have

$$p = a_0 \bmod 2.$$

Reading this equation modulo q we obtain

$$p' = 2^{-1} p \bmod q.$$

This shows that the sequence of numerators $(p, p', \cdots)$ is obtained by multiplying by γ and reducing mod q, and that the sequence of bits $(a_0, a_1, \cdots)$ is obtained by reducing the numerators modulo 2. Finally, the initial state is arbitrary and given by the choice of some $A \in Z_q$. □

It was remarked in [240, p. 125] that Peterson and Weldon [345] considered only the case where q is prime and 2 is a primitive element modulo q, and that their proof of Theorem 15.5 (p. 458) may be used in this situation to give another proof of Theorem 14.7.1.

In many cryptographic applications it is necessary to require the keystream sequences to have a large minimum period. It is well known that binary LFSRs of length m with primitive feedback polynomials can produce sequences with minimum period length $2^m - 1$, which is the maximum value. Such sequences are referred to as maximum-length sequences (briefly, m-sequences).

By Corollary 14.1.9, the maximum possible period for a FCSR with connection integer q is $T = q - 1$. This period is attained for any non-trivial loading of memory if and only if q is prime and 2 is a primitive root modulo q. In this case, for any initial loading of the register, the output sequence will either degenerate into all 0's or all 1's, or else it will ultimately drop into

Table 14.2: Values of q giving rise to ℓ-sequences for length ≤ 8.

Length	Values of q giving ℓ-sequences
1	3
2	5
3	11, 13
4	19, 29
5	37, 53, 59, 61
6	67, 83, 101, 107
7	131, 139, 149, 163, 173, 179, 181, 197, 211, 227
8	269, 293, 317, 347, 349, 373, 379, 389, 419, 421
8	443, 461, 467, 491, 509

the big periodic state (see Section 14.1). To emphasize the analogy with m-sequences, an *ℓ-sequence* is a periodic sequence (of period $T = q - 1$) which is obtained from a FCSR with prime connection integer q for which 2 is a primitive root.

By Propositions 14.1.3, 14.1.5, 14.1.6 and Corollary 14.1.9, such a sequence is (a shift of) the reverse of the binary expansion,

$$1/q = b_0 2^{-1} + b_1 2^{-2} + b_3 2^{-3} + \cdots$$

of the fraction $1/q$ [245, Section 4.1, ex. 31]. This binary expansion is called a *$1/q$-sequence* in [26], any single period of which is a codeword in the Barrows-Mandelbaum arithmetic code [13, 282]. These sequences have the following properties:

1. They are balanced [159].

2. They have the generalized de Bruijn property [282, 26, Theorem 1, p. 370]: In any given period of the sequence, every binary string of length $\lfloor \log_2(q) \rfloor$ occurs at least once and every binary string of length $\lfloor \log_2(q) \rfloor + 1$ occurs at most once.

We have seen in Chapter 5 that there are a number of techniques for finding large primes having primitive 2. Typical primes having primitive root 2 include Sophie German and Stern primes. With such a prime as connection integer the FCSR sequences have the maximum period.

Long pseudorandom sequences can also be generated by FCSR's with nonprime connection integer q. When $q = p^u$ is a power of a prime, with extremely high probability q has primitive 2 if p does. For details about primitive roots we refer to Chapters 3 and 5. A table of q's with which the FCSR produces ℓ-sequences are given in Table 14.2 [240].

For the distribution property of some ℓ-sequences we have the following result, which follows easily from the primitivity of 2.

Proposition 14.7.2 *Let q be a power of a prime p, say $q = p^e$, and suppose that 2 is primitive modulo q. Let a^∞ be any maximal period FCSR sequence, generated by a FCSR with connection integer q. The number of zeros and the number of ones in one period of a^∞ are equal.*

Regarding higher order distributions these sequences are close to having the deBruijn property that each subsequence of length log of the period occurs exactly once in each period. It was shown that for any two such subsequences, their numbers of occurrences can differ by at most two [240].

Theorem 14.7.3 *Let q be a power of a prime p, say $q = p^e$, and suppose that 2 is primitive modulo q. Let s be any nonnegative integer, and A and B be s bit subsequences. Let a^∞ be any maximal period, purely periodic FCSR sequence, generated by a FCSR with connection integer q. Then the numbers of occurrences of A and B in a^∞ with their starting positions in a fixed period of a^∞ differ by at most 2.*

Proof: The purely periodic FCSR sequences with connection integer q are precisely the 2-adic expansions of rational numbers $-x/q$, with $0 \le x < q$. Such a sequence has maximum period if and only if p does not divide x. Since 2 is primitive modulo q, the cyclic shifts of a^∞ correspond to the set of all rational numbers $-x/q$, with $0 \le x < q$. Thus an s bit subsequence A occurs in a^∞ if and only if it occurs as the first s bits in the 2-adic expansion of some rational number $-x/q$ with $0 \le x < q$ and p not dividing x. Two rational numbers $-x_1/q$ and $-x_2/q$ have the same first s bits if and only if $-x_1/q \equiv -x_2/q \pmod{2^s}$, if and only if $x_1 \equiv x_2 \pmod{2^s}$. Thus we want to count the number of x with a given first s bits, $0 \le x < q$, and x not divisible by p.

Let $2^r < q < 2^{r+1}$. If $s > r$, there is either zero or one such x, so the result follows. Thus we may assume $s \le r$.

We first count the number of x with the first s bits fixed and $0 \le x < q$, ignoring the divisibility condition. If $A = a_0, \cdots, a_{s-1}$, we let $\alpha = \sum_{i=0}^{s-1} a_i 2^i$. Let $q = \sum_{i=0}^{r} q_i 2^i$, and $q' = \sum_{i=0}^{s-1} q_i 2^i$. If $\alpha < q'$, then every choice of $a_s, \cdots, a_r$ with $\sum_{i=s}^{r} a_i 2^i \le \sum_{i=s}^{r} q_i 2^i$ gives a unique x in the right range. If $\alpha \ge q'$, then every choice of $a_s, \cdots, a_r$ with $\sum_{i=s}^{r} a_i 2^i < \sum_{i=s}^{r} q_i 2^i$ gives a unique x in the right range. Thus for different choices of A, the numbers of such x differ by at most one.

Next we consider those x for which $0 \le x < q$ and p divides x. That is, $x = py$ for some y, and $0 \le y < q/p = p^{e-1}$. As above, $x_1 = py_1$ and $x_2 = py_2$ have the same first s bits if and only if the same is true of y_1 and

y_2. The preceding paragraph shows that the numbers of such y for different choices of the first s bits differ by at most one. But if $x = py$, then $y \equiv A \pmod{2^s}$ if and only if $x \equiv pA \pmod{2^s}$, so for any B and C, the number of x divisible by p with first s bits equal to B differs from the number of x divisible by p with first s bits equal to C by at most 1. We have

$$\begin{aligned} & |\{x : 0 \le x < q,\, p \not| x, \text{ and } x \equiv \alpha \pmod{2^s}\}| \\ = \ & |\{x : 0 \le x < q \text{ and } x \equiv \alpha \pmod{2^s}\}| \\ & -|\{x : 0 \le x < q,\, p|x, \text{ and } x \equiv \alpha \pmod{2^s}\}|. \end{aligned}$$

As α varies the two terms on the right hand side differ by at most one from their values for any fixed choice of α. Thus the difference varies by at most 2. It is easy to check that the difference can be as large as 2. □

For recent results on the arithmetic cross-correlation of FCSR sequences and the distinctness of decimations of l-sequences, the reader is refered to Klapper and Goresky [241], and Goresky, Klapper and Murty [173]. Goresky and Klapper have recently given Fibonacci and Galois representations of feedback with carry shift registers [172].

14.8 Blum-Blum-Shub Sequences & Class Numbers

One of the cryptographically interesting number-theoretic generators is the *Blum-Blum-Shub generator* [26]. This generator can be described as follows. Let $p \equiv 3 \pmod 4$ and $q \equiv 3 \pmod 4$ be primes. Such an integer $N = pq$ is called a *Blum integer* [88]. Let x_0 be an integer which is a quadratic residue modulo N, i.e., $x_0 \equiv u^2 \pmod N$ for some integer u and $\gcd(x_0, N) = 1$. The Blum-Blum-Shub generator is then defined by

$$b_i = x_i \bmod 2,$$

where

$$x_i = x_{i-1}^2 \bmod N, \quad i = 1, 2, ... \tag{14.17}$$

As before, $x \bmod N$ is defined to be the least nonnegative integer congruent to x modulo N.

Blum, Blum and Shub [26] proved that the least period of the sequence x^∞ defined by (14.17) divides $\lambda(\lambda(N))$ if x_0 is a quadratic residue modulo a Blum integer N, where λ is the lambda function defined in Section 3.2. Thus, the least period of the binary sequence b^∞ must divide $\lambda(\lambda(N))$. If the Blum integer $N = pq$ is chosen such that

$$p = 2p_1 + 1,\ p_1 = 2p_2 + 1,\ q = 2q_1 + 1,\ q_1 = 2q_2 + 1,$$

where p, p_1, p_2, q, q_1, q_2 are all odd primes, then

$$\lambda(\lambda(N)) = 2p_2q_2.$$

Such primes p and q are called *special* [26, 88]. Note that a prime p is special if and only if $(p-1)/2$ and $(p-3)/4$ are both Sophie German primes. If p and q are special and $N = pq$, then the least period must be one of $p_2, q_2, 2p_2, 2q_2, 2p_2q_2, p_2q_2$. Thus, it must be no less than $\min\{p_2, q_2\}$. With such a special Blum integer the least period of the binary sequence b^∞ is controllable. In addition, we have the following conclusion about the linear complexity of the sequence b^∞ whose proof is similar to those of Theorems 3.3.5 and 3.3.6.

Proposition 14.8.1 [126] *Let $N = pq$, where p, q are special, and let x_0 be a quadratic residue modulo N. Then for the sequence b^∞ over $GF(q)$*

$$\mathrm{L}(b^\infty) \geq \min\{\mathrm{ord}_{(p-3)/4}(2), \mathrm{ord}_{(q-3)/4}(2)\}.$$

Thus, with a special Blum integer the linear complexity and its stability of the Blum-Blum-Shub sequence can be controlled by controlling the orders of 2 modulo $(p-3)/4$ and $(q-3)/4$.

By using exponential sum estimates, it is proved in [158] that if its period is large enough, then the sequence x^∞ defined by (14.17) is uniformly distributed modulo m.

Another cryptographically interesting property of the Blum-Blum-Shub generator is its unpredictability under the hypothesis that any efficient procedure for guessing the quadratic residuacity of a given m modulo N will be incorrect for a positive fraction of the inputs [26]. We note that the unpredictability problem for sequences defined by Blum, Blum and Shub [26] is similar to undecidability problems of formal languages, where the Church Hypothesis is needed (see Rozenberg and Salomaa [374] and Salomaa [379]).

It is important that the imbalance between 0's and 1's of cryptographic binary sequences is controlled. It is not strange that results about the imbalance of Blum-Blum-Shub sequences are obtained ten years after the proposing of the generator, since the imbalance problem of Blum-Blum-Shub sequences seems to be related to some quite advanced topics in number theory [88].

Substantial progress on this problem has been made by Cusick who proved that the average imbalance for these sequences is no worse than what would be expected in a truly random bit string of the same length [88]. However, the imbalance problem for each individual Blum-Blum-Shub sequence still remains open. Solving this problem might involve many more results in number theory. In this section we follow Cusick [88] to see how the average imbalance problem of Blum-Blum-Shub sequences is related to Gauss' class number problem for imaginary quadratic fields, the lambda function and the Kronecker symbol.

Let d and n be integers with $d \equiv 0$ or $1 \pmod 4$ and not a square, $n > 0$. The *Kronecker symbol* is defined by

1. $\left(\frac{d}{n}\right) = 0$ if $\gcd(d, n) > 1$,
2. $\left(\frac{d}{1}\right) = 1$,
3. if d is odd, $(d/2) = (2/|d|)$, a Jacobi symbol, so

$$\left(\frac{d}{2}\right) = \begin{cases} +1, & d \equiv 1 \text{ or } 7 \pmod 8, \\ -1, & d \equiv 3 \text{ or } 5 \pmod 8, \end{cases} \tag{14.18}$$

4. if $n = \prod_{i=1}^{r} p_i$ then $(d/n) = \prod_{i=1}^{r}(d/p_i)$, a product of Legendre symbols and, if n is even, the symbol $(d/2)$.

By the above definition the following basic properties are easily verified (see Hua [213, pp. 304-306] or Rosen [372, pp. 65-66]).

1. $\left(\frac{d}{n}\right) = \left(\frac{n}{|d|}\right)$ if d is odd.
2. $\left(\frac{d}{mn}\right) = \left(\frac{d}{m}\right)\left(\frac{d}{n}\right)$.
3. $\left(\frac{d_1 d_2}{n}\right) = \left(\frac{d_1}{n}\right)\left(\frac{d_2}{n}\right)$.
4. $\left(\frac{d}{m}\right) = \left(\frac{d}{n}\right)$ if $m \equiv n \pmod{|d|}$ and $\left(\frac{d}{m}\right) = \left(\frac{d}{n}\right)\text{sign}(d)$ if $m \equiv -n \pmod{|d|}$.

To go further, we need some results about quadratic fields. Any extension of the rational number field Q of degree 2 is called a *quadratic field.* It is easily seen that any quadratic field K is of the form $Q(\theta)$, where θ is a root of a polynomial $x^2 - d$ with $d \neq 1$ and d a square-free rational integer (positive or negative). The field is usually written as $Q(\sqrt{d})$.

If d and d' are not equal to 1 and square-free, then $Q(\sqrt{d}) \neq Q(\sqrt{d'})$. The basic invariant of a quadratic field is its *discriminant*, which is defined to be

$$D_K = \begin{cases} d, & \text{if } d \equiv 1 \pmod 4, \\ 4d, & \text{otherwise.} \end{cases}$$

Since $D_K \equiv 0$ or $1 \pmod 4$ and $K = Q(\sqrt{D_K})$, a quadratic field is determined by its discriminant.

It is easy to prove that any element of $Q(\sqrt{d})$ can be uniquely expressed as

$$\alpha = x + y\sqrt{d},$$

where x and y are rationals. The *conjugate* of α, written $\overline{\alpha}$, is defined to be $\overline{\alpha} = x - y\sqrt{d}$, and the *norm* of α is $\mathrm{N}(\alpha) = \alpha\overline{\alpha} = x^2 - dy^2$.

An element α of a quadratic field is called an *algebraic integer* or *integer* if α satisfies a polynomial equation

$$x^2 + bx + c = 0,$$

where b and c are rational integers. It is also easy to verify that the set of integers O_K of a quadratic field $K = Q(\sqrt{d})$ forms a ring with respect to the addition and multiplication of the quadratic field K and is described by

$$O_K = \begin{cases} Z[(1+\sqrt{d})/2], & \text{if } d \equiv 1 \pmod 4, \\ Z[\sqrt{d}], & \text{otherwise.} \end{cases}$$

An ideal I of the ring O_K is called a *principal ideal* if there exists an integer α such that $I = \{\lambda\alpha : \lambda \in O_K\}$. Two ideals I_1 and I_2 are said to be *equivalent* if there is a principal ideal (α) such that $I_1 = (\alpha)I_2$. These ideals are *narrowly equivalent* if the norm of α is positive. The *class number*, written $h(D_K)$, is the number of ideal classes in the narrow sense in a quadratic field $K = Q(\sqrt{d})$. We can also define the class number with respect to the usual equivalence relation. But for an imaginary quadratic field $K = Q(\sqrt{d})$, i.e., $d < 0$, the two kinds of equivalence relations are the same, since $\mathrm{N}(\alpha) > 0$ for any nonzero α.

The Dirichlet class number formula for the imaginary quadratic field $K = Q(\sqrt{d})$, where $d < 0$, is described by the following lemma (for proof, see Davenport [98]).

Lemma 14.8.2 *Suppose $D < 0$, $D \equiv 0$ or $1 \pmod 4$, D not a squure. Then the class number $h(D)$ of the imaginary quadratic field with discriminant D is given by*

$$h(D) = -\frac{w(D)}{2|D|}\sum_{j=1}^{|D|}\left(\frac{D}{j}\right)j,$$

where

$$w(D) = \begin{cases} 6, & D = -3, \\ 4, & D = -4, \\ 2, & D < -4. \end{cases}$$

There is an intimate relation between the theory of quadratic forms and that of quadratic fields. Thus the class number defined in Section 12.5 is closely related to the class number here. For details we refer to Buell [45], Cox [86], and Borevich and Shafarevich [28]. Now we turn back to the imbalance problem of Blum-Blum-Shub sequences and class numbers.

Let A denote the $\phi(N)/2$ by $\lambda(\lambda(N))$ array whose ith row is the ith sequence in the list of sequences s_a of length $\lambda(\lambda(N))$, where

$$s_a = \{c_j = a^{2^j} \bmod N : \ j = 1, 2, ..., \lambda(\lambda(N))\}$$

and a runs through the integers satisfying $1 \leq a < N/2$ and $\gcd(a, N) = 1$. Each of the integers c_i is a quadratic residue modulo N and so is a possible seed x_0 for the Blum-Blum-Shub generator.

Cusick [88] observed the following three properties of the array A:

1. The array A includes exactly two copies of the first $\lambda(\lambda(N))$ terms of each sequence x^∞ which can be produced by the $x^2 \bmod N$ generator, since each quadratic residue modulo N has two square roots a modulo N in the interval $1 \leq a < N/2$.

2. Each row of A contains at least one period of the corresponding sequence x^∞, since the period of x^∞ divides $\lambda(\lambda(N))$.

3. Each column of A contains some permutation of two copies of the set

$$S_N = \{a^2 \bmod N : \ \gcd(a, N) = 1, 1 \leq a < N/2\}$$

of the $\phi(N)/4$ quadratic residues modulo N, since the square-modulo-N operation gives a one-to-one correspondence between the set of $\phi(N)/4$ quadratic residues modulo N and itself.

These three key observations play an important role in relating the average imbalance of Blum-Blum-Shub sequences to class numbers, the lambda function and the Kronecker symbol.

Let B denote the $\phi(N)/2$ by $\lambda(\lambda(N))$ array of bits which is obtained by taking the elements in A modulo 2. The imbalance of the set S_N is defined by

$$\begin{aligned} I(S_N) &= \text{number of even elements of } S_N - \\ &\quad \text{number of odd elements of } S_N \end{aligned}$$

and the imbalance $I(B)$ of the array B by

$$I(B) = |\text{number of 1's in } B - \text{number of 0's in } B|.$$

The integer $I(S_N)$ is referred to as the *signed imbalance*. Since each column of B is made up of two permutations of S_N, then

$$I(B) = 2\lambda(\lambda(N))|I(S_N)|. \tag{14.19}$$

Thus the following theorem holds [88].

Theorem 14.8.3 *The average imbalance of Blum-Blum-Shub sequences is*

$$4\lambda(\lambda(N))|I(S_N)|/\phi(N)|.$$

By this theorem it is clear that the average imbalance is related to the lambda function, Euler function, and $I(S_N)$. Now let us see how $I(S_N)$ can be related to the class number of imaginary quadratic fields. To this end, we need the following two lemmas due to Cusick (for proof, see [88]).

Lemma 14.8.4 *Suppose $d > 0$, $d \equiv 0$ or $1 \pmod 4$, d not a square. Then*

$$\sum_{j=1}^{d} \left(\frac{d}{j}\right) j = 0.$$

Lemma 14.8.5 *If $q \equiv 3 \pmod 4$ is prime, then*

$$\sum_{j=1,\ j\ odd} \left(\frac{j}{q}\right) = \left(\frac{2}{q}\right)\left(\left(\frac{2}{q}\right) - 2\right)\frac{2}{w(-q)}h(-q).$$

It is easy to see that k is a quadratic residue modulo a Blum integer $N = pq$ if and only if $(k/p) = (k/q) = 1$, where $(\cdot/\cdot)$ is the Legendre or Kronecker symbol. To derive the relation between the signed imbalance $I(S_N)$ and class numbers, Cusick introduced the term antiresidue. An integer k is called an *antiresidue* modulo a Blum integer $N = pq$ if and only if

$$\left(\frac{k}{q}\right) = \left(\frac{k}{p}\right) = -1.$$

The key idea needed to establish the relation between the signed imbalance and class numbers is the set up of the following equation.

$$\begin{aligned} \Sigma &:= \textstyle\sum_{k=1}^{pq}\left(1 + (-1)^{k+1}\left(\frac{k}{p}\right)\right) \times \\ &\quad \left(1 + (-1)^{k+1}\left(\frac{k}{q}\right)\right)\left(\frac{pq}{k}\right) k \\ &= 4(\text{sum of odd residues } k \bmod N + \\ &\quad \text{sum of even residues } k \bmod N) \\ &= 4pq|\{\text{odd residues } k \bmod N,\ 1 \le k \le pq\}|. \end{aligned} \tag{14.20}$$

Note

$$\sum_{k=1,\gcd(k,pq)=1}^{pq} k = \frac{1}{2}pq\phi(pq)$$

and

$$\left(\frac{k}{p}\right) = \left(\frac{-p}{k}\right), \quad \left(\frac{k}{q}\right) = \left(\frac{-q}{k}\right).$$

Then expanding the sum in (14.20) gives

$$\begin{aligned}\Sigma &= \textstyle\sum_{k=1}^{pq} \left(\frac{(pq)^2}{k}\right) k + \sum_{k=1}^{pq} \left(\frac{pq}{k}\right) k + \\ &\quad \textstyle\sum_{k=1}^{pq} (-1)^{k+1} \left(\frac{pq}{k}\right) \left(\left(\frac{-p}{k}\right) + \left(\frac{-q}{k}\right)\right) k \\ &= \textstyle\frac{1}{2}pq\phi(pq) + \sum_{k=1}^{pq} (-1)^{k+1} \left(\frac{pq}{k}\right) \left(\frac{-p}{k}\right) k + \\ &\quad \textstyle\sum_{k=1}^{pq} (-1)^{k+1} \left(\frac{pq}{k}\right) \left(\frac{-q}{k}\right) k.\end{aligned} \tag{14.21}$$

Here Lemma 14.8.4 is used. Then we break up the final sum in (14.21) into two parts, one for k odd and one for k even. We find from basic properties of the Kronecker symbol that

$$\begin{aligned}&\textstyle\sum_{k=1,k\ even}^{pq} (-1)^{k+1} \left(\frac{pq}{k}\right) \left(\frac{-p}{k}\right) k \\ &= -\textstyle\sum_{j=1,j\ odd}^{pq} \left(\frac{pq}{pq-j}\right) \left(\frac{-p}{pq-j}\right) (pq - j) \\ &= \textstyle\sum_{j=1,j\ odd}^{pq} \left(\frac{pq}{j}\right) \left(\frac{-p}{j}\right) (pq - j).\end{aligned} \tag{14.22}$$

Note that

$$\sum_{j=1,j\ odd}^{2q} \left(\frac{j + 2tq}{q}\right) = 0, \quad \text{for } t = 0, 1, ...,$$

so

$$\sum_{j=1,j\ odd}^{pq} \left(\frac{j}{q}\right) = \sum_{j=1,j\ odd}^{q} \left(\frac{j}{q}\right). \tag{14.23}$$

Then combining (14.22) and (14.23) yields

$$\begin{aligned}&\textstyle\sum_{k=1}^{pq} (-1)^{k+1} \left(\frac{pq}{j}\right) \left(\frac{-p}{j}\right) k \\ &= pq \textstyle\sum_{j=1,j\ odd}^{pq} \left(\frac{pq}{j}\right) \left(\frac{-p}{j}\right) \\ &= \textstyle\sum_{j=1,j\ odd}^{pq} \left(\frac{p}{j}\right)^2 \left(\frac{-q}{j}\right) \\ &= \textstyle\sum_{j=1,j\ odd}^{pq} \left(\frac{j}{p}\right) - \sum_{j=1,j\ odd}^{q} \left(\frac{pj}{q}\right) \\ &= \left(1 - \left(\frac{p}{q}\right)\right) \textstyle\sum_{j=1,j\ odd}^{q} \left(\frac{j}{q}\right).\end{aligned} \tag{14.24}$$

Combining (14.20), (14.21), (14.24) and Lemma 14.8.5 yields

$$\begin{aligned} & 4pq|\{\text{odd residues } k \bmod N,\ 1 \le k \le pq\}| \\ = \ & \tfrac{1}{2}\phi(pq) + \left(1 - \left(\tfrac{p}{q}\right)\right)\left(\tfrac{2}{q}\right)\left(\left(\tfrac{2}{q}\right) - 2\right)\tfrac{2h(-q)}{w(-q)} + \\ & \left(1 - \left(\tfrac{q}{p}\right)\right)\left(\tfrac{2}{p}\right)\left(\left(\tfrac{2}{p}\right) - 2\right)\tfrac{2h(-p)}{w(-p)}. \end{aligned} \tag{14.25}$$

Since $p \equiv q \equiv 3 \pmod 4$, the Law of Quadratic Reciprocity implies

$$\left(\frac{p}{q}\right) = -\left(\frac{q}{p}\right),$$

so exactly one of the summands on the right-hand side of (14.25) is nonzero. It follows from the basic properties of the Kronecker symbol that

$$\left(\frac{2}{q}\right)\left(\left(\frac{2}{q}\right) - 2\right) = \begin{cases} -1, & q \equiv 7 \pmod 8, \\ +3, & q \equiv 3 \pmod 8. \end{cases} \tag{14.26}$$

Combining (14.25) and (14.26) gives

$$4|\{\text{odd residues } k \bmod N,\ 1 \le k \le pq\}| = \frac{1}{2}\phi(pq) - C_r h(-r) \tag{14.27}$$

and

$$4|\{\text{even residues } k \bmod N,\ 1 \le k \le pq\}| = \frac{1}{2}\phi(pq) + C_r h(-r), \tag{14.28}$$

where

$$r = \begin{cases} p, & \text{if } (p/q) = 1, \\ q, & \text{otherwise,} \end{cases} \tag{14.29}$$

and

$$C_r = \begin{cases} +2, & r \equiv 7 \pmod 8, \\ -6, & r \equiv 3 \pmod 8. \end{cases} \tag{14.30}$$

Finally by (14.27), (14.28) and the definition of $I(S_N)$ we have the following relation

$$I(S_N) = \frac{1}{2}C_r h(-r), \tag{14.31}$$

which shows the relation between the singed imbalance and the class number for imaginary quadratic fields.

Note that $h(-r) \to \infty$ as $r \to \infty$ (for proof, see [161]). By formulas (14.31), (14.29), and (14.30) we have the following result [88].

Theorem 14.8.6 *As N = Blum integer tends to infinity through any sequence of $N = pq$ with both p and q tending to infinity, we have $|I(S_N)| \to \infty$. Furthermore, the signed imbalance $I(S_N)$ takes on both positive and negative values infinitely often.*

By Theorem 14.8.6 the average imbalance of the Blum-Blum-Shub sequences will tend to infinity as N runs through any sequence of Blum integers, such that p and q tend to infinity and $\lambda(\lambda(N))/\phi(N)$ is bounded below.

To control the period of the sequences, Blum, Blum and Shub suggested using special primes. However, this way to ensure large period leads to also large imbalances due to Theorem 14.8.6 and the following lemma, which is easy to prove [88].

Lemma 14.8.7 *If p and q are special primes and $N = pq$, then*

$$\lambda(\lambda(N))/\phi(N) = \frac{1}{8} - \frac{1}{4}\frac{p+q-8}{(p-1)(q-1)}.$$

It then follows from Theorem 14.8.6 and Lemma 14.8.7 that the following conclusion is true.

Corollary 14.8.8 *If N runs through a sequence of special Blum integers such that p and q both tend to infinity, then the average imbalance of the Blum-Blum-Shub generator is asymptotic to $|I(S_N)|/2$.*

With some further arguments, Cusick was able to prove that the average imbalance of these sequences has order no larger than $\sqrt{\lambda(\lambda(N))}$, which is no worse than what would be expected in a random bit string of the same length [88].

Chapter 15

Prime Ciphering Algorithms

Traditional stream ciphers are usually based on shift registers that are hardware-oriented. There are now a number of software-oriented stream ciphers, such as RC4 [383, p. 397], SEAL [371], and WAKE [446]. But we do not even know the least period of the keystream sequences of these algorithms, let alone their linear complexity. However, for additive synchronous stream ciphers the linear complexity of their keystream sequences must be controlled. In this chapter we describe two fast stream ciphering algorithms for which some security aspects can be proved.

15.1 Prime-32: A Description

TWOPRIME is a a fast stream ciphering algorithm described in [133]. However, it is weak. The *Prime-32* algorithm described in this section is a variant of TWOPRIME, which was strengthend by C. Ding. The design of Prime-32 is based on the following considerations:

- its structure should be different from those in the public domain;
- the key size should be large enough;
- it is for 32-bit computers;
- it should work on blocks of bytes;
- it should be fast in software;
- it should be analyzable;
- it should be easily modified for 64-bit computers;

- it is expected to be secure.

Prime-32 is an additive stream cipher that works on blocks of bytes. The keystream generator produces an 8-byte keystream block at each time unit, and this keystream block is then bytewise xored with the 8-byte block of input.

The key of the algorithm has 16 bytes, denoted by $k_0k_1 \cdots k_{15}$, which are divided into four parts. Let these parts be

$$\begin{aligned} K_0 &= k_8 + k_9 2^8 + k_{10} 2^{16} + k_{11} 2^{24}, \\ K_1 &= k_{12} + k_{13} 2^8 + k_{14} 2^{16} + k_{15} 2^{24}, \\ K_2 &= (k_0, k_1, k_2, k_3), \\ K_3 &= (k_4, k_5, k_6, k_7). \end{aligned}$$

The algorithm has seventeen layers. The first layer consists of two (p, a) cyclic counters. A (p, a) *cyclic counter* has an internal register that can store any integer between 0 and $p - 1$, thus the register has $\lceil \log_2 p \rceil$ bits of memory. The initial value of the register is an integer k, where $0 \le k \le p-1$. The value of the register at time unit i is defined to be

$$r_i = (ai + k \bmod p),$$

where $z \bmod m$ denotes the least nonnegative integer that is congruent to z modulo m, where $\gcd(a, p) = 1$. When $a = 1$, it counts the numbers $k, (k+1) \bmod p, \cdots, (k+p-1) \bmod p$ cyclically. For any $a \bmod p$, we call it a cyclic counter with step a and period p, in other words, a (p, a) cyclic counter.

In the two (p_i, a_i) cyclic counters, p_0 and p_1 are two distinct primes having 32 bits, and a_0 and a_1 are two constants between 0 and $p_i - 1$ respectively. The largest two 32-bit primes are $p_1 = 4294967291$ and $p_0 = 4294967279$. Note that

$$p_1 - 1 = 2 \times 5 \times 19 \times 22605091, \; p_0 - 1 = 2 \times 7 \times 17 \times 18046081.$$

It has been computed that

$$\begin{aligned} \mathrm{ord}_{p_0}(2) &= (p_0 - 1)/2 = 2147483639 = 2^{31} - 9, \\ \mathrm{ord}_{p_1}(2) &= p_1 - 1 = 2^{32} - 6, \end{aligned}$$

where $\mathrm{ord}_{p_i}(2)$ denotes the multiplicative order of 2 modulo p_i. Thus, 2 is a primitive root modulo p_1. We have that the two primes are almost equal to $2^{32} - 1$ and that the orders of 2 modulo them are close to 2^{31}.

The two constants a_i are chosen such that

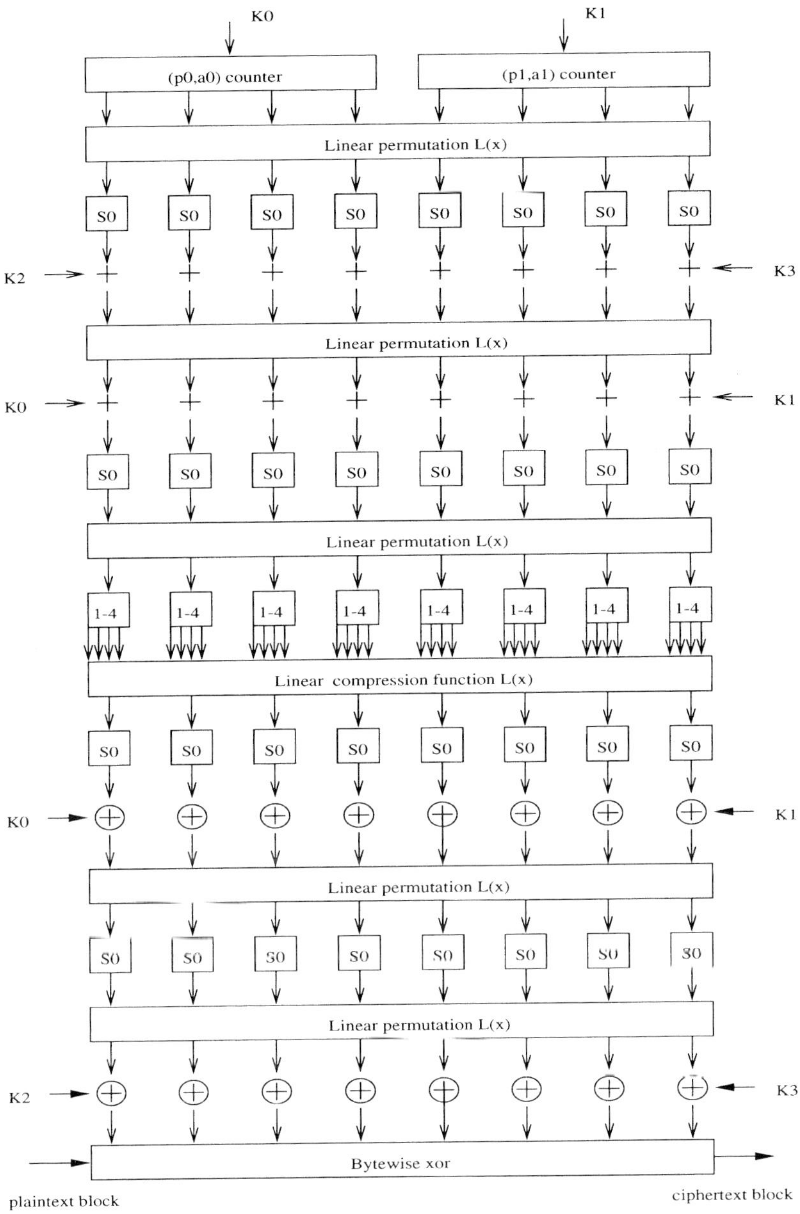

Figure 15.1: Structure of the ciphering algorithm.

1. $a_i > (p_i - 1)/2$ for $i = 0$ and 1, in order that in every two consecutive updatings of the two registers of the two cyclic counters in the keystream generator there is at least one modulo-p_i reduction;

2. they are different and the difference is large enough;

3. each a_i is not too close to p_i;

4. they are primes (this leads to $\gcd(a_1, a_2) = 1$).

Based on the above considerations we suggest the following two constants:

$$a_0 = 2345986071, \quad a_1 = 3124567807.$$

Of course, there are many such choices for the constants a_i. The first layer is intended to control the least period and linear complexity of the keystream sequence.

The second layer is a linear one that is for diffusion. The first, second, third, and fourth bytes of the contents of the two registers of the two cyclic counters are used as the inputs of this second layer. If $X_0, \cdots, X_7$ are the eight input bytes of this layer, then its eight output bytes are defined by

$$Y_j = \sum_{i=0}^{7} X_i - X_j, \ j = 0, 1, \cdots, 7, \tag{15.1}$$

where " + " and " − " denote the addition and subtraction of Z_{256}. It is clear that the change of one byte leads to a change of seven of the eight output bytes of this layer. We use $L(X)$ to denote this linear permutation.

The third layer of the algorithm consists of eight S-boxes S_0, each of which is a permutation of Z_{256} with good nonlinearity with respect to the addition of the residue class ring Z_{256}. This is the first nonlinear layer.

The nonlinear permutation S_0 is defined by

$$S_0(x) = [(x^{255} \bmod 257) \bmod 256], \quad x \in Z_{256}.$$

The permutation $x^{255} = x^{-1}$ has good nonlinearity with respect to the addition of Z_{257}. Computation proves that the above permutation S_0 also has good nonlinearity with respect to the addition of Z_{256}. The approach to finding a good nonlinear permutation of Z_{256} here follows that used by Massey [293].

The fourth layer is the bytewise addition of the outputs of the third round and the partial keys K_2 and K_3. The outputs of the first four S-boxes S_0 are added to the four bytes of K_2, and those of the second four

S-boxes S_0 are added to the four bytes of K_3, where all additions are integer addition modulo 256.

The fifth layer is a linear one that is exactly the same as the second layer. The sixth layer is again a key-addition layer, but this time the partial keys K_0 and K_1 are added. This is expected to make it difficult to find some key-equivalence classes, by which we mean that they determine the same encryption transformation.

The seventh layer is a nonlinear layer which is the same as the third one. The eighth layer is a linear layer which is the same as the second and fifth layers. The ninth layer is nonlinear and also for data expansion. It has eight-byte inputs, but 32-byte outputs. Each box containing a symbol 1-4 denotes an array of four S-boxes in the order S_1, S_2, S_3, S_4. The four S-boxes are defined by

$$\begin{aligned} S_1(x) &= [x^3 \bmod 257] \bmod 256, \\ S_2(x) &= [x^{171} \bmod 257] \bmod 256, \\ S_3(x) &= [45^x \bmod 257] \bmod 256, \\ S_4(x) &= \begin{cases} [\log_{45} x \bmod 257] \bmod 256, & \text{if } x \neq 0; \\ 128, & \text{if } x = 0. \end{cases} \end{aligned}$$

S_3 and S_4 are the two S-boxes used in SAFER [293]. As far as nonlinearity is concerned, S_3 and S_4 are good nonlinear permutations of Z_{256} with respect to the addition of Z_{256}. S_1 and S_2 have also good nonlinearity, but not as good as S_3 and S_4. In fact, S_1 and S_2 have the same nonlinearity as S_0. However, it should be mentioned that the nonlinearity with respect to the bytewise xor of S_1 and S_2 is much better than that of S_3 and S_4.

The *nonlinearity* of a permutation $P(x)$ of Z_{256} with respect to the addition of Z_{256} is measured by the probability

$$\Pr(P(x+a) - P(x) = b).$$

When $a = 0$, this probability is 1 or 0 and it is not interesting in any attack. So we are only interested in the case $a \neq 0$. Note that if $P(x)$ is a permutation, the equation $P(x+a) - P(x) = 0$ has no solution. So for any fixed $a \neq 0$ we have

$$\max_{b \neq 0} \Pr(P(x+a) - P(x) = b) \geq 2/256 = 1/128.$$

Hence

$$\begin{aligned} \tfrac{1}{128} &\leq \max\nolimits_{b \neq 0} \Pr(P(x+a) - P(x) = b) \leq \\ &\leq \max\nolimits_{a \neq 0} \max\nolimits_{b \neq 0} \Pr(P(x+a) - P(x) = b). \end{aligned} \tag{15.2}$$

For some cryptographic applications the smaller the $\max_{a \neq 0} \max_{b \neq 0} \Pr(P(x+a) - P(x) = b)$ the better the security with respect to some attacks. If equality holds in both places in (15.2), we say that the permutation $P(x)$ has the best nonlinearity with respect to the addition of Z_{256}.

The two permutations $S_3(x)$ and $S_4(x)$ have the best nonlinearity with respect to the addition of Z_{256}. This has been proved by a C program.

It should be noted that the two permutations $S_3(x)$ and $S_4(x)$ have relatively bad nonlinearity with respect to the bytewise xor operation [133].

The tenth layer is a linear compression one, which has 32-byte inputs and eight-byte outputs. We denote the inputs of this layer from the left to the right by $X_0, X_1, \cdots, X_{31}$, and the outputs from the left to the right by $Y_0, Y_1, \cdots, Y_7$. Then the linear compression function $C(x)$ is defined by

$$\begin{cases} Y_0 &= X_0 + X_5 + X_{10} + X_{15} + X_{16} + X_{22} + X_{24} + X_{30} + X_{31}, \\ Y_1 &= X_1 + X_6 + X_{11} + X_{12} + X_{17} + X_{23} + X_{25} + X_{31}, \\ Y_2 &= X_2 + X_7 + X_8 + X_{13} + X_{18} + X_{20} + X_{26} + X_{28}, \\ Y_3 &= X_3 + X_4 + X_9 + X_{14} + X_{19} + X_{21} + X_{27} + X_{29}, \\ Y_4 &= X_0 + X_6 + X_8 + X_{14} + X_{16} + X_{21} + X_{26} + X_{31}, \\ Y_5 &= X_3 + X_5 + X_{11} + X_{13} + X_{17} + X_{22} + X_{27} + X_{28}, \\ Y_6 &= X_2 + X_4 + X_{10} + X_{12} + X_{18} + X_{23} + X_{24} + X_{29}, \\ Y_7 &= X_1 + X_7 + X_9 + X_{15} + X_{19} + X_{20} + X_{25} + X_{30}, \end{cases} \tag{15.3}$$

Thus, every output byte depends on 8 input bytes except that the first byte depends on 9 inputs, and every input byte affects two output bytes except that the last input byte affects three output bytes. This linear compression function is surjective, and is mainly for compression, but it also plays an important role in diffusion. The data expansion and compression are designed to prevent one from inverting the whole system backwards. This also makes each output byte dependent on as many S-boxes and inputs of the expansion layer as possible.

The eleventh layer is a nonlinear one, where eight S-boxes S_0 are applied. The twelfth layer is a key addition layer, where the first part (K_0, K_1) of the key is xored with the outputs of the eleventh layer. The thirteenth layer is a linear one that is the same as the second and fifth layers. The fourteenth layer is a nonlinear one with eight S-boxes S_0. The fifteenth layer is a linear one that is the same as the second and fifth layers. The sixteenth layer is again a key-addition one, but here the addition is bytewise xor. This is also designed to prevent one from going backwards to the front of the keystream generator. The last layer is the bytewise xor of the keystream block and the plaintext block.

15.2 Theoretical Results about Prime-32

Let R be a commutative ring with multiplicative identity 1, and let

$$s^N = s_0 s_1 \cdots s_{N-1}$$

be a sequence of length N over R, where $s_i \in R$. If s^N satisfies a linear recurrence relation

$$s_i = a_1 s_{i-1} + a_2 s_{i-2} + \cdots + a_l s_{i-l}, \quad i \geq l, \ a_i \in R,$$

then there exists a shortest such linear recurrence relation, and the shortest l is called the linear complexity or linear span of the sequence and is denoted by $\mathrm{L}(s^N)$.

If the linear complexity of a sequence over a field is l, then $2l$ successive characters of the sequence can be used to determine a linear recurrence relation of length l satisfied by the sequence by using the Berlekamp-Massey algorithm [291], which has complexity $O(l^2)$. Thus, $2l$ successive characters of the sequence are sufficient to determine the whole sequence. Thus, sequences over fields for additive stream ciphers should have large linear complexity.

For sequences over Z_m, which is the ring $\{0, 1, ..., m-1\}$ with integer addition modulo m and multiplication modulo m, the Berlekamp-Massey algorithm does not work, but the Reeds-Sloane algorithm works. The latter is an analog of the Berlekamp-Massey algorithm, and it is also efficient [360]. Thus, it is necessary to control the linear complexity of sequences over Z_m for additive stream ciphering.

Proposition 15.2.1 *Concerning the keystream generator we have the following conclusions:*

1. *Each output sequence of bytes has least period* p_0, p_1 *or* $p_0 p_1$.

2. *Each output sequence of bytes over the ring* Z_{256} *has linear complexity at least* $\min\{\mathrm{ord}_{p_0}(2), \mathrm{ord}_{p_1}(2)\} = 2^{31} - 9$.

3. *The elements of* Z_{256} *are almost equally likely distributed in a cycle of each output sequence of bytes.*

4. *All the above conclusions hold for each output bit sequence.*

Proof: Note that the output sequence of the register of the (p_0, a_0) (resp. (p_1, a_1)) cyclic counter has least period p_0 (resp. p_1). Let X_1, X_2, X_3, X_4 be the four output bytes of the (p_0, a_0) cyclic counter at each time unit, and let X_5, X_6, X_7, X_8 be the four output bytes of the (p_1, a_1) cyclic counter. It follows that the semi-infinite sequences X_i^∞ have least period p_0 for $i = 1, 2, 3, 4$, and p_1 for $i = 5, 6, 7, 8$.

Consider $Y_1 = X_2 + X_3 + X_4 + X_5 + X_6 + X_7 + X_8 \bmod 256$. Then the semi-infinite sequence Y_1^∞ has period p_0p_1. It follows that its least period must be one of $1, p_0, p_1, p_0p_1$. Obviously, Y_1^∞ is not a constant sequence. Thus, its least period cannot be 1. Suppose that the least period of Y_1^∞ is p_0. Then the semi-infinite sequence $(X_5 + X_6 + X_7 + X_8)^\infty$ must have a period p_0, but it has a period p_1. This is impossible since p_0 and p_1 are distinct primes. Hence, the semi-infinite sequence Y_1^∞ must have least period p_0p_1. The same conclusion holds for Y_i^∞, where $2 \le i \le 8$. Since each output byte (bit) sequence cannot be a constant sequence, the least period of each output sequence should be one of $\{p_0, p_1, p_0p_1\}$.

We have already proven that each output bit sequence has a period (not necessary the least one) p_0p_1. By Basic Theorem 3.3.1, the linear complexity of each output bit sequence is at least $\min\{\mathrm{ord}_{p_0}(2), \mathrm{ord}_{p_1}(2)\}$.

Let $z_1^\infty = Z_1^\infty \bmod 2$, where Z_1^∞ is the output byte sequence of the first output byte position of the keystream generator. It is easily seen that the linear complexity of the semi-infinite sequence z_1^∞ over Z_{256} is no less than that of Z_1^∞. Thus, we have proved the second claim.

If $p_0 = p_1 = 2^{32}$, then each X_i takes on elements of Z_{256} with equal probability, and so does each Y_i. However, since $p_0 = 2^{32} - 17$ and $p_1 = 2^{32} - 5$, each output byte Y_i takes on elements of Z_{256} with almost equal probability. Since each layer is either a permutation layer or a linear layer, each keystream byte sequence has an almost uniform distribution of the elements of Z_{256}, so each bit sequence of the keystream block sequence has an almost uniform distribution of ones and zeroes. □

Remark: It should be extremely unlikely that the least period of a byte (bit) sequence is p_0 or p_1.

15.3 Security Arguments

A cipher must be secure against ciphertext-only attacks if it is secure against known plaintext attacks. So in the sequel we shall argue some security aspects of the algorithm only with respect to some known plaintext attacks. When doing so, we assume the cryptanalyst has sufficiently many keystream blocks. As with other practical ciphers, it is hard to prove the security of a ciphering algorithm since we cannot sort out all possible attacks on a cipher.

With respect to brute-force attack

An attack that applies to every cipher is the brute-force attack by trying all possible keys. Since the number of possible keys of our ciphering algorithm is 2^{128}, this attack should not work. On the other hand, it might be possible that a number of keys determine the same encryption transformation, but we do not see a way to prove their existence, let alone to determine them if they exist.

With respect to linear complexity attacks

Since this is an additive synchronous stream cipher, it is necessary to control the least period (cycle length) of the keystream sequences and its component bit sequences. As proved before, the least period of the output sequence and its component bit sequences all have least period $\geq \min\{p_0, p_1\}$, and the linear complexities of the output sequence and its component bit sequences are at least

$$\min\{\text{ord}_{p_0}(2), \text{ord}_{p_1}(2)\} = 2^{31} - 9.$$

Thus, any attack based on the Berlekamp-Massey algorithm [291] or Reeds-Sloane algorithm [360] should not work. We can also prove that the linear complexities of the output sequence and its component bit sequences have ideal stability, thus, it is hard to construct an LFSR to approximate the output sequence of the generator [138].

With respect to inverting attacks

One basic question is whether this keystream generator is invertible. All layers except the data-expansion layer are permutation layers, when the key is fixed. But without the key it could be impossible to invert the keystream generator.

Let $X_i^{(j)}$ denote the output bytes of the jth layer. It follows that

$$X_i^{(16)} = X_i^{(15)} \oplus k_i, \quad i = 0, 1, \cdots, 7,$$

where the addition is the bytewise xor. Assuming that the key is randomly chosen, the information about $(X_0^{(15)}, \cdots, X_7^{(15)})$ provided by the keystream block $(X_0^{(15)}, \cdots, X_7^{(15)})$ is zero as the partial key $(k_0, k_1, \cdots, k_7)$ is unknown. Thus, it is impossible to use one keystream block to go backwards.

Since the last layer of the keystream generator is linear with respect to bytewise xor, one may consider the difference of two keystream blocks at time t_1 and t_2, in order to get rid of the partial keys K_2 and K_3 added.

Let $X_i^{(j,t)}$ denote the output bytes of the jth layer at time t. It follows that

$$X_i^{(16,t_1)} \oplus X_i^{(16,t_2)} = X_i^{(15,t_1)} \oplus X_i^{(15,t_2)} \tag{15.4}$$

for $i = 0, 1, \cdots, 7$. This might be a useful relation.

Note that the linear permutation $L(x)$ is linear with respect to the integer addition modulo 256, but nonlinear with respect to bytewise xor operation. It is hard to get information about $X_i^{(14,t_1)} \oplus X_i^{(14,t_2)}$ and $X_i^{(13,t_1)} + X_i^{(13,t_2)}$.

With respect to correlation attacks

There are different kinds of correlation attacks, but they are only for special stream ciphers. It is impossible to sort out all correlation attacks on a system, but the essential idea of a correlation attack would be to find a relation between output keystream characters and some part or the whole of the key, or a relation between some intermediate variables. The purpose of finding such a relation is to get information about the key from known keystream blocks. A way to protect a stream cipher from correlation attacks is to use correlation-immune functions in the system in a proper way.

There are six layers that could protect the system from such an attack: the five linear permutation layers described by (15.1), where each function is correlation-immune of order 6 [402], and the linear compression layer described by (15.3), where each function is correlation immune of order 7. These correlation-immune functions and layers are expected to protect the cipher from correlation attacks.

With respect to affine approximation attacks

The idea of an affine approximation attack would be to use the best affine approximation (BAA) of some nonlinear components of the system to replace the nonlinear parts, in order to construct a pseudo-keystream generator which produces an output sequence that matches the original keystream sequence with high probability or to recover the key of the original keystream generator. Such an attack, carried out for two kinds of stream ciphers in [138], should not work on this algorithm, due to the high nonlinearity of the S-boxes and the five diffusion layers.

The most reasonable affine approximation of the nonlinear S-boxes is to use affine functions $ax + b$ over Z_{256} to approximate the five S-boxes. However, with a simple C program we have obtained the following result.

Proposition 15.3.1 *Let* Pr *denote the probability. Then*

$$\begin{aligned}
\max_{0\le a\le 255}\ \max_{0\le b\le 255} \Pr(S_0(x) = ax+b) &= \frac{2}{256}\\
\max_{0\le a\le 255}\ \max_{0\le b\le 255} \Pr(S_1(x) = ax+b) &= \frac{3}{256}\\
\max_{0\le a\le 255}\ \max_{0\le b\le 255} \Pr(S_2(x) = ax+b) &= \frac{3}{256}\\
\max_{0\le a\le 255}\ \max_{0\le b\le 255} \Pr(S_3(x) = ax+b) &= \frac{3}{256}\\
\max_{0\le a\le 255}\ \max_{0\le b\le 255} \Pr(S_4(x) = ax+b) &= \frac{4}{256}.
\end{aligned}$$

This result shows that affine approximations of the S-boxes with $ax+b$ over Z_{256} are very poor. Clearly, every permutation $P(x)$ of Z_{256} can be identified as a permutation $P'(y)$ of Z_2^8, and one might therefore be concerned with the affine approximation of $P'(y)$ with respect to Boolean affine functions over Z_2^8. However, as our operations in the ciphering algorithm are almost totally based on those of Z_{256}, such an affine approximation will probably not work.

15.4 Performance of Prime-32

On a Pentium (75 MHz) an initial C code (Borland C++ compiler, version 1991) of the Prime-32 runs at 4.5 Mbits/sec. The test is done with a self-feeding 4 Mbyte input data. An optimized code should run faster.

15.5 Prime-32 with a 192-Bit Key

The key length of Prime-32 can be 192-bits. To this end, the key addition at the twelfth layer uses the remaining 64 bits of key, instead of using (K_0, K_1). The algorithm itself need not be changed.

15.6 Prime-64

A variant of Prime-32 is the Prime-64 for 64-bit machines. Prime-64 has only one (p, a) cyclic counter, in which p is the closest 64-bit prime to $2^{64}-1$ such that $\mathrm{ord}_p(2) \ge 2^{32}$, and a is any prime that is approximately $3p/4$. Thus, only the first layer of Prime-32 is modified and others remain the same. The content of the register of the (p, a) cyclic counter is similarly divided into 8 bytes which are used as the input of the next layer.

A possible choice of the prime p is

$$p = 18446744073709551557 = 2^{64} - 59.$$

Thus,

$$p - 1 = 2^2 \times 11 \times 137 \times 547 \times 5594472617641.$$

The order of 2 modulo p can be computed with a 64-bit computer, based on this factorization.

We have about the same theoretical results for Prime-64, but have not tested Prime-64 for performance. However it is clearly much faster than Prime-32, as the first layer of Prime-64 is much faster than that of Prime-32.

The security of Prime-32 and Prime-64 should be at the same level. The choice between the two algorithms depends on the machines used.

Chapter 16

Cryptographic Problems and Philosophies

There are many unsolved cryptographic problems. Some have been attacked by cryptographers for many years without much success. One example is the definition and measure of security for ciphers. This makes cryptology very different from many other sciences. This chapter is intended mainly to discuss some cryptographic problems and philosophies, but not to solve them.

16.1 Nonlinearity and Linearity

Both nonlinear and linear functions are of significance for block and stream ciphers as well as for hash functions. Nonlinear functions are usually used to achieve confusion, while linear functions are employed to achieve diffusion. Nonlinear functions are useful in protecting a cipher from a differential cryptanalysis [257, 334, 19, 122], from determining the key by solving equations and/or by approximation and so forth. One example of the application of linear functions to achieve diffusion is the cipher algorithm SAFER K-64 developed by Massey [293], where pseudo-Hadamard transforms have been employed. Another important role that linear functions can play is the control of the density of cryptographic transformations, which will be introduced in Section 16.3.

Let $f(x_0, x_1, \cdots, x_{n-1})$ be a Boolean function. If

$$f(x) = x_0 + g(x_1, x_2, \cdots, x_{n-1}),$$

then

$$f(x_0 + 1, x_1, \cdots, x_{n-1}) = 1 + f(x_0, x_1, \cdots, x_{n-1})$$

for all $x \in Z_2^n$. Conversely, if

$$f(x_0 + 1, x_1, \cdots, x_{n-1}) = 1 + f(x_0, x_1, \cdots, x_{n-1})$$

for all $x \in Z_2^n$, then it is easy to prove that

$$f(x_0, x_1, \cdots, x_{n-1}) = x_0 + g(x_1, \cdots, x_{n-1}).$$

Furthermore,

$$f(x_0, x_1, \cdots, x_{i-1}, x_i + 1, x_i, \cdots, x_{n-1}) = 1 + f(x_0, x_1, \cdots, x_{n-1})$$

for each i and all x if and only if

$$f(x) = x_0 + x_1 + \cdots + x_{n-1}$$

or

$$f(x) = x_0 + x_1 + \cdots + x_{n-1} + 1.$$

This clearly shows why linear functions play the role in achieving diffusion, by which we usually mean the extent of changes in the ciphertext when a small number of changes in the plaintext or key occur.

Recall that the nonlinearity of a function $f(x)$ from an Abelian group $(G, +)$ to another Abelian group $(H, +)$ is measured by

$$P_f = \max_{0 \neq a \in G} \max_{\neq b \in H} \Pr(f(x + a) - f(x) = b). \tag{16.1}$$

The minimum distance between a mapping and all linear functions is also a rational nonlinearity measure in some cases [307], but this measure does not make sense in many cases, as there are only trivial affine mappings from some Abelian groups to some others. Thus, the most suitable measure for nonlinearity may be the quantity P_f.

These definitions show that the nonlinearity of a function (mapping) is relative to the two operations concerned. To see the relativity, we consider the function $f(x) = (x^{(p-1)/2} \bmod p) \bmod 2$ from Z_p to Z_2. As proved in Section 4.3.1, $f(x)$ is almost linear with respect to $(Z_p, \times)$ and $(Z_2, +)$, but has the best nonlinearity with respect to $(Z_p, +)$ and $(Z_2, +)$.

We consider now another example. Let $G = GF(2)^5$ and $(H, +) = (GF(2), \oplus)$, where $\oplus$ is modulo-2 addition. In $GF(2)^5$ we define two kinds of operations. The first operation $+'$ is bitwise modulo-2 addition and the second one is defined by

$$(x_0 x_1 \cdots x_4) +'' (y_0 y_1 \cdots y_4) = (z_0 z_1 \cdots z_4),$$

where

$$z_0 + z_1 2 + \cdots + z_4 2^4 = [x_0 + \cdots + x_4 2^4 + y_0 + \cdots + y_4 2^4] \bmod 2^5.$$

Let $V_0 = \{x = (x_0, \cdots, x_4) : \text{WH}(x) \text{ even}\}$ and let $f(x)$ be the characteristic function of V_0, where $\text{WH}(x)$ is the Hamming weight of x. We have then

$$\Pr(f(x +' a) \oplus f(x) = 1) = \begin{cases} 0, & a \in V_0 \\ 1, & \text{otherwise} \end{cases}$$

and

$$\Pr(f(x +' a) \oplus f(x) = 0) = \begin{cases} 1, & a \in V_0 \\ 0, & \text{otherwise} \end{cases}$$

Actually $f(x)$ is linear with respect to $(+', \oplus)$. But with respect to $(+'', \oplus)$ we have the result described in Table 16.1, where if $a \geq 17$, we have

$$32 \Pr(f(x +'' a) \oplus f(x) = 0) = 32 \Pr(f(x +'' (-a)) \oplus f(x) = 0).$$

This shows that $f(x)$ has relatively much better nonlinearity with respect to $(+'', \oplus)$ than to $(+', \oplus)$.

It is possible for one function to have the same local nonlinearity with respect to many binary operations of the input and output Abelian groups. Let $(G, +')$ and $(H, +)$ be two finite Abelian groups. For any $b \in G$, we define another binary operation $+''$ of G by

$$x +'' y = x +' y +' b.$$

Then it is easy to see that $(G, +'')$ is an Abelian group. Let $f(x)$ be any function from G to H, then we have

$$\Pr(f(x +'' a) - f(x) = r) = \Pr(f(x +' a +' b) - f(x) = r).$$

This means that the local nonlinearity of $f(x)$ at a with respect to the $(+'', +)$ is the same as that at $a +' b$ with respect to $(+', +)$. Thus, every function from G to H has the same nonlinearity with respect to $(+'', +)$ and $(+', +)$.

One of the most interesting general results about linear functions is Theorem 13.2.6, i.e.,

> For every nonzero linear function $L(x)$ from $F = GF(q^m)$ to $K = GF(q)$ with respect to $(F, +)$ and $(K, +)$, its nonlinearity with respect to $(F^*, \times)$ and $(K, +)$ is optimal.

Table 16.1: An example of the relativity of nonlinearity.

a	$32\,\Pr(f(x +'' a) \oplus f(x) = 0)$	$32\,\Pr(f(x +'' a) \oplus f(x) = 1)$
1	10	22
2	12	20
3	22	10
4	8	24
5	18	14
6	20	12
7	14	18
8	16	16
9	18	14
10	12	20
11	14	18
12	22	10
13	10	22
14	20	12
15	22	10
16	0	32

This shows another cryptographic significance of linear functions. It tells us that these linear functions are very good nonlinear functions provided they are used properly.

Summarizing this section, we see that linear functions are cryptographically important in

1. achieving "diffusion";
2. controlling the density of cryptographic transformations specified by keys;
3. serving as good nonlinear functions in a suitable context.

16.2 Stability and Instability

Stability problems are everywhere and all around us. Every one has to stabilize his/her relations with most of the people around him/her. We have the problem of stabilizing the world. Every country has its own stability problems, which include the stability of the general welfare by taxing, the stability of political rights and social order by law, and stability among

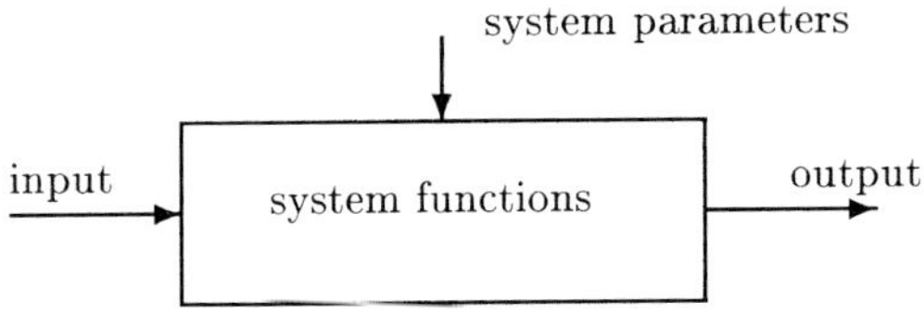

Figure 16.1: A description of cryptographic systems.

individuals and among different social classes. Similarly, every family has also its own stability problems. Generally, we may say that every system has its stability problems, so do cipher systems.

16.2.1 Stability and Diffusion

Many systems can be described with Figure 16.1, where whenever an input is given, a corresponding output is produced by the system, using functions which are controlled by the system parameters. One important stability problem of such systems is the study of changes in the output when the input is changed. Suppose that each input is taken from a metric space $(I, +, |.|_1)$, and each output is from a metric space $(O, +, |.|_2)$, where $(I, +)$ and $(O, +)$ are Abelian groups with norms $|.|_1$ and $|.|_2$ respectively. Furthermore, suppose that the system has only one system mapping $F(k_1, \cdots, k_r, i)$, where $k = (k_1, \cdots, k_r)$ denotes the system parameter, which is supposed to be taken from another metric space $(K, +, |.|_3)$. Then there are two basic stability problems about the system. One is the study of the ratio

$$\Delta_F(i_1, i_2) = \frac{|F(k, i_1) - F(k, i_2)|_2}{|i_1 - i_2|_1}, \tag{16.2}$$

where k is a fixed element of K. This is a measure of the extent of change in the output relative to the change in the input, which is a stability problem for many such systems. Another stability problem is how sensitive the system is to parameter changes, which can be measured by

$$\Delta_F(k, k') = \frac{|F(k, i) - F(k', i)|_2}{|k - k'|_3}, \tag{16.3}$$

where i is fixed. For many such continuous systems, calculus can be used to treat the two stability problems. Derivatives are measures of such stabilities. For discrete systems some mathematical tools which are analogous to calculus are needed to treat these two stability problems. The stability of the solution of some linear systems is one example of such problems.

Every cipher system may be described by the system of Figure 16.1, where the inputs are plaintexts, the outputs are ciphertexts and the system parameters are keys and/or the initial values of the internal memory state. The system functions are those which give the encryption transformation. For block ciphers, inputs and outputs are blocks of digits, say $p = (p_0, p_1, \cdots, p_{m-1})$ is a plaintext block and $c = (c_0, c_1, \cdots, c_{n-1})$ is the corresponding ciphertext block. Let E be the block encryption algorithm and $E(k, .)$ the encryption transformation specified by a key k; then we have the relation

$$c = E(k, p). \tag{16.4}$$

For the sake of simplicity, we assume that both plaintext blocks and ciphertext blocks are taken from $(GF(2)^n, +, |.|)$, where $|.|$ denotes the Hamming weight.

To guide the design of practical ciphers, Shannon suggested two general principles, which he called diffusion and confusion [397]. By *diffusion*, he meant the spreading out of the influence of a single plaintext digit over many ciphertext digits so as to hide the statistical structure of the plaintext. An extension of this idea is to spread the influence of a single key digit over many digits of ciphertext so as to frustrate a piecemeal attack on the key.

Thus, the concept of diffusion suggested by Shannon and its extension are in fact two kinds of instabilities which can be measured by (16.2) and (16.3) respectively, where F is the encryption algorithm. This means that Shannon's diffusion and its extension suggest designing ciphers which are not too stable with respect to both plaintext and keys. However, if we use (16.2) and (16.3) to measure the plaintext diffusion and key diffusion, it may be mathematically proven that there is a tradeoff between the extent of plaintext (key) diffusion and the nonlinearity of the encryption function $E(k, p)$ with respect to the additions of $(P, +)$ $((K, +))$ and $(C, +)$, where $(P, +)$, $(K, +)$ and $(C, +)$ are respectively the plaintext block space, key space and ciphertext block space. This tradeoff can be seen from the discussion of Section 16.3. It follows that we have to make a compromise between diffusion and nonlinearity of the encryption transformations when designing ciphers.

By *confusion*, Shannon meant the use of enciphering transformations that complicate the determination of how the statistics of the ciphertext depend on the statistics of the plaintext. Thus, the nonlinearity of the enciphering transformations may be used as a partial measure of confusion. If it is rational to do so, this means there is a tradeoff between diffusion and confusion within a block cipher.

Another cryptographic stability function is the linear complexity stability of sequences which is described in Section 2.3.4.

16.2.2 Correlation Stability and Pattern Stability

To see the conservation between correlations, we take the autocorrelation of binary periodic sequences as an example. Recall the definition of the autocorrelation function $\mathrm{AC}_s(l)$ of a binary sequence s^∞ of period N, which is defined by

$$\mathrm{AC}_s(l) = \sum_{i=0}^{N-1} (-1)^{s_i + s_{i+l}}/N.$$

Let n denote the number of 1's in one period of the sequence; then we have

$$\sum_{l=1}^{N-1} \mathrm{AC}_s(l) = (2n - N)^2/N - 1.$$

This means that the autocorrelations of sequences of period N with a fixed number of 1's in one period are conservative. Thus, keeping their stability is necessary.

As derived in Section 2.3.2, there is a conservation law of patterns in the period of a periodic sequence. Thus, keeping the stability of patterns in a periodic sequence is also necessary. The relation of autocorrelation stability and pattern stability has already been made clear in Section 2.3.2.

16.2.3 Mutual Information Stability

To show the stability of mutual information between keys and keystream digits, we consider now the binary NSG of Figure 2.5(b) [122]. Theoretically every bit of a keystream can give information about a generator's initial state and the key. Thus a basic requirement for stream ciphers is that every bit of keystream gives approximately the same amount of information. In our case, this yields balance requirements for the filter function $f(x)$. This *single bit analysis* is apparently applicable to all synchronous stream ciphers. Let $C_i = \{x \in Z_N : f(x) = i\}$ for $i = 0, 1$. If $n = \log_2 N$, we can write

$$I(k; h_0 = 0) = n - \log_2 |C_0| \text{ bits,}$$
$$I(k; h_0 = 1) = n - \log_2 |C_1| \text{ bits.}$$

Noticing that $|C_0| + |C_1| = N$, we get

$$2^{n - I(k; h_0 = 0)} + 2^{n - I(k; h_0 = 1)} = N.$$

This is the theoretical basis for keeping the mutual information stability of a keystream bit and the key as flat as possible.

If we now consider two bits h_i and h_j separately or arbitrarily, we may not obtain $I(k; h_i) + I(k; h_j)$ bits of information about the key. If the cipher is not properly designed, some combinations of bits may give much more information about the key than others. We call such combinations with their length $(h_i, h_j, |i-j|)$'s *bad patterns*. The idea behind the differential attack on this generator [122] is to look for bad patterns, and in particular for triples $(i, j; w)$ which give as much information about the key as possible. One may argue that we should design our cipher so that the mutual information $I(k; (i, j; w))$ is as small as possible for all $(i, j; w) \in Z_2 \times Z_2 \times Z_N$, but in fact we cannot achieve this: One pattern $(i, j; w) \in Z_2 \times Z_2 \times Z_N$ gives

$$I = n - \log_2 d_f(i, j; w) = n - \log_2 |C_i \cap (C_j - w)| \text{ bits}$$

of information about the key. Now consider the following theorems:

Theorem 16.2.1 (Conservation Law for Difference Parameters) *With the symbols as before, we have*

$$\sum_j d_f(i, j; w) = |C_i|, \ i \in Z_2, \ w \in Z_N;$$

$$\sum_i d_f(i, j; w) = |C_j|; \ j \in Z_2, \ w \in Z_N;$$

$$\sum_{(i,j) \in Z_2 \times Z_2} d_f(i, j; w) = N, \ w \in Z_N.$$

These are the laws of conservation between difference parameters which appear in three forms. By this theorem we have the following conclusion:

Theorem 16.2.2 (Conservation Law of Mutual Information) *With the symbols as before, we have*

$$\sum_j 2^{n - I(k;(i,j;w))} = |C_i|, \ i \in Z_2, \ w \in Z_N;$$

$$\sum_i 2^{n - I(k;(i,j;w))} = |C_j|; \ j \in Z_2, \ w \in Z_N;$$

$$\sum_{(i,j) \in Z_2 \times Z_2} 2^{n - I(k;(i,j;w))} = N, \ w \in Z_N;$$

$$\sum_{(i,j;w)} 2^{n - I(k;(i,j;w))} = N^2.$$

It is not difficult to prove the above theorems, which provide the theoretical basis for analyzing mutual information stability between two-bit patterns and the key. Generalizing the above theorems to the case of an arbitrary finite G is also straightforward.

For other generators there will usually also exist conservations of some mutual information. Thus, it is important to discover those conservations and to make compromises. Asking too much gain in one sense without considering the possible loss in another sense could be dangerous.

16.3 Localness and Globalness

One of the most troublesome problems in cryptography may be the control of some local cryptographic properties, such as local linear complexity, local sphere complexity and local density of cryptographic transformations and so forth. We analyze some cryptographic properties of the designed key stream within one period, which is usually very large, but we use only a very small part of each key stream. Thus, local properties are in general much more important than global ones.

We begin with densities of cryptographic transformations (briefly, transdensity or T-density [122]). Let M be the plaintext space, C the ciphertext space, K the key space and T_K the set of encryption or decryption transformations specified by the keys. Then the transdensities are defined by

$$D(T,K) = 1 - \sum_{k,k'} \frac{\Pr(t_k, t_{k'})}{\binom{|K|}{2}}$$

$$D_0(T,K) = 1 - \max_{k,k'} \Pr(t_k, t_{k'})/|K|,$$

where $\Pr(t_k, t_{k'})$ denotes the probability of agreement between the two encryption or decryption transformations specified by the two keys, which is usually replaced by $d(t_k, t_{k'})/|M|$ for simplicity, where $d(t_k, t_{k'})$ denotes the distance between t_k and $t_{k'}$. The introduction of transdensities was inspired by the following three cryptographic questions.

Question 16.3.1 *To break a cipher or to decipher a piece of ciphertext, do we have to recover the original key?*

Question 16.3.2 *Are the encryption transformations and decryption transformations specified by the keys really "different" from one another?*

Question 16.3.3 *When the answer to Question 16.3.2 is "yes", for a given key k, is there any key $k' \in K$ such that the probability of agreement*

$\Pr(t_k, t_{k'})$ *or the distance* $d(t_k, t_{k'})$ *is small enough? If there are such keys, which ones are they and how many are there?*

The importance of the questions is clear, as attacks may involve trying partial keys. That they are practical, follows from the fact that the M-209 cipher machine had large equivalence classes of keys. However, it seems that for most proposed ciphers the above three questions have not been answered.

The transdensity is related to partial-key attacks, key density, key size, message density, message and cryptogram residue classes, perfect secrecy, autocorrelation and crosscorrelation functions of sequences, difference sets, difference property of partitions, nonlinearity of cryptographic functions, affine approximation of functions, mutual information stability and source coding. Thus, the importance of transdensities is clear.

However, it follows from the definition that $D(T, K)$ is a global property of the cryptographic transformations. Theoretically, many enciphering transformations may be different, but practically they may be the same, since the block length of most proposed block ciphers is quite large. Intuitively, the larger the block length, the more difficult it is to control local properties of cryptographic systems.

The linear complexity of key streams is clearly a global property. The local linear complexities of a key stream are more important, but they are usually difficult to control. We have same problems for sphere complexity.

16.4 Goodness and Badness

Before discussing problems concerning goodness and badness, we should agree on what "goodness" and "badness" mean. Unfortunately, we cannot give strict mathematical definitions for goodness and badness. It is clear that these two concepts are relative. Nevertheless, this does not mean that there is no distinction between goodness and badness.

The first point of the relativity of goodness and badness is that they are comparative concepts. When we say that something is good, we mean that it is good relative to a specific thing or a set of specific things. Goodness and badness are also relative to

1. the aspect from which a thing is considered,

2. the context in which a thing occurs,

3. the time at which a thing occurs,

4. the purpose for which a thing is used.

These facts may be illustrated by the following cryptographic examples.

As shown in Section 16.1, it is very hard to decide whether linear functions are cryptographically good or bad. It depends on the system in which they are used and how they are used. This may indicate that it is necessary to discuss the cryptographic properties of some building blocks, but to conclude their cryptographic values without specific context may not be reasonable.

Now we turn to primes. Many ciphers are based on numbers and, in particular, on primes. Thus, primes are building blocks of many ciphers. Similarly, it may be impossible to say which primes are cryptographically good or better without giving specific cryptographic contexts. Mersenne primes are cryptographically bad when they are used as periods of binary sequences due to the fact the order of 2 modulo a Mersenne prime $2^m - 1$ is m, but they are good as periods of sequences over some other fields. Primes for RSA and those for stream ciphers are different in some aspects (see Section 5.10), and primes for different keystream generators are also required to have some special properties.

Summarizing the section, we conclude that many possible cryptographic building blocks have both good sides and bad sides with respect to specific contexts. What a cryptographer does is to find those good and bad sides with respect to some considerations and to use those good and bad sides in a proper way.

16.5 About Good plus Good

Before discussing the topic, let us agree that we have a measure of goodness for some cryptographic building blocks with respect to some cryptographic aspect. Given the definition of plus as some kind of combination of two cryptographic building blocks, then does "good plus good" give good? This depends on the measure of goodness and the definition of plus together with the two building blocks. The answer could be both "good" and "bad".

The bitwise modulo-2 sum of two sequences with large linear complexity may give a new sequence with very small linear complexity or with large linear complexity. Also let $f(x)$ be a mapping from an Abelian group $(A, +)$ to an Abelian group $(B, +)$, and let $g(x)$ be a mapping from $(B, +)$ to an Abelian group $(C, +)$. Assume that f and g have good nonlinearity with respect to $[(A, +), (B, +)]$ and $[(B, +), (C, +)]$ respectively. Then the composition function $h(x) = g(f(x))$ may have good or bad nonlinearity with respect to $[(A, +), (C, +)]$. Concerning the pattern distribution and difference property of sequences we have the same conclusion. Now the problem is how to develop techniques which ensure "good plus good =

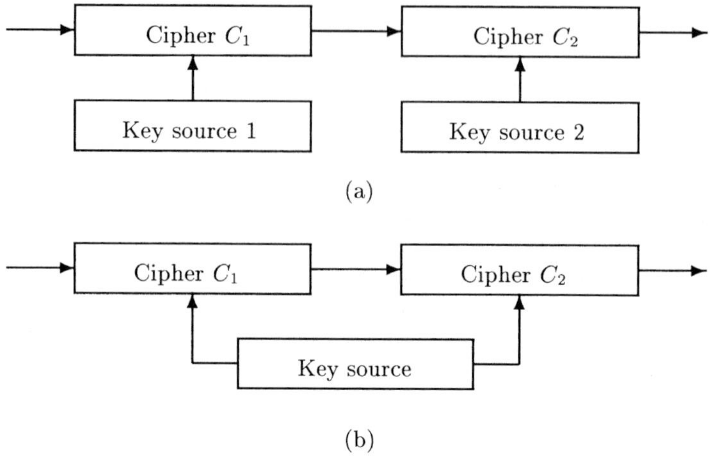

Figure 16.2: (a): a cascade of two ciphers. (b): a product of two ciphers.

good."

We now turn to the cascade of ciphers. Cascade ciphers and product ciphers can be depicted by Figure 16.2(a) and 16.2(b) respectively [297]. The distinction between *cascaded ciphers* and *product ciphers* [297, 397] is that in the latter the keys of the component ciphers need not be statistically independent, where they are in the former. Assume that ciphers C_1 and C_2 are good ciphers with respect to some specific security measures and cascade is considered as a kind of plus. Then one question is whether we have "good plus good =good", or equally whether the cascaded cipher is good with respect to those specific security measures. For details about this problem we refer to Maurer and Massey [297].

Iteration is the most used technique in designing block ciphers. Many block ciphers are based on the iteration of a round function several times. It is usually easy for us to control some cryptographic properties of the round function or that of the S-boxes of the round function, but very difficult to control those of the cryptographic transformations due to the iterations. For example, for many block ciphers, we do not even know how many fixed points their enciphering transformations have. This is one of the basic cryptographic problems we should solve, since ciphers whose cryptographic transformations have many fixed points are not secure.

Summarizing this section, we conclude that "good plus good" could be "good" or "bad", given the definitions of "good" and plus. The most important problem is to develop techniques that ensure "good plus good

= good," which is usually not easy. Of course, the techniques depend on what the measures of goodness and plus are. For example, if we consider the bitwise modulo-2 sum of two binary sequences and take the size of the linear complexity as a measure of goodness, then one technique to ensure "good plus good = good" is to ensure that the minimal polynomials of the two sequences are relative prime.

Consider now the composition of mappings, which is used in many ciphers. This is to say that here plus is defined to be the mapping composition. Taking the nonlinearity of mappings as the measure of goodness, we ask the question as to how to develop techniques for ensuring "good plus good = good," that is, how can we develop techniques to ensure that the composition mapping of two mappings with good nonlinearity has also good nonlinearity?

16.6 About Good plus Bad

At a first glance we may have the impression that this is very similar to "good plus good". This is however not true. In fact it is easy to give examples of "good plus bad = good", but difficult to find examples of "good plus bad = bad". One example for the latter case is the bitwise product of two binary sequences when the balance between 1's and 0's within one period is taken as the measure of goodness. Generally, whether "good plus bad" gives "good" depends on the measure of goodness and whether the "plus" has a tendency to preserve "goodness." However, it should be pointed out that we do not know whether "good plus bad" gives "good" in most cases. Thus, techniques for ensuring "good plus bad = good" are needed.

One of the main techniques employed in this book is the use of a mapping with good nonlinearity together with a (almost) linear function to get another mapping with good nonlinearity. This technique has been used frequently in Chapter 7, Chapter 10 and Chapter 13.

16.7 About Bad plus Good

It is not hard to see that "bad" plus "good" could give "bad" or "good". To illustrate this, we consider the composition of two mappings and take nonlinearity as a measure of goodness. Let α be a nontrivial linear onto mapping from an Abelian group $(F, +)$ to another one $(G, +)$, and let β be an onto mapping from $(G, +)$ to a third Abelian group $(H, +)$. Then the composition mapping defined by

$$\gamma(x) = \beta(\alpha(x)), \quad x \in F$$

is a mapping from F to H.

For any $g \in G$ define

$$F_g = \{x \in F : \alpha(x) = g\}.$$

Since α is a linear onto mapping, $(F_0, +)$ must be a subgroup of $(F, +)$ with $|F|/|G|$ elements. It is also easily seen that

$$\cup_{g \in G} F_g = F, \quad F_{g_1} \cap F_{g_2} = \emptyset \text{ for } g_1 \neq g_2.$$

Since α is constant on every F_g, for $f \in F$ and $h \in H$ we have

$$\begin{aligned} & |\{x \in F : \gamma(x + f) - \gamma(x) = h\}| \\ = & |\{x \in F : \beta(\alpha(x) + \alpha(f)) - \beta(\alpha(x)) = h\}| \\ = & \frac{|F|}{|G|} |\{y \in G : \beta(y + g) - \beta(y) = h\}|, \end{aligned}$$

where $g = \alpha(f)$. Hence

$$\Pr(\gamma(x + f) - \gamma(x) = h) = \Pr(\beta(y + g) - \beta(y) = h).$$

Thus, if $|F| = |G|$, then the nonlinearity of γ is the same as that of β. In this case, we have "bad plus good = good" and "bad plus bad = bad". However, if $|F|/|G|$ is quite large, the nonlinearity of γ is much worse than that of β. In this case, we have "bad plus bad = bad" and "bad plus good = bad". Thus if we are going to have goodness, we have to pay for it.

16.8 Hardware and Software Model Complexity

In stream ciphers the linear complexity (linear span), quadratic span, and 2-adic span are based on the linear feedback shift register model, the quadratic feedback shift register model, and the feedback with carry shift register model, respectively. These complexity measures are based on special hardware circuits which are usually quite efficient. These generators can be used as complexity models for producing sequences over a finite field (GF(2) in the case of 2-adic span) because every ultimately periodic sequence over the field can be generated by such a generator by choosing proper design parameters. However, to use the least amount of memory in such a special generator as a security measure, we have to have an "efficient" algorithm to determine the design parameters or initial loading with which the generator produces a given sequence. With respect to the LFSR model, we have the efficient Berlekamp-Massey algorithm. With respect to the FCSR model, we have also an efficient algorithm, the rational approximation algorithm.

Although every periodic sequence can be generated by the NSG in Section 2.2.1, it cannot be used as a security model since we haven't found an efficient algorithm to determine the initial loading with which it produces a given sequence.

For simple and efficient hardware models the main problem is the memory size, since in such a model the computational complexity is very small due to the speciality and simplicity of the models. Of course, the computational complexity of the model usually increases when the size of memory does.

It is also possible to produce every ultimately periodic binary sequence with the software algorithm of Section 14.2. Theoretically every ultimately periodic sequence can be produced with a similar algorithm by a proper selection of the input parameters p and q. To use the smallest memory, the input parameters should be reduced. Due to the 2-RA algorithm of Section 14.6, the index defined by

$$\Lambda_2(p, q) = \lceil \log |p| \rceil + \lceil \log |q| \rceil$$

should be a software complexity for the 2-adic expansion sequence of the rational number p/q, where p/q is reduced. The above measure $\Lambda_2(p, q)$ is really an analogue of the usual linear span, which behaves slightly differently from the 2-adic span. This software model of complexity for binary ultimately periodic sequences is with respect to the specific algorithm of Section 14.2, where the number of computations needed is very small. The 2-RA algorithm indicates that it should be a security measure. This is a software 2-adic span. In some cases a software complexity could be much more convenient and reasonable than a hardware one based on some awkward hardware model. With the advent of powerful computers, software model complexities with respect to some algorithms seem to be more promising. We should be aware of the fact that the linear span can be defined without the LFSR hardware model. It existed long before the electronic age.

Similar to hardware model complexities, software model complexities must be relative to an algorithm. The algorithm should usually be efficient in software; in this case the software complexity should be mainly based on the memory size.

There are also some other complexity models and security measures based on hardware models, which are used not to produce the original sequence, but to produce a sequence which is almost the same as the original sequence. The linear complexity (linear span) of a sequence could be very large, but it could be possible to use a very small linear feedback shift register to produce another sequence which is almost the same as the original sequence. The sphere complexity (see Section 2.3.4) is based on the LFSR hardware approximation model [137, 138].

Notes on Sequences

As this book is mainly about keystream sequences and number theory, we cannot cover other aspects of sequences. In this note we give some other information on sequences.

For linear complexity profiles of sequences we refer to Dai [91], Niederreiter [319, 320, 321, 322], Niederreiter and Vielhaber [321, 324, 326]. Information on the linear complexity and minimal polynomials of the products of sequences can be found in Zierler and Mills [473], Herlestam [203], and Göttfert and Niederreiter [175, 176], where the relation between Hasse-Teichmüller derivatives and products of sequences is established. The linear complexity of bent sequences is discussed by Kumar and Scholtz [251].

Information on integer sequences can be found in the two books by Sloane [407] and Sloane and Plouffe [408]. The book by Golomb [169] is devoted to shift-register sequences. Information on sequences with lower correlation can be found in Helleseth and Kumar [200], Sarwate [381], Klapper [234], No and Kumar [331], and No [329]. For clock-controlled sequences we refer to Gollmann [163, 164], Gollmann and Chambers [165, 166], and Smeets [409, 410] for detailed references.

Sequences over rings are interesting in both theory and applications. Information on this topic is available from Kløve [242, 243, 244], Dai, Beth, and Gollmann [92].

Design and analysis of geometric sequences are carried out by Chan and Games [68], and Klapper [233]. For the existence of secure keystream generators, see Klapper [240].

Appendix A

More About Cyclotomic Numbers

The cryptographic importance of cyclotomic numbers has been seen in some of the preceding chapters. Formulae for the cyclotomic numbers of orders 2, 3, 4, 5, 6 [106]; 7 [273]; 8 [264]; 9 [14]; 10 [456]; 11 [274]; 12 [455]; 14 [316]; 15 [44]; 16 [454, 148]; 18 [14]; 20 [317]; 24 [149], are already known. Some of these cyclotomic numbers have been already introduced in Chapter 4. Due to the cryptographic importance of cyclotomic numbers, we make some notes about those which have not been introduced in Chapter 4. Formulae for some cyclotomic numbers are also given here. Others are too long to present here.

A.1 Cyclotomic Numbers of Order 7

The cyclotomic numbers of order seven, calculated by Leonard and Williams [273], can be given in terms of the solutions of certain triple of Diophantine equations, analogous to the expressions for the cyclotomic numbers of order 5 in terms of the solutions of a pair of Diophantine equations (see for example [455]). To introduce the cyclotomic numbers, we need the following result about Diophantine equations due to Leonard and Williams [272, 273]:

Proposition A.1.1 *If $p \equiv 1 \pmod 7$ then there are exactly six integral simultaneous solutions of the triple of Diophantine equations*

$$\begin{cases} 72p = 2x_1^2 + 42(x_2^2 + x_3^2 + x_4^2) + 343(x_5^2 + 3x_6^2), \\ 12x_2^2 - 12x_4^2 + 147x_5^2 - 441x_6^2 + 56x_1x_6 \\ +24x_2x_3 - 24x_2x_4 + 48x_3x_4 + 98x_5x_6 = 0, \\ 12x_3^2 - 12x_4^2 + 49x_5^2 - 147x_6^2 + 28x_1x_5 \\ +28x_1x_6 + 48x_2x_3 + 24x_2x_4 + 24x_3x_4 + 490x_5x_6 = 0, \end{cases} \tag{A.1}$$

satisfying $x_1 \equiv 1 \pmod 7$, *distinct from the two "trivial" solutions*

$$(-6t, \pm 2u, \pm 2u, \mp 2u, 0, 0),$$

where t *is given uniquely and* u *is given ambiguously by*

$$p = t^2 + 7u^2, \ t \equiv 1 \pmod 7. \tag{A.2}$$

If $(x_1, \cdots, x_6)$ *is a nontrivial solution with* $x_1 \equiv 1 \pmod 7$ *then others are given by*

$$(x_1, -x_3, x_4, x_2, (-x_5 - 3x_6)/2, (x_5 - x_6)/2)$$

and

$$(x_1, -x_4, x_2, -x_3, (-x_5 + 3x_6)/2, (-x_5 - x_6)/2).$$

Each of the other three can be obtained from one given above by changing the signs of x_2, x_3, x_4.

The following well-known relations about cyclotomic numbers

$$\begin{aligned} (h,k) &= (h+ae, k+be) \text{ for any integers } a \text{ and } b, \\ (h,k) &= (k,h) \text{ if } f \text{ is even}, \\ (h,k) &= (e-h, k-b) \end{aligned}$$

yield the following matrix [414, 273]

$$\begin{bmatrix} A & B & C & D & E & F & G \\ B & G & H & I & J & K & H \\ C & H & F & K & L & L & I \\ D & I & K & E & J & L & J \\ E & J & L & J & D & I & K \\ F & K & L & L & I & C & H \\ G & H & I & J & K & H & B \end{bmatrix} \tag{A.3}$$

in which the letter in the hth row and kth column, $h, k = 0, 1, 2, \cdots, 6$, represents the value of (h, k). Thus the 49 cyclotomic numbers of order 7 reduce to the determination of the 12 quantities $A, B, C, D, E, F, G, H, I, J, K, L$.

By making use of the Jacobi sum $J(m, n)$ and the Dickson-Hurwitz sums of order 7, Leonard and Williams got the following results about the 12 constants.

Proposition A.1.2 *Let $p \equiv 1 \pmod 7$ be a prime. If $(x_1, \cdots, x_6)$ is any nontrivial solution of (A.1) with $x_1 \equiv 1 \pmod 7$ and (t, u) is the solution of (A.2) and the sign of u is chosen to satisfy*

$$u \equiv 3x_2 + 2x_3 \pmod 7.$$

Then for some primitive root g of p the cyclotomic numbers of order 7 are given by (A.3) and

$$\begin{aligned}
49A &= p - 20 - 12t + 3x_1, \\
588B &= 12p - 72 + 24t + 168u - 6x_1 + 84x_2 - 42x_3 + 147x_4 + 147x_6, \\
588C &= 12p - 72 + 24t + 168u - 6x_1 + 84x_3 - 42x_4 - 294x_6, \\
588D &= 12p - 72 + 24t - 168u - 6x_1 + 42x_2 + 84x_4 - 147x_5 + 147x_6, \\
588E &= 12p - 72 + 24t + 168u - 6x_1 - 42x_2 - 84x_4 - 147x_5 + 147x_6, \\
588F &= 12p - 72 + 24t - 168u - 6x_1 - 84x_3 - 42x_4 - 294x_6, \\
588G &= 12p - 72 + 24t - 168u - 6x_1 - 84x_2 + 42x_3 + 147x_5 + 147x_6, \\
588H &= 12p + 12 + 24t + 8x_1 - 196x_5, \\
588I &= 12p + 12 - 60t - 84u - 6x_1 + 42x_2 + 42x_3 - 42x_4, \\
588J &= 12p + 12 + 24t + 8x_1 + 98x_5 - 294x_6, \\
588K &= 12p + 12 - 60t + 84u - 6x_1 - 42x_2 + 42x_4, \\
588L &= 12p + 12 + 24t + 8x_1 + 98x_5 + 294x_6.
\end{aligned}$$

A.2 Cyclotomic Numbers of Orders 9, 18

The cyclotomic numbers of orders nine and eighteen were determined by Baumert and Fredickson in 1967 [14]. The relations between the 81 cyclotomic constants are given by Table A.1. Thus, the 81 possible cyclotomic numbers reduce to just 19 distinct ones. Each cyclotomic number of orders 9 and 18 is expressed as a constant plus a linear combination of p, L, M, $c_0, \cdots, c_5$ where $4p = L^2 + 27M^2$, $L \equiv 7 \pmod 9$ and (β being a primitive 9th root of unity)

$$p = \left(\sum_{i=0}^{5} c_i \beta^i\right)\left(\sum_{i=0}^{5} c_i \beta^{-i}\right)$$

is a factorization of p in the field of 9th roots of unity. The formulas for cyclotomic numbers of order 9 are relatively simple. But the tables of cyclotomic numbers of order 18 are too large to present [14], which were deposited in the unpublished mathematical tables file maintained by *Mathematics of*

Table A.1: The relations between the cyclotomic numbers of order 9.

k/h	0	1	2	3	4	5	6	7	8
0	00	01	02	03	04	05	06	07	08
1	01	08	12	13	14	15	16	17	12
2	02	12	07	17	24	25	26	24	13
3	03	13	17	06	16	26	36	25	14
4	04	14	24	16	05	15	25	26	15
5	05	15	25	26	15	04	14	24	16
6	06	16	26	36	25	14	03	13	17
7	07	17	24	25	26	24	13	02	12
8	08	12	13	14	15	16	17	12	01

Computation. However, the relations between the cyclotomic constants of order 18 and some selected cyclotomic numbers can be found in the tables in [14]. The application of those cyclotomic numbers to the determination of residue difference sets has also been discussed in that paper.

A.3 Cyclotomic Numbers of Order Eleven

The basic work for evaluating the cyclotomic numbers of order 11 was laid by Dickson [106, 108]. A complete treatment of the cyclotomic numbers of order eleven was given by Leonard and Williams [273]. Let $p = 11f + 1$ be a prime with f even. Based on the basic relations among cyclotomic numbers, the 121 cyclotomic constants are reduced to 26 quantities as in the following matrix, and the relations about the 121 constants are described by

$$\begin{bmatrix} A & B & C & D & E & F & G & H & I & J & K \\ B & K & L & M & N & O & P & Q & R & S & L \\ C & L & J & S & T & U & V & W & X & T & M \\ D & M & S & I & R & X & Y & Z & Y & U & N \\ E & N & T & R & H & Q & W & Z & Z & V & O \\ F & O & U & X & Q & G & P & V & Y & W & P \\ G & P & V & Y & W & P & F & O & U & X & Q \\ H & Q & W & Z & Z & V & O & E & N & T & R \\ I & R & X & Y & Z & Y & U & N & D & M & S \\ J & S & T & U & V & W & X & T & M & C & L \\ K & L & M & N & O & P & Q & R & S & L & B \end{bmatrix}.$$

The evaluation of the cyclotomic numbers of order 11 is based on the solutions of a set of Diophantine equations. We refer to [273].

A.4 On Other Cyclotomic Numbers

Based on the Jacobi sum, Muskat carried out the cyclotomic numbers of order fourteen, and investigated their application to residue difference sets [316].

The Jacobi sums of order 15 were evaluated by Dickson and Muskat. Based on these evaluations, Bucks, Smith, Spearman, and Williams obtained the Dickson-Hurwitz sum of order 15. Then they expressed each cyclotomic number in terms of the Dickson-Hurwitz sums, and finally obtained explicit formulas for the cyclotomic numbers of order 15 using the values for the Dickson-Hurwitz sum. For details we refer to [44].

The cyclotomic numbers of order sixteen were treated by Whiteman in [454], where a table of formulas for $(i, 0)$ was given. In [148] Evans and Hill gave a complete table of the formulas for the cyclotomic numbers of order sixteen. Each number is expressed as a linear combination of parameters of quartic, octic, and bioctic Jacobi sums. Applications of these formulas were also discussed.

Complete formulas for the cyclotomic numbers of order twenty were derived by Muskat and Whiteman [317]. The application of those cyclotomic constants to residue difference sets has also been discussed in that paper. The cyclotomic numbers of order 24 were calculated by Evans [149, 150]. According to [149], there are 48 tables, and each of the 48 tables contains 109 formulas .

A.5 Behind Cyclotomic Numbers

It is interesting to note that cyclotomic formulas have the same form. Behind this uniformity of known cyclotomic numbers is the Riemann Hypothesis for Curves over Finite Fields, which can be described as follows

Theorem A.5.1 *Suppose that $F(x, y)$ is a polynomial of total degree d, with coefficients in $GF(q)$ and with N zeros $(x, y) \in GF(q) \times GF(q)$. Suppose that $F(x, y)$ is absolutely irreducible, i.e., irreducible not only over $GF(q)$, but also over every algebraic extension thereof. Then*

$$|N - q| \leq 2g\sqrt{q} + c_1(d),$$

where g is the genus of the curve $F(x, y) = 0$ and $c_1(d)$ is a constant depending on d.

This theorem, proven by Weil [449, 448, 382], indicates the uniformity of the form of cyclotomic formulas . By the uniformity we mean that $|(i,j) - p/d^2|$ is of order $O(\sqrt{p})$. It can be proved that $g \leq (d-1)(d-2)/2$, hence

$$|N - q| \leq (d-1)(d-2)\sqrt{q} + c_1(d).$$

Weil's proof is far from elementary and uses deep techniques in algebraic geometry. An elementary proof was given by Stepanov in 1969. A complete elementary proof of the Theorem A.5.1 and a detailed account of the historical development is presented by Schmidt in [382].

As far as the above Theorem A.5.1 is concerned, we are much interested in the case $F(x,y) = y^d - f(x)$. Applying Theorem A.5.1 for such a curve may give a rough estimation on the distribution of kth power residues and nonresidues, and thus on the pattern distributions of some cyclotomic sequences. In order to be able to apply Theorem A.5.1 for such a curve, we have to know when $y^d - f(x)$ is absolutely irreducible. The following two results whose proofs can be found in [382] give some conditions which ensure the irreducibility.

Proposition A.5.2 *Suppose that the polynomial $y^d - f(x)$ has coefficients in a field K. Then the following three conditions are equivalent:*

1. *$y^d - f(x)$ is absolutely irreducible;*
2. *$y^d - cf(x)$ is absolutely irreducible for every $0 \neq c \in K$;*
3. *if $f(x) = c_1(x - x_1)^{d_1} \cdots (x - x_s)^{d_s}$ is the factorization of f in $\overline{K}$ with $x_i \neq x_j (i \neq j)$, where $\overline{K}$ is the algebraic closure of K, then* $\gcd(d, d_1, ..., d_s) = 1$.

A very practical result is the following.

Proposition A.5.3 *Suppose that deg(f) $= m$. Then $y^d - f(x)$ is absolutely irreducible if* $\gcd(m, d) = 1$.

Let g be the primitive root of p, and $\alpha = g^f$. Then

$$y^d = x^d - 1 = \prod_{i=0}^{d-1} (x - \alpha^i).$$

By Proposition A.5.2, $x^d - 1 = y^d$ is absolutely irreducible. Note that the genus of the curve $x^d - 1 = y^d$ is $(d-1)(d-2)/2$. Then by Theorem A.5.1

$$|N_d - p| \leq (d-1)(d-2)\sqrt{p} + c_1(d),$$

where N_d is the number of solutions of the equation $x^d - 1 = y^d$ over Z_p. Note that $x^d = 1$ has d solutions over Z_p. $y^d = x^d - 1$ has d solutions $(0, y)$ and d solutions $(x, 0)$. By definition it is easily seen that

$$(0,0) = (N_d - 2d)/d^2.$$

Hence

$$\left|(0,0) + \frac{2}{d} - \frac{p}{d^2}\right| \leq \frac{(d-1)(d-2)}{d^2}\sqrt{p} + \frac{c_1(d)}{d^2}.$$

This gives a lower and upper bound on $(0, 0)$. Similar bounds on other cyclotomic numbers may be established.

Theorem A.5.1 is interesting not only in cryptography, but also in coding theory. It could be interesting in any field where the number of solutions of a set of equations is concerned. Finally, we mention that Stepanov's elementary proof of Theorem A.5.1 is further simplified by Bombieri who gives a complete proof in five pages based on the Riemann-Roch theorem [27].

It is also interesting to think of the possibility of using Theorem A.5.1 to estimate or to control the number of fixed points of the cryptographic encryption (resp. decryption) transformation of some block ciphers since fixed points are solutions of equations. Other cryptographic measures on cipher systems which are based on the number of solutions of equations may also be controlled by this theorem. For example, the nonlinearity of round functions for iterated block ciphers is also related to the number of solutions of some equations. The same conclusion holds for any kind of correlation measures. Theorem A.5.1 is fairly general and thus better bounds are possible to develop.

Appendix B

Cyclotomic Formulae of Orders 6, 8 and 10

This appendix presents cyclotomic formulas of orders 6, 8 and 10. For explanations and meanings of these formulas , see Chapter 7.

Table B.1: The cyclotomic numbers of order 6 for even f.

	$m \equiv 0 \pmod 3$	$m \equiv 1 \pmod 3$	$m \equiv 2 \pmod 3$
36(0,0)	$p-17-20A$	$p-17-8A+6B$	$p-17-8A-6B$
36(0,1)	$p-5+4A+18B$	$p-5+4A+12B$	$p-5+4A+6B$
36(0,2)	$p-5+4A+6B$	$p-5+4A-6B$	$p-5-8A$
36(0,3)	$p-5+4A$	$p-5+4A-6B$	$p-5+4A+6B$
36(0,4)	$p-5+4A-6B$	$p-5-8A$	$p-5+4A+6B$
36(0,5)	$p-5+4A-18B$	$p-5+4A-6B$	$p-5+4A-12B$
36(1,2)	$p+1-2A$	$p+1-2A-6B$	$p+1-2A+6B$
36(1,3)	$p+1-2A$	$p+1-2A-6B$	$p+1-2A-12B$
36(1,4)	$p+1-2A$	$p+1-2A+12B$	$p+1-2A+6B$
36(2,4)	$p+1-2A$	$p+1+10A+6B$	$p+1+10A-6B$

Table B.2: The cyclotomic numbers of order 6 for odd f.

	$m \equiv 0 \pmod 3$	$m \equiv 1 \pmod 3$	$m \equiv 2 \pmod 3$
36(0,0)	$p-11-8A$	$p-11-2A$	$p-11-2A$
36(0,1)	$p+1-2A+12B$	$p+1+4A$	$p+1-2A-12B$
36(0,2)	$p+1-2A+12B$	$p+1-2A+12B$	$p+1-8A+12B$
36(0,3)	$p+1+16A$	$p+1+10A-12B$	$p+1+10A+12B$
36(0,4)	$p+1-2A-12B$	$p+1-8A-12B$	$p+1-2A-12B$
36(0,5)	$p+1-2A-12B$	$p+1-2A+12B$	$p+1+4A$
36(1,0)	$p-5+4A+6B$	$p-5-2A+6B$	$p-5+4A+6B$
36(1,1)	$p-5+4A-6B$	$p-5+4A-6B$	$p-5-2A-6B$
36(1,2)	$p+1-2A$	$p+1+4A$	$p+1+4A$
36(2,1)	$p+1-2A$	$p+1-8A-12B$	$p+1-8A+12B$

Table B.3: The cyclotomic numbers of order 8 in subcase I.

	If 2 is a quartic residue	If 2 is not a quartic residue
64(0,0)	$p-23-18x-24a$	$p-23+6x$
64(0,1)	$p-7+2x+4a+16y+16b$	$p-7+2x+4a$
64(0,2)	$p-7+6x+16y$	$p-7-2x-8a-16y$
64(0,3)	$p-7+2x+4a-16y+16b$	$p-7+2x+4a$
64(0,4)	$p-7-2x+8a$	$p-7-10x$
64(0,5)	$p-7+2x+4a+16y-16b$	$p-7+2x+4a$
64(0,6)	$p-7+6x-16y$	$p-7-2x-8a+16y$
64(0,7)	$p-7+2x+4a-16y-16b$	$p-7+2x+4a$
64(1,2)	$p+1+2x-4a$	$p+1-6x+4a$
64(1,3)	$p+1-6x+4a$	$p+1+2x-4a-16b$
64(1,4)	$p+1+2x-4a$	$p+1+2x-4a+16y$
64(1,5)	$p+1+2x-4a$	$p+1+2x-4a-16y$
64(1,6)	$p+1-6x+4a$	$p+1+2x-4a+16b$
64(2,4)	$p+1-2x$	$p+1+6x+8a$
64(2,5)	$p+1+2x-4a$	$p+1-6x+4a$

Table B.4: The cyclotomic numbers of order 8 in subcase II.

	If 2 is a quartic residue	If 2 is not a quartic residue
64(0,0)	$p-15-2x$	$p-15-10x-8a$
64(0,1)	$p+1+2x-4a-16y$	$p+1+2x-4a-16b$
64(0,2)	$p+1+6x+8a-16y$	$p+1-2x+16y$
64(0,3)	$p+1+2x-4a-16y$	$p+1+2x-4a-16b$
64(0,4)	$p+1-18x$	$p+1+6x+24a$
64(0,5)	$p+1+2x-4a+16y$	$p+1+2x-4a+16b$
64(0,6)	$p+1+6x+8a+16y$	$p+1-2x-16y$
64(0,7)	$p+1+2x-4a-16y$	$p+1+2x-4a+16b$
64(1,0)	$p-7+2x+4a$	$p-7+2x+4a+16y$
64(1,1)	$p-7+2x+4a$	$p-7+2x+4a-16y$
64(1,2)	$p+1-6x+4a+16b$	$p+1+2x-4a$
64(1,3)	$p+1+2x-4a$	$p+1-6x+4a$
64(1,7)	$p+1-6x+4a\quad 16b$	$p+1+2x-4a$
64(2,0)	$p-7-2x-8a$	$p-7+6x$
64(2,1)	$p+1+2x-4a$	$p+1\quad 6x+4a$

Table B.5: The first set of cyclotomic numbers of order 10 for even f.

100(0,0)	$p - 29 + 18x$
100(0,1)	$p - 9 - 2x + 25u + 50v - 25w$
100(0,2)	$p - 9 - 2x + 25v - 25w$
100(0,3)	$p - 9 - 2x - 50u + 25v + 25w$
100(0,4)	$p - 9 - 2x + 25u + 25w$
100(0,5)	$p - 9 - 2x$
100(0,6)	$p - 9 - 2x - 25u + 25w$
100(0,7)	$p - 9 - 2x + 50u - 25v + 25w$
100(0,8)	$p - 9 - 2x - 25v - 25w$
100(0,9)	$p - 9 - 2x - 25u - 50v - 25w$
200(1,2)	$2p + 2 + x + 25w$
200(1,3)	$2p + 2 + x + 75w$
200(1,4)	$2p + 2 + x - 75w$
200(1,5)	$2p + 2 + x + 25w$
200(1,6)	$2p + 2 + x + 25w$
200(1,7)	$2p + 2 + x - 75w$
200(1,8)	$2p + 2 + x + 75w$
200(2,4)	$2p + 2 + x - 25w$
200(2,5)	$2p + 2 + x - 25w$
200(2,6)	$2p + 2 + x + 25w$
200(2,7)	$2p + 2 + x - 25w$
200(3,6)	$2p + 2 + x - 25w$

Table B.6: The second set of cyclotomic numbers of order 10 for even f.

400(0,0)	$4p - 116 - 3x - 150u + 75w$
100(0,1)	$p - 9 - 2x + 50w$
400(0,2)	$4p - 36 + 17x + 50u - 25w$
200(0,3)	$2p - 18 - 4x + 25u - 25v + 25w$
200(0,4)	$2p - 18 - 4x + 25u - 25v + 25w$
400(0,5)	$4p - 36 + 17x + 50u - 25w$
100(0,6)	$p - 9 - 2x - 50w$
400(0,7)	$4p - 36 + 17x + 50u - 25w$
200(0,8)	$2p - 18 - 4x - 75u + 75v - 75w$
200(0,9)	$2p - 18 - 4x + 25u - 25v + 25w$
200(1,2)	$2p + 2 + x + 25u + 25v - 50w$
200(1,3)	$2p + 2 + x - 50v - 75w$
200(1,4)	$2p + 2 + x - 25u - 25v$
200(1,5)	$2p + 2 + x + 50v + 25w$
200(1,6)	$2p + 2 + x - 25u - 25v$
200(1,7)	$2p + 2 + x + 25u + 25v - 50w$
200(1,8)	$2p + 2 + x - 50u + 75w$
200(2,4)	$2p + 2 + x + 50u + 75w$
400(2,5)	$4p + 4 - 23x + 50u - 25w$
200(2,6)	$2p + 2 + x - 25u - 25v$
200(2,7)	$2p + 2 + x - 25u - 25v$
200(3,6)	$2p + 2 + x + 50v + 25w$

Table B.7: The first set of cyclotomic numbers of order 10 for odd f.

100(0,0)	$p - 19 + 8x$
200(0,1)	$2p + 2 + x + 50u + 50v - 25w$
200(0,2)	$2p + 2 + x - 50u + 50v - 75w$
200(0,3)	$2p + 2 + x - 50u + 50v + 25w$
200(0,4)	$2p + 2 + x + 50u + 50v + 75w$
100(0,5)	$p + 1 - 12x$
200(0,6)	$2p + 2 + x - 50u - 50v + 75w$
200(0,7)	$2p + 2 + x + 50u - 50v + 25w$
200(0,8)	$2p + 2 + x + 50u - 50v - 75w$
200(0,9)	$2p + 2 + x - 50u - 50v - 25w$
100(1,0)	$p - 9 - 2x + 25v$
100(1,1)	$p - 9 - 2x - 25v$
200(1,2)	$2p + 2 + x + 25w$
200(1,3)	$2p + 2 + x - 25w$
200(1,4)	$2p + 2 + x - 75w$
200(1,8)	$2p + 2 + x - 25w$
200(1,9)	$2p + 2 + x + 25w$
100(2,0)	$p - 9 - 2x + 25u$
200(2,1)	$2p + 2 + x - 75w$
100(2,2)	$p - 9 - 2x - 25u$
200(2,3)	$2p + 2 + x + 75w$
200(3,1)	$2p + 2 + x + 75w$

Table B.8: The second set of cyclotomic numbers of order 10 for odd f.

400(0,0)	$4p - 76 - 7x + 50u + 25w$
200(0,1)	$2p + 2 + x + 50v + 125w$
400(0,2)	$4p + 4 - 23x + 50u - 25w$
200(0,3)	$2p + 2 + x + 25u - 75v + 50w$
200(0,4)	$2p + 2 + x - 25u - 25v$
400(0,5)	$4p + 4 + 27x + 150u - 75w$
200(0,6)	$2p + 2 + x + 50v - 75w$
400(0,7)	$4p + 4 - 23x + 50u - 25w$
200(0,8)	$2p + 2 + x - 75u + 25v - 50w$
200(0,9)	$2p + 2 + x - 25u - 25v$
100(1,0)	$p - 9 - 2x$
200(1,1)	$2p - 18 - 4x + 25u - 25v + 25w$
200(1,2)	$2p + 2 + x - 25u - 25v$
200(1,3)	$2p + 2 + x + 50u - 25w$
200(1,4)	$2p + 2 + x - 25u - 25v$
200(1,8)	$2p + 2 + x + 50v + 125w$
200(1,9)	$2p + 2 + x + 25u + 25v - 50w$
400(2,0)	$4p - 36 + 17x + 50u - 25w$
200(2,1)	$2p + 2 + x + 25u + 25v - 50w$
200(2,2)	$2p - 18 - 4x - 25u + 25v - 25w$
200(2,3)	$2p + 2 + x - 50u - 25w$
200(3,1)	$2p + 2 + x - 50v + 25w$

Appendix C

Finding Practical Primes

For the design of some keystream generators, we may need primes of certain size. A good source for such primes is the book "Factorizations of $b^n \pm 1 : b = 2, 3, 5, 6, 7, 10, 11, 12$ up to Higher Powers" by J. Brillhart, D. H. Lehmer, J. L. Selfridge, Bryant Tuckerman, and S. S. Wagstaff, Jr [39], where many primes with more than 25 but less than 300 decimal digits are collected. Those primes are factors of $b^n \pm 1$ for $b = 2, 3, 5, 6, 7, 10, 11, 12$.

To design binary sequence generators, we may need primes p of certain size such that $\mathrm{ord}_p(2)$ is large enough. To this end, the following result is useful. Its proof is left as an exercise.

Proposition C.0.4 *If p is a prime factor of $2^n - 1$, then $\mathrm{ord}_p(2)$ must divide n. If p is a prime factor of $2^n + 1$, then $\mathrm{ord}_p(2)$ must divide $2n$.*

Hence, for all prime factors p of $2^n \pm 1$, we have $\mathrm{ord}_p(2) \leq 2n$. On the other hand, the factorization of $2^n \pm 1$ is generally very hard for $n \geq 4000$. Hence, almost all prime factors of $2^n \pm 1$ listed in the above book are not interesting from this point of view.

However, for prime factors of $b^n \pm 1$, where $b = 3, 5, 6, 7, 11$, the order of 2 modulo them could be very large. For the case $b = 3$, we have the following numerical examples:

Prime p	Order $\mathrm{ord}_p(2)$	$b^n \pm 1$
96656723	$p - 1$	$3^{47} - 1$
20381027	$p - 1$	$3^{29} - 1$
1001523179	$p - 1$	$3^{23} - 1$
4404047	$(p - 1)/2$	$3^{31} - 1$
21523361	$(p - 1)/8$	$3^{16} + 1$
42521761	$(p - 1)/40$	$3^{20} + 1$
22996651	$p - 1$	$3^{25} + 1.$

Further computation shows that for prime factors of $3^n \pm 1$, the probability of having a large $\text{ord}_p(2)$ is rather high.

For the case $b = 5$, we have the following numerical examples:

Prime p	Order $\text{ord}_p(2)$	$b^n \pm 1$
12207031	$(p-1)/10$	$5^{11} - 1$
305175781	$(p-1)/6$	$5^{13} - 1$
3981071	$(p-1)/10$	$5^{19} - 1$
41540861	$p-1$	$5^{17} + 1$
632133361	$(p-1)/2$	$5^{20} + 1$
38923	$p-1$	$5^{13} + 1.$

Further computation shows that for prime factors of $5^n \pm 1$, the probability of having a large $\text{ord}_p(2)$ is also rather high.

For the case $b = 7$, we have the following numerical examples:

Prime p	Order $\text{ord}_p(2)$	$b^n \pm 1$
293459	$p-1$	$7^{11} - 1$
2583253	$p-1$	$7^{27} - 1$
12323587	$(p-1)/11$	$7^{33} - 1$
10746341	$(p-1)/11$	$7^{11} + 1$
228511817	$(p-1)/2$	$7^{13} + 1$
59361349	$p-1$	$7^{34} + 1.$

Further computation shows that for prime factors of $7^n \pm 1$, the probability of having a large $\text{ord}_p(2)$ is also rather high.

For the case $b = 10$, we have the following numerical examples:

Prime p	Order $\text{ord}_p(2)$	$b^n \pm 1$
99990001	$(p-1)/2$	$10^{24} - 1, 10^{12+1}$
1058313049	$(p-1)/4$	$10^{26} - 1$
121499449	$(p-1)/2$	$10^{14} + 1.$

The above numerical examples show that to find primes of certain size such that $\text{ord}_p(2)$ is large enough, one can check prime factors of $b^n \pm 1$ for $b = 3, 5, 7, 10, 11$ listed in the above book.

Similarly, for all prime factors of $3^n \pm 1$, we have $\text{ord}_p(3) \leq 2n$. Thus, to find primes of certain size such that $\text{ord}_p(3)$ is large enough, one can check the prime factors of $b^n \pm 1$ for $b = 2, 5, 7, 10, 11$ listed in that book.

Appendix D

List of Research Problems

When setting up bridges between number theory and stream ciphers, we have proposed a number of research problems from the viewpoint of designing some sequences. Those problems might be interesting to some number theorists, cryptologists and computer scientists, so we present here a table of those research problems in order to make them easy to find. In addition, there are also quite a number of research problems which have been implied in our discussions, but have not been stated out.

1. Research Problem 2.3.5, about the relation between linear and quadratic span.
2. Research Problem 3.3.4, about $\text{ord}_{p^k}(a) = \text{ord}_p(a)$.
3. Research Problem 5.2.4, about primes of form $4p + 1$.
4. Research Problem 5.2.5, about primes of form $8p + 1$.
5. Research Problem 5.2.6, about primes of form $16p + 1$.
6. Research Problem 5.2.7, about primes of form $32p + 1$.
7. Research Problem 5.3.1, about the order of q modulo primes of the form $N = k2^n + 1$.
8. Research Problem 5.3.2, about the primitivity of 3 modulo primes of the form $N = k2^n + 1$.
9. Research Problem 5.4.1, about the primitive roots of Mersenne primes.
10. Research Problem 5.4.3, about the primitivity of 2 modulo some special primes.

11. Research Problem 5.4.4, about the primitivity and order 2 modulo some special primes.

12. Research Problem 5.4.6, about the order of an integer modulo $R317$ and $R1031$.

13. Research Problem 5.5.1, about the primality of two integers.

14. Research Problem 5.5.2, about the primality of two integers.

15. Research Problem 5.6.5, about twin primes.

16. Research Problem 5.6.6, about twin primes.

17. Research Problem 5.6.7, about twin primes.

18. Research Problem 5.6.8, about the primitivity and order of 2 modulo some twin primes.

19. Research Problem 5.7.2, about twin primes.

20. Research Problem 6.3.9, about difference sets.

21. Research Problem 6.3.14, about difference sets.

22. Research Problem 6.5.5, about perfect nonlinear functions.

23. Research Problem 7.3.2, about the linear complexity of ADSC sequences.

24. Research Problem 7.4.1, about Menon difference sets.

25. Research Problem 7.4.2, about the existence of Hadamard matrices.

26. Research Problem 8.1.1, about Stern primes.

27. Research Problem 8.1.2, about almost difference sets.

28. Research Problem 11.6.9, about permutations.

29. Research Problem 12.7.7, about large primes.

30. Research Problem 12.7.8, about algorithms for quadratic partition.

31. Research Problem 13.2.8, about generators.

32. Research Problem 14.6.4, about the relation between linear and 2-adic span of binary sequences.

33. Research Problem 14.6.5, about special generators.
34. Research Problem 9.4.10, about distributions of quadratic residues and nonresidues.

Appendix E

Exercises

Chapter 2

1. Prove Proposition 2.3.2.

2. Prove Proposition 2.3.3.

3. Show that the variant of the CFB mode for block ciphers described in the text gives a self-synchronous stream cipher.

Chapter 3

1. Prove Proposition 3.1.2.

2. Prove Proposition 3.2.1.

3. Derive the formula

$$a(x) = (2f + 1)(3 + (u - 1)x - (n + 2)x^2)$$

given in Section 4.3.2 for the polynomial $a(x)$ in the case $d = 3$.

4. Suppose $GF(q)$ is a finite field, r is an odd prime and q is a primitive root of r such that r^2 does not divide q^{r-1}. Prove that if $Q_r(x)$ is a cyclotomic polynomial, then $Q_r(x^r)$ is irreducible over $GF(q)$.

Chapter 4

1. Prove Properties (A)—(F) described in Section 4.1.

2. Prove Proposition 4.3.4.

3. Complete the proof of Proposition 4.3.6.

4. Complete the proof of Proposition 4.3.7.

5. Prove Properties (A)—(E) described in Section 4.4.1.

6. Prove Proposition 4.4.3.

7. Complete the proof of Proposition 4.4.7.

8. Calculate the number of solutions (x, y) of the equation $x^2 = y^6 + 1$ over Z_p^2, where $p \equiv 1 \pmod 6$, with the help of cyclotomic numbers of order 6. (Hint: Consider cyclotomic classes and numbers of order both 2 and 6, then find their relations and the relation among cyclotomic numbers of orders 2 and 6.)

Chapter 5

1. Prove Theorem 5.1.1.

2. Prove Proposition 5.2.1.

3. Prove Proposition 5.2.2.

4. Prove Proposition 5.2.3.

Chapter 6

1. Prove Theorem 6.3.10.

2. Prove Theorem 6.3.11.

3. Prove Theorem 6.3.13.

4. Prove Theorem 6.3.15.

5. Prove Theorem 6.3.17.

6. Prove Theorem 6.3.18.

Chapter 7

1. Let $K = GF(q)$, and let $F = GF(q^k)$ be an extension of K. Assume that $\theta \neq 0$ and α are elements of F. Define a sequence s^∞ by

$$s_n = \mathrm{Tr}_{F/K}(\theta\alpha^n), \quad n \geq 0. \tag{E.1}$$

(a) Prove that the least period of the sequence s^∞ is equal to the multiplicative order of α.

(b) Prove that the linear complexity of the sequence s^∞ is equal to the degree of the minimal polynomial of α over K.

2. If α is a generating element of F, the sequence s^∞ of (E.1) is called a maximum-length sequence (m-sequence for short). By the foregoing problem, this sequence has least period $q^k - 1$. For the case $q = 2$, calculate the autocorrelation values of this sequence.

Chapter 8

1. Prove Lemma 8.2.1.

2. Prove Lemma 8.2.6.

3. Attack Problem 8.3.1.

4. Attack Problem 8.3.2.

5. Suppose α is a primitive nth root of unity and $\alpha \in GF(2^m)$, where m is the order of 2 modulo n. Let

$$f(x) = \prod_{i \in S} (x - \alpha^i),$$

where S is a subset of $\{0, 1, \cdots, n-1\}$. Prove that $f(x)$ has all coefficients in $\{0, 1\}$ if and only if $k \in S$ implies $2k \bmod n \in S$.

Chapter 9

1. Prove the two Jacobsthal formulas described in Section 9.4.

2. Use the two Jacobsthal formulas to prove Proposition 9.4.1.

3. Given an odd prime p, prove that the Legendre symbol formula

$$\sum_{h=0}^{p-1} \left(\frac{h}{p}\right)\left(\frac{h+k}{p}\right) = -1 \tag{E.2}$$

holds for each $k = 1, 2, \cdots, p-1$.

4. Let the Legendre sequence s^∞ for an odd prime p be defined by

$$\begin{aligned} s_i &= \left(\frac{i}{p}\right)' \\ &= \begin{cases} 1 & \text{if } p|i, \\ \left(\frac{i}{p}\right) & \text{otherwise.} \end{cases} \end{aligned}$$

Prove that if $p \equiv 3 \pmod 4$ the autocorrelation function given by

$$A(k) = \sum_{i=0}^{p-1} s_i s_{i+k} \quad (0 \le k \le p-1) \tag{E.3}$$

satisfies

$$A(k) = \begin{cases} p & \text{if } k = 0, \\ -1 & \text{if } 1 \le k \le p-1. \end{cases}$$

Chapter 10

1. Let θ be a pth root of unity over $GF(r^m)$, where p and r are primes such that $p = rt + 1$. Let

$$S(x) = \sum_{i=1}^{r-1} i \sum_{u \in D_i} x^u \tag{E.4}$$

be defined in $GF(r)[x]$, where $D_0, D_1, \cdots, D_{r-1}$ are the cyclotomic classes of order r. Prove that $(S(\theta))^r = S(\theta^r)$.

2. For two different primes $p_i = rt_i + 1$, where r is also prime, calculate the linear complexity of the sum sequence of the output sequences of the two rth-order cyclotomic generators defined by the two primes.

Chapter 11

1. Prove Proposition 11.4.4.

2. Prove Proposition 11.4.5.

3. Prove Lemma 11.4.16.

4. Prove Proposition 11.5.1.

Chapter 12

1. Prove Proposition 12.2.1.
2. Prove Proposition 12.2.4.
3. Prove Proposition 12.2.5.
4. Prove Proposition 12.3.1.
5. Prove Proposition 12.5.1.

Chapter 13

1. Prove Proposition 13.1.1.
2. Prove Proposition 13.1.2.
3. Prove Proposition 13.2.1.
4. Prove Proposition 13.2.2.
5. Prove Corollary 13.2.3.
6. Prove Proposition 13.2.4.
7. Prove Proposition 13.2.5.

Chapter 14

1. Prove Lemma 14.1.4.
2. Prove Proposition 14.1.5.
3. Prove Corollary 14.1.9.
4. Prove Lemma 14.6.6.
5. Prove Corollary 14.8.1.

Appendix F

List of Mathematical Symbols

AAC(l)	Aperiodic autocorrelation function, see Section 2.3.3.
ACC(l)	Aperiodic crosscorrelation function, see Section 2.3.3.
AC(l)	Periodic autocorrelation function, see Section 2.3.3.
$a\|b$	a divides b.
CC(l)	Periodic crosscorrelation function, see Section 2.3.3.
$\mathrm{C}_f(.)$	Autocorrelation function of f, same as $\mathrm{AC}_f(.)$, see Chapter 5.
$\mathrm{C}_s(.)$	Autocorrelation function of s, same as $\mathrm{AC}_s(.)$, see Chapter 5.
$d_f(i,j,w)$	Difference parameters of f, see Section 2.4.
gcd	Greatest common divisor.
$GF(q)$	Galois field with q elements.
$I(A,B)$	Amount of mutual information between events A and B.
lcm	Least common multiple.
L(S)	Linear complexity or linear span, see Section 2.3.1.
$(m,n)_d$	Cyclotomic number of order d, see Chapter 7.
$\mathrm{nord}_n(a)$	Negative order of a modulo n, see Section 3.2.
$N(x)$	Norm of x.
$\mathrm{ord}_n(a)$	Multiplicative order of a modulo n, see Sections 3.1 and 3.2.
$Q_{[p]}$	Field of p-adic numbers, see Section 14.4.
Q	Rational number field (sometimes not).
$Q_n(x)$	Cyclotomic polynomial, see Section 3.1.

$\mathrm{SC}_k(S)$	Sphere complexity, see Section 2.3.4.
$\mathrm{Tr}(x)$	Trace of x.
$\mathrm{WC}_k(S)$	Weight complexity, see Section 2.3.4.
Z_n	Residue class ring modulo n.
$Z_{[p]}$	Ring of p-adic integers, see Section 14.4.
Z	Ring of rational integers.
$\phi(x)$	Euler function (sometimes a mapping).
$\lambda(x)$	Lambda or Carmichael function, see Section 3.2.
$\lambda_2(S)$	2-adic span of S, see Section 14.6.
$\Lambda_2(p, q)$	Software complexity of p/q, see Section 16.8.
$(\frac{n}{m})$	Legendre, Jacobi, Kronecker symbols, see Sections 5.1, 14.8 resp.

Bibliography

[1] L. M. Adleman, D. R. Estes, and K. S. McCurley, Solving bivariate quadratic congruences in random polynomial time, Math. Comput. 17 (1987), 17–28.

[2] A. S. Ambrosimov, Properties of bent functions of q-valued logic over finite fields, Discrete Math. Appl. 4(4) (1994), 341–350.

[3] W. Alexi, B. Chor, O. Goldreich and C. P. Schnorr, RSA and Rabin functions: Certain parts are as hard as the whole, SIAM J. Comput. 17 (1988), 194–209.

[4] R. J. Anderson, Solving a class of stream ciphers, Cryptologia 14 (1990), 285–288.

[5] R. J. Anderson, Fast attacks on certain stream ciphers, Electronics Letters 29 (1993), 1322–1323.

[6] T. M. Apostol, Introduction to Analytic Number Theory, Springer Verlag, 1976.

[7] K. T. Arasu, Recent results on difference sets, in: Coding Theory and Design Theory, Part II, D. Ray-Chaudhuri ed., Springer Verlag, 1990, 1–23.

[8] K. T. Arasu, J. A. Jedwab and S. Sehgal, New constructions of Menon difference sets, J. Comb. Theory A 64 (1993), 329–336.

[9] E. Bach, Comments on search procedures for primitive roots, Math. Comput. 66 (1997), 1719–1727.

[10] R. Baillie, New primes of the form $k \times 2^n + 1$, Math. Comput. 33 (1979), 1333–1336.

[11] R. Balasubramanian, J.-M. Deshouillers, F. Dress, Problem de Waring pour les bicarrés, C.R.A.S. 303 (1986), 85–86 and 161–163.

[12] R. H. Barker, Group synchronizing of binary digital systems, in: Communication Theory, W. Jackson, ed., Butterworths, London, 1953, 273–287.

[13] J. M. Barrows, Jr, A new method for constructing multiple error correcting linear residue codes, Rep. R-277, Coordinated Sci. Lab., University of Illinois, Urban, 1966.

[14] L. D. Baumert and H. Fredricksen, The cyclotomic number of order eighteen with applications to difference sets, Math. Comput. 21 (1967), 204–219.

[15] L. D. Baumert, Cyclic Difference Sets, Lecture Notes in Mathematics 182, Springer Verlag, 1971.

[16] H. Beker and F. Piper, Cipher Systems: The protection of communications, Northwood Books, London, 1982.

[17] J. W. Bergquist, Difference sets and congruences modulo a product of primes, Dissertation, University of Southern California, 1963.

[18] E. R. Berlekamp, Algebraic Coding Theory, McGraw-Hill, New York, 1968.

[19] T. Beth and C. Ding, On almost perfect nonlinear permutations, in: Advances in Cryptology - Eurocrypt'93, T. Helleseth ed., LNCS 765 (1993), Springer Verlag, 65–76.

[20] T. Beth, D. Jungnickel and H. Lenz, Design Theory, Mannheim 1985, Cambridge 1986.

[21] T. Beth and F. C. Piper, The stop-and-go generator, in: Advances in Cryptology - Eurocrypt'84, T. Beth, N. Cot and I. Ingemarsson, eds., LNCS 209 (1984), Springer Verlag, 88–92.

[22] E. Biham and A. Shamir, Differential cryptanalysis of DES-like cryptosystems, in: Advances in Cryptology - Crypto' 90, A. J. Menezes and S. A. Vanstone, eds., LNCS 537 (1991), Springer Verlag, 2–21.

[23] E. Biham and A. Shamir, Differential Cryptanalysis of the Data Encryption Standard, Springer, New York, 1993.

[24] R. E. Blahut, Fast Algorithms for Digital Signal Processing, Addison-Wesley Publishing Company, 1985.

[25] L. Blum and S. Micali, How to generate cryptographically strong sequences of pseudorandom bits, SIAM J. Comput. 13 (1984), 850–864.

[26] L. Blum, M. Blum and M. Shub, A simple unpredictable pseudorandom number generator, SIAM J. Comput. 15 (1986), 364–383.

[27] E. Bombieri, Counting points on curves over finite fields (d'après S.A. Stepanov), Sem. Bourbaki, Vol. 1972–73, Exposé 430, Lecture Notes in Mathematics 383 (1974), Springer Verlag, New York, 234–241.

[28] Z. I. Borevich and I. R. Shafarevich, Number Theory, Academic Press, New York, 1966.

[29] A. Borning, Some results for $k!+1$ and $2\cdot 3\cdot 5\cdots p+1$, Math. Comput. 26 (1972), 567–570.

[30] J. Boyar, Inferring sequences produced by pseudorandom number generators, J. ACM 36 (1989), 129–141.

[31] A. Brauer, On a class of Hadamard determinants, Math. Z. 58 (1953), 219–225.

[32] R. P. Brent, Irregularities in the distribution of primes and twin primes, Math. Compt. 29 (1975), 43–56.

[33] R. P. Brent, Tables concerning irregularities in the distribution of primes and twin primes to 10^{11}, Math. Comput. 30 (1976), 379.

[34] E. F. Brickell, A. M. Odlyzko, Cryptanalysis, In: Contemporary Cryptography: The Science of Information Integrity, G.J. Simmons, ed., IEEE Press, 1992.

[35] W. E. Briggs, An elementary proof of a theory about the representation of primes by quadratic forms, Canadian J. Math. 6 (1954), 353–363.

[36] J. Brillhart, J. Tonascia, P.J. Weiberger, On the Fermat quotient , in. Computers in Number Theory, A. O. L. Atkin and B.J. Birch, eds., Academic Press, New York, 1971, 213–222.

[37] J. Brillhart, Note on representing a prime as a sum of two squares, Math. Comput. 26 (1972), 1011–1013.

[38] J. Brillhart, D. H. Lehmer and J. L. Sefridge, New primality criteria and factorizations of $2^m \pm 1$, Math. Comput. 26 (1972), 567–570.

[39] J. Brillhart, D. H. Lehmer, J. L. Sefridge, B. Tuckerman and S. S. Wagstaff, Jr, Factorizations of $b^n \pm 1$: b=2, 3, 5, 6, 7, 10, 11, 12 up to high powers, Contemporary Mathematics 22, Second edition, Amer. Math. Soc., Providence, 1988.

[40] B. W. Brock, Hermitian congruence and the existence and completion of generalized Hadamard matrices, J. Combin. Theory A 49 (1988), 233–261.

[41] B. W. Brock, A new construction of circulant GH(p^2; $\mathbf{Z}_p$), Discrete Math. 112 (1993), 249–252.

[42] R. H. Bruck, Difference sets in a finite group, Trans. Amer. Math. Soc. 78 (1955), 464–481.

[43] R. H. Bruck, Computational aspects of certain combinatorial problems, Proceedings of Symposia in Applied Mathematics 6, McGraw-Hill, New York, 1956, 31–43.

[44] N. Buck, L. Smith, B. K. Spearman and K. S. Williams, The cyclotomic numbers of order fifteen, Math. Comput. 48 (1987), 67–83.

[45] D. A. Buell, Binary Quadratic Forms, Springer Verlag, New York, 1989.

[46] J. P. Buhler, R. E. Crandall and M. A. Penk, Primes of the form $n! \pm 1$ and $2 \cdot 3 \cdot 5 \cdots p \pm 1$, Math. Comput. 38 (1982), 639–643.

[47] D. A. Burgess, On character sums and primitive roots, Proc. London Math. Soc. (3) 12 (1962), 179–192.

[48] D. A. Burgess and P. D. T. A. Elliot, The average of the least primitive root, Mathematika 15 (1968), 39–50.

[49] C. K. Caldwell, On the primality of $n! \pm 1$ and $2 \cdot 3 \cdot 5 \cdots p \pm 1$, Math. Comput. 64 (1995), 889–890.

[50] C. Caldwell, Web page Mersenne primes: history, theorems and lists, http://www.utm.edu/research/primes/mersenne. shtml.

[51] P. Camion and A. Canteaut, Construction of t-resilient functions over a finite alphabet, in: Advances in Cryptology, EUROCRYPT'96, LNCS 1070, Springer, 1996, 283-293.

[52] P. Camion and A. Canteaut, Generalization of Siegenthaler inequality and Schnorr-Vaudenay multipermutations, In: N. Koblitz, ed., Advances in Cryptology - CRYPTO'96, LNCS 1109, Springer-Verlag, 1996, 372–386.

[53] A. Canteaut, C. Carlet, P. Charpin and C. Fontaine, Propagation characteristics and correlation-immunity of highly nonlinear Boolean functions, in: Proceedings of Eurocrypt'00, LNCS 1807, Springer Verlag, 2000, 507–520.

[54] A. Canteaut, P. Charpin and H. Dobbertin, Weight divisibility of cyclic codes, highly nonlinear functions on F_{2^m}, and crosscorrelation of maximum-length sequences, SIAM J. Discrete Math. 13(1) (2000), 105–138.

[55] C. Carlet, Codes de Reed-Muller, Codes de kerdock et de Preparata, thesis, Publication of LITP, Institut Blaise Pascal, Université Paris 6, 90.59 (1990).

[56] C. Carlet, Two new classes of bent functions, in: Advances in Cryptology – Eurocrypt'93, LNCS 765, Heidelberg, Springer Verlag, 1994, 77–101.

[57] C. Carlet, A construction of bent functions, in: Finite Fields and Applications, London Mathematical Society Lecture Notes Series 233, Cambridge, Cambridge University Press, 1996, 47–58.

[58] C. Carlet, Recent results on bent functions, in: Proceedings of the International Conference on Combinatorics, Information Theory and Statistics, 1999, 275-291.

[59] C. Carlet, On cryptographic propagation criteria for Boolean functions, Information and Computation 151 (1999), 32–56.

[60] C. Carlet and C. Ding, Highly nonlinear mappings, J. Complexity, to appear in 2004.

[61] C. Carlet and S. Dubuc, On generalized bent and q-ary perfect nonlinear functions, in: D. Jungnickel and H. Niederreiter Eds., Finite Fields and Applications, Proceedings of Fq5, Springer Verlag, 2000, 81-94.

[62] C. Carlet and P. Guillot, A characterization of binary bent functions, Designs, Codes and Cryptography 14 (1998), 130–140.

[63] C. Carlet and P. Guillot, An alternate characterization of the bentness of binary functions with uniqueness, J. Comb. Theory A 76 (1996), 328–335.

[64] C. Carlet and P. Guillot, A new characterization of Boolean functions, in: Proceedings of AAECC'13, Lecture Notes in Computer Science, vol. 1719, Springer Verlag, 94–103.

[65] L. Carlitz, Distribution of primitive roots in a finite field, Quart. J. Math. 4 (1953), 4–10.

[66] R. D. Carmichael, On sequences of integers defined by recurrence relations, Quart. J. Pure Appl. Math. 48 (1920), 343–372.

[67] F. Chabaud and S. Vaudenay, Links between differential and linear cryptanalysis, in: Advances in Cryptology - Eurocrypt'94, A. De Santis, ed., LNCS 950 (1995), Springer Verlag, 356–365.

[68] A. H. Chan and R. A. Games, On the linear span of binary sequences obtained from finite geometries, in: Advances in Cryptology - Crypto '86, A. M. Odlyzko, ed., LNCS 263 (1987), Springer Verlag, 405–417.

[69] A. H. Chan and R. A. Games, On the quadratic span of DeBruijn sequences, IEEE Trans. Info. Theory 36 (1990), 822–829.

[70] K. Chandrasekharan, Elliptic Functions, Grundlehren der Mathematischen Wissenschaften 281, Springer Verlag, 1985.

[71] J. R. Chen, Waring's problem for $g(5) = 37$, Sci. Sinica 13 (1964), 1547–1568.

[72] P. H. Chen, Multisequence linear shift register synthesis and its application to BCH decoding, IEEE Trans. Commun. 24 (1976), 438–440.

[73] Y. Q. Chen, On the existence of abelian Hadamard difference sets and a new family of difference sets, Finite Fields Appl. 3 (1997), 234–256.

[74] U. Cheng, On the continued fraction and Berlekamp's algorithm, IEEE Trans. Info. Theory 30 (1984), 541–544.

[75] S. Chowla, A property of biquadratic residues, Proc. Nat. Acad. Sci. India Sec. A 14 (1944), 45–46.

[76] S. Chowla, Contributions to the theory of the construction of balanced incomplete block designs, Math. Student 12 (1945), 82–85.

[77] S. Chowla and H. J. Ryser, Combinatorial problems, Canad. J. Math. 2 (1950), 93–99.

[78] S. Chowla, The Riemann Hypothesis and Hilbert's Tenth Problem, Gordon and Breach, New York, 1965, Chapters IV, V.

[79] H. Chung and P. V. Kumar, A general construction for generalized bent functions, IEEE Trans. Info. Theory 35 (1989), 206–209.

[80] H. Cohn, A Classical Invitation to Algebraic Numbers and Class Fields, Springer Verlag, New York, Berlin, Heidelberg, London, Paris, Tokyo, 1984.

[81] S. D. Cohn, Exceptional polynomials and the reducibility of substitution polynomials, L'Enseignement Mathématique 36 (1990), 53–65.

[82] C. J. Colbourn and W. de Launey, Difference matrices, in: C. Colbourn and J. H. Dinitz eds., Handbook of Combinatorial Designs, New York, CRC Press, 1996, Chapter IV.11, pp. 287–297.

[83] W. N. Colquitt and L. Welsh, Jr, New Mersenne prime $2^{110503} - 1$, Math. Comput. 56 (1991), 867–870.

[84] C. M. Cordes, Permutations mod m in the form x^n, Amer. Math. Monthly 83 (1976), 32–33.

[85] R. S. Coulter and R. Matthews, Planar functions and plans of the Lenz-Barlotti class II, Designs, Codes and Cryptography 10 (1997), 165–195.

[86] D. A. Cox, Primes of the Form $x^2 + ny^2$: Fermat, Class Field Theory, and Complex Multiplication, John Wiley & Sons, 1989.

[87] R. Crandall, J. Doenias, C. Norrie and J. Young, The twenty-second Fermat number is composite, Math. Comput. 64 (1995), 863–868.

[88] T. W. Cusick, Properties of the X^2 mod N generator, IEEE Trans. Info. Theory 41 (1995), 1155–1159.

[89] T. W. Cusick, Constructing differentially uniform permutations via crosscorrelation functions, Preprint, Feb. 1995.

[90] T. W. Cusick and H. Dobbertin, Some new 3-valued crosscorrelation functions of binary sequences, IEEE Trans. Info. Theory 42 (1996), 1238–1240.

[91] Z. Dai, Proof of Rueppel's linear complexity conjecture, IEEE Trans. Info. Theory 32 (1986), 440–443.

[92] Z. Dai, T. Beth and D. Gollmann, Lower bounds for the linear complexity of sequences over residue rings, in: Advances in Cryptology - Eurocrypt'90, I. Damgard, ed., LNCS 473 (1990), Springer Verlag, 189–195.

[93] Z. Dai, J. Yang, G. Gong and P. Wang, On the linear complexity of generalized Legendre sequences, in: Sequences and their Applications, T. Helleseth, P.V. Kumar, K. Yang eds., Springer Verlag, 2001, 145–153.

[94] Z. Dai and K. C. Zeng, Continued fractions and the Berlekamp-Massey algorithm, in: Advances in Cryptology - Auscrypt'90, J. Seberry and J. Pieprzyk, eds., LNCS 453 (1990), Springer Verlag, 24–31.

[95] I. Damgard, On the randomness of Legendre and Jacobi sequences, in: Advances in Cryptology - Crypto '88, S. Goldwasser, ed., LNCS 403 (1990), Springer Verlag, 163–172.

[96] H. Davenport, On the distribution of quadratic residues (mod p), J. London Math. Soc. 6 (1931), 49–54.

[97] H. Davenport, The Higher Arithmetic, 5th ed., Cambridge University Press, 1982.

[98] H. Davenport, Multiplicative Number Theory, Springer Verlag, New York, 1980.

[99] J. A. Davis, Almost difference sets and reversible difference sets, Arch. Math. 59 (1992), 595–602.

[100] W. de Launey, Square GBRDs over non-abelian groups, Ars Combin. 27 (1989), 40–49.

[101] W. de Launey, Generalized Hadamard matrices which are developed modulo a group, Discrete Math. 104 (1992), 49–65.

[102] W. de Launey, Circulant GH($p^2, \mathbf{Z}_p$) exist for all primes p, Graphs Combin. 8 (1992), 317–321.

[103] D. E. Denning, Cryptography and Data Security, Addison-Wesley, 1983.

[104] J.-M. Deshouillers, Waring's problem and the circle-method, in: Number Theory and Applications, R. A. Mollin, ed., Kluwer Academic Publishers, 1989, 37–44.

[105] P. Diaconis, Average running time of the fast Fourier transform, J. Algorithms 1 (1980), 197–208.

[106] L. E. Dickson, Cyclotomy, higher congruences, and Waring's problem, Amer. J. Math. 57 (1935), 391–424, and 463–474.

[107] L. E. Dickson, Cyclotomy and trinomial congruences, Trans. Amer. Math. Soc. 37 (1935), 363–380.

[108] L. E. Dickson, Cyclotomy when e is composite, Trans. Amer. Math. Soc. 38 (1935), 187–200.

[109] L. E. Dickson, Solution of Waring's problem, Amer. J. Math. 58 (1936), 530–535.

[110] L. E. Dickson, History of the Theory of Numbers, Vol. 1–3, Chelsea Publishing Company, 1952.

[111] W. Diffie and M. E. Hellman, New directions in cryptography, IEEE Trans. Info. Theory 22 (1976), 644–654.

[112] W. Diffie and M. Hellman, Privacy and authentication: An introduction to cryptography, Proc. IEEE 67(3) (1979), 397–427.

[113] W. Diffie, The first ten years of public key cryptology, in: Contemporary Cryptology: The Science of Information Integrity, G. J. Simmons, ed., IEEE Press, 1992.

[114] J. F. Dillon, A survey of bent functions, NSA Technical Journal, Special Issue, 1972, 191–215.

[115] J. F. Dillon, Elementary Hadamard Difference sets, Ph. D. Thesis, Univ. of Maryland, 1974.

[116] J. F. Dillon, Multiplicative difference sets via additive characters, Designs, Codes and Cryptography 17 (1999), 225–235.

[117] J. F. Dillon and H. Dobbertin, Cyclic difference sets with Singer Parameters, Manuscript, 1999.

[118] C. Ding, Proof of Massey's conjectured algorithm, in: Advances in Cryptology - Eurocrypt'88, C. G. Günther, ed., LNCS 330 (1989), Springer-Verlag, 345–349.

[119] C. Ding, Lower Bounds on the weight complexity of cascaded binary sequences, in: Proceed. Auscrypt'90, LNCS 453 (1991), Springer-Verlag, 39–43.

[120] C. Ding, Pattern distributions of Legendre sequences, IEEE Trans. Info. Theory 44 (1998), 1693–1698.

[121] C. Ding, Autocorrelation values of the generalized cyclotomic sequences of order 2, IEEE Trans. Info. Theory 44 (1998), 1698–1702.

[122] C. Ding, The differential cryptanalysis and design of the natural stream ciphers, Fast Software Encryption, LNCS 809 (1994), Springer-Verlag, 101–115.

[123] C. Ding, Binary cyclotomic generators, Fast Software Encryption, LNCS 1008 (1995), Springer, 29–61.

[124] C. Ding, Linear complexity of generalized cyclotomic binary sequences of order 2, Finite Fields and Their Applications 3 (1997), 159–174.

[125] C. Ding, Cryptographic Counter Generators, TUCS Series in Dissertation **4**, Turku Centre for Computer Science, 1997, ISBN 951-650-929-0.

[126] C. Ding, Blum-Blum-Shub generator, Electron. Lett. 33 (1997), 677.

[127] C. Ding, Linear complexity of some generalized cyclotomic sequences, International Journal on Algebra and Computation 8 (1998), 431–442.

[128] C. Ding and T. Helleseth, Cyclotomic generator of order r, Information Processing Letters 66 (1998), 21–25.

[129] C. Ding and T. Helleseth, New generalized cyclotomy and its applications, Finite Fields & Their Applications 4 (1998), 21–25.

[130] C. Ding, T. Helleseth and K. Y. Lam, Several classes of binary sequences with three-level autocorrelation, IEEE Trans. Information Theory 45 (1999), 2606–2612.

[131] C. Ding, T. Helleseth and H. M. Martinsen, New families of binary sequences with optimal three-level autocorrelation, IEEE Trans. Info. Theory 47(1) (2001), 428–433.

[132] C. Ding, T. Helleseth and W. Shan, On the linear complexity of Legendre sequences, IEEE Trans. Info. Theory 44 (1998), 1276–1278.

[133] C. Ding, V. Niemi, A. Renvall and A. Salomaa, TWOPRIME: A fast stream ciphering algorithm, in: Fast Software Encryption, LNCS 1267, Springer Verlag, 1997, 88–102.

[134] C. Ding, P. Pei and A. Salomaa, Chinese Remainder Algorithm: Applications in Computing, Coding, Cryptography, World Scientific, Singapore, 1996.

[135] C. Ding and A. Salomaa. Cooperatively distributed hashing and ciphering, Computers and Artificial Intelligence 15 (1996), 233–245.

[136] C. Ding and G. Xiao, Stream Ciphers and Their Applications (in Chinese), National Defense Press, Beijing, 1994.

[137] C. Ding, G. Xiao and W. Shan, New measure indexes on the security of stream ciphers, in Chinese, Proc. Third Chinese National Workshop on Cryptology, Xian, China, 1988, 5–15.

[138] C. Ding, G. Xiao and W. Shan, The Stability Theory of Stream Ciphers, LNCS 561 (1991), Springer-Verlag.

[139] H. Dobbertin, Construction of bent functions and balanced Boolean functions with high nonlinearity, in: B. Preneel ed., Fast Software Encryption, LNCS 1008, Heidelberg, Springer Verlag, 1995, 61–74.

[140] H. Dobbertin, One-to-one highly nonlinear functions on finite fields with characteristic 2, Appl. Algebra Engrg. Comm. Comput. 9 (1998), 139–152.

[141] H. Dobbertin, Almost perfect nonlinear power functions on $GF(2^n)$: The Welch case, IEEE Trans. Info. Theory 45 (1999), 1271–1275.

[142] H. Dobbertin, Almost perfect nonlinear power functions on $GF(2^n)$: The Niho case, Information and Computation 151 (1999), 57–72.

[143] K. Dörge, Zur Verteilung des Quadratischen Reste, Jahresbericht Deutschen Math. Vereinigung 88 (1929), 41–49.

[144] S. Eliahou and M. Kervaire, Barker sequences and difference sets, L'Enseign. Math. 38 (1992), 345–382.

[145] W. J. Ellison, Waring's problem, Amer. Math. Month. 78 (1971), 10–36.

[146] P. D. T. A. Elliott, The distribution of primitive roots, Canad. J. Math. 21 (1969), 822–844.

[147] H. T. Engstrom, On sequences defined by linear recurrence relations, Trans. Amer. Math. Soc. 33 (1931), 210–218.

[148] R. J. Evans and J. R. Hill, The cyclotomic numbers of order sixteen, Math. Comput. 33 (1979), 827–835.

[149] R. J. Evans, The cyclotomic numbers of order twenty-four, Math. Comput. 35 (1980), 1036–1038.

[150] R. J. Evans, Twenty-fourth power residue difference sets, Math. Comput. 40 (1983), 677–683.

[151] D. R. Evertse, L. M. Adleman, K. Kompella, K. S. McCurley and G. Miller, Breaking the Ong-Schnorr-Shamir signature scheme for quadratic number fields, in: Advances in Cryptology - Crypto'85, H. C. Williams, ed., LNCS 218 (1986), Springer Verlag, 3–13.

[152] J. H. Evertse, Linear structures in block ciphers, in: Advances in Cryptology - Eurocrypt'87, LNCS 304 (1988), Springer Verlag, 249–266.

[153] J. Feigenbaum, Overview of Interactive proof systems and zero-knowledge, in: Contemporary Cryptology: The Science of Information Integrity, G.J. Simmons, ed., IEEE Press, 1992, 423–440.

[154] G. Feng and K. K. Tzeng, An iterative algorithm for the multi-sequences synthesis with a shortest LFSR, Sci. Sinica (Science in China), August 1985, 740–749.

[155] G. L. Feng and K. K. Tzeng, A generalized Euclidean algorithm for multisequence shift-register synthesis, IEEE Trans. Info. Theory 35 (1989), 584–594.

[156] G. L. Feng and K. K. Tzeng, A generalization of the Berlekamp-Massey algorithm for multisequence shift-register synthesis with applications to cyclic codes, IEEE Trans. Infom. Theory 37 (1991), 1274–1287.

[157] H. Fredricksen, A survey of full length nonlinear shift register cyclic algorithms, SIAM Review 24 (1982), 195–221.

[158] J. B. Friedlander and I. E. Shparlinski, On the distribution of the power generator, Math. Comput. 70 (2001), 1575–1589.

[159] C. F. Gauss, Disquisitiones Arithmeticae, Leipzig, 1801. English translation, Yale, New Haven, 1966. (Reprint by Springer Verlag, Berlin, Heidelberg, and New York, 1986).

[160] R. Gold, Maximal recursive sequences with 3-valued recursive cross-correlation functions, IEEE Trans. Info. Theory 14 (1968), 154–156.

[161] D. Goldfeld, Gauss' class number problem for imaginary quadratic fields, Bull. Amer. Math. Soc. 13 (1985), 23–37.

[162] S. Goldwasser, The search for provably secure cryptosystems, in: Cryptology and Computational Number Theory, C. Pomerance, ed., Proc. of Symposia in Applied Mathematics, American Mathematical Society, 1990.

[163] D. Gollmann, Linear recursions of cascaded sequences, in: Contributions to General Algebra 3, Proc. of the Vienna Conference, June 1984, Verlag Hölder-Pichler-Tempsky, Verlag BG Teubner, Stuttgart, Wien, 1985.

[164] D. Gollmann, Pseudo random properties of cascade connections of clock controlled shift registers, in: Advances in Cryptology - Eurocrypt'84, T. Beth, N. Cot and I. Ingemarsson, eds., LNCS 209 (1985), Springer Verlag, 93–98.

[165] D. Gollmann and W. G. Chambers, Lock-in effect in cascades of clock-controlled shift registers, in: Proc. Eurocrypt'88, LNCS 330 (1988), Springer Verlag, 331–343.

[166] D. Gollmann, W. G. Chambers, Clock-controlled shift registers: A review, IEEE J. on Selected Areas in Communications 7 (1989), 525–533.

[167] S. Golomb, Sequences with randomness properties, Glenn L. Martin Co. Report, Baltimore 1955. (Reprinted in [169]).

[168] S. Golomb, et al., Digital Communications with Space Applications, Prentice-Hall, Englewood Cliffs, New Jersey, 1964.

[169] S. W. Golomb, Shift Register Sequences, Aegean Park Press, Laguna Hills, California, 1982.

[170] J. Gordon, Strong primes are easy to find, in: Advances in Cryptology - Eurocrypt'84, T. Beth, N. Cot and I. Ingemarsson, eds., LNCS 209 (1985), Springer Verlag, 216–223.

[171] B. Gordon, W. H. Mills and L. R. Welch, Some new difference sets, Canadian J. Math. 14 (1962), 614–625.

[172] M. Goresky and A. Klapper, Fibonacci and Galois representations of feedback-with-carry shift registers, IEEE Trans. Info. Theory 48 (2002), 2826-2836.

[173] M. Goresky, A. Klapper and R. Murty, On the distinctness of decimations of l-sequences, in: Sequences and their Applications, T. Helleseth, P. V. Kumar and K. Yang Eds., Springer Verlag, 2001, 197–208.

[174] M. Goreski, A. Klapper and L. Washington, Fourier transforms and the 2-adic span of periodic binary sequences, IEEE Trans. Info. Theory 46 (2000), 687-691.

[175] R. Göttfert and H. Niederreiter, Hasse-Teichmüller derivatives and products of linear recurring sequences, in: Contemporary Mathematics 168 (1994), 117–125.

[176] R. Göttfert and H. Niederreiter, On the minimal polynomial of the product of linear recurring sequences, Finite Fields and Their Applications 1 (1995), 204–218.

[177] R. T. Gregory and E. V. Krishnamurthy, Methods and Applications of Error-Free Computation, Springer Verlag, N. Y., 1984.

[178] E. Grosswald, Topics from the Theory of Numbers, The Macmillan Company, New York, 1966.

[179] E. Grosswald, Representations of Integers as Sums of Squares, Springer Verlag, 1985.

[180] E. J. Groth, Generation of binary sequences with controllable complexity, IEEE Trans. Info. Theory, 17 (1971).

[181] W. Gruner, Einlagerung des regulären n-Simplex in den n-dimensionalen Würfel, Comment. Math. Helv. 12 (1939–40), 149–152.

[182] C. G. Günther, On some properties of the sum of two pseudorandom sequences, Advances in Cryptology - Eurocrypt'86, Linköping, Sweden, May 1986.

[183] C. G. Günther, Alternating step generators controlled by de Bruijn sequences, in; Advances in Cryptology - Eurocrypt' 87, LNCS 309 (1988), Springer Verlag, 5–14.

[184] R. K. Guy, Unsolved Problems in Number Theory, Springer Verlag, New York, 1982.

[185] M. Hall, Divisibility sequences of third order, Amer. J. Math. 58 (1936) 577–584.

[186] M. Hall, Divisors of second order sequences, Bull. Amer. Math. Soc. 43 (1937), 78–80.

[187] M. Hall, An isomorphism between linear recurring sequences and algebraic rings, Trans. Amer. Math. Soc. 44 (1938), 196–218.

[188] M. Hall, Equidistribution of residues in sequences, Duke Math. J. 4 (1938), 691–695.

[189] M. Jr. Hall, A survey of difference sets, Proc. Amer. Math. Soc. 7 (1956), 975–986.

[190] M. Jr. Hall, Difference sets in combinatorics, eds. M. Hall, Jr. and J. H. van Lint, Dordrecht: D. Reidel, 1975, 321–346.

[191] A. R. Hammons Jr., P. V. Kumar, A. R. Calderbank, N. J. A. Sloane and P. Solé, The Z_4-linearity of Kerdock, Preparata, Goethals and related codes, IEEE Trans. info. Theory 40(2) (1994), 301–319.

[192] G. H. Hardy, S. Ramanujan, Une formule asymptotique pour le nombre des partitions de n, C.R.A.S. 164 (1917), 35–38.

[193] G. H. Hardy and J. E. Littlewood, Some problems of 'Partitio Numerorum': VI. Further Researches in Waring's Problem, Math Zeit. 23 (1925), 1–37.

[194] H. Hasse, Vorlesungen über Zahlentheorie, Springer Verlag, Berlin, 1964.

[195] D. R. Heath-Brown, Artin's conjecture for primitive roots, Quart. J. Math. Oxford (2) 37 (1986), 27–38.

[196] T. Helleseth, Some results about the cross-correlation function between two maximal linear sequences, Discr. Math. 16 (1976), 209–232.

[197] T. Helleseth, A note on the cross-correlation function between two binary maximal length linear sequences, Discr. Math. 23 (1978), 301–307.

[198] T. Helleseth, Legendre sums and codes related to QR codes, Discr. Appl. Math. 35 (1992), 107–113.

[199] T. Helleseth, On the correlation of m-sequences and related sequences, in: Sequences and their Applications, T. Helleseth, P.V. Kumar and K. Yang eds., Springer Verlag, 2001, 34–45.

[200] T. Helleseth and P. V. Kumar, Sequences with low correlation in: Handbook of Coding Theory, Pless, Brualdi, and Huffman, eds., Elsevier, 1998.

[201] T. Helleseth and D. Sandberg, Some power mappings with low differential uniformity, Applicable Algebra in Engin., Commun. and Computing 8 (1997) 363–370.

[202] T. Helleseth, C. Rong and D. Sandberg, New families of almost perfect nonlinear power mappings, IEEE Trans. Infom. Theory 45(2) (1999), 475–485.

[203] T. Herlestam, On functions of linear shift register sequences, in: LNCS, vol. 219, Advances in Cryptology, Springer Verlag, 1986, 119–129.

[204] C. Hermite, Note au sujet de l'article précédent, J. Math. Pures Appl. 13 (1848), 15; also: Note sur un théorème rélatif aux nombers entières Oeuvres. vol. 1, 264.

[205] E. Hewitt and K. Ross, Abstract Harmonic Analysis, Springer, Heidelberg, 1970.

[206] D. Hilbert, Beweis für die Darstellbarkeit der ganzen Zahlen durch eine feste Anzahl n-ter Potenzen Waringsches problem, Math. Ann. 67 (1909), 281–300.

[207] I. Honkala, and A. Tietäväinen, Codes and Number Theory, In: Handbook of Coding Theory, Pless, Huffman, eds., Elsevier, 1998.

[208] H. Hopf, Über die Verteilung quadratischer Reste, Math. Zeitschrift 32 (1930), 222–231.

[209] C. Hooley, On Artin's conjecture, J. für die reine und angewandte Mathematik 225 (1967), 209–220.

[210] X. D. Hou, q-ary bent functions constructed from chain rings, Finite Fields and their Applications 4 (1998), 55–61.

[211] X. D. Hou, Bent functions, Partial difference sets, and quasi-Frobenius local rings, Designs, Codes and Cryptography 20 (2000), 251–268.

[212] X. D. Hou and P. Langevin, Results on bent functions, J. Comb. Theory A 80 (1997), 232–246.

[213] L. K. Hua, Introduction to Number Theory, Springer Verlag, 1982.

[214] R. H. Hudson, On the first occurrence of certain patterns of quadratic residues and non-residues, Israel J. Math. 44 (1983), 23–32.

[215] K. Ireland and M. Rosen, A classical Introduction to Modern Number Theory, Springer Verlag, Berlin, Heidelberg, and New York, 1982.

[216] Information technology – Data cryptographic techniques – Modes of operation for a 64-bit block cipher algorithm, IS 8372, ISO/IEC, 1987.

[217] C. G. J. Jacobi, Fundamenta Nova Theoriae Functionum Ellipticarum, 1829.

[218] E. Jacobsthal, Über die Darstellung der Primzahlen der Form $4n + 1$ als Summe Zweier Quadrate, J. für die reine und angewandte Mathematik 132 (1907), 238–245.

[219] C. J. A. Jansen and D. E. Boekee, The shortest feedback shift register that can generate a given sequence, in: Advances in Cryptology - Crypto' 89, LNCS 435 (1989), Springer, 90–99.

[220] H. Janwa and R. Wilson, Hyperplane sections of Fermat varieties in P^3 in char. 2 and some applications to cyclic codes, in: Proceedings AAECC-10, LNCS 673, Berlin, Springer-Verlag, 1993, 180–194.

[221] S. M. Jennings, Multiplexed sequences: some properties of the minimal polynomial, in: LNCS 149 (1983), Springer Verlag, 189–206.

[222] R. R. Jueneman, Analysis of certain aspects of output feedback mode, in: Advances in Cryptology, Proc. Crypto' 82, D. Chaum, R. L. Rivest and A. T. Sherman, eds., Santa Barbara, CA, Plenum Press, New York, 1983, 99–127.

[223] D. Jungnickel, Finite Fields, Structure and Arithmetics, Bibliographishes Institut and F.A. Brockhaus AG, Mannheim, 1993.

[224] D. Jungnickel, Difference sets, in: J. Dinitz and D. R. Stinson eds., Contemporary Design Theory: A Collection of Surveys, John Wiley & Sons, 1992.

[225] D. Jungnickel and A. Pott, Perfect and almost perfect sequences, Discrete Applied Mathematics 95 (1999), 331–359.

[226] D. Jungnickel and A. Pott, Difference sets: an introduction, in: A. Pott, P.V. Kumar, T. Helleseth and D. Jungnickel eds., Difference Sets, Sequences and their Correlation Properties, Amsterdam, Kluwer, 1999, 259–295.

[227] T. Kaida, S. Uehara and K. Imamura, An algorithm for the k-error linear complexity of sequences over $GF(p^m)$ with period p^n, p a prime, Information and Computation 151 (1999), 134–147.

[228] T. Kasami, Weight enumerators for several class of subcodes of the 2nd order binary Reed-Muller codes, Information and Control 18 (1971), 369–394.

[229] A. M. Kerdock, A class of low-rate nonlinear codes, Information and Control 20 (1972), 182-187.

[230] P. Kesave Menon, Certain Hadamard designs, Proc. Amer. Math. Soc. 13 (1962), 524–531.

[231] E. L. Key, An analysis of the structure and complexity of nonlinear binary sequence generators, IEEE Trans. Info. Theory 22 (1976), 732–763.

[232] K. Kjeldsen and E. Andersen, Some random properties of cascaded sequences, IEEE Trans. Info. Theory 26 (1982), 854–862.

[233] A. Klapper, The vulnerability of geometric sequences based on fields of odd characteristic, J. Cryptology 7 (1994), 33-51.

[234] A. Klapper, D-form sequences: families of sequencews with optimal correlation values and linear span, IEEE Trans. Info. Theory 41 (1995), 1–9.

[235] A. Klapper, Feedback with carry shift registers over finite fields, in: Fast Software Encryption, LNCS 1008, Springer Verlag 1995, 170–178.

[236] A. Klapper, On the existence of secure keystream generators, J. Cryptology 14 (2001), 1–15.

[237] A. Klapper and M. Goresky, 2-adic shift registers, in: Fast Software Encryption: Proc. of the 1993 Cambridge Security Workshop, R. Anderson, ed., LNCS 809, Springer Verlag, 1994, 174–178.

[238] A. Klapper and M. Goresky, Cryptanalysis based on 2-adic rational approximation, in: Advances in Cryptology - Crypto' 95, D. Coppersmith, ed., LNCS 963 (1995), Springer Verlag, 262–273.

[239] A. Klapper and M. Goresky, Large period nearly deBruijn FCSR sequences, in: Advances in Cryptology–Eurocrypt 1995, LNCS 921, Springer Verlag, 1995, 263-273.

[240] A. Klapper and M. Goresky, Feedback shift registers, 2-adic span, and combiners with memory, Journal of Cryptology 10 (1997), 111–147.

[241] A. Klapper and M. Goresky, Arithmetic cross-correlation of FCSR sequences, IEEE Trans. Info. Theory 43 (1997), 1342–1346.

[242] T. Kløve, Periodicity of recurring sequences in rings, Math. Scand. 32 (1972), 165–168.

[243] T. Kløve, Linear recurring sequences in Boolean rings, Math. Scand. 33 (1973), 5–12.

[244] T. Kløve, On exponential recurring sequences, Math. Scand. 34 (1974), 44–50.

[245] D. Knuth, The Art of Computer Programming, Vol. 2. Seminumerical Algorithms, Addison-Wesley, Reading MA, 1981.

[246] N. Koblitz, Introduction to Elliptic Curves and Modular Forms, Springer, New York, 1984.

[247] N. Koblitz, A Course in Number Theory and Cryptography, Springer, New York, 1988.

[248] R. G. Kraemer, Proof of a conjecture on Hadamard 2-groups, J. Comb. Theory A 63 (1993), 1–10.

[249] E. Kranakis, Primality and Cryptography, Wiley-Teubner Series in Computer Science, B. G. Teubner and John Wiley & Sons, 1986.

[250] E. V. Krishnamurthy and R. T. Gregory, Mapping integers and Hensel codes onto Farey fractions, BIT vol. 23 (1983), 9–20.

[251] P. V. Kumar and R. A. Scholtz, Bounds on the linear span of bent sequences, IEEE Trans. Info. Theory 29 (1983), 854–862.

[252] P. V. Kumar, R. A. Scholtz and L. R. Welch, Generalized bent functions and their properties, J. Combinatorial Theory, Series A, 40 (1985), 90–107.

[253] K. Kurosawa, F. Sato, T. Sakata and W. Kishimoto, A relation between the linear complexity and k-error linear complexity, IEEE Trans. Info. Theory 46(2) 2000, 694–698.

[254] G. Lachaud and J. Wolfmann, The weights of the orthogonal of the extended quadratic binary Goppa codes, IEEE Trans. info. Theory 36 (1990), 686–692.

[255] J. C. Lagarias and J. Reeds, Unique extrapolation of polynomial recurrences, SIAM J. Computing 17, 342–260.

[256] J. C. Lagarias, Pseudorandom number generators in cryptography and number theory, in: Cryptography and Computational Number Theory, C. Pomerance, ed., Proceedings of Symposia in Applied Mathematics, vol. 42, Amer. Mathematical Society, 1990, 115–143.

[257] X. Lai, J. L. Massey and S. Murphy, Markov ciphers and differential cryptanalysis, in: Advances in Cryptology - Eurocrypt' 91, LNCS 547 (1991), Springer Verlag, 17–38.

[258] E. Landau, Über die Klassenzahl der binären quadratischen Formen von negativer Discriminante, Math. Annalen 56 (1903), 671–676.

[259] E. S. Lander, Symmetric Design: An algebraic approach, London Math. Society Lecture Note Series 74, 1983.

[260] P. Langevin, On generalized bent functions, in: CISM Courses and Lectures 339 (Eurocode), 1992, 147–157.

[261] A. G. B. Lauder and K. G. Paterson, Computing the error linear complexity spectrum of a binary sequence of period 2^n, IEEE Trans. Info. Theory 49(1) 2003, 273–280.

[262] D. H. Lehmer, An extended theory of Lucas' functions, Ann. of Math. (2) 31 (1930), 419–448.

[263] E. Lehmer, On residue difference sets, Canad. J. Math. 5 (1953), 425–432.

[264] E. Lehmer, On the number of solutions of $u^k + D \equiv w^2 \pmod{p}$, Pacific J. Math. 5 (1955), 103–118.

[265] D. H. Lehmer, Computer technology applied to the theory of numbers, Studies in Number Theory, Math. Assoc. Amer. (distributed by Prentice-Hall, Englewood Cliffs, N.J.), 1969, 117–151.

[266] D.H. Lehmer, On Fermat's quotient base two, Math. Comput. 36 (1981), 289–290.

[267] A. Lehmpel, M. Cohn, Maximal families of bent sequences, IEEE Trans. Info. Theory 28 (1982), 865–868.

[268] A. Lempel, M. Cohn and W. L. Eastman, A class of binary sequences with optimal autocorrelation properties, IEEE Trans. Info. Theory 23(1) (1977), 38–42.

[269] H. W. Lenstra, Factoring with elliptic curves, Ann. Math. 126 (1987), 649–673.

[270] A. Lenstra, Primality Testing, in: Cryptology and Computational Number Theory, C. Pomerance, ed., Proc. of Symp. in Appli. Math., American Mathm. Society, 1990, 13–26.

[271] A. K. Lenstra, H. W. Lenstra, Jr., M. S. Manasse and J. M. Pollard, The number field sieve, in: Proc. 22nd ACM Symp. Theory of Computing, 1990, 461–572.

[272] P. A. Leonard and K. S. Williams, A diophantine system of Dickson, Rend. Accad. Naz. Lincei 56 (1974), 145–150.

[273] P. A. Leonard and K. S. Williams, The cyclotomic numbers of order seven, in: Proc. Amer. Math. Soc. 51 (1975), 295–300.

[274] P. A. Leonard and K. S. Williams, The cyclotomic numbers of order eleven, Acta Arith. 26 (1975), 367–383.

[275] W. J. LeVeque, Topics in Number Theory, vol. 1, Reading, Massachusetts, Addison-Wesley, 1956.

[276] R. Lidl, H. Niederreiter, Finite Fields, in Encyclopedia of Mathematics and Its Applications, vol. 20, Addison-Wesley, 1983.

[277] O. A. Logachev, A. A. Salnikov and V. V. Yashchenko, Bent functions on a finite Abelian group, Discrete Math. Appl. 7(6) (1997) 547-564.

[278] D. L. Long and A. Wigderson, The discrete log hides $O(\log n)$ bits, SIAM J. Computing 17, 363–372.

[279] F. J. MacWilliams and N. J. A. Sloane, The Theory of Error-Correcting Codes, North-Holland Publishing Company, 1977.

[280] K. Mahler, On a geometrical representation of p-adic numbers, Ann. of Math. 41 (1940), 8–56.

[281] K. Mahler, P-adic Numbers and Their Functions, Second edition, Cambridge University Press, 1981.

[282] D. Mandelbaum, Arithmetic codes with large distance, IEEE Trans. Info. Theory 13 (1967), 237–242.

[283] D. Mandelbaum, An approach to an arithmetic analog of Berlekamp's algorithm, IEEE Trans. Info. Theory 30 (1984), 758–762.

[284] H. B. Mann, Addition Theorems, Interscience Publishers, 1965.

[285] H. B. Mann, Difference sets in elementary Abelian groups, J. Math. 9 (1965), 212–219.

[286] H. B. Mann, Recent advances in difference sets, Amer. Math. Monthly 74 (1967), 229–235.

[287] Jr. H. Marshall, Combinatorial Theory, Blaisdell Publishing Company, 1967.

[288] G. Marsaglia, A current view of random number generators, in: Proc. Comput. Sci. Statistics: Sixteenth Symp. Interface, Keynote address.

[289] G. Marsaglia and A. Zaman, A new class of random number generators, Annals of Applied Probability 1 (1991), 462–480.

[290] A. Maschietti, Difference sets and hypherovals, Designs, Codes and Cryptography 14 (1998), 89–98.

[291] J. L. Massey, Shift-register synthesis and BCH decoding, IEEE Trans. Info. Theory 15 (1969), 122–127.

[292] J. L. Massey and T. Schaub, Linear complexity and applications, in: Coding Theory and Applications, G. Cohn, P. Godlewski, eds., LNCS 311 (1987), Springer Verlag, 19–31.

[293] J. L. Massey, SAFER K-64: A byte-oriented block-ciphering algorithm, in: Fast Software Encryption, R. Anderson, ed., LNCS 809, Springer Verlag, 1994, 1–17.

[294] M. Matsui, Linear cryptanalysis method for DES cipher, in: Advances in Cryptology - EUROCRYPT'93, LNCS 765. Springer-Verlag, 1994, 386–397.

[295] G. Matthew and H. C. Williams, Some new primes of the form $k \times 2^n + 1$, Math. Comput. 31 (1977), 797–798.

[296] U. M. Maurer, A provably-secure strongly randomized cipher, in Advances in Cryptology - Eurocrypt'90, I. Damgard, ed., LNCS 473 (1991), Springer Verlag, 361–373.

[297] U. M. Maurer and J. L. Massey, Cascaded ciphers: the importance of being first, J. Cryptology 6 (1993), 55–61.

[298] J. H. McClellan and C. H. Rader, Number Theory in Digital Signal Processing, Prentice-Hall, Inc. Englewood Cliffs, New Jersey, 1979.

[299] K. S. McCurley, The discrete logarithm problem, in: Cryptography and Computational Number Theory, C. Pomerance, ed., Proc. of Symposia in Applied Mathematics, vol. 42, Amer. Mathematical Society, 1990, 49–74.

[300] K. S. McCurley, Odds and ends from cryptography and computational number theory, in: Cryptography and Computational Number Theory, C. Pomerance, ed., Proc. of Symposia in Applied Mathematics, vol. 42, Amer. Mathematical Society, 1990, 145–166.

[301] R. J. McEliece and H. Rumsey Jr, Euler products, cyclotomy, and coding, J. Number Theory 4 (1972), 302–311.

[302] R. L. McFarland, A family of difference sets in noncyclic groups, J. Comb. Theory, Series A 15 (1973), 1–10.

[303] W. Meidl and H. Niederreiter, On the expected value of the linear complexity and the k-error linear complexity of periodic sequences, IEEE Trans. Info. Theory 48 (2002), 2817–2825.

[304] W. Meidl and H. Niederreiter, Linear Complexity, k-error linear complexity, and the discrete Fourier transform, J. Complexity 18 (2002), 87–103.

[305] W. Meidl and H. Niederreiter, Counting functions and expected values for the k-error linear complexity, Finite Fields and Their Applications 8 (2002), 142–154.

[306] W. Meier and O. Staffelbach, Fast correlation attacks on certain stream ciphers, J. Cryptology 1(3) (1989), 159–176.

[307] W. Meier and O. Staffelbach, Nonlinearity criteria for cryptographic functions, in: Advances in Cryptology – Crypto '90, LNCS 434 (1990), Springer Verlag, 1990, 549–562.

[308] A. Menezes, P. van Oorschot, and S. Vanstone, Handbook of Applied Cryptography, CRC Press Series on Discrete Mathematics and Its Applications, 1996.

[309] P. K. Menon, On difference sets whose parameters satisfy a certain relation, in: Proc. AMS 13 (1986), 739–745.

[310] R. C. Merkle and M. E. Hellman, Hiding information and signatures in trapdoor knapsacks, IEEE Trans. Info. Theory 24 (1978), 523–530.

[311] W. H. Mills, Continued fractions and linear recurrences, Math. Comput. 29 (1975), 173–180

[312] C. J. Mitchell, F. Piper and P. Wild, Digital signatures, in: Contemporary Cryptography, G. J. Simmons, ed., IEEE Press, 1992, 325–378.

[313] P. L. Montgomery, New solutions of $a^{p-1} \equiv 1 \pmod{p^2}$, Math. Comput. 61 (1993), 361–363.

[314] Gary L. Mullen, Permutation polynomials over finite fields, in: Finite Fields, Coding Theory with Advances in Comm. and Computing, Proc. of Las Vegas Conference, August, 1991, Lecture Notes in Pure and Applied Math. 141 (1993), 131-151, Marcel Dekker, Inc.

[315] L. Murata, On the magnitude of the least prime primitive root, J. Number Theory 37 (1991), 47–66.

[316] J. B. Muskat, The cyclotomic numbers of order fourteen, Acta Arith. 11 (1966), 263–279.

[317] J. B. Muskat and A. L. Whiteman, The cyclotomic numbers of order twenty, Acta Arith. 17 (1970), 185–216.

[318] J. Nechvatal, Public key cryptography, in: Contemporary Cryptology: The Science of Information Integrity, G. J. Simmons, ed., IEEE Press, 1992.

[319] H. Niederreiter, Sequences with almost perfect linear complexity profile, in: Advances in Cryptology - Proc. Eurocrypt'87, LNCS 304, Springer, 1988, 37–51.

[320] H. Niederreiter, Keystream sequences with a good linear complexity profile for every starting point, in: Advances in Cryptology - Proc. Eurocrypt'89, LNCS 434, Springer, 1990, 523–532.

[321] H. Niederreiter, A combinatorial approach to probabilistic results on the linear complexity profile of random sequences, J. of Cryptology 2 (1990), 105–112.

[322] H. Niederreiter, The linear complexity profile and the jump complexity of keystream sequences, in: Advances in Cryptology - Proc. Eurocrypt'90, LNCS 473, Springer, 1991, 174–188.

[323] H. Niederreiter, Periodic sequences with large k-error linear complexity, IEEE Trans. Info. Theory 49 (2003), 501–505.

[324] H. Niederreiter and M. Vielhaber, On the fractal nature of the set of all binary sequences with almost perfect linear complexity profile, Communications and Multimedia Security, R. Posch, ed., Chapman & Hall, London, 1995, 214–221.

[325] H. Niederreiter and M. Vielhaber, Tree complexity and a doubly exponential gap between structured and random sequences, J. Complexity 12 (1996), 187–198.

[326] H. Niederreiter and M. Vielhaber, Linear complexity profiles: Hausdorff dimensions for almost perfect profiles and measure for general profiles, J. Complexity 13 (12997), 353-383.

[327] Y. Niho, Multi-valued cross-correlation functions between two maximal linear recursive sequences, Ph.D thesis, Elec. Eng., Southern Calif. (USCEE Report 409), 1972.

[328] I. Niven and H. S. Zuckerman, An Introduction to the Theory of Numbers, Third Edition, John Wiley and Sons Inc., 1972.

[329] J. S. No, Generalization of GWM sequences and No sequences, IEEE Trans. Info. Theory 35 (1989), 371–379.

[330] J. S. No, S. W. Golomb, G. Gong, H. K. Lee and P. Gaal, Binary pseudorandom sequences of period $2^m - 1$ with ideal autocorrelation generated by the polynomial $z^d + (z+1)^d$, IEEE Trans. Information Theory 44(3) (1998), 1278-1282.

[331] J. S. No and P. V. Kumar, A new family of binary pseudorandom sequences having optimal periodic correlation properties and large linear span, IEEE Trans. Info. Theory 35 (1989), 371–379.

[332] K. Nyberg, Perfect nonlinear S-boxes, in: Advances in Cryptology - Eurocrypt'91, D. W. Davies, ed., LNCS 547 (1991), Springer Verlag, 378–386.

[333] K. Nyberg, Differentially uniform mappings for cryptography, in: Advances in Cryptology - Eurocrypt'93, T. Helleseth, ed., LNCS 765 (1994), Springer Verlag, 55–64.

[334] K. Nyberg and L. R. Knudsen, Provable security against differential cryptanalysis, in: Advances in Cryptology - Crypto' 92, E. F. Brickell, ed., LNCS 740 (1993), Springer Verlag, 566–574.

[335] J. D. Olsen, R. A. Scholtz and L. R. Welch, Bent-function sequences, IEEE Trans. Info. Theory 28 (1982), 858–864.

[336] H. Ong, C. P. Schnorr and A. Shamir, An efficient signature based on quadratic equations, in: Proc. of 16th Annual ACM Symposium on Theory of Computing, 208–216.

[337] P. C. van Oorschot, A comparison of practical public key cryptosystems based on integer factorization and discrete logarithms, in: Contemporary Cryptology: The Science of Information Integrity, G. J. Simmons, ed., IEEE Press, 1992.

[338] R. E. A. C. Paley, On orthogonal matrices, J. Math. and Phys. 12 (1933), 311–320.

[339] B. K. Parady, J. F. Smith and S. E. Zarantonello, Largest known twin primes, Math. Comput. 55 (1990), 381–382.

[340] Y.-H. Park, D. Hong and E. Chun, On the linear complexity of some generalized cyclotomic sequences, International J. of Algebra and Computation, to appear.

[341] D. Pei, Personal communications Jan. 1994.

[342] O. Perron, Die Lehre von den Kettenbrüchen, 2nd ed., Chelsea, New York, 1950, 32–34. MR 12, 254.

[343] F. R. Pichler, Finite state machine modeling of cryptographic systems in Loops, in: Proc. Eucrocrypt'87, LNCS 304, Springer Verlag, 1988.

[344] R. Peralta, On the distribution of quadratic residues and nonresidues modulo a prime number, Math. Comput. 58 (1992), 433–440.

[345] W. W. Peterson and E. J. Weldon, Jr., Error-Correcting Codes, second edition, MIT Press, Cambridge MA, 1972.

[346] J. P. Pieprzyk, Nonlinearity of exponent permutations, in: Advances in Cryptology - Eurocrypt'89, Springer Verlag, 1990.

[347] F. Piper, Stream Ciphers, Elektrotechnik und Maschinenbau 104 (12) (1987), 564–568.

[348] V. S. Pless, Encryption schemes for computer confidentiality, IEEE Trans. Comput., vol. C-26 (1977), 1133–1136.

[349] V. S. Pless and W. C. Huffman, Handbook of Coding Theory, Amsterdam, Elsevier, 1998.

[350] J. Pollard, The fast Fourier transform in a finite field, Math. Comput. 25 (1971), 365–374.

[351] J. M. Pollard and C. P. Schnorr, An efficient solution of the congruence $x^2 + ky^2 \equiv m \pmod{n}$, IEEE Trans. Info. Theory 33 (1987), 702–709.

[352] C. Pomerance, Analysis and comparison of some integer factoring algorithm, in: Computational Methods in Number Theory, H. W. Lenstra, Jr., and R. Tijdeman, eds., Math. Centrum Tract 154 (1982), 89–139.

[353] C. Pomerance, The quadratic sieve factoring algorithm, in: LNCS 209; Advances in Cryptology - Eurocrypt'84, T. Beth, N. Cot, and I. Ingemarsson, eds., LNCS 209 (1985), Springer Verlag, 169–182.

[354] C. Pomerance, J. W. Smith and R. Tuler, A pipeline architecture for factoring large integers with the quadratic sieve algorithm, SIAM J. Computing 17(2) (1988), 387–403.

[355] C. Pomerance, Factoring, in: Cryptography and Computational Number Theory, C. Pomerance, ed., Proceed. of Symp. in Applied Math. vol. 42, American Mathematical Society, 1990, 27–48.

[356] A. Pott, Finite Geometry and Character Theory, Lecture Notes in Mathematics, vol. 1601, Berlin, Springer Verlag, 1995.

[357] B. Preneel, Design and Analysis of Cryptographic Hash Functions, Ph.D thesis, Katholieke Universiteit Leuven, 1993.

[358] F. Proth, Théorèmes sur les nombres premiers, C. R. Acad. Sci. Paris 87 (1878), p. 926.

[359] M. O. Rabin, Digital signatures and public-key functions as intractable as factorization, MIT Laboratory for Computer Science, TR-212, 1979.

[360] J. A. Reeds and N. J. A. Sloane, Shift-register synthesis (modulo m), SIAM J. Comput. 14(3) (1985), 505–513.

[361] P. Ribenboim, The Book of Prime Number Records, Springer Verlag, 1988.

[362] P. Ribenboim, The Little Book of Big Primes, Springer Verlag, 1991.

[363] H. Riesel, A note on the primes of the numbers of the form $N = (6a+1)2^{2n-1} - 1$ and $M = (6a-1)2^{2n} - 1$, Ark. Math. 3 (1956), 245–253. MR 17, 945.

[364] H. Riesel, Lucasian criteria for the primality of $h2^{2n} - 1$, Math. Comput. 23 (1969), 869–875.

[365] H. Riesel, Primes Numbers and Computer Methods for Factorization, Progress in Mathematics 57, Birkhäuser, 1985.

[366] M. P. Ristembatt and J. L. Daws, Jr, Performance criteria for spread Communications, IEEE Trans. Comm. 25(8) (1977), 756–763.

[367] R. L. Rivest, A. Shamir and L. Adleman, A method for obtaining signature and public-key cryptosystems, Comms. of ACM 21(2) (1978), 120–126.

[368] R. L. Rivest, Remarks on a proposed cryptanalytic attack on the M.I.T. public-key cryptosystem, Cryptologia 2 (1978).

[369] R. M. Robinson, The converse of Fermat's theorem, Amer. Math. Monthly 64 (1957), 703–710. MR 55 #4520.

[370] R. M. Robinson, A report on primes of the form $k2^n+1$ and on factors of Fermat numbers, Proc. Amer. Math. Soc. 9 (1958), 673–681. MR 20 # 3079.

[371] P. Rogaway and D. Coppersmith, A software-oriented encryption algorithm, in: Fast Software Encryption, LNCS **809**, Springer Verlag, 1994, 56–63.

[372] H. E. Rose, A Course in Number Theory, Clarendon Press, Oxford, 1988.

[373] O. S. Rothaus, On bent functions, J. Combinatorial Theory 20 (1976), 300–305.

[374] G. Rozenberg and A. Salomaa, Cornerstones of Undecidability, Prentice Hall, New York, 1994.

[375] R. Y. Rubinstein, Simulation and the Monte Carlo method, John Wiley & Sons, New York, 1982.

[376] R. A. Rueppel, Analysis and Design of Stream Ciphers, Springer Verlag, 1986.

[377] H. J. Ryser, Combinatorial Mathematics, Carus Mathematical Monograph, no. 14, 1963.

[378] A. Salomaa, Computation and Automata, Cambridge University Press, Cambridge, 1985.

[379] A. Salomaa, Formal Languages, Academic Press, New York, 1973.

[380] A. Salomaa, Public-key Cryptography, EATCS Monographs on Theoretical Computer Science, vol. 23, Springer Verlag, 1990.

[381] D. V. Sarwate and M. B. Pursley, Crosscorrelation properties of pseudorandom and related sequences, Proc. IEEE 5 (1980), 593–619.

[382] W. M. Schmidt, Equations over Finite Fields: An Elementary Approach, Lecture Notes in Mathematics, vol. 536, Springer Verlag, 1976.

[383] B. Schneier, Applied Cryptography, 2nd Edition John Wiley & Sons, 1996.

[384] C. P. Schnorr, On the construction of random number generators and random function generators, in: Advances in Cryptology - Eurocrypt'88, C. G. Günther, ed., LNCS 330 (1989), Springer Verlag, 225–232.

[385] R. A. Scholtz, The origins of spectrum communications, IEEE Trans. Commun. 30 (1982), 882–854.

[386] A. Schönhage and V. Strassen, Schnelle Multiplikation grosser Zahlen, Computing 7 (1971), 281–292.

[387] M. R. Schröder, Number Theory in Science and Communication, Springer Series in Information Sciences 7, Springer Verlag, 1984.

[388] L. von Schrutka, Eine Beweis für die Zerlegbarkeit der Primezahlen von der Form $6n + 1$ in ein einfaches und ein dreifaches Quadrat, J. für die reine und angewandte Mathematik 140 (1911), 252–265.

[389] M. P. Schutzenberger, A nonexistence theorem for an infinite family of symmetrical block designs, Ann. Eugenics, 14 (1949), 286–287.

[390] E. S. Selmer, Linear Recurrence Relations over Finite Fields, Department of Mathematics, University of Bergen, Norway, 1966.

[391] J. A. Serret, Sur un théorème rélatif aux nombers entières, J. Math. Pures Appl. 13 (1848), 12–14.

[392] A. Shamir, On the generation of cryptographically strong pseudorandom sequences, in: Proceed. of the 8th Int. Colloquium on Automata, Languages and Programming, LNCS 62, Springer Verlag, 1981.

[393] D. Shanks, Review of "A table of Gaussian primes," by L. G. Diehl and J. H. Jordan , Math. Comput. 21 (1967), 260–262.

[394] D. Shanks, J. W. Wrench, Brun's constant, Math. Comput. 28 (1974), 293–299.

[395] D. Shanks, Review of Brent UMT 21, Math. Compt. 30 (1976), p. 379.

[396] D. Shanks, Solved and Unsolved Problems in Number Theory, second edition, Chelsea Publishing Company, 1978.

[397] C. E. Shannon, Communication theory of secrecy systems, Bell Sys. Tech. J. 28 (1949), 657–715.

[398] D. E. Shippee, Four new factors of Fermat numbers, Math. Comput. 32 (1978), p. 941.

[399] V. Shoup, Searching for primitive roots in finite fields, Math. Comput. 58(197) (1992), 369–380.

[400] T. Siegenthaler, Correlation-immunity of nonlinear combining functions for cryptographic applications, IEEE Trans. Info. Theory 30 (1984), 776–780.

[401] T. Siegenthaler, Decrypting a class of stream ciphers using ciphertext only, IEEE Trans. Comput. 34 (1985), 81–85.

[402] T. Siegenthaler, Cryptanalyst's representation of nonlinearly filtered ml-sequences, in: LNCS, vol. 219, Advances in Cryptology - Proc. Eurocrypt'85, F. Pichler, ed., Springer Verlag, 1986.

[403] T. Siegenthaler, Methoden für den Entwurf von Stream-Cipher Systemen, Diss. ETH Nr. 8185, ADAG Zürich, 1986.

[404] M. K. Simon, J. K. Omura, R. A. Scholtz and B. K. Levitt, Spread Spectrum Communications, vol. I, Rockville, MD: Computer Science Press, 1985.

[405] J. Singer, A theorem in finite projective geometry and some applications to number theory, Trans. Amer. Math. Soc. 43 (1938), 377–385.

[406] Th. Skolem, S. Chowla and D. J. Lewis, The Diophantine equation $2^{n+2}-7=x^2$ and related problems, Proc. Amer. Math. Soc. 10 (1959), 663–669.

[407] N. J. A. Sloane, A Handbook of Integer Sequences, Academic Press, N.Y., 1973.

[408] N. J. A. Sloane and S. Plouffe, The Encyclopedia of Integer Sequences, Academic Press, San Diego, 1973.

[409] B. J. M. Smeets, On the autocorrelation function of some sequences generated by clock-controlled shift-registers, in Proc. 2nd Joint Swedish-Soviet Int. Workshop on Inf. Theory, 1985, Linköping, Sweden.

[410] B. Smeets, A note on sequences generated by clock-controlled shift registers, in Advances in Cryptology - Proc. Eurocrypt'85, F. Pichler, ed., LNCS 219, Springer Verlag, 1986, 142–148.

[411] R. G. Stanton and D. A. Sprott, A family of difference sets, Canad. J. Math. 10 (1958), 73–77.

[412] M. Stamp and C. F. Martin, An algorithm for the k-linear complexity of binary sequences with period 2^n, IEEE Trans. Info. Theory 39 (1993), 1393–1401.

[413] H. Stichtenoth, Algebraic Function Fields and Codes, Springer Verlag, 1986.

[414] T. Storer, Cyclotomy and Difference Sets, Marham, Chicago, 1967.

[415] T. Storer, Cyclotomies and difference sets modulo a product of two distinct odd primes, Michigan Math. J. 14 (1967), 117–127.

[416] M. Szalay, On the distribution of the primitive roots of a prime, J. Number Theory 7 (1975), 184–188.

[417] H. Tarnanen, A. Tietäväinen, A simple method to estimate the maximum nontrivial correlation of some sets of sequences, AAECC 5 (1994), 123–128.

[418] M. Templer, On the primality of $k! + 1$ and $2 \cdot 3 \cdot 5 \cdot \cdots \cdot p + 1$, Math. Comput. 34 (1980), 303–304.

[419] A. Tietäväinen, On the cardinality of sets of sequences with given maximum correlation, Discr. Math. 106/107 (1992), 471–477.

[420] A. Tietäväinen, On the correlation of sequences Algebraic Coding, in: LNCS 573 (1992), 1–4.

[421] R. C. Titsworth, Optimal ranging codes, IEEE Trans. Space Electronics and Telemetry, March 1964, 19–30.

[422] J. F. Traub, Computational complexity of interactive process, SIAM J. Comput. 1 (1972), 167–179.

[423] J. F. Traub, G. W. Wasilkowski and H. Wozniakowski, Information, Uncertainty, Complexity, Addison-Wesley, Reading, Mass., 1983.

[424] R. J. Turyn, A special class of Williamson matrices and difference sets, J. Comb. Theory A 36 (1984), 111–115.

[425] T. W. Tze, S. Chanson, C. Ding, T. Helleseth and M. Parker, Logarithm authentication codes, Information and Computation 184(1) (2003), 93–108

[426] R. C. Vaughan, A remark on the divisor function $d(n)$, Glasgow Math. J. 14 (1973), 54–55.

[427] R. C. Vaughan, The Hardy-Littlewood Method, Cambridge Tract in Math., no. 8, 1981.

[428] E. Vegh, Pairs of consecutive roots modulo a prime, Proc. Amer. Math. Soc. (19) 2 (1968), 1169–1170.

[429] E. Vegh, Arithmetic progressions of primitive roots of a prime, J. reine angew. Math. 244 (1970), 108–111.

[430] E. Vegh, A note on the distribution of the primitive roots of a prime, J. Number Theory 3 (1971), 13–18.

[431] H. Walum, A recurrent pattern in the list of quadratic residues mod a prime in the values of the Liouville *lambda* function, J. Number Theory 12 (1980), 53–56.

[432] Y. Wang, On the least primitive root of a prime, Sci. Sinica 10 (1961), 1–14.

[433] M. Ward, Some arithmetical properties of sequences satisfying a linear relation, Ann. Math. (2) 32 (1931), 734–738.

[434] M. Ward, The distribution of residues in a sequence satisfying a linear recursion relation, Trans Amer. Math. Soc. 33, 166–190.

[435] M. Ward, The algebra of recurring series, Ann. of math. (2) 32 (1931), 1–9.

[436] M. Ward, The characteristic number of a sequence of integers satisfying a linear recursion relation, Trans. Amer. Math. Soc. 33 (1931), 153–165.

[437] M. Ward, Some arithmetical properties of sequences satisfying a linear recursion relation, Ann. of Math. (2) 32 (1931), 734–738.

[438] M. Ward, The arithmetical theory of linear recurring sequences, Transactions of Americ. Math. Soc. vol. 35 (1933), 600–628.

[439] M. Ward, Some arithmetical theory of linear recurring series, Trans. Amer. Math. Soc. 35 (1935), 600–628.

[440] M. Ward, An arithmetical property of recurring series of the second order, Bull. Amer. Math. Soc. 40 (1934), 825–828.

[441] M. Ward, Note on an arithmetical property of recurring series, Math. Z. 39 (1935), 211–214.

[442] M. Ward, The null divisor of linear recurring series, Duke Math. J. 2 (1936), 472–476.

[443] M. Ward, Linear divisibility sequences, Trans. Amer. Math. Soc. 41 (1937), 276–286.

[444] M. Ward, Arithmetical properties of sequences in rings, Ann. of Math. (2) 39 (1938), 210–219.

[445] M. Ward, Memoir on elliptic divisibility sequences, Amer. J. Math. 70 (1948), 31–74.

[446] D. Wheeler, A Bulk Data Encryption Algorithm, in: Fast Software Encryption, LNCS 809, Springer Verlag, 1994, 127–134.

[447] A. Weil, Sur les courbes algèbriques et les variètes qui s'en dèduisent, Actualitès Math. Sci., No. 1041 (Paris, 1945), deuxiéme partie, Section IV.

[448] A. Weil, Number of solutions of equations in a finite field, Bull. Am. Math. Soc. 55 (1949), 497–508.

[449] A. Weil, Sur les courbes algébriques et les variétés qui s'en dduisent Actualits sci ind. No. 1041.

[450] A. Weil, Number Theory: An Approach Through History, Birkhäuser, Boston, Basel, and Stuttgart, 1984.

[451] H. Weber, Beweis des Satzes, daß jede endlich primitive quadratische Form unendlich viele Primzahlen darzustellen fährig ist, Math. Annalen 20 (1882), 301–329.

[452] B. M. M. de Weger, Approximation lattices of p-adic numbers, J. Num. Th. 24 (1986), 70–88.

[453] L. R. Welch and R. A. Scholtz, Continued fractions and Berlekamp's algorithm, IEEE Trans. Info. Theory 25 (1979), 19–27.

[454] A. L. Whiteman, The cyclotomic numbers of order sixteen, Trans. Amer. Math. Soc. 86 (1957), 401–413.

[455] A. L. Whiteman, The cyclotomic numbers of order twelve, Acta Arith. 6 (1960), 53–76.

[456] A. L. Whiteman, The cyclotomic numbers of order ten, in: Proc. Sympos. in Appl. Math. 10, Amer. Math. Soc., 1960, 95–111.

[457] A. L. Whiteman, A family of difference sets, Illinois J. Math. 6 (1962), 107–121.

[458] H. C. Williams and C. R. Zarnke, A note on the prime numbers of the form $N = (6a+1)2^{2n-1} - 1$ and $M = (6a-1)2^{2n-1} - 1$, Math. Comput. 22 (1968) 420–422.

[459] H. C. Williams and C. R. Zarnke, Some prime numbers of the form $2A3^n + 1$ and $2A3^n - 1$, Math. Comput. 26 (1972), 995–998.

[460] H. C. Williams, Some primes with interesting digit patterns, Math. Comput. 32 (1978), 1306–1310.

[461] H. C. Williams and E. Seah, Some primes of the form $(a^n - 1)/(a-1)$, Math. Comput. 33(148) (1979), 1337–1342.

[462] H. C. Williams and H. Dubner, The primality of R1031, Math. Comput. 47 (1986), 703–712.

[463] J. Wolfmann, Bent functions and coding theory, in: A. Pott, P. V. Kumar, T. Helleseth and D. Jungnickel eds., Difference Sets, Sequences and their Correlation Properties, Amsterdam, Kluwer, 1999, pp. 393–417.

[464] M. Xia, Some infinite class of Williamson matrices and difference sets, J. Comb. Theory A 61 (1992), 230–242.

[465] Q. Xiang, Recent results on difference sets with classical parameters, in: A. Pott, P. V. Kumar, T. Helleseth and D. Jungnickel eds., Difference Sets, Sequences and their Correlation Properties, Amsterdam, Kluwer, 1999, 419–434.

[466] G. Z. Xiao, J. L. Massey, A spectral characterization of correlation-immune functions, IEEE Trans. Info. Theory 34 (1988), 569–571.

[467] G. Xiao and S. Wei, Fast algorithm for determining the linear complexity of periodic sequences, in: Proceed. of INDOCRYPT 2002, LNCS 2551, Springer Verlag, 2002, 12–21.

[468] A. Yao, Theory and application of trapdoor functions, in: Proc. of the 23th IEEE Symposium on Foundations of Computer Science, Chicago, IL, 1982, 80–91.

[469] A. Yao, Computational information theory, in Complexity in Information Theory, Y. Abu-Mostafa, ed., Springer Verlag, New York, 1–15.

[470] D. Zagier, Die ersten 50 Millionen Primzahlen, in: Lebendige Zahlen: Fünf Exkursionen, Mathematische Miniaturen 1, W. Borho et al., Birkhäuser Verlag, 1981.

[471] K. C. Zeng, C. H. Yang and T. R. N. Rao, On the linear consistency test (LCT), in: cryptanalysis and its applications Advances in Cryptology, Crypto '89, LNCS 435 (1990), Springer Verlag, 164–174.

[472] N. Zierler, Linear recurring sequences, J. Soc. Ind. Appli. Math. 7 (1959), 31–48.

[473] N. Zierler and W. H. Mills, Products of linear recurring sequences, J. Algebra 27 (1973), 147–157.

Index